中国大洋海底地理实体名录

Chinese Gazetteer of Undersea Features
on the International Seabed

(2017)

中国大洋矿产资源研究开发协会办公室　著

海洋出版社

2018年·北京

图书在版编目(CIP)数据

中国大洋海底地理实体名录. 2017 : 汉、英 / 中国大洋矿产资源研究开发协会办公室著. — 北京 : 海洋出版社, 2018.8

ISBN 978-7-5210-0175-4

Ⅰ. ①中… Ⅱ. ①中… Ⅲ. ①海底 – 海洋地理学 – 中国 – 2017 – 名录 – 汉、英 Ⅳ. ①P737.2-62

中国版本图书馆CIP数据核字(2018)第186471号

责任编辑：王　溪
责任印制：赵麟苏

海洋出版社 出版发行
http://www.oceanpress.com.cn
北京市海淀区大慧寺路 8 号　　邮编：100081
北京朝阳印刷厂有限责任公司印刷　　新华书店北京发行所经销
2018年11月第1版　　2018年11月第1次印刷
开本：880mm×1230mm　　1／16　　印张：35.75
字数：626千字　　图幅数：490幅　　定价：390.00元
发行部：010-62132549　　邮购部：010-68038093　　总编室：010-62114335
海洋版图书印、装错误可随时退换

《中国大洋海底地理实体名录（2017）》
作者名单

组　长：李　波

副组长：宋成兵　　石学法　　何高文　　韩喜球　　陶春辉　　李传顺
　　　　马维林　　朱本铎

主要成员（按姓氏笔画排序）：

王叶剑　　邓显明　　石丰登　　刘丽强　　孙思军　　许津德

吴自银　　宋　超　　张国堙　　张瑞端　　李四海　　李守军

李春菊　　李洪林　　李鹏飞　　杨永平　　杨克红　　罗　祎

姜　静　　赵荻能　　盖增喜　　黄文星　　程永寿

《中国大洋海底地理实体名录（2017）》
顾问组

林绍花　　李裕伟

前言

　　为体现我国在深海海底地理实体的发现、命名方面对国际社会的重要贡献，彰显我国海洋科技工作者对国际深海的认知水平，中国大洋矿产资源研究开发协会办公室在 2016 年组织编写出版了《中国大洋海底地理实体名录（2016）》（以下简称《名录（2016）》），是目前我国最为齐全的深海海底地理实体名录，全书共收录了由我国命名和翻译的深海地理实体标准地名 230 个，《名录（2016）》出版后，受到广泛关注。

　　为满足统筹协调深海活动的需求，依据我国颁布实施的《中华人民共和国深海海底区域资源勘探开发法》有关精神，我们组织有关人员开展了《名录（2017）》的编写工作。在《名录（2016）》的基础上，新收录了地名 22 个，使得《名录（2017）》收录标准地名数量达到 252 个；同时，根据《名录（2016）》出版以来有关人员反馈的意见对相关内容进行了针对性修改，使新出版的《名录（2017）》内容更加准确和完善。其中，新收录的 22 个标准地名中，包含 2017 年经国际海底地理实体命名分委会（SCUFN）审议通过的标准地名 15 个，2018 年经 SCUFN 审议通过的标准地名 6 个以及经中国地名主管部门批准使用的标准地名 1 个。我们相信，《名录（2017）》的出版，将持续推动我国深海地理实体命名工作的开展，提升我国在国际海洋合作领域的话语权，同时在提高我国的国际地位和影响力方面继续发挥积极作用。

Foreword

In 2016, the Office of China Ocean Mineral Resources Research and Development Association organized and published the *Chinese Gazetteer of Undersea Features on the International Seabed* (2016) (hereinafter referred to as *Gazetteer* 2016), in order to demonstrate the significant contribution of China to the international community in the discovery and naming of undersea features, and reveal the research level in international seabed of Chinese marine scientists and engineers. This *Gazetteer* is currently the most complete directory of undersea feature in China. There have been 230 standard geographical names of deep-sea geographic entities named and translated by China. Extensive attention has been received after the publication of the *Gazetteer* 2016.

In order to meet the needs of coordinating the deep-sea survey activities, we organized relevant personnel to carry out the compilation of the *Gazetteer* 2017 in accordance to the spirit of the Law of the Exploration and Development of Deep seabed Resources issued by China. Based on the *Gazetteer* 2016, 22 new undersea feature names have been included resulting a total of 252 undersea feature names in the *Gazetteer* 2017. In the meantime, the relevant contents have been revised according to the feedback since the publication of the *Gazetteer* 2016 which makes the newly published *Gazetteer* 2017 more accurate. Among the 22 new standard feature names, there are 15 standard undersea feature names approved by SCUFN in 2017, 6 standard place names approved by SCUFN in 2018, and 1 standard place name approved by Chinese toponymical authorities.

We believe that the publication of the *Gazetteer* 2017 will continue to promote the naming of deep-sea geographic entities in China, enhance China's voice in the field of international maritime cooperation, and continue to play an active role in improving the international status and influence of China.

深海海底地理实体的命名工作是人类开展深海探测活动成果的重要体现，对深海海底地理实体的发现和命名并予以公布，是对国际社会的重要贡献，同时也彰显出命名者的科技实力和认识水平，提升在此领域的影响力，并在某种程度上体现出其对命名实体的潜在权益。因此，自20世纪50年代以来，美、英、德、法、俄、日等海洋强国开始重视深海海底命名工作并一直主导着该项工作的发展；同时，自20世纪90年代以来，新兴的海洋国家，如巴西、阿根廷和韩国等国，也加强了海底命名工作，进步明显。

根据国际海底地理实体命名分委会（Sub-Committee on Undersea Feature Names，SCUFN）的规定，海底地理实体是海底可测量并可划分界限的地貌单元，赋予其标准名称的行为称为"海底命名"。海底地名包含通名和专名两部分，通名是用来区分地理实体类型的词语，专名是用来区分每个地理实体的词语。如地名"织女平顶海山"，"平顶海山"为海底地理实体的一种类型，是通名；"织女"为该地理实体的专有名称，即专名。在国际海域的地名需要得到SCUFN审议批准，截至2015年年底，SCUFN共审议批准了国际海域3 940个海底地名。

自中国大洋矿产资源研究开发协会（以下简称"大洋协会"）经国务院批准成立20多年来，已组织开展国际海域调查航次39个（截至2016年10月），发现了大量海底地理实体，但仅对小部分海底地理实体以字母代码表示，给调查研究工作带来诸多不便。2011年，根据国家海洋局部署，大洋协会办公室成立了大洋地理实体命名工作组，正式启动了国际海域（本书中"国际海域"与"深海"含义一致）海底命名工作。根据需要，工作组编制完成了《国际海底区域地理实体命名管理规定（试行）》，对海底命名的内容、程序和技术方法进行了确定，建立了地名数据库。截至 2016 年 10 月，工作组共新命名了 163 个国际海域海底地理实体，已由国务院批准对外公布使用。我们每年从中选取并编制一定数量的地名提案提交 SCUFN 审议，目前已有 63 个得到 SCUFN 审议核准。同时，除新命名的地理实体之外，考虑到大洋航次调查的实际需要，工作组在遵守国际惯例的基础上对我国开展过调查但已被命名的 18 个海底地理实体进行了整理和名称翻译，还参照海底命名的技术要求对我国已经发现的 42 个海底热液区的名称进行了标准化处理。这样，工作组在国际海域内共新命名、翻译和标准化处理 223 个名称。我们将这 223 个标准化名称和海军出版社、国家海洋信息中心命名的 7 个国际海域海底地名汇总编制形成《中国大洋海底地理实体名录（2016）》（以下简称《名录》），以中英文对照的方式予以出版，供海洋科技工作者使用。这 230 个国际海域的地名蕴含着中国传统文化元素，符合我国相关法律和规范的要求，也符合 SCUFN 的规定，体现了我国海洋科技工作者对世界海洋认知水平的进步，展示了我国海洋科技实力。同时，《名录》所含地名的水深数据主要来源于我国大洋航次所获得的多波束测深数据，质量和精度满足海底命名要求，并具有全局性、代表性和控制性特征，均通过了中国地名分委会的技术审查。

《名录》由中国大洋协会办公室组织编写，在编写过程中，国内熟悉海底命名工作的专家作为顾问给予了具体指导。大洋协会办公室李波副主任和周宁总工亲自参与了重要环节的编写工作，如专名体系框架确定、通名技术要求等方面，在编写后期也给予了悉心指导。在具体编写过程中，主要分工如下：西太平洋海山区主要由何高文、黄文星、朱本铎完成；中太平洋海山区主要由马维林、杨克红完成；东太平洋海盆区主要由朱本铎、黄文星、刘丽强完成；东太平洋海隆区主要由程永寿、邓显明、张瑞端完成；大西洋中脊区主要由石学法、陶春辉、石丰登完成；西南印度洋洋中脊区主要由陶春辉、邓显明、张国堙完成；西北印度洋洋中脊区主要由韩喜球、王叶剑、李洪林、吴招才完成。《名录》后期图件制作由程永寿、黄文星、杨克红、孙思军、张瑞端、刘丽强完成，专名主要由李鹏飞、盖增喜审核；

《名录》提交审查、报批及后期统稿主要由宋成兵、孙思军、黄文星、盖增喜完成；《名录》中文稿初稿完成后，主要由盖增喜、宋超等承担了翻译、校对等工作；《名录》编写和出版过程中，宋成兵、盖增喜、孙思军具体承担了大量组织协调工作。

在《名录》公开出版之际，我们衷心感谢为中国大洋事业做出贡献的人们，没有他们的辛劳工作，就没有《名录》编写和出版的基础。我们也非常感谢在《名录》报批过程中给予支持的民政部、国家海洋局有关职能部门的同志，他们的责任心也非常值得我们钦佩。感谢海军出版社、国家海洋信息中心同意其编制并经过 SCUFN 核准的名称也纳入《名录》。感谢海洋出版社付出的心血和提供的帮助，特别感谢国际地名分委会中国籍委员林绍花提出了诸多宝贵建议。

《名录》的公开出版，恰逢我国颁布实施《深海海底区域资源勘探开发法》。这些蕴含中华文化元素的地名将载入世界海底地名词典，在全球范围内推广使用，这将进一步提升我国在 SCUFN 乃至国际海洋合作领域的话语权，提高我国的国际地位和影响力；同时，也以实际行动体现了依据《深海海底区域资源勘探开发法》统筹、有序管理国内力量开展深海海底活动的精神。我们相信《名录》的公开出版，将引起国内外的广泛关注，不仅填补了我国在国际海域海底命名工作的空白，更直接体现了我国作为负责任的大国在海洋科学领域对全球的重大贡献。

《名录》所收录的资料和信息均截止到 2016 年 10 月，我们将随着工作的进展及时修订和补充有关国际海域地理实体命名的信息。《名录》不足之处，恳请斧正。

作　者

2016 年 10 月

Foreword of the edition 2016

Naming of undersea features is the important reflection of achievements of human beings in deep sea exploration. Discovery, naming and publication of undersea features are important contributions to international society, as well as the demonstration of scientific and cognitive level of nomenclatures, which will improve their impact in the field and represent potential rights to the named feature to certain extent. Therefore, starting from 1950s, traditional maritime powers including US, UK, Germany, Russia and Japan have started to lay emphasis on deep undersea feature naming and dominated the development of this work. Moreover, since 1990s, newly emerged marine countries such as Brazil, Argentina and Korea have also enhanced undersea feature naming and achieved significant progresses.

According to the definition of the Sub-Committee on Undersea Feature Names (SCUFN), undersea features are undersea geomorphic units that can be measured and delimited by relief, the behavior of giving them standard names is known as undersea feature naming. An undersea feature name consists of a generic term and a specific term, the former is to define the type of geographic features, while the latter is to distinguish geographic features from each other. For instance, in the name "Zhinyu Guyot", "Guyot" is a type of undersea geographic feature, the generic term, and "Zhinyu" is the distinguished name of the geographic feature, the specific term.

Over twenty years since establishment, China Ocean Mineral Resources Research and Development Association (COMRA) has organized 39 research cruises to international seas by the end of October 2016 and discovered numerous undersea features. However, only a small part of them have been marked with alphabetic codes, resulting in much inconvenience in future researches and investigations. In 2011, according to the deployment of State Oceanic Administration of China, COMRA Office established the group for naming of undersea features, officially launching the naming work in international sea area (same as deep sea in this book). The experts of the group compiled *the Provisions on Administration of the Naming of International Undersea Geographical Feature (Trial)*, hence defined the contents, processes and technical methods of undersea feature naming, and established undersea feature name database. By the end of

October 2016, the group has named totally 163 undersea features in international sea area, which were published pursuant to the approval of the State Council. The group has been compiling a certain amount of undersea features naming proposals each year for SCUFN to review, and by the end of October 2016, 63 proposals have been approved by SCUFN. In addition to the newly named undersea features, considering the actual needs of oceanic survey cruises, the group has summarized and translated 18 already named undersea features that were investigated by China, on the basis of abiding by international practices. By referring to the technical requirements for undersea features naming, it also standardized the names of 42 seafloor hydrothermal fields that were already discovered by Chinese scientists. In this way, the group has newly named, translated and standardized 223 names in international seabed. We incorporated 223 names and 7 undersea feature names named by China Navigation Publications and National Marine Data and Information Service into the *Chinese Gazetteer of Undersea Features on the International Seabed (2016)*, for the use of marine scientists and engineers. In order to further expand the scope of application and impact of the *Gazetteer*, it is published in both English and Chinese. With the publication of the *Gazetteer*, those names implying elements of traditional Chinese culture will go down in the *GEBCO Gazetteer of Undersea Feature Names* and be used worldwide, which will further improve China's discursive power in SCUFN and even international marine cooperation, thus enhancing China's international status and impact. Bathymetric data included in the *Gazetteer* mainly came from the multi-beam echo sounding data gained in China's oceanic cruises. It was in accordance with requirements of undersea feature naming for quality and precision. Featuring globality, representativeness and controllability, it has passed the technical review of China Committee on Geographical Names.

The compilation of the *Gazetteer* was organized by the COMRA Office under the specific consultancy guidance of specialists that were familiar with undersea feature naming. The Deputy Director of the Association, Li Bo and Chief Engineer, Zhou Ning participated in the compilation in person to have defined the framework of specific term system and technical requirements for generic terms and provided guidance in later stage of compilation. During compilation, the division of work was as follows: seamount area in Western Pacific Ocean was mainly accomplished by He Gaowen, Huang Wenxing and Zhu Benduo; seamount area in Middle Pacific Ocean by Ma Weilin and Yang Kehong; oceanic basin area in Eastern Pacific Ocean by Zhu Benduo, Huang Wenxing and Liu Liqiang; Eastern Pacific Rise by Cheng Yongshou, Deng Xianming and Zhang Ruiduan; Mid-ocean ridge area in Atlantic Ocean by Shi Xuefa, Tao Chunhui and Shi Fengdeng; Mid-ocean ridge area in Southwestern Indian Ocean by Tao Chunhui, Deng Xianming and Zhang Guoyin; Mid-ocean ridge area of Northwestern Indian Ocean by Han Xiqiu, Wang Yejian, Li Honglin and Wu Zhaocai. The production of maps in the *Gazetteer* in later stage was completed by Cheng Yongshou, Huang Wenxing, Yang Kehong, Sun Sijun, Zhang Ruiduan and Liu Liqiang, while the specific terms were reviewed by Li Pengfei and Ge Zengxi. Submission for review, report for approval and verification of the *Gazetteer* in later stage were assumed by Song Chengbing, Sun Sijun, Huang Wenxing, Ge Zengxi. Ge Zengxi and Song Chao undertook translation and proofreading work after completing the first draft of Chinese edition of the *Gazetteer*. During compilation and publication of the *Gazetteer*, Song Chengbing, Ge Zengxi and Sun Sijun have taken the responsibilities for organization and coordination.

On the occasion of publication of the *Gazetteer*, we are sincerely grateful to people who have contributed to China's oceanic cause. Without their diligent efforts, there won't be any foundation for the compilation and publication of the *Gazetteer*. We are also thankful to comrades from Ministry of Civil Affairs, related functional departments of the State Oceanic Administration of China, who deserve admiration for their senses of responsibility. We are gratitude to China navigation Publications and National Marine Data and Information Service for the names compiled by them and approved by SCUFN being incorporated into the *Gazetteer*. Thanks to the efforts and help of the China Ocean Press, and special thanks to the numerous valuable suggestions provided by Lin Shaohua, the Chinese member of SCUFN.

As China has issued and implemented *Law of the People's Republic of China on Resource Exploration and Exploitation in the Deep Seabed Area*, publication of the *Gazetteer* demonstrated the spirit of this law in coordination and organized management of domestic power to carry out deep undersea activities. At the same time, the *Gazetteer* has intensively presented achievements of China in investigations and researches on international seas for the past twenty years, and highlighted outstanding contributions of China in ocean science in international seas as a responsible power.

Data and information included in the *Gazetteer* are up to the end of October 2016. As the progress continues, we will promptly supplement naming information of related geographic features in international seas, and amend the *Gazetteer*.

The Authors

October 2016

目录

Contents

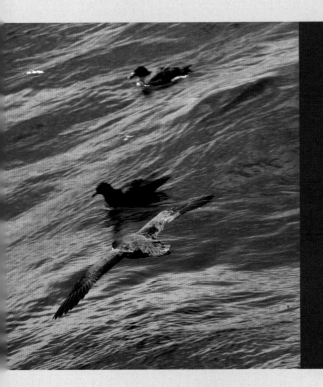

第1章
术语与定义
Chapter 1
Terms and Definitions

　　《中国大洋海底地理实体名录（2017）》（以下简称《名录》）中的地理实体类型涉及海山、平顶海山、海丘、圆海丘、海山群、平顶海山群、海丘群、海脊、海山链、海岭、海渊、海底洼地、裂谷和断裂带等，《名录》中还涉及热液区等，其定义[①]分述如下。

　　The undersea features in this catalog include Seamount, Guyot, Hill, Knoll, Seamounts, Guyots, Hills, Ridge, Seamount Chain, Deep, Depression, Gap and Fracture Zone, etc. The hydrothermal fields are also included in the catalog. The definitions of these terms are listed below.

■ 海　山

孤立的海底高地，地形高差大于 1 000 m，如魏源海山（图 1-1）。

Seamount

An isolated undersea elevation, terrain relief greater than 1 000 m as measured from the deepest isobath that surrounds most of the feature, characteristically of conical form. E.g., Weiyuan Seamount (Fig.1-1).

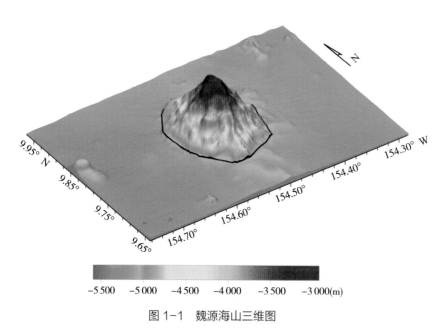

图 1-1　魏源海山三维图

Fig.1-1　3-D topographic map of the Weiyuan Seamount

[①] 定义根据《海底地名命名》（GB29432-2012）和《海底地名命名标准》（B-6 出版物）修订。
　　The definitions are revised from *Nomenclature of undersea feature names* (GB29432-2012) and *Standardization of Undersea Feature Names* (Publication B-6).

■ 平顶海山

顶部发育较大规模平台的海山，如白驹平顶海山（图 1-2）。

Guyot

A seamount with a comparatively smooth flat top. E.g., Baiju Guyot (Fig.1–2).

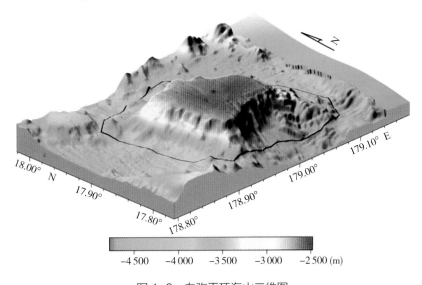

图 1-2　白驹平顶海山三维图

Fig.1-2　3-D topographic map of the Baiju Guyot

■ 海　丘

孤立的海底高地，地形高差小于 1 000 m，如景福海丘（图 1-3）。

Hill

An isolated undersea elevation generally of irregular shape, terrain relief less than 1 000 m as measured from the deepest isobath that surrounds most of the feature. E.g., Jingfu Hill (Fig.1–3).

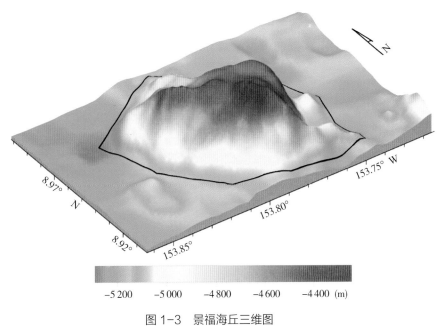

图 1-3　景福海丘三维图

Fig.1-3　3-D topographic map of the Jingfu Hill

■ 圆海丘

轮廓呈圆滑状的孤立海底高地，地形高差小于 1 000 m，如巩珍圆海丘（图 1-4）。

Knoll

An isolated undersea elevation with a rounded profile, terrain relief less than 1 000 m as measured from the deepest isobath that surrounds most of the feature. E.g., Gongzhen Knoll (Fig.1−4).

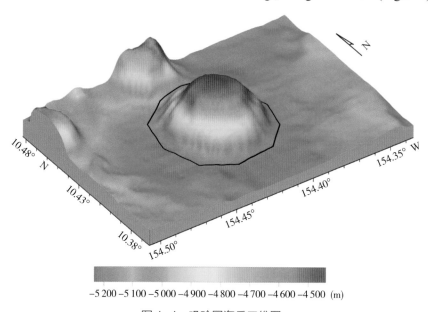

图 1-4　巩珍圆海丘三维图

Fig.1-4　3-D topographic map of the Gongzhen Knoll

■ 海山群

由多个相对聚集的海山、海丘构成的大型地理实体，如柔木海山群（图 1-5）。

Seamounts

A large undersea feature consisting of several relatively gathering seamounts and/or hills. E.g., Roumu Seamounts (Fig.1−5).

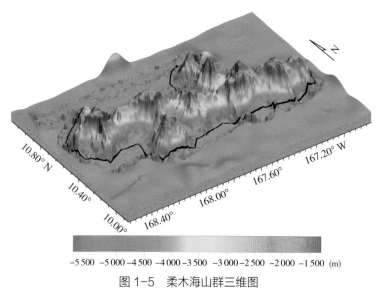

图 1-5　柔木海山群三维图

Fig.1-5　3-D topographic map of the Roumu Seamounts

■ 平顶海山群

由多个相对聚集的平顶山构成的大型地理实体，如斯干平顶海山群（图1-6）。

Guyots

A large undersea feature consisting of several relatively gathering guyots. E.g., Sigan Guyots (Fig.1-6).

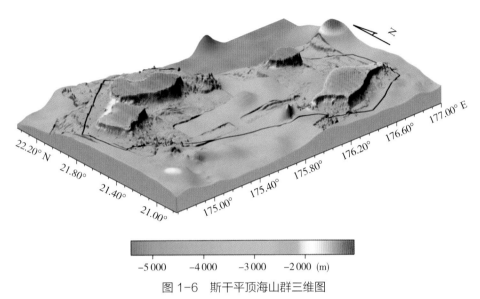

图1-6　斯干平顶海山群三维图

Fig.1-6　3-D topographic map of the Sigan Guyots

■ 海丘群

由多个相对聚集的海丘构成的大型地理实体，如莺萝海丘群（图1-7）。

Hills

A large undersea feature consisting of several relatively gathering hills and/or knolls. E.g., Niaoluo Hills (Fig.1-7).

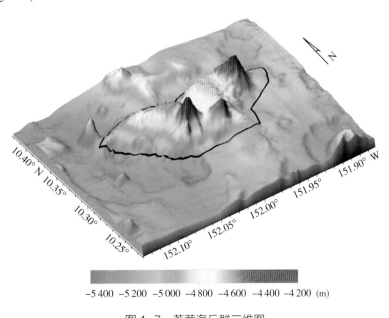

图1-7　莺萝海丘群三维图

Fig.1-7　3-D topographic map of the Niaoluo Hills

■ 海 脊

形态孤立且狭长的海底高地，呈线状延伸，如思文海脊（图 1-8）。

Ridge

An isolated and elongated undersea elevation of varying complexity, size and gradient, extending linearly. E.g., Siwen Ridge (Fig.1-8).

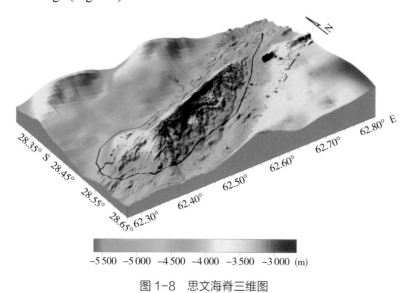

图 1-8 思文海脊三维图

Fig.1-8 3-D topographic map of the Siwen Ridge

■ 海山链

由多个互不相连的海山、海丘按一定方向排列构成的大型地理实体，如楚茨海山链（图 1-9）。

Seamount Chain

A large undersea feature consisting of several disconnected seamounts and/or hills arranged in a particular direction. E.g., Chuci Seamount Chain (Fig.1-9).

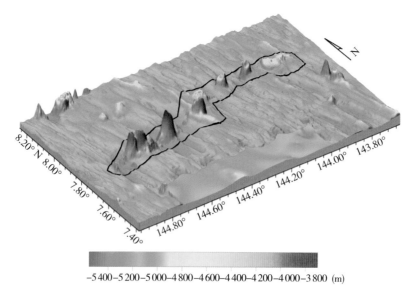

图 1-9 楚茨海山链三维图

Fig.1-9 3-D topographic map of the Chuci Seamount Chain

■ 海 岭

沿一定方向延伸、规模较大的海底高地，由多个山体互相连接的海山、海丘构成，如鉴真海岭（图 1-10）。

Ridge

A comparatively large and elongated undersea elevation, extending to a particular direction consisting of several connected seamounts and/or hills. E.g., Jianzhen Ridge (Fig.1-10).

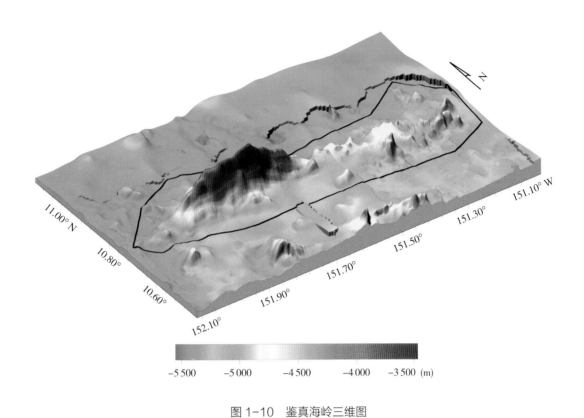

图 1-10　鉴真海岭三维图

Fig.1-10　3-D topographic map of the Jianzhen Ridge

■ 海 渊

海槽、海盆或海沟中出现的局部凹陷地貌单元，如太白海渊（图 1-11）。

Deep

A localized depressed geomorphic unit within the confines of a larger feature, such as a trough, basin or trench. E.g., Taibai Deep (Fig.1-11).

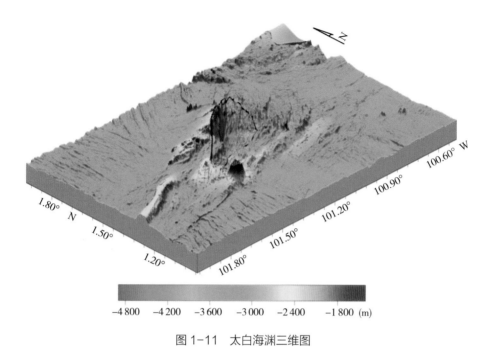

图 1-11 太白海渊三维图

Fig.1-11 3-D topographic map of the Taibai Deep

■ 海底洼地

发育于海盆、海底平原中的局部凹陷地貌单元，如天祜南海底洼地（图 1-12）。

Depression

A localized depressed geomorphic unit developed in a basin or plain. E.g., Tianhunan Depression (Fig.1-12).

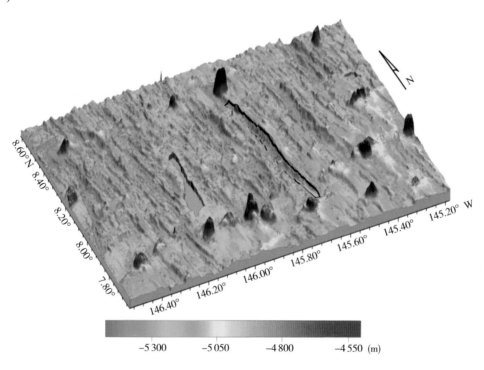

图 1-12 天祜南海底洼地三维图

Fig.1-12 3-D topographic map of the Tianhunan Depression

■ 裂　谷

海脊、海隆或其他海底高地中发育的狭窄且断裂的地貌单元，如德音裂谷（图 1-13）。

Gap

A narrow break developed in a ridge, rise or other undersea elevation. E.g., Deyin Gap (Fig.1-13).

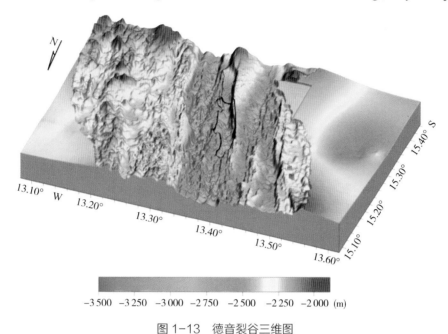

图 1-13　德音裂谷三维图

Fig.1-13　3-D topographic map of the Deyin Gap

■ 断裂带

因海底扩张脊轴位移而形成的地貌单元，形态狭长而不规则，通常由两翼地形陡峭且不对称的小型海脊、洼地或陡崖构成，如竺可桢断裂带（图 1-14）。

Fracture Zone

A long narrow zone of irregular topography formed by the movement of tectonic plates associated with an offset of a spreading ridge axis, usually consisting of steep-sided and/or asymmetrical small ridges, depressions or escarpments. E.g., Chukochen Fracture Zone (Fig.1-14).

■ 热液区

海底热液区一般分布在洋中脊、弧后盆地等区域，由若干个热液喷口组成，它的成因在于海水从地壳裂缝渗入地下，遇到熔岩被加热，溶解了周围岩层中的金属元素后又从地下喷出，有些活动的热液喷口温度高达 400℃，如龙旂热液区（图 1-15）。

Hydrothermal Field

A hydrothermal field is commonly found near areas such as mid-ocean ridges and backarc basins. It usually consists of several hydrothermal vents. A hydrothermal vent is a fissure in a planet's surface from which geothermally heated water issues. The sea water permeates underground through cracks in the crust. Then it is heated by encountered lava. Dissolving metal elements from surrounding rock formation, it issues from underground again. In contrast to the approximately 2 °C ambient water temperature at these depths, water emerges from these vents at temperatures can be as high as 400 °C. E.g., Longqi Hydrothermal Field (Fig.1–15).

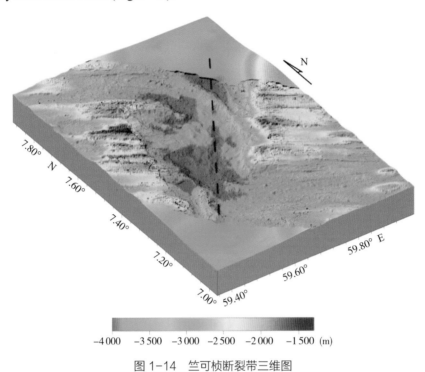

图 1-14　竺可桢断裂带三维图

Fig.1–14　3-D topographic map of the Chukochen Fracture Zone

图 1-15　龙旂热液区活动热液喷口

Fig.1–15　Active vents in the Longqi Hydrothermal Field

第2章
太平洋地理实体
Chapter 2
Undersea Features in the
Pacific Ocean

第1节 命名概况

太平洋位于亚洲、大洋洲、北美洲、南美洲和南极洲之间，是世界上规模最大、水深最深，而且边缘海和岛屿最多的一个大洋。太平洋海底地貌分为大陆架、大陆坡、海沟、大洋盆地和东太平洋海隆等5大类基本地貌单元。

太平洋的大陆架面积约938万平方千米，占太平洋总面积的5.2%，主要分布在太平洋西部、西南部和北部。西部大陆架较宽，渤海、黄海和东海的海底地形几乎全部为大陆架。太平洋东部大陆架狭窄，呈不连续的条带状，在北美洲宽仅18～20km，南美洲最宽处达70km。太平洋的大陆坡地形坡度大，其上发育海岭、盆地、峡谷、断崖和阶地等次级地理实体。在西太平洋大陆坡与洋盆交界处发育一系列岛弧和海沟，岛弧自北而南由阿留申群岛至克马德克群岛等10个群岛构成，长达9520km。岛弧外侧伴生一系列海沟，如千岛-堪察加海沟等，其中马里亚纳海沟的最大水深超过11000m。西太平洋岛弧内侧分布着一系列宽阔的边缘海盆地，形成西太平洋边缘海，主要有白令海、鄂霍次克海、日本海、黄海、东海、南海、爪哇海、苏拉威西海和珊瑚海等。大洋东岸的南北美大陆山系直逼海岸，大部分海岸线平直，无岛弧和边缘海盆地，但发育数个海沟，包括北部的阿拉斯加海沟、中部的中美洲海沟和南部的秘鲁-智利海沟。

太平洋的大洋盆地宽阔而深邃，其中部隆起一系列呈NW—SE走向的海底山脉，北起堪察加半岛，向南直抵土阿莫土群岛，绵延1万多千米，将太平洋海盆分为东、西两部分，两部分海域地貌特征截然不同。东太平洋海底地形相对比较平坦，最显著的地理实体为东太平洋海隆和一系列纬向断裂带以及大规模的东北太平洋深海盆地。纬向断裂带长达数千千米，宽约100～200km，自北向南发育门多西诺和加拉帕戈斯等22条。深海盆地水深为4000～6000m，盆底地形平缓，发育一些低缓的小型海山。西太平洋海域地形复杂，发育大量海山且以平顶型海山为主，海山基底直径一般大于20km，有60多座海山基底直径大于60km，山体高差超过2000m。这些海山按一定的规律排列构成多个海山区或海岭，包括皇帝-夏威夷海山链、莱恩海岭、吉尔伯特海山区、马绍尔海山区、麦哲伦海山区、中太平洋海山区和马尔库斯-威克海山区，部分海山出露海面成为环礁或海岛。

东太平洋海隆，是世界大洋中脊系的一部分，其始于60°S，60°W处，向东至130°W附近转向北，大致平行于美洲海岸向北延伸，直至阿拉斯加湾，长达15000km，高差2000～3000m、宽2000～4000km。东太平洋海隆以东发育着次一级的分支海岭，如智利海岭、纳斯卡海岭和加拉帕戈斯海岭等。

我国在菲律宾海盆、麦哲伦海山区、马尔库斯-威克海山区、马绍尔海山区、中太平洋海山区、莱恩海岭、克拉里昂-克拉伯顿区和东太平洋海隆等国际海底区域命名171个地理实体（含翻译地名）和18个热液区（图2-1）。

Section 1　Overview of the Naming

The Pacific Ocean is located between Asia, Oceania, the Americas and Antarctica. It is the world's largest and deepest ocean, with most marginal seas and islands. The Pacific seabed geomorphology can be divided into 5 categories of fundamental geomorphic units, which are continental shelf, continental slope, trench, ocean basin and the East Pacific Rise.

The area of Pacific continental shelf is about 9.38 million square kilometers, accounting for 5.2% of the total area of the Pacific Ocean, mainly distributed in the west, southwest and north of the Pacific Ocean. The western part of the continental shelf is wider. The seabed topography of Bohai Sea, Huanghai Sea and East China Sea are almost all of the continental shelves. The East Pacific continental shelf is narrow and in a discontinuous stripped shape. Its width at North America is only 18–20 km, while the widest at South America is 70 km. The topography of Pacific continental slope has great angles, on which develops ridges, basins, canyons, cliffs, terraces and other secondary undersea features. At the junction of the West Pacific continental slope and the ocean basin, it develops a series of island arcs and trenches. The island arcs consist of 10 islands, which are Aleutian Islands, Kermadec Islands and other from north to south, up to 9 520 km. The outer parts of island arcs are associated with a series of trenches, such as the Kuril-Kamchatka Trench. Among them, the maximum depth of the Mariana Trench is over 11 000 m. The inner parts of West Pacific island arcs distribute a series of wide marginal sea basins which form the West Pacific Marginal Sea, mainly the Bering Sea, Sea of Okhotsk, Sea of Japan, The Huanghai Sea, East China Sea, South China Sea, Java Sea, Celebes Sea, Coral Sea, etc. The North and South America Continental Mountains east to Pacific are encroaching the coast. Most of the coastline is straight with no arcs or marginal sea basins, but develops several trenches, including Alaska Trench in the north, Central America Trench in the middle, and Peru-Chile Trench in the south.

The basin of Pacific Ocean is wide and deep with a series of NW to SE running seamount ranges uplift in the central part. From Kamchatka Peninsula in the north to as far as Tuamotu Archipelago in the south, they stretch over 10 000 km, which divide the Pacific Basin into western and eastern parts with entirely different geomorphic characteristics. The East Pacific seabed topography is relatively flat. The most significant undersea features are the East Pacific Rise and a series of latitudinal fracture zones, and large-scale Northeast Pacific deep-sea basins. The latitudinal fracture zones have a length of thousands of kilometers and a width of about 100–200 km, with a development of 22 zones from north to south, such as Mendocino, Galapagos, etc. The water depth of deep-sea basin is about 4 000–6 000 m, the pelvic floor's terrain is smooth, with a development of low and flat small-scale seamounts. The topography of West Pacific is quite complex and develops a lot of seamounts, mainly guyots. The base diameters of these seamounts are generally greater than 60 km, while the total reliefs are more than

2 000 m. These seamounts constitute several seamount areas or ridges according to certain arrangement rules, including the Hawaiian-Emperor Seamount Chain, Line Islands Chain, Gilbert Seamount Area, Marshall Seamount Area, Magellan Seamount Area, Central Pacific Seamount Area and Marcus-Wake Seamount Area, while parts of the seamounts can reach out the sea surface becoming atolls or islands.

The East Pacific Rise, is a part of the world's mid-ocean ridge system. Beginning from 60° S and 60° W, it runs eastwards to around 130° W, then turns to north and extends 15 000 km to the Gulf of Alaska, roughly parallel to the coast of the America. The total relief is 2 000–3 000 m and the width is 2 000–4 000 km. Secondary ridge branches, such as the Chile Ridge, the Nazca Ridge, the Galapagos Ridge, etc. are developed to the east of the East Pacific Rise.

In total, 171 undersea features (including translations) and 18 hydrothermal fields have been named by China within the region of Philippine Basin, Magellan Seamount Area, Marcus-Wake Seamount Area, Marshall Seamount Area, Central Pacific Seamount Area, Line Islands Chain, Clarion-Clipperton Area, East Pacific Rise, etc., in the international seabed area of the Pacific Ocean (Fig. 2-1).

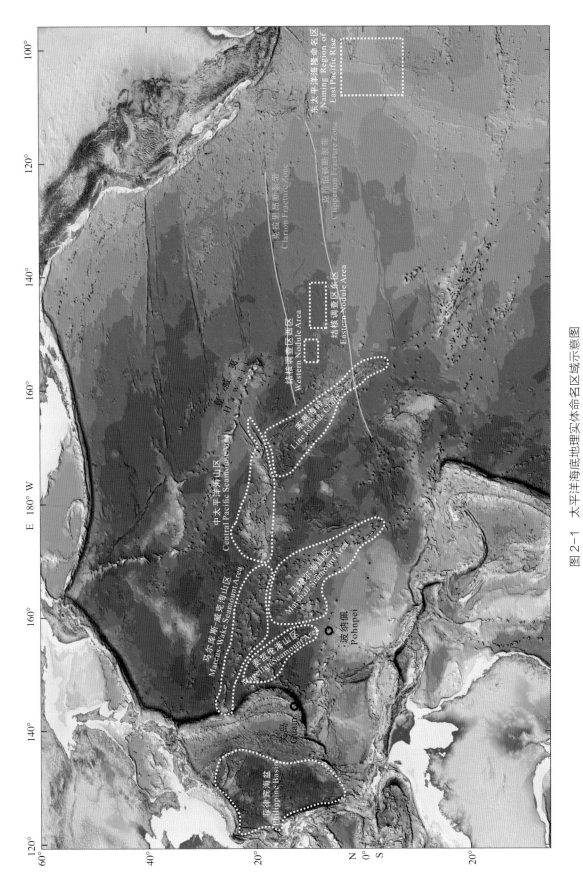

图 2-1　太平洋海底地理实体命名区域示意图

Fig.2-1　The Pacific Ocean undersea features naming regions

第 2 节　菲律宾海盆地理实体

菲律宾海盆位于太平洋西部边缘，台湾岛和菲律宾群岛以东海域。此海盆为大型深海盆地，被琉球海沟、菲律宾海沟和九州－帕劳海岭所环绕，水深达 5 000 ～ 6 500 m。海盆中部被 NW—SE 向延伸的长条状中央裂谷带分隔，形成多个次级海盆、海底高原、海台、海岭等地貌单元，在此基础上又发育海山、海丘、小型海脊众多小型地理实体。

西菲律宾海盆地貌单元的形成主要受海底扩张控制，大约在 55 Ma 至 33 ～ 30 Ma 期间，盆地内部发生以中央裂谷为中心的海底扩张，随着岩浆活动性质的海底扩张，深海平原地貌逐渐形成。随着扩张脊的发育，断裂带、海台、海山等地貌单元也逐步形成，并随着扩张的进行不断推移，分布于盆地的深海平原中。

在菲律宾海盆我国共命名 18 个地理实体。其中平顶海山 1 个，为日昇平顶海山；海山 10 个，包括海东青海山、天保海山、槐序海山、首阳海山、立春海山、惊蛰海山、清明海山、谷雨海山、静好海山和飒沓海山；海丘 4 个，包括日潭海丘、翠翘海丘、春分海丘和小满海丘；海脊 1 个，为月潭海脊；海山群 1 个，为睢鸠海山群；海盆 1 个，为芒种海盆（图 2-2）。

Section 2　Undersea Features in the Philippine Basin

The Philippine Basin is located in the western edge of the Pacific Ocean, east to Taiwan Island and the Philippine Islands. This sea basin is a large deep-sea basin, surrounded by Ryukyu Trench, Philippine Trench and Kyushu-Palau Ridge, with the water depth being up to 5 000–6 500 m. The central part of the basin is separated by an NW to SE running, elongated central rift zone, forming several secondary geomorphic units such as basins, plateaus, ridges, etc. Numerous small undersea features are developed in these features such as seamounts, knolls, small ridges, etc.

The formation of the West Philippine Basin geomorphic units are mainly controlled by the seafloor spreading. During around 55 Ma to 33–30 Ma, the seafloor spreading occurred with the central rift being the center inside the basin. Along with the magmatism featured seafloor spreading, geomorphology of abyssal plains gradually evolved. With the development of spreading ridges, fracture zones, plateaus, seamounts and other geomorphic units also gradually formed. Moreover, with the continuous progress of spreading, they distributed in the abyssal plains of the basin.

In total, 18 undersea features have been named by China in the region of the Philippine Basin including 1 guyot, which is Risheng Guyot; 10 seamounts, including Haidongqing Seamount, Tianbao Seamount, Huaixu Seamount, Shouyang Seamount, Lichun Seamount, Jingzhe Seamount, Qingming Seamount, Guyu Seamount, Jinghao Seamount and Sata Seamount; 4 hills, including Ritan Hill, Cuiqiao Hill, Chunfen Hill and Xiaoman Hill; 1 ridge, which is Yuetan Ridge; 1 seamounts, which is Jujiu Seamounts; 1 basin, which is Mangzhong Basin (Fig. 2-2).

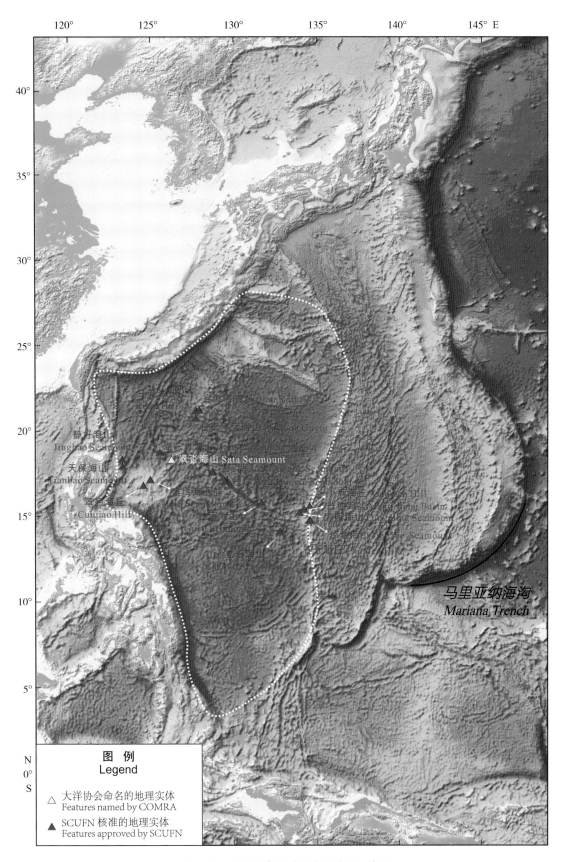

图 2-2　菲律宾海盆地理实体位置示意图

Fig.2-2　Locations of the undersea features in the Philippine Basin

2.2.1 日昇平顶海山
Risheng Guyot

中文名称 Chinese Name	日昇平顶海山 Risheng Pingdinghaishan	英文名称 English Name	Risheng Guyot	所在大洋 Ocean or Sea	西太平洋 West Pacific Ocean
发现情况 Discovery Facts	此平顶海山于 2004 年 10 月由中国科考船"大洋一号"在执行测量任务时调查发现。 This guyot was discovered by the Chinese R/V *Dayang Yihao* during the survey in October, 2004.				
命名历史 Name History	由我国命名为日昇平顶海山,于 2012 年提交 SCUFN 审议通过。 This feature was named Risheng Guyot by China and the name was approved by SCUFN in 2012.				
特征点坐标 Coordinates	20°42.60′N,127°44.10′E			长(km)×宽(km) Length (km) × Width (km)	23 × 21
最大水深（m） Max Depth（m）	5 200	最小水深（m） Min Depth（m）	3 147	高差（m） Total Relief（m）	2 053
地形特征 Feature Description	该平顶海山俯视平面形态呈近圆形,长宽分别为 23 km 和 21 km。山顶平台水深约 3 400 m,最浅处水深 3 147 m,山麓水深 5 200 m,高差 2 053 m,山顶平坦、边坡陡峭(图 2–3)。 The overlook plane shape of this guyot is nearly round. The length is 23 km and the width is 21 km. The depth of the top platform is about 3 400 m. The minimum depth is 3 147 m while the piedmont depth is 5 200 m. The total relief is 2 053 m. It has a flat top and steep slopes (Fig.2–3).				
命名释义 Reason for Choice of Name	"日昇"一词源于我国著名的世界文化遗产福建土楼其中的一个楼名,寓意太阳升起,蒸蒸日上。日昇楼在福建土楼中属于特色鲜明的圆形土楼。该海底地理实体山顶浑圆,犹如太阳从海底升起,形状又与日昇楼类似,故以"日昇"命名。 The word "Risheng" comes from the name of a traditional building Tulou in Fujian Province of China. Risheng Tulou is one of characteristic round Tulou. The feature has a round top like the sun rising from seabed, its shape is similar to the Risheng Tulou. Furthermore, "Risheng" in Chinese language means sunrise. So it was named Risheng.				

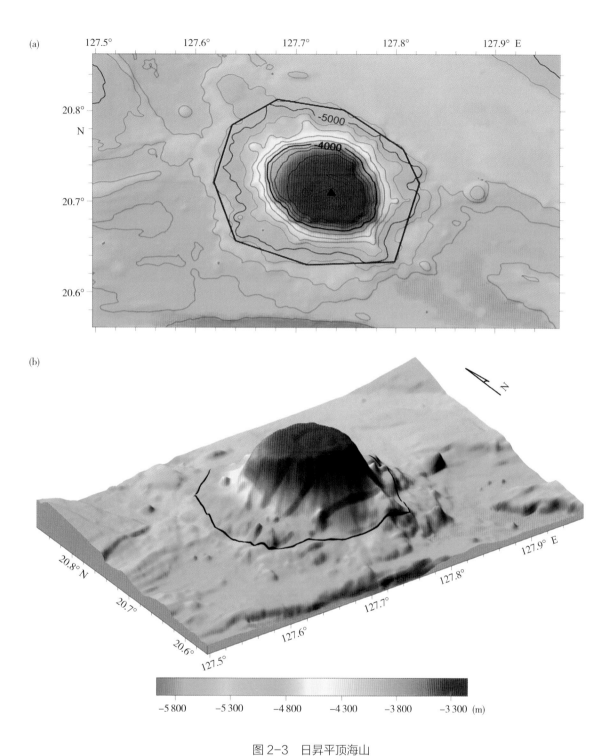

图 2-3　日昇平顶海山

(a) 地形图（等深线间隔 200 m）；(b) 三维图

Fig.2-3　Risheng Guyot

(a) Bathymetric map (the contour interval is 200 m); (b) 3-D topographic map

2.2.2 日潭海丘
Ritan Hill

中文名称 Chinese Name	日潭海丘 Ritan Haiqiu	英文名称 English Name	Ritan Hill	所在大洋 Ocean or Sea	西太平洋 West Pacific Ocean
发现情况 Discovery Facts	此海丘于 2004 年 10 月由中国科考船"大洋一号"在执行测量任务时调查发现。 This hill was discovered by the Chinese R/V *Dayang Yihao* during the survey in October, 2004.				
命名历史 Name History	由我国命名为日潭海丘，于 2012 年提交 SCUFN 审议通过。 This feature was named Ritan Hill by China and the name was approved by SCUFN in 2012.				
特征点坐标 Coordinates	21°09.40′N，127°45.20′E			长（km）× 宽（km） Length (km) × Width (km)	14 × 9
最大水深（m） Max Depth（m）	4 250	最小水深（m） Min Depth（m）	3 808	高差（m） Total Relief（m）	442
地形特征 Feature Description	位于日昇平顶海山东北侧，月潭海脊以北，海丘形状不规则，北高南低，长宽分别为 14 km 和 9 km。最浅处水深 3 808 m，山麓水深 4 250 m，高差 442 m（图 2–4）。 It is located in the northwest of Risheng Guyot, north to Yuetan Ridge. The shape of Ritan Hill is irregular, with depth less in the northern area than that in the southern area. The length is 14 km and the width is 9 km. The minimum depth of Ritan Hill is 3 808 m while the piedmont depth is 4 250 m. The total relief is 442 m (Fig.2–4).				
命名释义 Reason for Choice of Name	日月潭为著名风景名胜区，分为南北两部分，北部形如圆日，也称"日潭"，南部形如弯月，也称"月潭"。该海底地理实体形如圆日，且位于另一海底地理实体月潭海丘北部，故以"日潭"为名。 Riyue Lake is a famous scenic area in Taiwan, China, divided into northern and southern parts. The northern part shapes like sun, called Ritan (Sun Lake), whereas the southern part shapes like a half-moon, called as Yuetan (Moon Lake). The main part of the feature shapes like a sun, thus it is named Ritan Hill after the Sun Lake in Chinese language.				

(a)

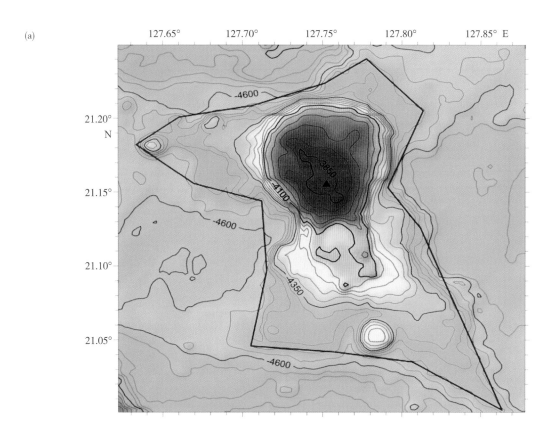

(b)

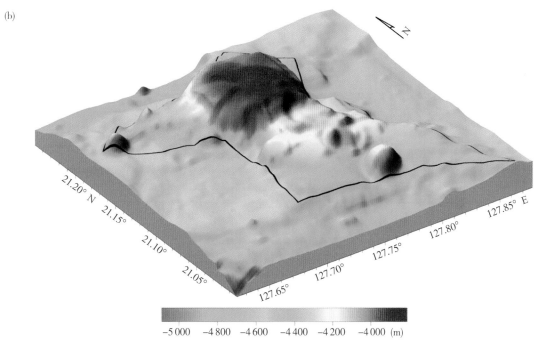

图 2-4　日潭海丘

(a) 地形图（等深线间隔 50 m）；(b) 三维图

Fig.2-4　Ritan Hill

(a) Bathymetric map (the contour interval is 50 m)；(b) 3-D topographic map

2.2.3 月潭海脊
Yuetan Ridge

中文名称 Chinese Name	月潭海脊 Yuetan Haiji	英文名称 English Name	Yuetan Ridge	所在大洋 Ocean or Sea	西太平洋 West Pacific Ocean
发现情况 Discovery Facts	此实体于 2004 年 10 月由中国科考船 "大洋一号" 在执行测量任务时调查发现。 This ridge was discovered by the Chinese R/V *Dayang Yihao* during the survey in October, 2004.				
命名历史 Name History	由我国命名为月潭海脊，于 2012 年提交 SCUFN 审议通过。 This feature was named Yuetan Ridge by China and the name was approved by SCUFN in 2012.				
特征点坐标 Coordinates	20°54.16′N，128°06.01′E			长(km)× 宽(km) Length (km) × Width (km)	130×30
最大水深（m） Max Depth（m）	5 600	最小水深（m） Min Depth（m）	4 050	高差（m） Total Relief（m）	1 550
地形特征 Feature Description	该海脊位于日昇平顶海山东北侧，日潭海丘以南。长宽分别为 130 km 和 30 km。最浅处水深 4 050 m，山麓水深 5 600 m，高差 1 550 m（图 2–5）。 Yuetan Ridge is located in the northeast of Risheng Guyot, south to Ritan Hill. The length and width are 130 km and 30 km respectively. The minimum depth is 4 050 m while the piedmont depth is 5 600 m. The total relief is 1 550 m (Fig.2–5).				
命名释义 Reason for Choice of Name	日月潭为著名风景名胜区，分为南北两部分，北部形如圆日，也称 "日潭"，南部形如弯月，也称 "月潭"。该海底地理实体形如一轮弯月，且位于另一海底地理实体日潭海丘以南，故以 "月潭" 为名。 Riyue Lake is a famous scenic area in Taiwan, China, divided into northern and southern part. The northern part shapes like sun, called Ritan (Sun Lake), whereas the southern part shapes like a half-moon, called as Yuetan (Moon Lake). The main part of this feature shapes like a half-moon, thus it is named Yuetan Ridge after the Moon Lake in Chinese language.				

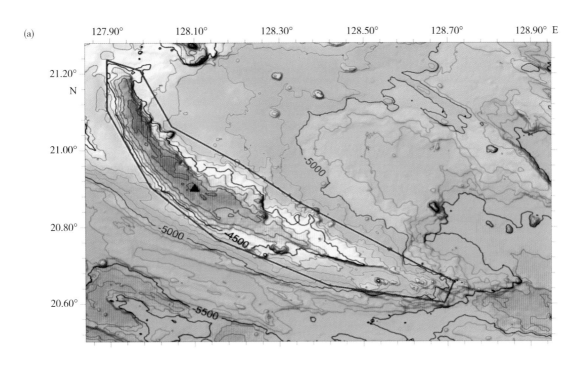

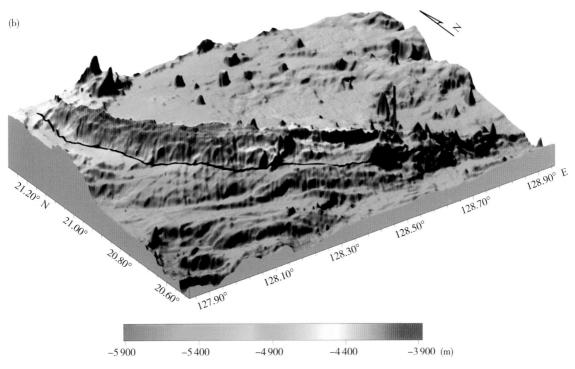

图 2-5　月潭海脊

(a) 地形图（等深线间隔 100 m）；(b) 三维图

Fig.2-5　Yuetan Ridge

(a) Bathymetric map (the contour interval is 100 m); (b) 3-D topographic map

2.2.4 海东青海山
Haidongqing Seamount

中文名称 Chinese Name	海东青海山 Haidongqing Haishan	英文名称 English Name	Haidongqing Seamount	所在大洋 Ocean or Sea	西太平洋 West Pacific Ocean
发现情况 Discovery Facts	此海山于 2004 年 6—9 月由中国测量船"李四光号"在执行测量任务时调查发现。 This seamount was discovered by the Chinese R/V *Lisiguanghao* during the survey from June to September, 2004.				
命名历史 Name History	由我国命名为海东青海山，于 2014 年提交 SCUFN 审议通过。 This feature was named Haidongqing Seamount by China and the name was approved by SCUFN in 2014.				
特征点坐标 Coordinates	18°41.70′N，125°34.40′E			长(km)×宽(km) Length (km)× Width (km)	23×20
最大水深（m） Max Depth（m）	5 319	最小水深（m） Min Depth（m）	2 697	高差（m） Total Relief（m）	2 622
地形特征 Feature Description	此海山位于吕宋海底高原北部，山体呈圆锥状，顶部水深 2 697 m（图 2–6）。 This seamount is located in the northern part of Luzon Plateau, like a cone in its formation. The top depth is 2 697 m (Fig.2–6).				
命名释义 Reason for Choice of Name	海东青，属于隼形目、隼科，是"矛隼"的一种，分布在北欧、北亚、中国吉林省、黑龙江省和辽宁省，有"万鹰之神"的含义，是中华满洲族系的最高图腾，代表勇敢、智慧、坚忍、正直、强大、开拓、进取和永不放弃的精神。以此命名海山，表现中华民族文化的精彩性和多样性。 "Haidongqing", belonging to falcon category of falconiformes, is a kind of bird scattering in North Europe, North Asia, and Jilin, Heilongjiang and Liaoning provinces of China. Haidongqing, suggesting the spirit of courage, wisdom, endurance, integrity, mightiness, creativity, progress and never surrendering, implicates "god of falcons". As a result, Haidongqing is revered as the sovereign totem of an ethnic group named Manchuria in China. The reason why the seamount is named after Haidongqing is to demonstrate the cultural brilliance and diversity of Chinese.				

(a)

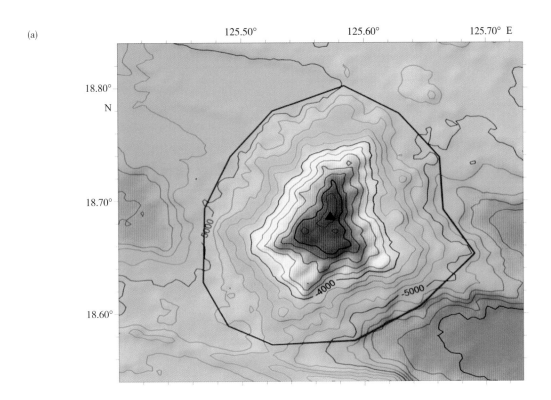

(b)

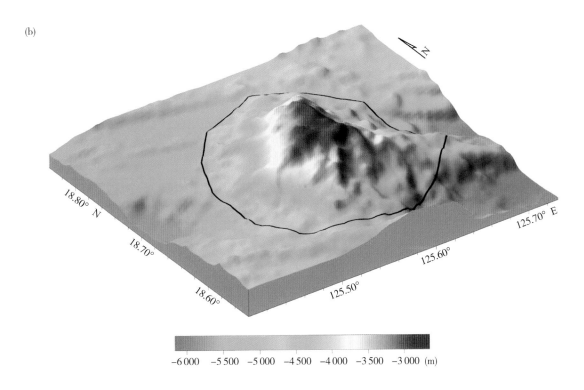

图 2-6　海东青海山

(a) 地形图（等深线间隔 200 m）；(b) 三维图

Fig.2-6　Haidongqing Seamount

(a) Bathymetric map (the contour interval is 200 m); (b) 3-D topographic map

2.2.5 天保海山
Tianbao Seamount

中文名称 Chinese Name	天保海山 Tianbao Haishan	英文名称 English Name	Tianbao Seamount	所在大洋 Ocean or Sea	西太平洋 West Pacific Ocean
发现情况 Discovery Facts	此海山于 2004 年 6—9 月由中国测量船"李四光号"在执行测量任务时调查发现。 This seamount was discovered by the Chinese R/V *Lisiguanghao* during the survey from June to September, 2004.				
命名历史 Name History	由我国命名为天保海山，于 2015 年提交 SCUFN 审议通过。 This feature was named Tianbao Seamount by China and the name was approved by SCUFN in 2015.				
特征点坐标 Coordinates	17°54.10′N，123°17.20′E			长 (km)× 宽 (km) Length (km) × Width (km)	97×59
最大水深（m） Max Depth（m）	4 230	最小水深（m） Min Depth（m）	2 005	高差（m） Total Relief（m）	2 225
地形特征 Feature Description	此海山位于吕宋海底高原北部，山体俯视平面形态呈椭圆形，顶部水深 2 005 m（图 2-7）。 This seamount is located in the northern part of Luzon Plateau, with an elliptic overlook plane shape. The top depth is 2 005 m (Fig.2–7).				
命名释义 Reason for Choice of Name	天保，上天保佑平安的意思，出自《诗经·小雅·天保》（《诗经》是公元前 11 世纪至公元前 6 世纪的中国诗歌总集）"天保定尔，亦孔之固"，指上天保佑安定，江山稳固太平。以此命名海山，表示古人对国家稳定平安的向往。 "Tianbao", meaning peace blessing, comes from a poem named *Tianbao* in *Shijing · Xiaoya*. *Shijing* is a collection of ancient Chinese poems from 11th century B.C. to 6th century B.C. "Heaven blesses you, protects you, ensures security for you." The poem means the providence blesses the peace and quietness, This title for the seamount expresses the expectation for peace and stabilization of ancient Chinese people.				

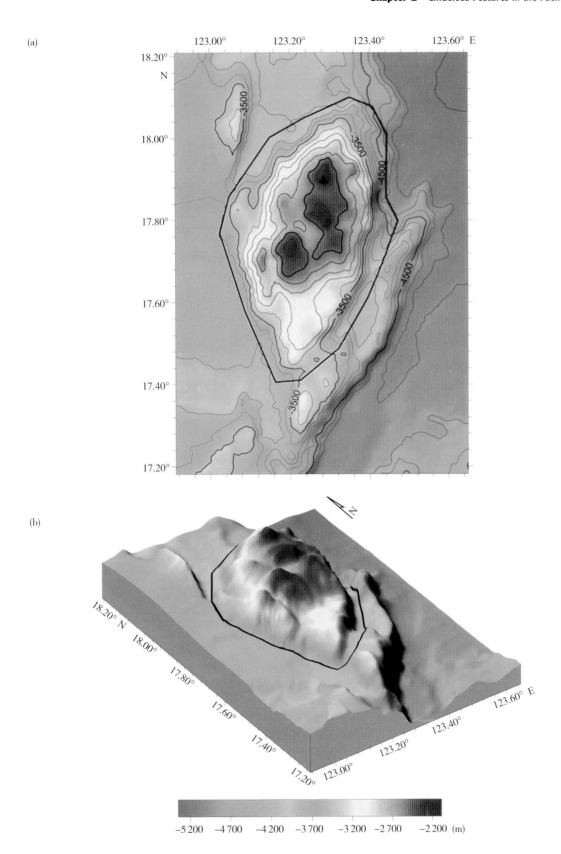

图 2-7 天保海山

(a) 地形图（等深线间隔 200 m）；(b) 三维图

Fig.2-7 Tianbao Seamount

(a) Bathymetric map (the contour interval is 200 m); (b) 3-D topographic map

2.2.6 静好海山
Jinghao Seamount

中文名称 Chinese Name	静好海山 Jinghao Haishan	英文名称 English Name	Jinghao Seamount	所在大洋 Ocean or Sea	西太平洋 West Pacific Ocean
发现情况 Discovery Facts	此海山于2004年6—9月由中国测量船"李四光号"在执行测量任务时调查发现。 This seamount was discovered by the Chinese R/V *Lisiguanghao* during the survey from June to September, 2004.				
命名历史 Name History	由我国命名为静好海山，于2015年提交SCUFN审议通过。 This feature was named Jinghao Seamount by China and the name was approved by SCUFN in 2015.				
特征点坐标 Coordinates	18°18.50′N，123°16.60′E			长(km)×宽(km) Length (km)× Width (km)	38×26
最大水深（m） Max Depth（m）	4 418	最小水深（m） Min Depth（m）	2 646	高差（m） Total Relief（m）	1 772
地形特征 Feature Description	此海山位于吕宋海底高原北部，山体形状不规则，顶部水深2 646 m（图2–8）。 This seamount is located in the northern part of Luzon Plateau, with an irregular shape. The top depth is 2 646 m (Fig.2–8).				
命名释义 Reason for Choice of Name	静好，宁静美好的意思，出自《诗经·郑风·女曰鸡鸣》。"琴瑟在御，莫不静好"，指夫妻弹琴鼓瑟相唱和，生活宁静美好。以此命名海山，表示人们对美好生活的歌颂和向往。 "Jinghao", meaning quiet and graceful literally, comes from a poem named *Nüyuejiming* in *Shijing · Zhengfeng. Shijing* is a collection of ancient Chinese poems from 11th century B.C. to 6th century B.C. "In peace and love we'll stay, ever happy, ever gay." "Jinghao" is usually quoted to depict a scene or a mood of quiet, graceful and exquisite in family life. This title is used to express the laud and yearning for an ideal life.				

(a)

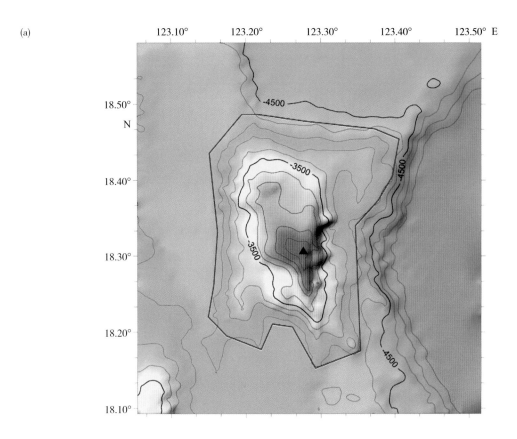

(b)

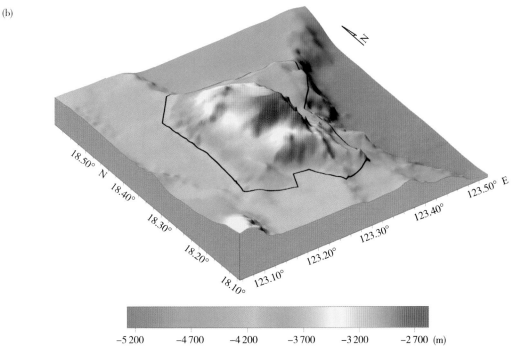

图 2-8 静好海山

(a) 地形图（等深线间隔 200 m）；(b) 三维图

Fig.2-8 Jinghao Seamount

(a) Bathymetric map (the contour interval is 200 m); (b) 3-D topographic map

2.2.7 翠翘海丘
Cuiqiao Hill

中文名称 Chinese Name	翠翘海丘 Cuiqiao Haiqiu	英文名称 English Name	Cuiqiao Hill	所在大洋 Ocean or Sea	西太平洋 West Pacific Ocean
发现情况 Discovery Facts	此海丘于 2004 年 9 月由中国测量船"李四光号"在执行测量任务时调查发现。 This hill was discovered by the Chinese R/V *Lisiguanghao* during the survey in September, 2004.				
命名历史 Name History	由我国命名为翠翘海丘，于 2016 年提交 SCUFN 审议通过。 This feature was named Cuiqiao Hill by China and the name was approved by SCUFN in 2016.				
特征点坐标 Coordinates	16°45.5′N, 124°34.3′E			长(km)× 宽(km) Length (km) × Width (km)	10.1 × 8.8
最大水深（m） Max Depth（m）	2 008	最小水深（m） Min Depth（m）	1 016	高差（m） Total Relief（m）	992
地形特征 Feature Description	此海丘位于吕宋海底高原北部，山体形状呈三角形，顶部水深 1 016 m（图 2–9）。 This hill lies in the northern part of Luzon Plateau, with triangle shape and top depth 1 016 m (Fig.2–9).				
命名释义 Reason for Choice of Name	翠翘，翠玉般的簪子。出自蔡伸的《一剪梅》。"堆枕乌云堕翠翘。午梦惊回，满眼春娇。"该诗句描写美丽的女子，日日对前线征战的丈夫的思念之情。 "Cuiqiao", means a hairpin like a piece of jade in its appearance. The expression is originated from a song *Yijianmei* by Caishen in Song Dynasty of China (A.D.960—1276). The verse is: Off her silky hair the hairpin slipped; From the afternoon doze she was aroused; And caught by the Spring flourish. The song depicts a young woman's incessant lovesickness for her husband served far in the frontier battlefield.				

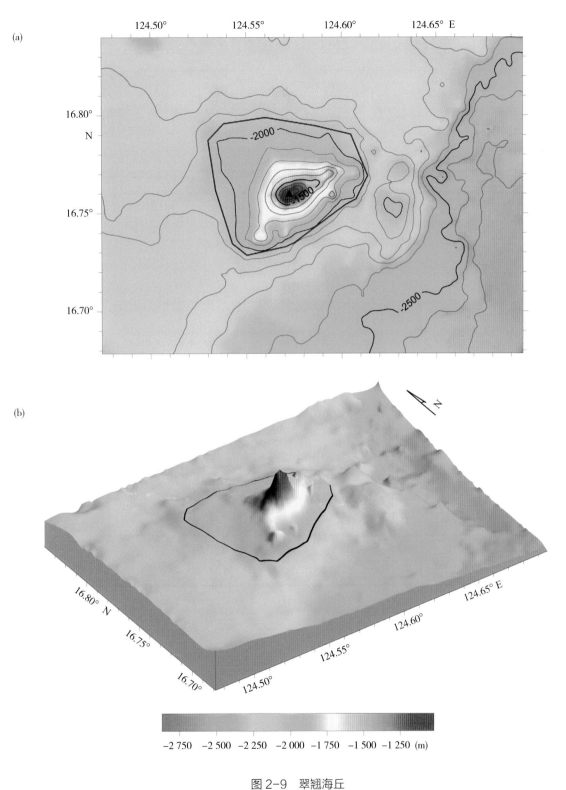

图 2-9 翠翘海丘

(a) 地形图（等深线间隔 100 m）；(b) 三维图

Fig.2-9 Cuiqiao Hill

(a) Bathymetric map (the contour interval is 100 m); (b) 3-D topographic map

2.2.8　雎鸠海山群
Jujiu Seamounts

中文名称 Chinese Name	雎鸠海山群 Jujiu Haishanqun	英文名称 English Name	Jujiu Seamounts	所在大洋 Ocean or Sea	西太平洋 West Pacific Ocean
发现情况 Discovery Facts	此海山群于 2004 年 9 月由中国测量船"李四光号"在执行测量任务时调查发现。 The seamounts were discovered by the Chinese R/V *Lisiguanghao* during the survey in September, 2004.				
命名历史 Name History	由我国命名为雎鸠海山群，于 2016 年提交 SCUFN 审议通过。 This feature was named Jujiu Seamounts by China and the name was approved by SCUFN in 2016.				
特征点坐标 Coordinates	17°05.5′N, 124°59.2′E 17°09.5′N, 124°00.2′E			长(km)×宽(km) Length (km) × Width (km)	26.6×19.8
最大水深（m） Max Depth（m）	2 850	最小水深（m） Min Depth（m）	1 804	高差（m） Total Relief（m）	1 046
地形特征 Feature Description	此海山群位于吕宋海底高原北部，山体形状呈"L"形，顶部水深 1 804 m（图 2–10）。 The seamounts lie in the northern part of Luzon Plateau, with "L" shape and top depth 1 804 m (Fig.2-10).				
命名释义 Reason for Choice of Name	雎鸠，水鸟名，即鱼鹰。出自《诗经·关雎》，"关关雎鸠，在河之洲。"《关雎》是描写古代男女恋情社会风俗习尚的诗篇，为《诗经》首篇，反应古代人对伦理的思想。 "Jujiu", name of a kind of waterfowl, selected from *Shijing* (the Book of Poetry of China). The original text is: The waterfowl would coo upon a sandbar in the brook. The song, as the first of The Book of Poetry, with the subject of courtship in ancient China, reflects certain attitude to social life at that time.				

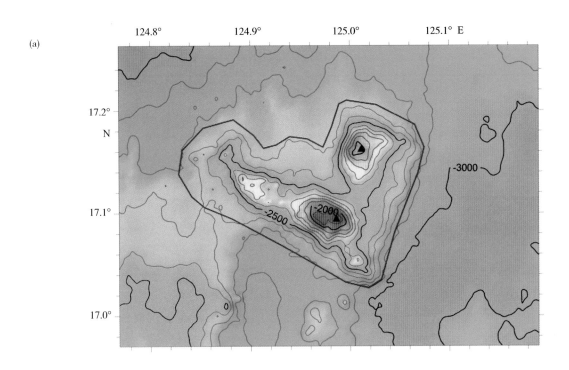

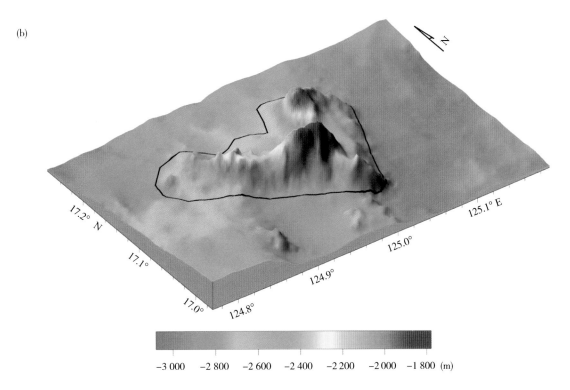

图 2-10　睢鸠海山群

(a) 地形图（等深线间隔 100 m）；(b) 三维图

Fig.2-10　Jujiu Seamounts

(a) Bathymetric map (the contour interval is 100 m); (b) 3-D topographic map

2.2.9 槐序海山
Huaixu Seamount

中文名称 Chinese Name	槐序海山 Huaixu Haishan	英文名称 English Name	Huaixu Seamount	所在大洋 Ocean or Sea	西太平洋 West Pacific Ocean
发现情况 Discovery Facts	此海山于 2004 年 9 月由中国测量船"李四光号"在执行测量任务时调查发现。 This seamount was discovered by the Chinese R/V *Lisiguanghao* during the survey in September, 2004.				
命名历史 Name History	由我国命名为槐序海山，于 2017 年提交 SCUFN 审议通过。 This feature was named Huaixu Seamount by China and the name was approved by SCUFN in 2017.				
特征点坐标 Coordinates	16°59.9′N, 129°29.3′E			长(km)×宽(km) Length (km) × Width (km)	39.05×22.21
最大水深（m） Max Depth（m）	5 130	最小水深（m） Min Depth（m）	3 554	高差（m） Total Relief（m）	1 576
地形特征 Feature Description	此海山位于菲律宾海盆东部，俯视平面形状呈椭圆，顶部水深 3 554 m（图 2–11）。 This Seamount is located in the eastern part of Philippine Basin, with ellipsoid form and minimum depth 3 554 m (Fig.2–11).				
命名释义 Reason for Choice of Name	槐序，即中国农历四月份的别称，中国农历四月是夏初，槐树开始开花的时令。古人将一年的月令与天气、万物的变化相结合，创造出诗情画意的月份别称，体现古人的智慧与情趣。 "Huaixu", another name for April in Chinese lunar calendar, i.e. the beginning of the summer when Chinese scholartrees blossom. The poetic and pictorial inspiring name, created by associating month, climate and the changes of great nature, conveys the wisdom and interests of people living in the ancient world.				

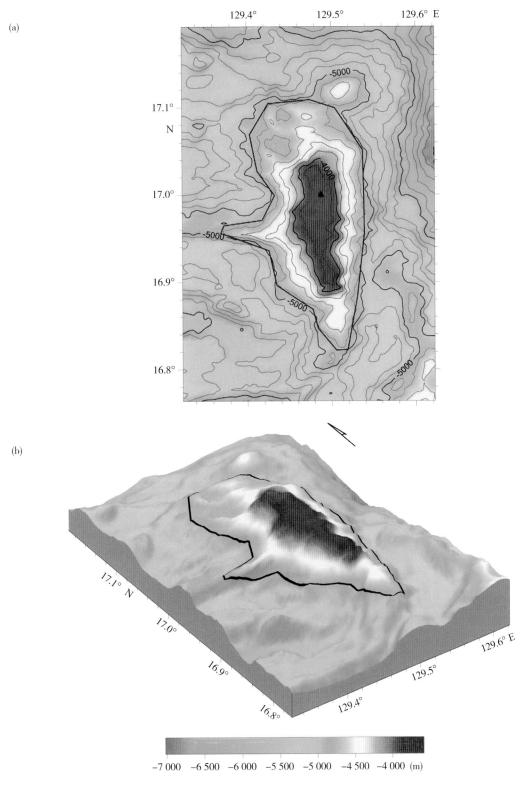

图 2-11　槐序海山

(a) 地形图（等深线间隔 200 m）；(b) 三维图

Fig.2-11　Huaixu Seamount

(a) Bathymetric map (the contour interval is 200 m); (b) 3-D topographic map

2.2.10　首阳海山
Shouyang Seamount

中文名称 Chinese Name	首阳海山 Shouyang Haishan	英文名称 English Name	Shouyang Seamount	所在大洋 Ocean or Sea	西太平洋 West Pacific Ocean
发现情况 Discovery Facts	此海山于2004年9月由中国测量船"李四光号"在执行测量任务时调查发现。 This seamount was discovered by the Chinese R/V *Lisiguanghao* during the survey in September, 2004.				
命名历史 Name History	由我国命名为首阳海山，于2017年提交SCUFN审议通过。 This feature was named Shouyang Seamount by China and the name was approved by SCUFN in 2017.				
特征点坐标 Coordinates	17°31.6′N, 128°45.5′E			长(km)×宽(km) Length (km) × Width (km)	16.0×14.7
最大水深（m） Max Depth（m）	5 228	最小水深（m） Min Depth（m）	3 520	高差（m） Total Relief（m）	1 708
地形特征 Feature Description	此海山位于吕宋岛东部约280 n mile，山体形状呈叉子状，顶部水深3 520 m（图2-12）。 This Seamount is located in the eastern part of Philippine Basin, with fork shape and minimum depth 3 520 m (Fig.2-12).				
命名释义 Reason for Choice of Name	首阳，即中国农历一月份的别称，农历一月是初春的时候，寒气逐渐退去，大地开始回暖。古人将一年的月令与天气、万物的变化相结合，创造出诗情画意的月份别称，体现古人的智慧与情趣。 "Shouyang", another name for January in Chinese lunar calendar, i.e. the beginning of the spring when the grim cold air gives way to the all encompassing warmth imperceptibly. The poetic and pictorial inspiring appellation, created by associating month, climate and the changes of great nature, conveys the wisdom and wisdom of people living in the ancient world.				

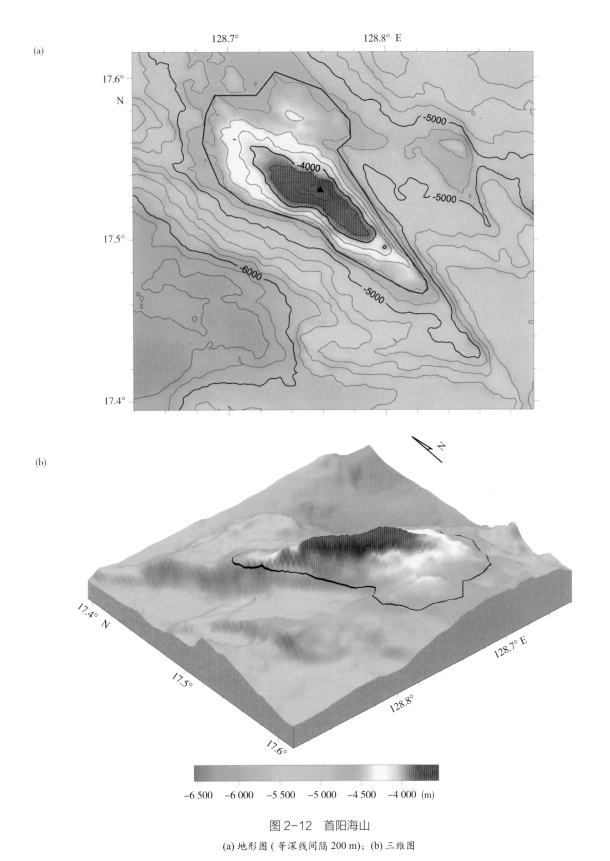

图 2-12　首阳海山

(a) 地形图（等深线间隔 200 m）；(b) 三维图

Fig.2-12　Shouyang Seamount

(a) Bathymetric map (the contour interval is 200 m); (b) 3-D topographic map

2.2.11 立春海山
Lichun Seamount

中文名称 Chinese Name	立春海山 Lichun Haishan	英文名称 English Name	Lichun Seamount	所在大洋 Ocean or Sea	西太平洋 West Pacific Ocean
发现情况 Discovery Facts	此海山于 2014 年 9 月由中国科考船"向阳红 10"船在执行测量任务时调查发现。 This seamount was discovered by the Chinese R/V *Xiangyanghong* 10 during the survey in September, 2014.				
命名历史 Name History	由我国命名为立春海山，于 2017 年提交 SCUFN 审议通过。 This feature was named Lichun Seamount by China and the name was approved by SCUFN in 2017.				
特征点坐标 Coordinates	14°08.0'N, 132°45.2'E		长(km)×宽(km) Length (km) × Width (km)		10×10
最大水深（m） Max Depth（m）	5 800	最小水深（m） Min Depth（m）	4 070	高差（m） Total Relief（m）	1 730
地形特征 Feature Description	该海山俯视平面形态近圆形，基座直径 10 km 左右，顶部水深 4 070 m，山麓水深 5 800 m，海山北坡地形较缓，南坡地形较陡（图 2–13）。 This seamount has a nearly round overlook plane shape. The base diameter is about 10km. The water depth is about 4 070 m to the top and about 5800 m to foothills. And the northern slope of the terrain is slow while the southern slope is steep (Fig.2–13).				
命名释义 Reason for Choice of Name	立春，农历二十四节气中的第一个节气，也是中国民间重要的传统节日之一。"立"是"开始"的意思，自秦代以来，中国就一直以立春作为孟春时节的开始。 The UN Educational, Scientific, and Cultural Organization (UNESCO) adopted a decision that China's "the 24 Solar Terms" be inscribed on the Representative List of the Intangible Cultural Heritage of Humanity on 30 November, 2016 in Ethiopia's capital Addis Ababa. "The 24 Solar Terms" is the Chinese heritage and knowledge in China of time and practices developed through observation of the sun's annual motion. The ancient Chinese divided the sun's annual circular motion into 24 segments. Each segment was called a specific Solar Term. "Lichun", the first term of the 24 Solar Terms, means spring commences.				

(a)

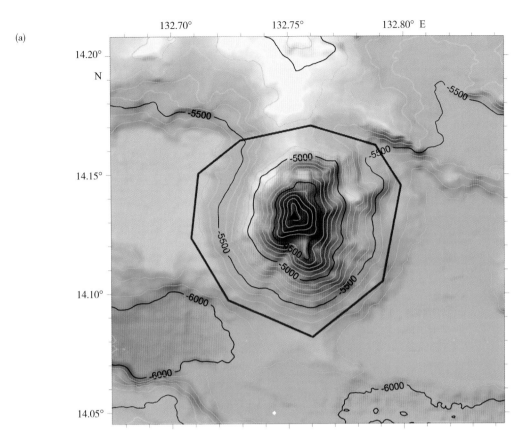

(b)

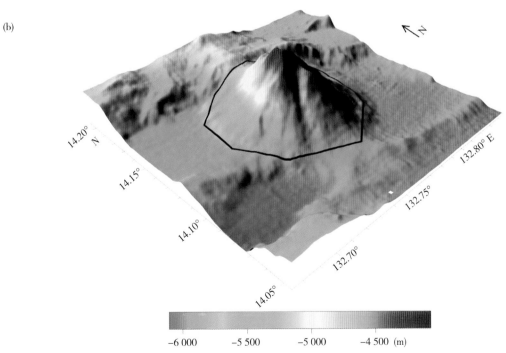

图 2-13　立春海山

(a) 地形图 (等深线间隔 100 m)；(b) 三维图

Fig.2-13　Lichun Seamount

(a) Bathymetric map (the contour interval is 100 m); (b) 3-D topographic map

2.2.12 惊蛰海山
Jingzhe Seamount

中文名称 Chinese Name	惊蛰海山 Jingzhe Haishan		英文名称 English Name	Jingzhe Seamount	所在大洋 Ocean or Sea	西太平洋 West Pacific Ocean
发现情况 Discovery Facts	此海山于 2014 年 9 月由中国科考船"向阳红 10"船在执行测量任务时调查发现。 This seamount was discovered by the Chinese R/V *Xiangyanghong* 10 during the survey in September, 2014.					
命名历史 Name History	由我国命名为惊蛰海山，于 2017 年提交 SCUFN 审议通过。 This feature was named Jingzhe Seamount by China and the name was approved by SCUFN in 2017.					
特征点坐标 Coordinates	14°41.7'N, 134°34.1'E			长(km)× 宽(km) Length (km) × Width (km)		20×9
最大水深（m） Max Depth（m）	3 670	最小水深（m） Min Depth（m）	1 900	高差（m） Total Relief（m）		1 770
地形特征 Feature Description	该海山位于九州－帕劳海脊之上，基座大小 20 km。顶部水深 1 900 m，山麓水深 3 670 m，海山北坡地形较缓，南坡地形较陡（图 2-14）。 This seamount is located on the Kyushu-Palau ridge. with a base size of 20 km. The water depth is about 1 900 m to the top and about 3 670 m to foothills. And the seamount northern slope is relatively slow yet southern slope is steep (Fig.2-14).					
命名释义 Reason for Choice of Name	惊蛰，是二十四节气之中的第三个节气。惊蛰节气在农忙上有着相当重要的意义。自古以来，我国劳动人民很重视惊蛰节气，把它视为春耕开始的日子。 The UN Educational, Scientific, and Cultural Organization (UNESCO) adopted a decision that China's "the 24 Solar Terms" be inscribed on the Representative List of the Intangible Cultural Heritage of Humanity on 30 November, 2016 in Ethiopia's capital Addis Ababa. "The 24 Solar Terms" is the Chinese heritage and knowledge in China of time and practices developed through observation of the sun's annual motion. The ancient Chinese divided the sun's annual circular motion into 24 segments. Each segment was called a specific Solar Term. "Jingzhe", the third term of the 24 Solar Terms, means "insets waken". "Jingzhe " plays a very important role in the busy farming season. In China, people have always been high valuing "Jingzhe", and regard it as the beginning of the spring farming season.					

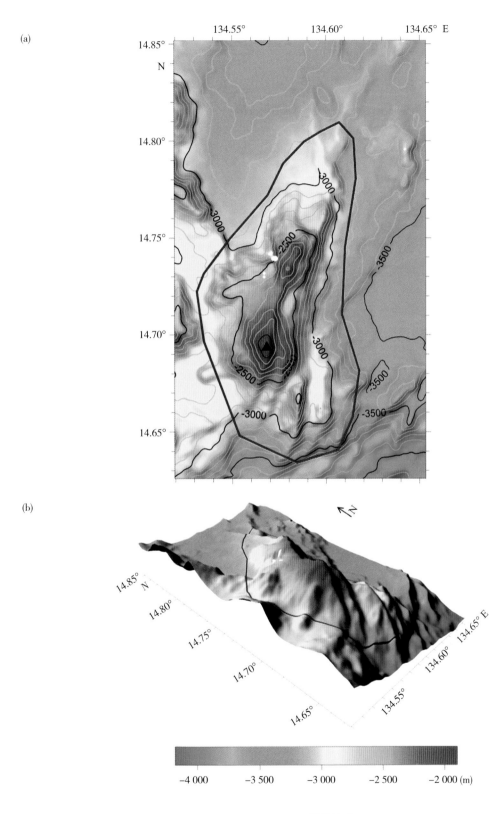

图 2-14　惊蛰海山

(a) 地形图 (等深线间隔 100 m)；(b) 三维图

Fig.2-14　Jingzhe Seamount

(a) Bathymetric map (the contour interval is 100 m); (b) 3-D topographic map

2.2.13 春分海丘
Chunfen Hill

中文名称 Chinese Name	春分海丘 Chunfen Haiqiu	英文名称 English Name	Chunfen Hill	所在大洋 Ocean or Sea	西太平洋 West Pacific Ocean
发现情况 Discovery Facts	此海丘于 2014 年 9 月由中国科考船"向阳红 10"船在执行测量任务时调查发现。 This hill was discovered by the Chinese R/V *Xiangyanghong* 10 during the survey in September, 2014.				
命名历史 Name History	由我国命名为春分海丘,于 2017 年提交 SCUFN 审议通过。 This feature was named Chunfen Hill by China and the name was approved by SCUFN in 2017.				
特征点坐标 Coordinates	14°42.1'N, 134°31.0'E		长(km)×宽(km) Length (km)× Width (km)		5×4
最大水深(m) Max Depth(m)	3 150	最小水深(m) Min Depth(m)	2 300	高差(m) Total Relief(m)	850
地形特征 Feature Description	该海丘位于九州 – 帕劳海脊之上,俯视平面形态近圆形,基座直径 5 km。海丘顶部水深 2 300 m,海丘底部水深 3 150 m(图 2–15)。 This hill is located on the Kyushu-Palau ridge. It has a nearly round overlook plane shape with a base size of 5 km. The water depth is about 2 300 m to the top and about 3 150 m to foothills (Fig.2–15).				
命名释义 Reason for Choice of Name	春分,是二十四节气之中的第四个节气,是春季九十天的中分点。春分是节日和祭祀庆典,古代帝王有春天祭日,秋天祭月的礼制。 The UN Educational, Scientific, and Cultural Organization (UNESCO) adopted a decision that China's "the 24 Solar Terms" be inscribed on the Representative List of the Intangible Cultural Heritage of Humanity on 30 November, 2016 in Ethiopia's capital Addis Ababa. "The 24 Solar Terms" is the Chinese heritage and knowledge in China of time and practices developed through observation of the sun's annual motion. The ancient Chinese divided the sun's annual circular motion into 24 segments. Each segment was called a specific Solar Term. "Chunfen", the fourth term of the 24 Solar Terms, means vernal equi-nox. The ancient emperors offered sacrifices to the God of Sun in "Chunfen".				

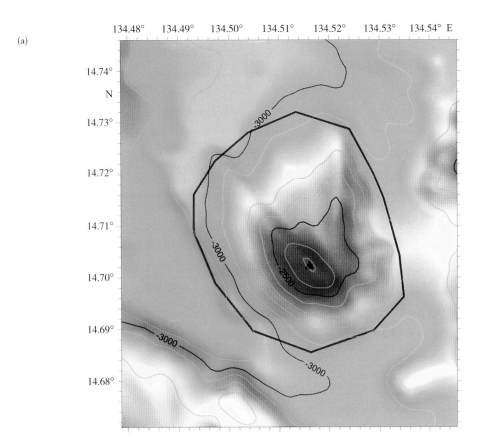

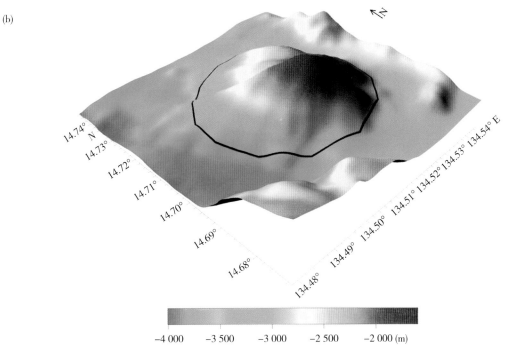

图 2-15　春分海丘

(a) 地形图 (等深线间隔 100 m)；(b) 三维图

Fig.2-15　Chunfen Hill

(a) Bathymetric map (the contour interval is 100 m); (b) 3-D topographic map

2.2.14 清明海山
Qingming Seamount

中文名称 Chinese Name	清明海山 Qingming Haishan	英文名称 English Name	Qingming Seamount	所在大洋 Ocean or Sea	西太平洋 West Pacific Ocean
发现情况 Discovery Facts	此海山于 2014 年 9 月由中国科考船"向阳红 10"船在执行测量任务时调查发现。 This seamount was discovered by the Chinese R/V *Xiangyanghong* 10 during the survey in September, 2014.				
命名历史 Name History	由我国命名为清明海山，于 2017 年提交 SCUFN 审议通过。 This feature was named Qingming Seamount by China and the name was approved by SCUFN in 2017.				
特征点坐标 Coordinates	15°10.1′N, 133°55.2′E		长(km)×宽(km) Length (km)× Width (km)	26.5×9	
最大水深（m） Max Depth（m）	4 850	最小水深（m） Min Depth（m）	3 410	高差（m） Total Relief（m）	1 440
地形特征 Feature Description	该海山位于芒种海盆东侧，基座大小约为 26 km，顶部水深 3 410 m，山麓水深 4 850 m，高差 1 440 m，海山东西坡地形较缓，南北坡地形较陡（图 2–16）。 This seamount is adjacent to the east side of the "Mangzhong" basin. Its base size is about 26 km. The water depth is about 3 410 m to the top and about 4 850 m to foothills. The eastern and western slope terrain of the seamount is very slow, while the terrain of the north and south slope is steep (Fig.2–16).				
命名释义 Reason for Choice of Name	清明，是二十四节气之中的第五个节气，清明节是中国重要的"时年八节"之一，与端午节、春节、中秋节并称为中国四大传统节日。 The UN Educational, Scientific, and Cultural Organization (UNESCO) adopted a decision that China's "the 24 Solar Terms" be inscribed on the Representative List of the Intangible Cultural Heritage of Humanity on 30 November, 2016 in Ethiopia's capital Addis Ababa. "The 24 Solar Terms" is the Chinese heritage and knowledge in China of time and practices developed through observation of the sun's annual motion. The ancient Chinese divided the sun's annual circular motion into 24 segments. Each segment was called a specific Solar Term. "Qingming", the fifth term of the 24 Solar Terms, means pure brightness. Qingming is also a festival when people commemorate their passed family members and ancestors at grave sites.				

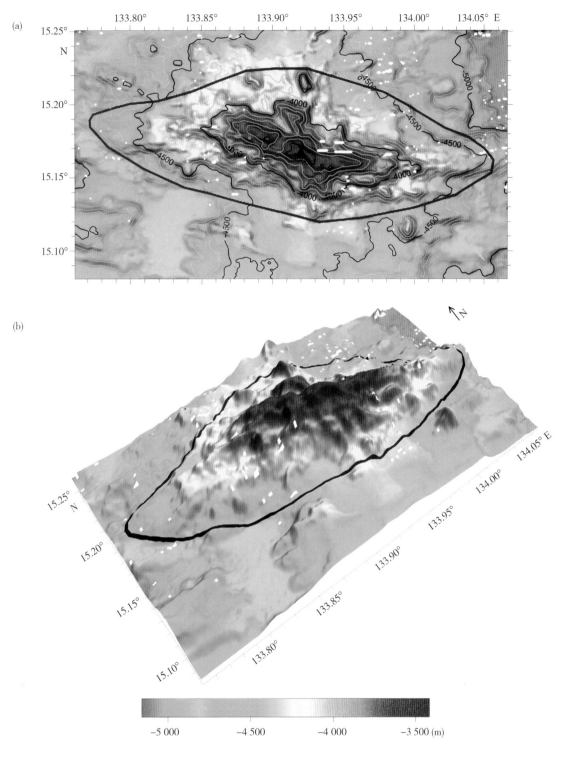

图 2-16　清明海山

(a) 地形图 (等深线间隔 100 m)；(b) 三维图

Fig.2-16　Qingming Seamount

(a) Bathymetric map (the contour interval is 100 m); (b) 3-D topographic map

2.2.15 谷雨海山
Guyu Seamount

中文名称 Chinese Name	谷雨海山 Guyu Haishan	英文名称 English Name	Guyu Seamount	所在大洋 Ocean or Sea	西太平洋 West Pacific Ocean
发现情况 Discovery Facts	此海山于 2014 年 9 月由中国科考船"向阳红 10"船在执行测量任务时调查发现。 This seamount was discovered by the Chinese R/V *Xiangyanghong* 10 during the survey in September, 2014.				
命名历史 Name History	由我国命名为谷雨海山，于 2017 年提交 SCUFN 审议通过。 This feature was named Guyu Seamount by China and the name was approved by SCUFN in 2017.				
特征点坐标 Coordinates	15°43.6'N, 131°40.8'E			长(km)×宽(km) Length (km) × Width (km)	22 × 18
最大水深（m） Max Depth（m）	5 900	最小水深（m） Min Depth（m）	4 180	高差（m） Total Relief（m）	1 720
地形特征 Feature Description	该海山位于菲律宾海，其西侧 250 km 处为清明海山，俯视平面形态近圆形，基座直径 22 km，海山顶部水深 4 180 m，海山底部水深 5 900 m（图 2–17）。 This seamount is located in Philippine Sea, with Qingming Seamount in its west direction about 250 km. It has a nearly circle overlook plane shape with a base size of 22 km. The water depth is about 4 180 m to the top and about 5 900 m to foothills. (Fig.2–17).				
命名释义 Reason for Choice of Name	谷雨，农历二十四节气中的第六个节气，也是春季最后一个节气。意味着寒潮天气基本结束，气温回升加快，有利于谷类农作物的生长。 The UN Educational, Scientific, and Cultural Organization (UNESCO) adopted a decision that China's "the 24 Solar Terms" be inscribed on the Representative List of the Intangible Cultural Heritage of Humanity on 30 November, 2016 in Ethiopia's capital Addis Ababa. "The 24 Solar Terms" is the Chinese heritage and knowledge in China of time and practices developed through observation of the sun's annual motion. The ancient Chinese divided the sun's annual circular motion into 24 segments. Each segment was called a specific Solar Term. "Guyu", the sixth term of the 24 Solar Terms, means grain rain. During the time, the cold weather is basically over and the temperature will speedy picked up. It is conducive to the growth of cereal crops".				

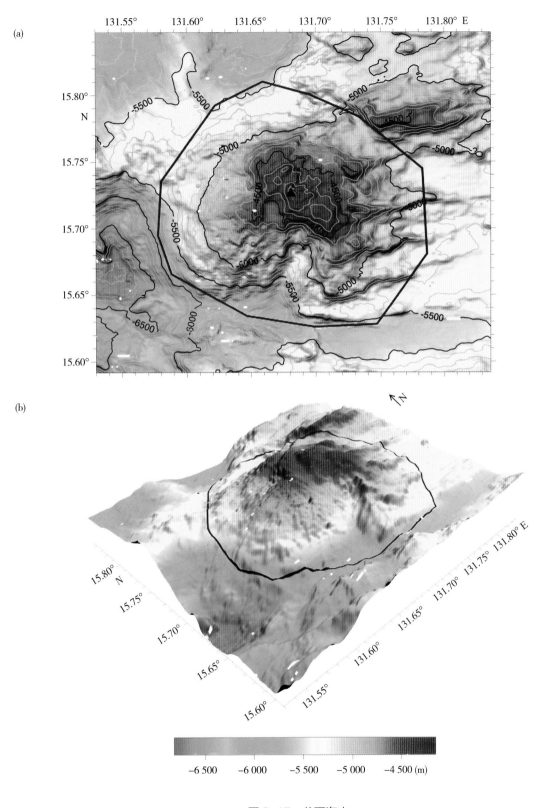

图 2-17　谷雨海山

(a) 地形图 (等深线间隔 100 m)；(b) 三维图

Fig.2-17　Guyu Seamount

(a) Bathymetric map (the contour interval is 100 m); (b) 3-D topographic map

2.2.16 小满海丘
Xiaoman Hill

中文名称 Chinese Name	小满海丘 Xiaoman Haiqiu	英文名称 English Name	Xiaoman Hill	所在大洋 Ocean or Sea	西太平洋 West Pacific Ocean
发现情况 Discovery Facts	此海丘于 2014 年 9 月由中国科考船"向阳红 10"船在执行测量任务时调查发现。 This hill was discovered by the Chinese R/V *Xiangyanghong* 10 during the survey in September, 2014.				
命名历史 Name History	由我国命名为小满海丘，于 2017 年提交 SCUFN 审议通过。 This feature was named Xiaoman Hill by China and the name was approved by SCUFN in 2017.				
特征点坐标 Coordinates	15°32.6′N, 134°32.8′E		长(km)×宽(km) Length (km)× Width (km)		9×8.5
最大水深（m） Max Depth（m）	3 400	最小水深（m） Min Depth（m）	2 600	高差（m） Total Relief（m）	800
地形特征 Feature Description	该海丘位于九州－帕劳海脊之上，基座大小 9 km，顶部最小水深 2 600 m，山麓水深 3 400 m，海丘西北坡地形较缓，东南坡地形较陡（图 2–18）。 This hill is located on the Kyushu-Palau ridge, The base size is about 9 km. The water depth is about 2 600 m to the top and about 3 400 m to foothills. And the northwestern slope is slow yet southeastern slope is steep (Fig.2–18).				
命名释义 Reason for Choice of Name	小满，农历二十四节气中的第八个节气，小满其含义是夏熟作物的籽粒开始灌浆饱满，但还未成熟。 The UN Educational, Scientific, and Cultural Organization (UNESCO) adopted a decision that China's "the 24 Solar Terms" be inscribed on the Representative List of the Intangible Cultural Heritage of Humanity on 30 November, 2016 in Ethiopia's capital Addis Ababa. "The 24 Solar Terms" is the Chinese heritage and knowledge in China of time and practices developed through observation of the sun's annual motion. The ancient Chinese divided the sun's annual circular motion into 24 segments. Each segment was called a specific Solar Term. "Xiaoman", the eighth sixth term of the 24 Solar Terms, means the summer ripe grain begins to full , but not mature yet.				

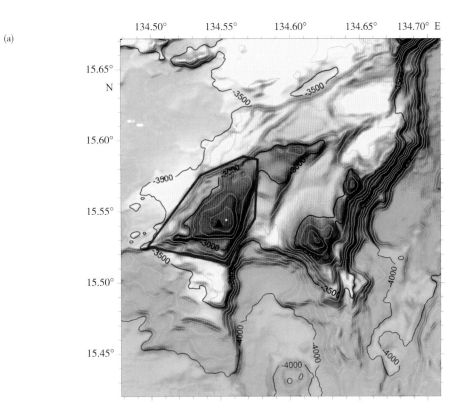

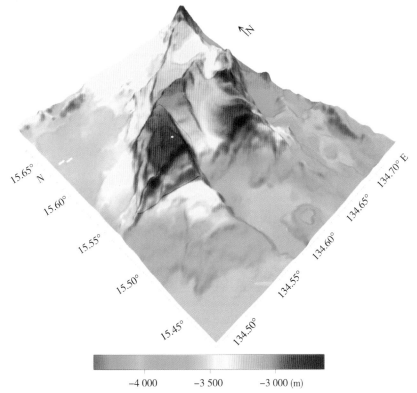

图 2-18　小满海丘

(a) 地形图（等深线间隔 100 m）；(b) 三维图

Fig.2-18　Xiaoman Hill

(a) Bathymetric map (the contour interval is 100 m); (b) 3-D topographic map

2.2.17　芒种海盆
Mangzhong Basin

中文名称 Chinese Name	芒种海盆 Mangzhong Haipen	英文名称 English Name	Mangzhong Basin	所在大洋 Ocean or Sea	西太平洋 West Pacific Ocean
发现情况 Discovery Facts	此海盆于 2014 年 9 月由中国科考船"向阳红 10"船在执行测量任务时调查发现。 This basin was discovered by the Chinese R/V *Xiangyanghong* 10 during the survey in September, 2014.				
命名历史 Name History	由我国命名为芒种海盆，于 2017 年提交 SCUFN 审议通过。 This feature was named Mangzhong Basin by China and the name was approved by SCUFN in 2017.				
特征点坐标 Coordinates	15°13.7′N, 134°07.3′E		长(km)×宽(km) Length (km) × Width (km)		32×22
最大水深（m） Max Depth（m）	5 200	最小水深（m） Min Depth（m）	4 500	高差（m） Total Relief（m）	700
地形特征 Feature Description	该海盆位于清明海山西侧，俯视平面形态近长方形，海盆长约为 32 km，最浅处水深 4 500 m，最深处水深 5 200 m，海盆内的平均坡度为 1.7°（图 2–19）。 This ocean basin is adjacent to the west side of Qingming Seamount. It has a nearly rectangle overlook plane shape, with length of 32 km. The shallowest water depth is about 4 500 m, and the deepest depth is about 5 200 m. The average slope is 1.7° in this basin (Fig.2–19).				
命名释义 Reason for Choice of Name	芒种，农历二十四节气中的第九个节气，夏季的第三个节气，表示仲夏时节的正式开始。芒种字面的意思是"有芒的麦子快收，有芒的稻子可种"。 The UN Educational, Scientific, and Cultural Organization (UNESCO) adopted a decision that China's "the 24 Solar Terms" be inscribed on the Representative List of the Intangible Cultural Heritage of Humanity on 30 November, 2016 in Ethiopia's capital Addis Ababa. "The 24 Solar Terms" is the Chinese heritage and knowledge in China of time and practices developed through observation of the sun's annual motion. The ancient Chinese divided the sun's annual circular motion into 24 segments. Each segment was called a specific Solar Term. "Mangzhong", the ninth term of the 24 Solar Terms, declares the midsummer officially begins. "Mangzhong" literally means Grain in Beard".				

(a)

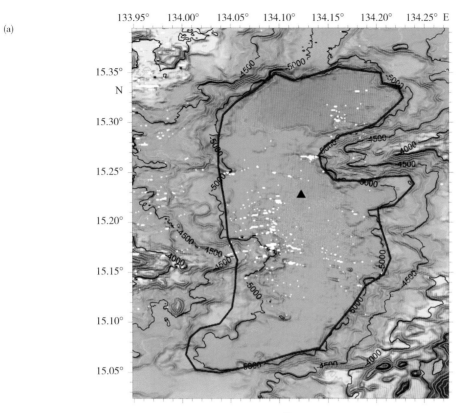

(b)

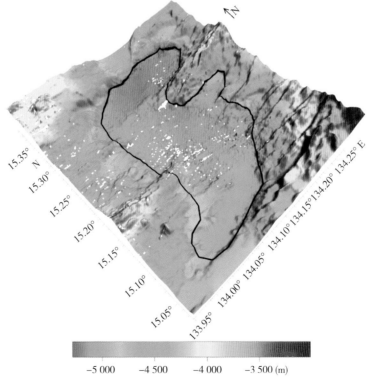

图 2-19　芒种海盆

(a) 地形图 (等深线间隔 100 m)；(b) 三维图

Fig.2-19　Mangzhong Basin

(a) Bathymetric map (the contour interval is 100 m); (b) 3-D topographic map

2.2.18 飒沓海山
Sata Seamount

中文名称 Chinese Name	飒沓海山 Sata Haishan	英文名称 English Name	Sata Seamount	所在大洋 Ocean or Sea	西太平洋 West Pacific Ocean
发现情况 Discovery Facts	此海山于 2004 年由中国测量船 "李四光号" 在执行测量任务时调查发现。 This seamount was discovered by the Chinese R/V *LiSiguangHao* during the survey in 2004.				
命名历史 Name History	在我国大洋航次调查报告和 GEBCO 地名辞典中未命名。 This seamount has not been named in Chinese cruise reports or GEBCO gazetteer.				
特征点坐标 Coordinates	18°14.8′N, 126°17.5′E			长(km)× 宽(km) Length (km) × Width (km)	40×35
最大水深（m） Max Depth（m）	5 552	最小水深（m） Min Depth（m）	2 432	高差（m） Total Relief（m）	3 120
地形特征 Feature Description	此海山位于吕宋深海高原北部，山体形状呈三角形，顶部水深 2 432 m（图 2–20）。 This seamount lies in the northern part of Luzon Plateau, with triangle shape and top depth 2 432 m (Fig.2–20).				
命名释义 Reason for Choice of Name	飒沓：群飞的样子，形容马跑得快。出自李白的诗句《侠客行》"银鞍照白马，飒沓如流星。"该诗句抒发作者对侠客的倾慕，对拯危济难、用世立功生活的向往。 Sata, describing the state of flying in swarm or horse galloping, is taken from Ode to Gallantry by Li Bai. The original text is: The silver saddle illuminated the white horse; Its galloping was like a shooting star. The poet showed his admiration for chivalrous swordsmen and longing for military exploits.				

(a)

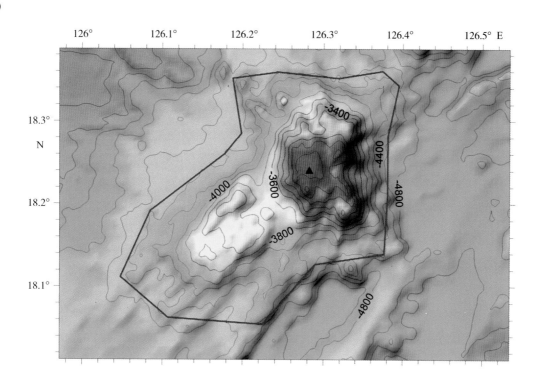

(b)

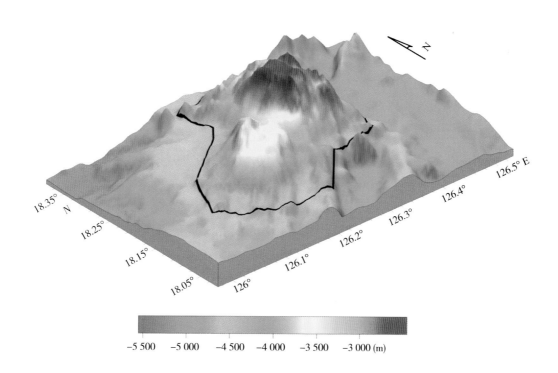

图 2-20　飒沓海山

(a) 地形图（等深线间隔 100 m）；(b) 三维图

Fig.2-20　Sata Seamount

(a) Bathymetric map (the contour interval is 100 m); (b) 3-D topographic map

第 3 节 麦哲伦海山区地理实体

麦哲伦海山区位于西太平洋马里亚纳海盆以北，西邻马里亚纳海沟，由近 20 个海山 / 海山群组成。海山区大致上呈链状排列，NW—SE 向延伸，长度超过 1 500 km。此海山区内的地理实体以大型平顶海山为主，平顶海山由地形平坦的山顶平台和陡峭的山坡构成，平台水深 1 400 ～ 2 200 m，山麓水深 5 000 ～ 6 000 m。

麦哲伦海山区是库拉板块内部的规模较大的海底火山活动区域，其展布受一条规模较大的断裂控制，岩浆沿断裂带喷发或溢出形成海山。火山岩浆喷发活动主要发生于晚侏罗纪至第三纪期间，此后，海山随板块漂移的同时曾隆起出露或接近海平面，遭受风化剥蚀，此后再度下沉至海面之下并接受沉积，发育成现今的平顶海山。

在此海域我国命名地理实体 12 个，翻译地名 8 个。其中海山群 3 个，包括骐骆平顶海山群、采薇海山群和嘉偕平顶海山群；海山 1 个，为采菽海山；平顶海山 16 个，包括维骐平顶海山、维骆平顶海山、劳里平顶海山、鹿鸣平顶海山、赫姆勒平顶海山、戈沃罗夫平顶海山、斯科尔尼亚科瓦平顶海山、戈尔金平顶海山、伊利切夫平顶海山、佩加斯平顶海山、采薇平顶海山、采杞平顶海山、维嘉平顶海山、维祯平顶海山、维偕平顶海山和布塔科夫平顶海山（图 2–21）。

Section 3　Undersea Features in the Magellan Seamount Area

The Magellan Seamount Area is located in the north of Mariana Basin in West Pacific, adjacent to Mariana Trench to the west, consisting of nearly 20 seamounts. The seamounts arranged generally in chain, and runs NW to SE with the length of over 1 500 km. The undersea features in this seamount area are mainly large guyots, which are made up of flat top platforms and steep slopes. The water depth of the platform is 1 400–2 200 m, while the piedmont depth is 5 000–6 000 m.

The Magellan Seamount Area is a relatively large-scale undersea volcanic chain inside the Kula Plate, whose extension is controlled by a large-scale fault. The lava erupts or overflows along the fracture zone, forming seamounts. Volcanic eruption activity occurs mainly during the Late Jurassic to the Tertiary period. After that, these seamounts uplifted out of or near the sea surface while drifting with plates. They suffered from weathering and denudation, then again to sink beneath the sea surface and received deposits, developing into current guyots.

In total, 12 undersea features have been named and 8 have been translated by China in the region of the Magellan Seamount Area. Among them, there are 3 seamounts, including Qiluo Guyots, Caiwei Seamounts and Jiaxie Guyots; 1 seamount, which is Caishu Seamount; 16 guyots, including Weiqi Guyot, Weiluo Guyot, Lowrie Guyot, Luming Guyot, Hemler Guyot, Govorov Guyot, Skornyakova Guyot, Gordin Guyot, Il'ichev Guyot, Pegas Guyot, Caiwei Guyot, Caiqi Guyot, Weijia Guyot, Weizhen Guyot, Weixie Guyot and Butakov Guyot (Fig. 2–21).

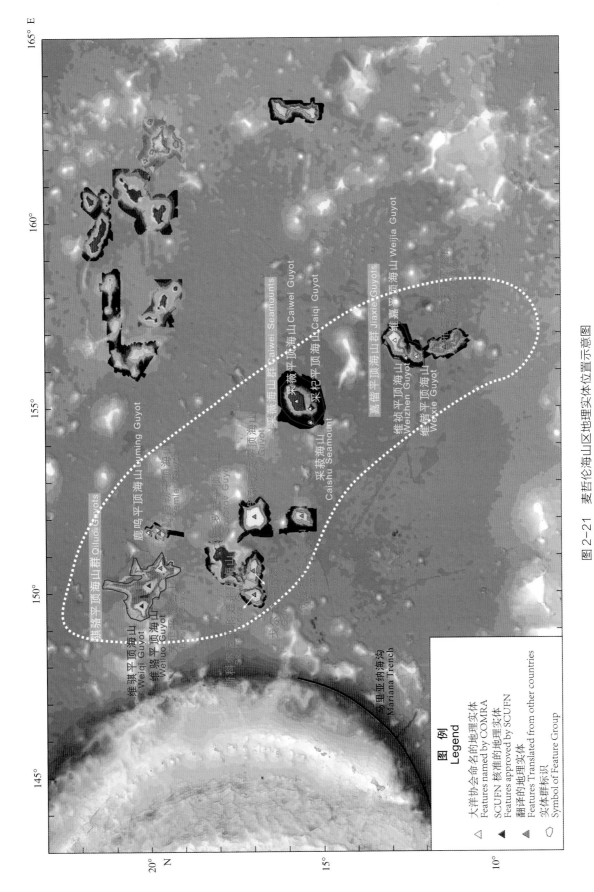

图 2-21　麦哲伦海山区地理实体位置示意图

Fig.2-21　Locations of the undersea features in the Magellan Seamount Area

2.3.1　骐骆平顶海山群
Qiluo Guyots

中文名称 Chinese Name	骐骆平顶海山群 Qiluo Pingdinghaishanqun	英文名称 English Name	Qiluo Guyots	所在大洋 Ocean or Sea	西太平洋 West Pacific Ocean
发现情况 Discovery Facts	此平顶海山群于 2001 年 6—7 月由中国科考船"大洋一号"在执行 DY105-11 航次时调查发现。 Qiluo Guyots were discovered by the Chinese R/V *Dayang Yihao* during the DY105-11 cruise from June to July, 2001.				
命名历史 Name History	骐骆平顶海山群曾经在中国大洋航次调查报告中以代号命名。它由维骐平顶海山、维骆平顶海山、劳里平顶海山以及 1 个小型海山和 1 个海脊组成。维骐平顶海山和维骆平顶海山由我国命名，于 2014 年提交 SCUFN 审议通过。劳里平顶海山在 GEBCO 地名辞典中称为"Lowrie Guyot"。 A code was used to name Qiluo Guyots temporarily in Chinese cruise reports. The guyots consist of Weiqi Guyot, Weiluo Guyot and Lowrie Guyot as well as a small seamount and a ridge. Both Weiqi Guyot and Weiluo Guyot were named by China and approved by SCUFN in 2014.				
特征点坐标 Coordinates	20°15.50′N，149°39.20′E（维骐平顶海山 / Weiqi Guyot） 20°00.80′N，150°09.20′E（维骆平顶海山 / Weiluo Guyot） 19°40.15′N，150°46.65′E（劳里平顶海山 / Lowrie Guyot）			长(km)×宽(km) Length (km)× Width (km)	195×120
最大水深（m） Max Depth（m）	5 500	最小水深（m） Min Depth（m）	1 460	高差（m） Total Relief（m）	4 040
地形特征 Feature Description	3 个平顶海山北西向排列组成，发育于水深 3 000 m 的平台之上，构成此平顶海山群的主体。平顶海山群的坡麓水深 5 500 m，最高峰为西北部的维骐平顶海山，最小水深 1 460 m（图 2–22）。 The guyots consist of three guyots arranged in NW direction, developing on the same platform with the depth of 3 000 m. The piedmont depth of this guyots is 5 500 m while the minimum depth is 1 460 m. The highest summit is Weiqi Guyot in the northwest (Fig.2–22).				
命名释义 Reason for Choice of Name	利用组合"维骐平顶海山"和"维骆平顶海山"名称的方式命名。维骐、维骆均指骏马，出自《诗经·小雅·皇皇者华》，具体释义见"维骐平顶海山"和"维骆平顶海山"。 Qiluo guyots consist of the last character of Weiqi and Weiluo in Chinese respectively. Both Weiqi and Weiluo mean fine horse and comes from a poem named *Huanghuangzhehua* in *Shijing · Xiaoya*, please refer to Weiqi Guyot and Weiluo Guyot for their specific meanings.				

(a)

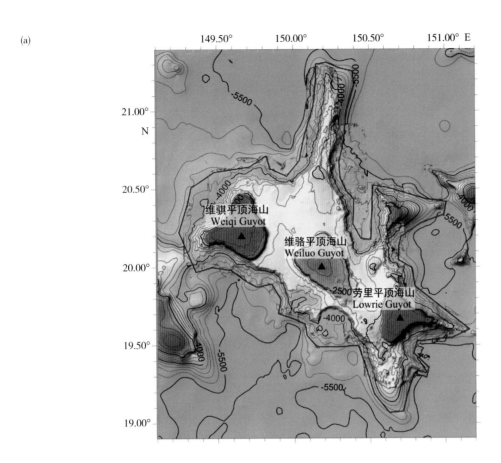

(b)

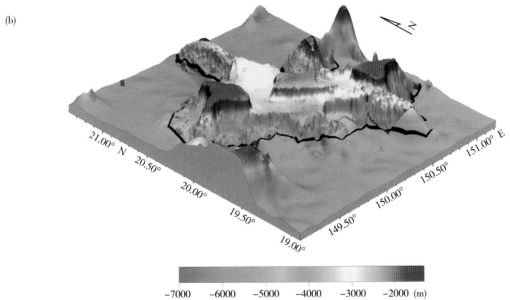

图 2-22　骐骆平顶海山群

(a) 地形图（等深线间隔 300 m）；(b) 三维图

Fig.2-22　Qiluo Guyots

(a) Bathymetric map (the contour interval is 300 m); (b) 3-D topographic map

2.3.2 维骐平顶海山
Weiqi Guyot

中文名称 Chinese Name	维骐平顶海山 Weiqi Pingdinghaishan	英文名称 English Name	Weiqi Guyot	所在大洋 Ocean or Sea	西太平洋 West Pacific Ocean
发现情况 Discovery Facts		此平顶海山于 2001 年 7 月由中国科考船 "大洋一号" 在执行 DY105-11 航次时调查发现。 This guyot was discovered by the Chinese R/V *Dayang Yihao* during the DY105-11 cruise in July, 2001.			
命名历史 Name History	由我国命名为维骐平顶海山，于 2014 年提交 SCUFN 审议通过。 This feature was named Weiqi Guyot by China and the name was approved by SCUFN in 2014.				
特征点坐标 Coordinates	20°15.50′N，149°39.20′E			长 (km) × 宽 (km) Length (km) × Width (km)	60 × 50
最大水深（m） Max Depth（m）	4 650	最小水深（m） Min Depth（m）	1 460	高差（m） Total Relief（m）	3 190
地形特征 Feature Description	此平顶海山位于骐骆平顶海山群的西北部，由地形平缓的山顶平台和陡峭的山坡组成，长宽分别为 60 km 和 50 km。山顶平台水深约 1 600 m，最浅处 1 460 m，山麓水深 4 650 m，高差 3 190 m（图 2-23）。 Weiqi Guyot is located in the northwest of Qiluo Guyots. It has a flat top platform and steep slopes. The length is 60 km and the width is 50 km. The top platform depth is about 1 600 m and the minimum depth is 1 460 m while the piedmont depth is 4 650 m, which make the total relief being 3 190 m (Fig.2-23).				
命名释义 Reason for Choice of Name	"维骐" 出自《诗经·小雅·皇皇者华》"我马维骐，六辔如丝"，指黑色条纹的马。本篇寓意使者出使途中不忘君王所教，忠于职守的高尚品德。以 "维骐" 命名此海山，赞誉中国古代使臣铭记职责，不辞辛劳的工作作风。 "Weiqi" comes from a poem named *Huanghuangzhehua* in *Shijing · Xiaoya*. *Shijing* is a collection of ancient Chinese poems from 11th century B.C. to 6th century B.C. "Black and blue are my horses, six silky reins guiding the courses." Weiqi means horse with black stripes. The poem implies that the envoy keep king's instructions in mind. This guyot is named Weiqi Guyot to praise ancient Chinese envoys' noble morality of dedication to responsibility.				

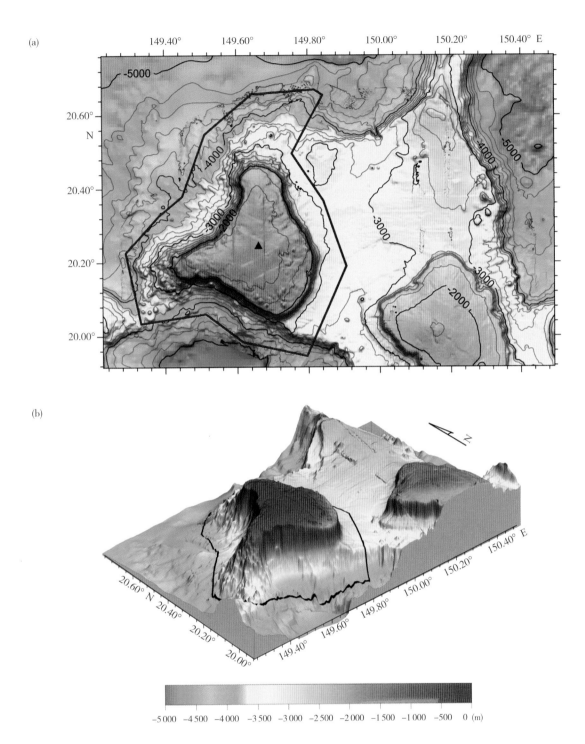

图 2-23　维骐平顶海山

(a) 地形图（等深线间隔 200 m）；(b) 三维图

Fig.2-23　Weiqi Guyot

(a) Bathymetric map (the contour interval is 200 m); (b) 3-D topographic map

2.3.3 维骆平顶海山
Weiluo Guyot

中文名称 Chinese Name	维骆平顶海山 Weiluo Pingdinghaishan	英文名称 English Name	Weiluo Guyot	所在大洋 Ocean or Sea	西太平洋 West Pacific Ocean
发现情况 Discovery Facts	此平顶海山于 2001 年 7 月由中国科考船"大洋一号"在执行 DY105-11 航次时调查发现。 This guyot was discovered by the Chinese R/V *Dayang Yihao* during the DY105-11 cruise in July, 2001.				
命名历史 Name History	由我国命名为维骆平顶海山，于 2014 年提交 SCUFN 审议通过。 This feature was named Weiluo Guyot by China and the name was approved by SCUFN in 2014.				
特征点坐标 Coordinates	20°00.80′N，150°09.20′E			长(km)× 宽(km) Length (km) × Width (km)	60 × 30
最大水深（m） Max Depth（m）	3 250	最小水深（m） Min Depth（m）	1 750	高差（m） Total Relief（m）	1 500
地形特征 Feature Description	维骆平顶海山位于骐骆平顶海山群的中部，由地形平缓的山顶平台和陡峭的山坡组成，NW—SE 走向，长宽分别为 60 km 和 30 km。山顶平台水深约 2 000 m，最浅处 1 750 m，山麓水深 3 250 m，高差 1 500 m（图 2-24）。 Weiluo Guyot is located in the center of Weiluo Guyots and consists of flat top platform and steep slopes. It runs NW to SE. The length is 60 km and the width is 30 km. The top platform depth is about 2 000 m. The minimum depth is 1 750 m. The piedmont depth is 3 250 m, which makes the total relief being 1 500 m (Fig.2–24).				
命名释义 Reason for Choice of Name	"维骆"出自《诗经·小雅·皇皇者华》"我马维骆，六辔沃若"，指黑鬣的白马。本篇寓意使者出使途中不忘君王所教，忠于职守的高尚品德。以"维骆"命名此海山，赞誉中国古代使臣铭记职责，四处精查细问的工作作风。 "Weiluo" comes from a poem named *Huanghuangzhehua* in *Shijing · Xiaoya*. *Shijing* is a collection of ancient Chinese poems from 11th century B.C. to 6th century B.C. "Black and white are my horses, six glossy reins guiding the courses." "Weiluo" means white horse with black hairs. The poem implies that the envoy keep king's instructions in mind. This guyot is named Weiluo Guyot to praise ancient Chinese envoys' noble morality of dedication to responsibility.				

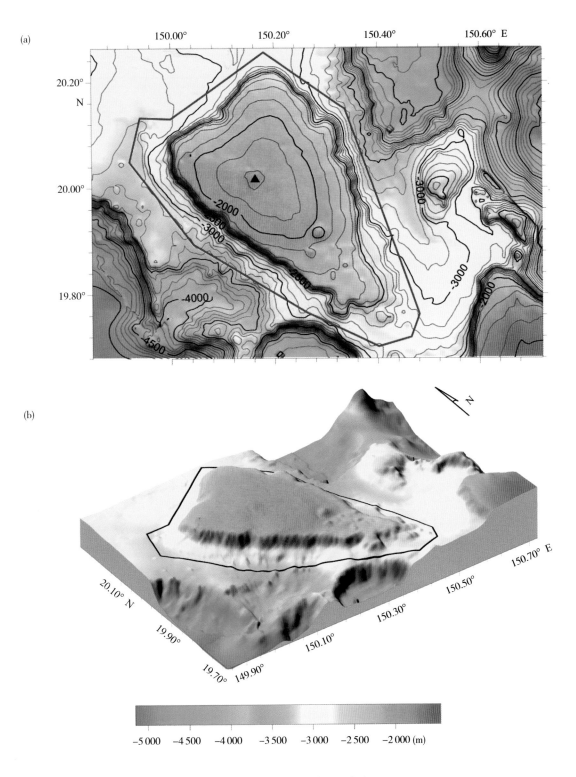

图 2-24　维骆平顶海山

(a) 地形图 (等深线间隔 100 m)；(b) 三维图

Fig.2-24　Weiluo Guyot

(a) Bathymetric map (the contour interval is 100 m); (b) 3-D topographic map

2.3.4 劳里平顶海山
Lowrie Guyot

中文名称 Chinese Name	劳里平顶海山 Laoli Pingdinghaishan	英文名称 English Name	Lowrie Guyot	所在大洋 Ocean or Sea	西太平洋 West Pacific Ocean
发现情况 Discovery Facts	此平顶海山于 2001 年 6—7 月由中国科考船"大洋一号"在执行 DY105-11 航次时调查发现。 The guyot was discovered by the Chinese R/V *Dayang Yihao* during the DY105-11 cruise from June to July, 2001.				
命名历史 Name History	该平顶海山在 GEBCO 地名辞典中称为"Lowrie Guyot",中文译名为"劳里平顶海山",提名国和提名时间不详。 This guyot is called Lowrie Guyot in GEBCO gazetteer. Its translation is Laoli Pingdinghaishan in Chinese. The country and time of the proposal was not listed in the gazetteer.				
特征点坐标 Coordinates	19°40.15′N,150°46.65′E			长 (km) × 宽 (km) Length (km) × Width (km)	70 × 55
最大水深（m） Max Depth（m）	5 500	最小水深（m） Min Depth（m）	1 470	高差（m） Total Relief（m）	4 030
地形特征 Feature Description	劳里平顶海山位于骐骆平顶海山群东南部,由地形平缓的山顶平台和陡峭的山坡组成,最长 70 km,最窄处 55 km。山顶平台水深约 1 700 m,最浅处 1 470 m,山麓水深 5 500 m,高差 4 030 m(图 2-25)。 Lowrie Guyot is located in the southeast of Qiluo Guyots and consists of a flat top platform and steep slopes. The length is 70 km and the width is 55 km. The depth of the platform is about 1 700 m. The minimum depth is 1 470 m. The piedmont depth is 5 500 m, which makes the total relief being 4 030 m (Fig.2-25).				
命名释义 Reason for Choice of Name	以 19 世纪 70 年代 ACUF 水道测量组织（U.S.H.O）的制图员 Allen Lowrie（生卒年月不详）的名字命名。 The guyots is named after Allen Lowrie (birth and death date unknown), a cartographer in U.S.H.O of ACUF in the 1870s.				

(a)

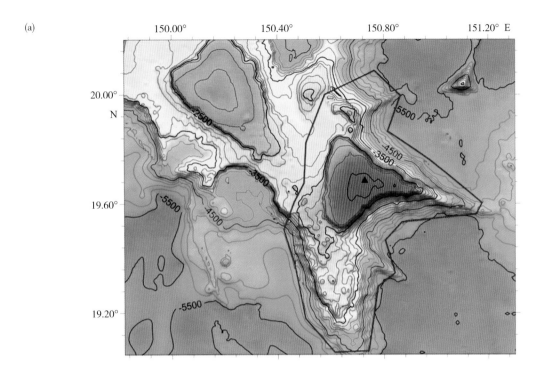

(b)

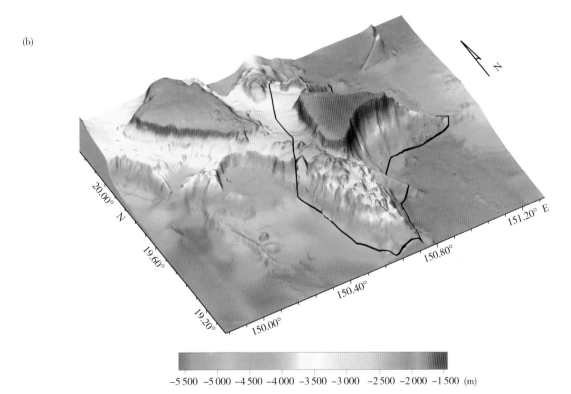

−5 500　−5 000　−4 500　−4 000　−3 500　−3 000　−2 500　−2 000　−1 500　(m)

图 2-25　劳里平顶海山

(a) 地形图（等深线间隔 200 m）；(b) 三维图

Fig.2-25　Lowrie Guyot

(a) Bathymetric map (the contour interval is 200 m); (b) 3-D topographic map

2.3.5 鹿鸣平顶海山
Luming Guyot

中文名称 Chinese Name	鹿鸣平顶海山 Luming Pingdinghaishan	英文名称 English Name	Luming Guyot	所在大洋 Ocean or Sea	西太平洋 West Pacific Ocean
发现情况 Discovery Facts	此平顶海山于 2001 年 6—7 月由中国科考船"大洋一号"在执行 DY105-11 航次时调查发现。 This guyot was discovered by the Chinese R/V *Dayang Yihao* during the DY105-11 cruise from June to July, 2001.				
命名历史 Name History	此平顶海山在我国大洋航次调查报告中曾暂以代号命名。 A code was used to name this guyot temporarily in Chinese cruise reports.				
特征点坐标 Coordinates	20°02.20′N，151°38.64′E			长(km)×宽(km) Length (km)× Width (km)	30×30
最大水深（m） Max Depth（m）	5 130	最小水深（m） Min Depth（m）	1 270	高差（m） Total Relief（m）	3 860
地形特征 Feature Description	该平顶海山由地形平缓的山顶平台和陡峭的山坡组成，俯视平面形态近圆形，基座直径 30 km。最浅处 1 270 m，山麓水深 5 130 m，高差 3 860 m。此平顶海山与赫姆勒平顶海山组成平顶海山群，两者之间以鞍部分隔，相距 40 km，鞍部水深 3 300 m（图 2-26）。 Luming Guyot consists of a flat top platform and steep slopes. It has a nearly round overlook plane shape with the base diameter of 30 km. The minimum depth is 1 270 m, while the piedmont depth is 5 130 m, which makes the total relief being 3 860 m. Luming Guyot and Hemler Guyot, located 40 km apart, constitute guyots together, which are separated by a saddle, whose depth is 3 300 m (Fig.2-26).				
命名释义 Reason for Choice of Name	"鹿鸣"出自《诗经·小雅·鹿鸣》"呦呦鹿鸣，食野之苹。我有嘉宾，鼓瑟吹笙"。此篇诗歌是古人宴请宾客时所唱，展现了原野上群鹿呦呦鸣叫，尊贵的客人弹琴吹笙奏乐，宾主共欢的场景。鹿鸣指鹿在鸣叫，选"鹿鸣"作为专名，引出该篇古人待宾客热情有道，和谐快乐的情景。 "Luming" comes from a poem named *Luming* in *Shijing · Xiaoya*. *Shijing* is a collection of ancient Chinese poems from 11th century B.C. to 6th century B.C. "The bleating deer seems to sing, as it nibbles the wormwood green. I have renowned guests to salute, with music of the lute and flute." The poem was played when feting guests, expressing the pleasure. "Luming" means deer bleating. We name this guyot "Luming" to show the hospitality of ancient Chinese.				

(a)

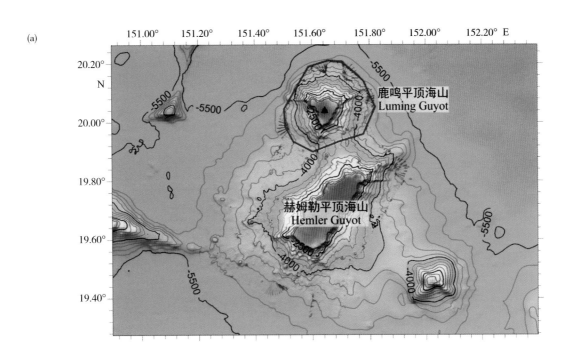

(b)

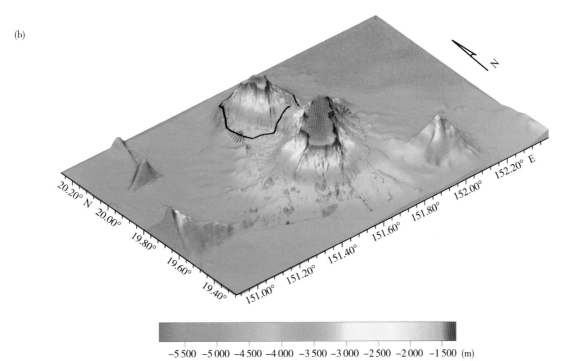

−5 500　−5 000　−4 500　−4 000　−3 500　−3 000　−2 500　−2 000　−1 500　(m)

图 2-26　鹿鸣平顶海山

(a) 地形图（等深线间隔 300 m）；(b) 三维图

Fig.2-26　Luming Guyot

(a) Bathymetric map (the contour interval is 300 m); (b) 3-D topographic map

2.3.6　赫姆勒平顶海山
Hemler Guyot

中文名称 Chinese Name	赫姆勒平顶海山 Hemule Pingdinghaishan	英文名称 English Name	Hemler Guyot	所在大洋 Ocean or Sea	西太平洋 West Pacific Ocean
发现情况 Discovery Facts	此平顶海山于 2001 年 6—7 月由中国科考船"大洋一号"在执行 DY105-11 航次时调查发现。 The guyot was discovered by the Chinese R/V *Dayang Yihao* during the DY105-11 cruise from June to July, 2001.				
命名历史 Name History	该平顶海山在 GEBCO 地名辞典中称为"Hemler Guyot"，中文译名为"赫姆勒平顶海山"，提名国和提名时间不详。此平顶海山在我国大洋航次调查报告中曾暂以代号命名。 The guyot is called Hemler Guyot in GEBCO gazetteer. Its translation is Hemule Pingdinghaishan in Chinese. Its country and time of proposal is unknown. A code was used to name this guyot temporarily in Chinese cruise reports.				
特征点坐标 Coordinates	19°45.77′N，151°45.00′E			长(km)×宽(km) Length (km) × Width (km)	60 × 40
最大水深（m） Max Depth（m）	5 120	最小水深（m） Min Depth（m）	1 300	高差（m） Total Relief（m）	3 820
地形特征 Feature Description	赫姆勒平顶海山由地形平缓的山顶平台和陡峭的山坡组成，SW—NE 走向，长宽分别为 60 km 和 40 km。山顶平台水深约 1 500 m，最浅处 1 300 m，山麓水深 5 120 m，高差 3 820 m（图 2–27）。 Hemler Guyot consists of a flat top platform and steep slopes and runs NE to SW. The length is 60 km and the width is 40 km. The depth of top platform is about 1 500 m and the minimum depth is 1 300 m. The piedmont depth is 5 120 m, which makes the total relief being 3 820 m (Fig.2–27).				
命名释义 Reason for Choice of Name	不详。 The reason remains unknown.				

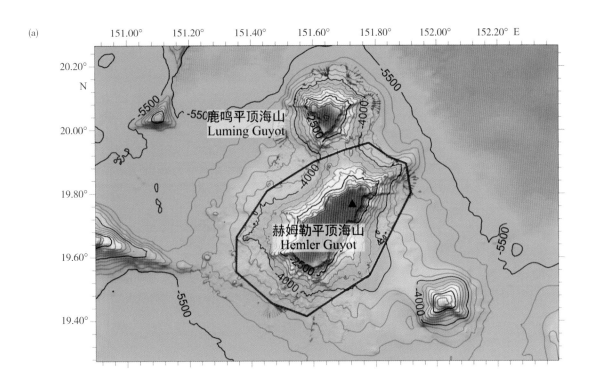

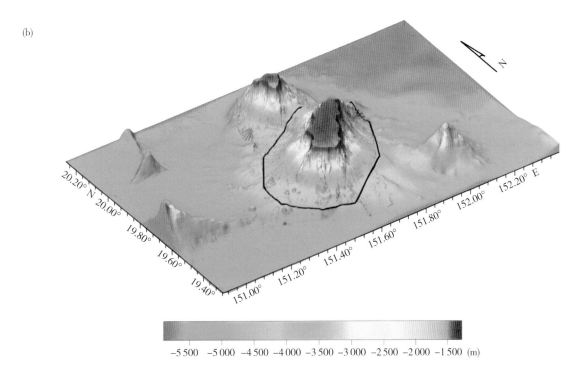

图 2-27　赫姆勒平顶海山

(a) 地形图（等深线间隔 300 m）；(b) 三维图

Fig.2-27　Hemler Guyot

(a) Bathymetric map (the contour interval is 300 m); (b) 3-D topographic map

2.3.7 戈沃罗夫平顶海山
Govorov Guyot

中文名称 Chinese Name	戈沃罗夫平顶海山 Gewoluofu Pingdinghaishan	英文名称 English Name	Govorov Guyot	所在大洋 Ocean or Sea	西太平洋 West Pacific Ocean
发现情况 Discovery Facts	此平顶海山于 1999 年 5—6 月由中国科考船"海洋四号"在执行 DY95-10 航次时调查发现。俄罗斯从 1987 年开始多次调查，其中在 2005 年利用"Gelendzhik"号科考船进行了多波束地形调查。 This guyot was discovered by the Chinese R/V *Haiyang Sihao* during the DY95-10 cruise from May to June in 1999. Russia had several investigations since 1987 and used R/V *Gelendzhik* to carry on multibeam topographic investigations in 2005.				
命名历史 Name History	该平顶海山在 2006 年由俄罗斯提交 SCUFN 审议通过，命名为"Govorov Guyot"，中文译名为"戈沃罗夫平顶海山"。此平顶海山在我国大洋航次调查报告中曾暂以代号命名。 This guyot was proposed to SCUFN by Russia and approved in 2006. It is called Govorov Guyot with the translation being Gewoluofu Pingdinghaishan in Chinese. A code was temporarily used to name it in Chinese cruise reports.				
特征点坐标 Coordinates	17°58.85′N，151°05.88′E		长(km)×宽(km) Length (km)× Width (km)		177×168
最大水深（m） Max Depth（m）	5 600	最小水深（m） Min Depth（m）	1 320	高差（m） Total Relief（m）	4 280
地形特征 Feature Description	此平顶海山由地形平缓的山顶平台和陡峭的山坡组成，俯视平面形态不规则，整体呈 NE—SW 走向，长宽分别 177 km 和 168 km。平顶海山西坡较陡，东坡较缓（图 2–28）。 The guyot consists of a flat top platform and steep slopes. It has an irregular overlook plane shape, running NE to SW overall. The length is 150 km and the width is 110 km. The west slope of the guyot is steep while the east slope is flat (Fig.2–28).				
命名释义 Reason for Choice of Name	以俄罗斯科学家 I. N. Govorov（1920—1997 年）的名字命名。Govorov 是俄罗斯的太平洋地磁专家。他的主要研究方向包括火山学、地质学和构造学。他对马尔库斯－威克海山区、麦哲伦海山区和 h 笠原脊等地的磁场进行了研究。 It is named after I. N. Govorov (1920–1997), a Russian specialist on magnetism of the Pacific Ocean. His main research fields were volcanism, geology, and tectonics. He studied the magmatism of the Marcus-Wake Seamount Area, the Magellan Seamount Area, and the Ogasawara Rise.				

(a)

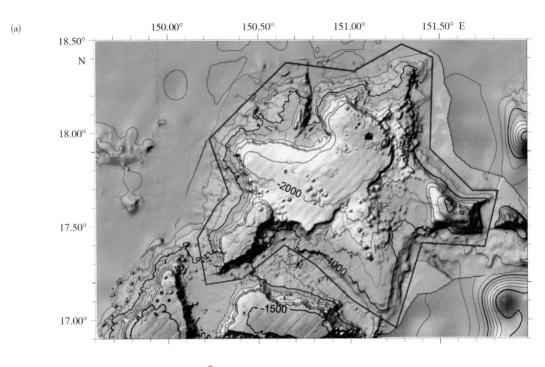

(b)

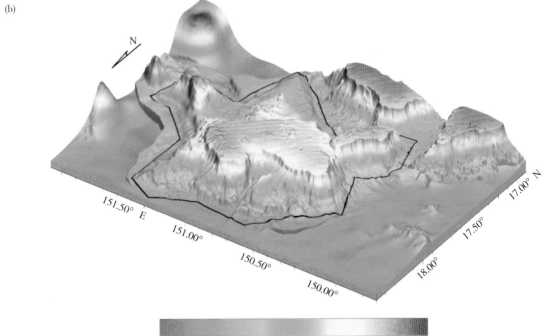

图 2-28　戈沃罗夫平顶海山

(a) 地形图 （等深线间隔 500 m）；(b) 三维图

Fig.2-28　Govorov Guyot

(a) Bathymetric map (the contour interval is 500 m); (b) 3-D topographic map

2.3.8　斯科尔尼亚科瓦平顶海山
Skornyakova Guyot

中文名称 Chinese Name	斯科尔尼亚科瓦 平顶海山 Sikeerniyakewa Pingdinghaishan	英文名称 English Name	Skornyakova Guyot	所在大洋 Ocean or Sea	西太平洋 West Pacific Ocean
发现情况 Discovery Facts	此平顶海山于 2001 年 6—7 月由中国科考船"大洋一号"在执行 DY105-11 航次时调查发现。俄罗斯使用"Gelendzhik"号科考船于 2006 年调查。 This guyot was discovered by the Chinese R/V *Dayang Yihao* during the DY105-11 cruise from June to July, 2001. Russia had investigated it using R/V *Gelendzhik* in 2006.				
命名历史 Name History	该平顶海山在 2007 年由俄罗斯提交 SCUFN 审议通过，命名为"Skornyakova Guyot"，中文译名为斯科尔尼亚科瓦平顶海山。该平顶海山在我国大洋航次调查报告中曾暂以代号命名。 This guyot was proposed to SCUFN by Russia and approved in 2007. It is called Skornyakova Guyot with translation being Sikeerniyakewa Pingdinghaishan in Chinese. A code was temporarily used to name it in Chinese cruise reports.				
特征点坐标 Coordinates	16°51.92′N，149°53.80′E			长(km)×宽(km) Length (km)× Width (km)	90×50
最大水深（m） Max Depth（m）	5 450	最小水深（m） Min Depth（m）	1 227	高差（m） Total Relief（m）	4 223
地形特征 Feature Description	此平顶海山由地形平缓的山顶平台和陡峭的山坡组成，俯视平面形态呈梨形，北窄南宽，南北长 90 km，东西宽 50km。山顶平台水深约 1 500 m，最浅处 1 227 m，山麓水深 5 450 m，高差 4 223 m（图 2–29）。 This guyot consists of a flat top platform and steep slopes. It has a pear-like overlook plane shape, with a narrow end in the north and a broad end in the south. The length is 90 km and the width is 50 km. The top platform depth is about 1 500 m and the minimum depth is 1 227 m. The piedmont depth is 5 450 m, which makes the total relief being 4 223 m (Fig.2–29).				
命名释义 Reason for Choice of Name	以俄罗斯海洋地质学家 N. S. Skornyakova（1924—1995 年）的名字命名。N. S. Skornyakova 是俄罗斯海洋地质学家，地质学博士学位，多次参加了太平洋和印度洋科学考察，致力于多金属结核研究，发表了 150 余篇论文。 It is named after N. S. Skornyakova (1924–1995), a Russian marine geologist, doctor of geological sciences, and participant in Pacific Ocean and Indian Ocean expeditions. He was a specialist in the study of polymetallic nodules, and author of more than 150 scientific publications.				

(a)

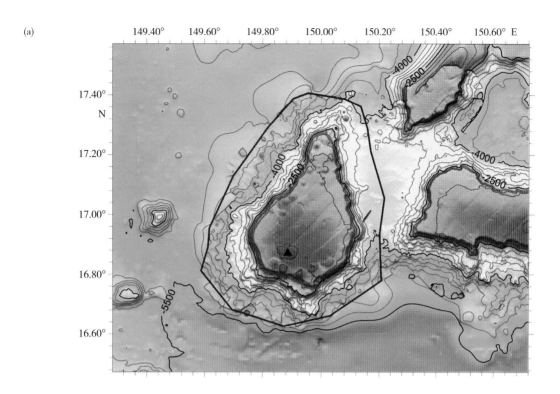

(b)

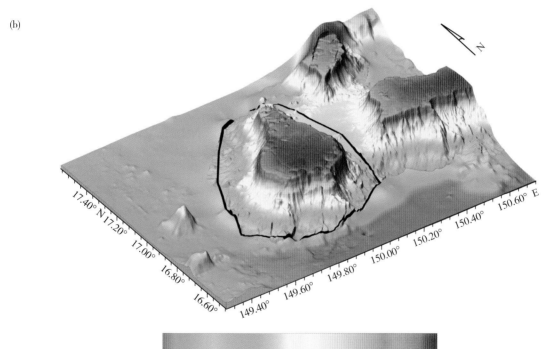

$-5\,500\quad-5\,000\quad-4\,500\quad-4\,000\quad-3\,500\quad-3\,000\quad-2\,500\quad-2\,000\quad-1\,500\quad$ (m)

图 2-29　斯科尔尼亚科瓦平顶海山

(a) 地形图（等深线间隔 300 m）；(b) 三维图

Fig.2-29　Skornyakova Guyot

(a) Bathymetric map (the contour interval is 300 m); (b) 3-D topographic map

2.3.9 戈尔金平顶海山
Gordin Guyot

中文名称 Chinese Name	戈尔金平顶海山 Geerjin Pingdinghaishan	英文名称 English Name	Gordin Guyot	所在大洋 Ocean or Sea	西太平洋 West Pacific Ocean
发现情况 Discovery Facts	此平顶海山于 2001 年 6—7 月由中国科考船"大洋一号"在执行 DY105-11 航次时调查发现。俄罗斯使用"Gelendzhik"号科考船于 2006 年调查。 This guyot was discovered by the Chinese R/V *Dayang Yihao* during the DY105-11 cruise from June to July, 2001. Russia had investigated it using R/V *Gelendzhik* in 2006.				
命名历史 Name History	该平顶海山在 2007 年由俄罗斯提交 SCUFN 审议通过，命名为"Gordin Guyot"，中文译名为"戈尔金平顶海山"。该平顶海山在我国大洋航次调查报告中曾暂以代号命名。 This guyot was proposed to SCUFN by Russia and approved in 2007. It is called Gordin Guyot with the translation being Geerjin Pingdinghaishan in Chinese. A code was temporarily used to name it in Chinese cruise reports.				
特征点坐标 Coordinates	16°56.78′N，150°47.82′E			长 (km)× 宽 (km) Length (km) × Width (km)	90 × 50
最大水深（m） Max Depth（m）	5 450	最小水深（m） Min Depth（m）	1 227	高差（m） Total Relief（m）	4 223
地形特征 Feature Description	此平顶海山位于戈沃罗夫平顶海山以南 20 km，斯科尔尼亚科瓦平顶海山以东 20 km。海山由地形平缓的山顶平台和陡峭的山坡组成，俯视平面形态狭长，东西长 90 km，南北宽 60 km。山顶平台水深约 1 500 m，最浅处 1 312 m，山麓水深 5 900 m，高差 4 588 m（图 2–30）。 Gordin Guyot is located in 20 km south to Govorov Guyot, 20 km east to Skornyakova Guyot. It consists of a flat top platform and steep slopes. It has an elongated overlook plane shape. The length is 90 km from east to west and the width is 50 km from north to south. The top platform depth is about 1 500 m and the minimum depth is 1 312 m. The piedmont depth is 5 900 m, which makes the total relief being 4 588 m (Fig.2–30).				
命名释义 Reason for Choice of Name	以俄罗斯海洋地球物理学家 V. M.Gordin（1942—2002）博士的名字命名。他参加了太平洋和印度洋的科考工作。他是海洋调查、地磁数据分析和理论方面的专家，著有《海洋地磁学》一书，并发表了 130 余篇科学论文。 It is named after V. M. Gordin (1942–2002), a Russian marine geophysicist, doctor of sciences, and participant in Pacific Ocean and Indian Ocean expeditions. He was a specialist in marine surveying, the theory and analysis of geomagnetic data, and the author of book *Marine Magnetometry* and more than 130 additional scientific publications.				

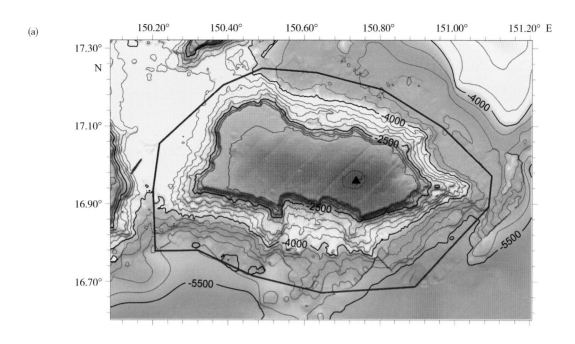

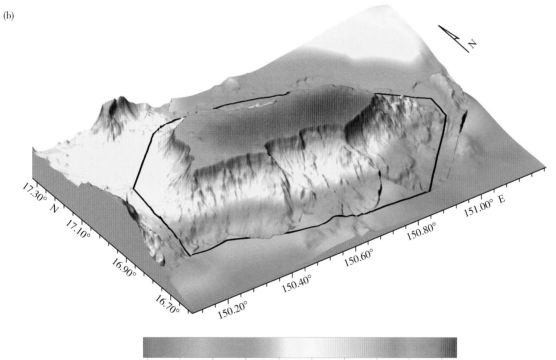

图 2-30　戈尔金平顶海山

(a) 地形图 (等深线间隔 300 m)；(b) 三维图

Fig.2-30　Gordin Guyot

(a) Bathymetric map (the contour interval is 300 m); (b) 3-D topographic map

2.3.10 伊利切夫平顶海山
Il'ichev Guyot

中文名称 Chinese Name	伊利切夫平顶海山 Yiliqiefu Pingdinghaishan	英文名称 English Name	Il'ichev Guyot	所在大洋 Ocean or Sea	西太平洋 West Pacific Ocean
发现情况 Discovery Facts	此平顶海山于1998年5—6月由中国科考船"海洋四号"在执行DY95-9航次时调查发现。俄罗斯从1986年开始多次调查，其中在2005年利用"Gelendzhik"号科考船进行了多波束地形调查。 This guyot was discovered by the Chinese R/V *Haiyang Sihao* during the DY95-9 cruise from May to June, 1998. Russia had several investigations since 1986 and used R/V *Gelendzhik* to carry on multibeam topographic investigations in 2005.				
命名历史 Name History	该平顶海山在2006年由俄罗斯提交SCUFN审议通过，命名为"Il'ichev Guyot"，中文译名为"伊利切夫平顶海山"。此平顶海山在我国大洋航次调查报告中曾暂以代号命名。 This guyot was proposed to SCUFN by Russia and approved in 2006. It is called Il'ichev Guyot with the translation being Yiliqiefu Pingdinghaishan in Chinese. A code was temporarily used to name it in Chinese cruise reports.				
特征点坐标 Coordinates	16°55.00′N，152°04.98′E			长(km)×宽(km) Length (km)× Width (km)	55×55
最大水深（m） Max Depth（m）	6 100	最小水深（m） Min Depth（m）	1 372	高差（m） Total Relief（m）	4 728
地形特征 Feature Description	此平顶海山由地形平缓的山顶平台和陡峭的山坡组成。山顶平台水深约1 500 m，最浅处1 372 m，山麓水深6 100 m，高差4 728 m（图2-31）。 This guyot consists of a flat top platform and steep slopes. The top platform depth is about 1 500 m and the minimum depth is 1 372 m. The piedmont depth is 6 100 m, which makes the total relief being 4 728 m (Fig.2–31).				
命名释义 Reason for Choice of Name	以俄罗斯海洋学家和声学专家 V. I. Il'ichev（1932—1994）的名字命名。他主要研究西北太平洋地区，曾于1974年至1994年期间担任俄罗斯科学院远东区太平洋海洋研究所所长。 It is named after professor V. I. Il'ichev (1932–1994), a Russian oceanographer and acoustic specialist. His research focused on the Northwest Pacific Ocean. He was the director of the Pacific Oceanology Institute, Far Eastern Branch of the Russian Academy of Sciences from 1974 to 1994.				

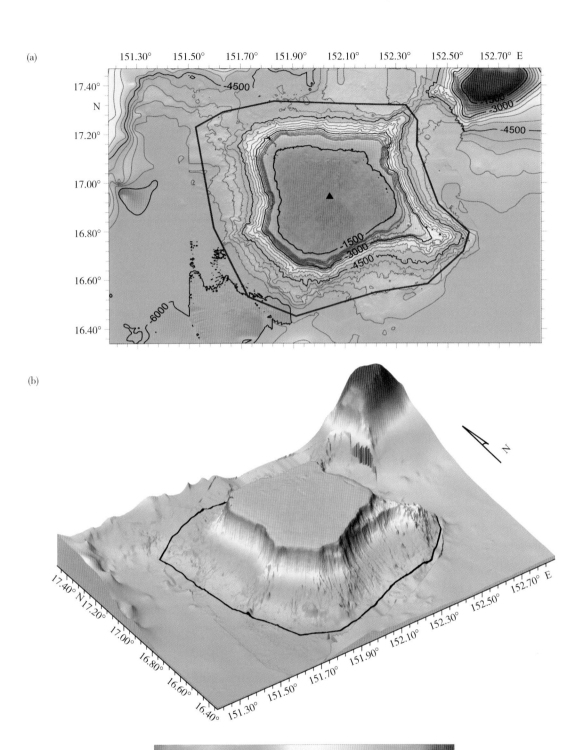

图 2-31 伊利切夫平顶海山

(a) 地形图（等深线间隔 300 m）；(b) 三维图

Fig.2-31 Il'ichev Guyot

(a) Bathymetric map (the contour interval is 300 m); (b) 3-D topographic map

2.3.11 佩加斯平顶海山
Pegas Guyot

中文名称 Chinese Name	佩加斯平顶海山 Peijiasi Pingdinghaishan	英文名称 English Name	Pegas Guyot	所在大洋 Ocean or Sea	西太平洋 West Pacific Ocean
发现情况 Discovery Facts	此平顶海山于1998年5—6月由中国科考船"海洋四号"在执行DY95-9航次时调查发现。俄罗斯从1983年开始多次调查，其中在2005年利用"Gelendzhik"号科考船进行了多波束地形调查。 This guyot was discovered by the Chinese R/V *Haiyang Sihao* during the DY95-9 cruise from May to June, 1998. Russia had several investigations since 1983 and used R/V *Gelendzhik* to carry on multibeam topographic investigations in 2005.				
命名历史 Name History	该平顶海山在2006年由俄罗斯提交SCUFN审议通过，命名为"Pegas Guyot"，中文译名为"佩加斯平顶海山"。此平顶海山在我国大洋航次调查报告中曾暂以代号命名。 This guyot was proposed to SCUFN by Russia and approved in 2006. It is called Pegas Guyot with the translation being Peijiasi Pingdinghaishan in Chinese. A code was temporarily used to name it in Chinese cruise reports.				
特征点坐标 Coordinates	15°36.56′N，152°05.28′E			长(km)×宽(km) Length (km)× Width (km)	100×80
最大水深（m） Max Depth（m）	6 120	最小水深（m） Min Depth（m）	1 320	高差（m） Total Relief（m）	4 800
地形特征 Feature Description	此平顶海山由地形平缓的山顶平台和陡峭的山坡组成，NNE—SSW走向，长宽分别为100 km和80 km。山顶平台水深约1 500 m，最浅处1 320 m，山麓水深6 120 m，高差4 800 m（图2–32）。 This guyot consists of a flat top platform and steep slopes. It runs NNE to SSW. The length is 100 km and the width is 80 km. The top platform depth is about 1 500 m and the minimum depth is 1 320 m. The piedmont depth is 6 120 m, which makes the total relief being 4 800 m (Fig.2–32).				
命名释义 Reason for Choice of Name	以俄罗斯海洋调查船"Pegas"号的名字命名。在1975—1976期间，"Pegas"号主要在东马里亚纳海盆、麦哲伦海山区以及马尔库斯–威克海山区等地进行地质与地球物理考察。 It is named after the Russian research vessel *Pegas*, which conducted regional geologic-geophysical investigations in the area of the East Mariana Basin, Magellan Seamount Area and Marcus-Wake Seamount Area in 1975–1976.				

(a)

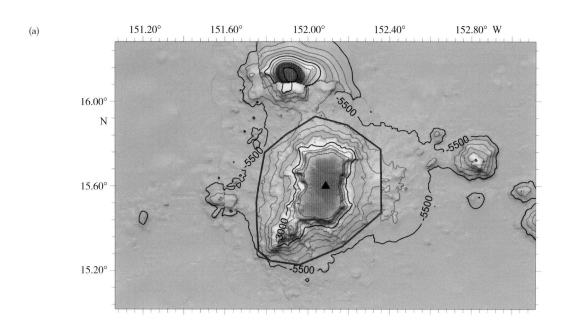

(b)

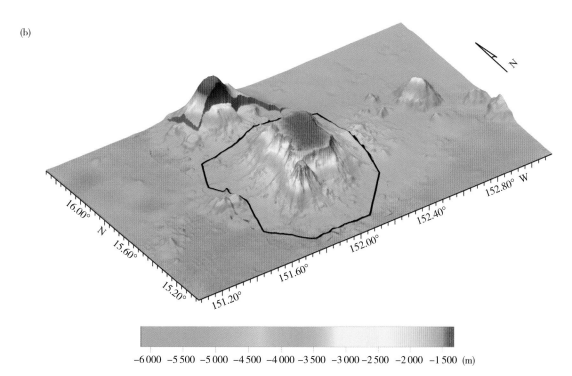

图 2-32　佩加斯平顶海山

(a) 地形图 （等深线间隔 500 m）；(b) 三维图

Fig.2-32　Pegas Guyot

(a) Bathymetric map (the contour interval is 500 m); (b) 3-D topographic map

2.3.12 采薇海山群
Caiwei Seamounts

中文名称 Chinese Name	采薇海山群 Caiwei Haishanqun	英文名称 English Name	Caiwei Seamounts	所在大洋 Ocean or Sea	西太平洋 West Pacific Ocean
发现情况 Discovery Facts	此海山群于1997年6月由中国科考船"海洋四号"在执行DY95-7航次时调查发现。 The seamounts were discovered by the Chinese R/V *Haiyang Sihao* during the DY95-7 cruise in June, 1997.				
命名历史 Name History	该海山群由3个海山或平顶海山组成，在我国大洋航次调查报告中曾暂以代号命名，其中规模最大的平顶海山在GEBCO地名辞典中称为"Pallada Guyot"，我国现命名为采薇平顶海山。 The seamounts consist of 3 seamounts or guyots. A code was temporarily used to name it in Chinese cruise reports. The biggest guyot is named after "Pallada Guyot" in GEBCO gazetteer. It is named after Caiwei Guyot by China now.				
特征点坐标 Coordinates	15°40.00′N，155°10.00′E（采薇平顶海山 / Caiwei Guyot） 15°18.00′N，155°00.00′E（采杞平顶海山 / Caiqi Guyot） 15°21.07′N，154°36.05′E（采菽海山 / Caishu seamount）			长 (km) × 宽 (km) Length (km) × Width (km)	156 × 115
最大水深（m） Max Depth（m）	5 830	最小水深（m） Min Depth（m）	1 230	高差（m） Total Relief（m）	4 600
地形特征 Feature Description	此海山群由采薇平顶海山、采杞平顶海山和采菽海山组成。采薇平顶海山规模最大，山顶平台水深约1 500 m，最浅处1 230 m，山麓水深5 830 m，高差4 600 m（图2-33）。 The seamounts consist of Caiwei Guyot, Caiqi Guyot and Caishu Seamount. Caiwei Guyot is the largest one, its top platform depth is about 1 500 m with the minimum depth of 1 230 m. The piedmont depth is 5 830 m, which makes the total relief being 4 600 m (Fig.2-33).				
命名释义 Reason for Choice of Name	以采薇平顶海山命名。采薇来源于《诗经·小雅·采薇》，采集野菜之意。此海山群的3个海山采用群组化命名方法，分别命名为采薇、采杞、采菽，均取词于《诗经·小雅·采薇》。 It is named after the largest guyot, Caiwei Guyot in this group. Caiwei comes from a poem named *Caiwei* in *Shijing · Xiaoya*. *Shijing* is a collection of ancient Chinese poems from 11th century B.C. to 6th century B.C. "Caiwei" means picking potherbs. There are three seamounts named using group method in this group, and all their names, Caiwei, Caiqi and Caishu, come from the same poem.				

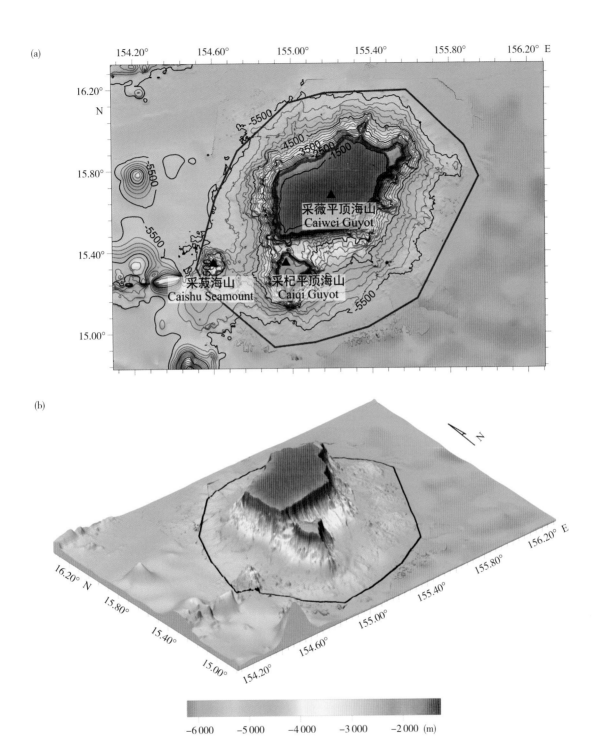

图 2-33　采薇海山群

(a) 地形图（等深线间隔 200 m）；(b) 三维图

Fig.2-33　Caiwei Seamounts

(a) Bathymetric map (the contour interval is 200 m); (b) 3-D topographic map

2.3.13 采薇平顶海山
Caiwei Guyot

中文名称 Chinese Name	采薇平顶海山 Caiwei Pingdinghaishan	英文名称 English Name	Caiwei Guyot (Pallada Guyot)	所在大洋 Ocean or Sea	西太平洋 West Pacific Ocean
发现情况 Discovery Facts	此平顶海山于 1997 年 6 月由中国科考船"海洋四号"在执行 DY95-7 航次时调查发现。 This guyot was discovered by the Chinese R/V *Haiyang Sihao* during the DY95-7 cruise in June, 1997.				
命名历史 Name History	该平顶海山在 GEBCO 地名辞典中称为"Pallada Guyot", 2004 年由俄罗斯提名, SCVFN 审议通过。该平顶海山在我国大洋航次调查报告中曾暂以代号命名, 2011 年由中国大洋协会命名为采薇平顶海山。 This guyot is named Pallada Guyot in GEBCO gazetteer, which was proposed by Russia and approved by SCVFN in 2004. A code was temporarily used to name it in Chinese cruise reports. It was named Caiwei Guyot by COMRA in 2011.				
特征点坐标 Coordinates	15°40.00′N，155°10.00′E			长(km)×宽(km) Length (km)× Width (km)	110×95
最大水深（m） Max Depth（m）	4 830	最小水深（m） Min Depth（m）	1 230	高差（m） Total Relief（m）	3 600
地形特征 Feature Description	此平顶海山为采薇海山群中规模最大的平顶海山，由地形平缓的山顶平台和陡峭的山坡组成，NE—SW 走向，长宽分别为 110 km 和 95 km。山顶平台水深约 1 450 m，最浅处 1 230 m，山麓水深 4 830 m，高差 3 600 m（图 2–34）。 This guyot consists of a flat top platform and steep slopes, running NE to SW. It is the largest guyot among Caiwei Seamounts. The length is 110 km and the width is 95 km. The top platform depth is 1 450 m and the minimum depth is 1 230 m. The piedmont depth is 4 830 m, which makes the total relief being 3 600 m (Fig.2–34).				
命名释义 Reason for Choice of Name	"采薇"出自《诗经·小雅·采薇》"采薇采薇，薇亦作止"，意为采集一种可食用的野菜。此篇诗歌是士兵在返乡途中所唱，表达其对故乡的思念与对和平的向往。 GEBCO 地名辞典中称为 Pallada Guyot，以俄罗斯的"Pallada"号护卫舰的名字命名。 "Caiwei" comes from a poem named *Caiwei* in *Shijing · Xiaoya*. *Shijing* is a collection of ancient Chinese poems from 11th century B.C. to 6th century B.C. "We pick the vetches, the vetches; Springing up are the vetches." "Caiwei" means collecting a kind of potherbs for food. The poem was sung by soldiers going home showing their eager for peace. It is also named Pallada Guyot after the Russian frigate, *Pallada*, in GEBCO gazetteer.				

(a)

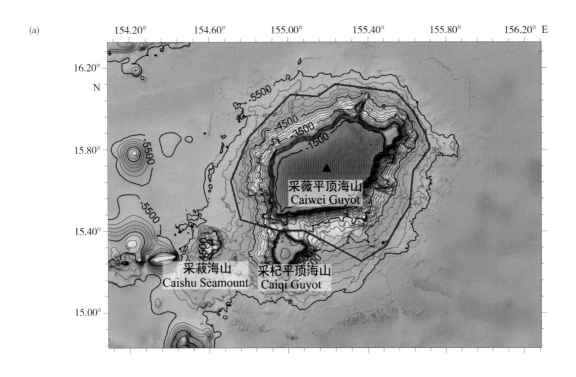

(b)

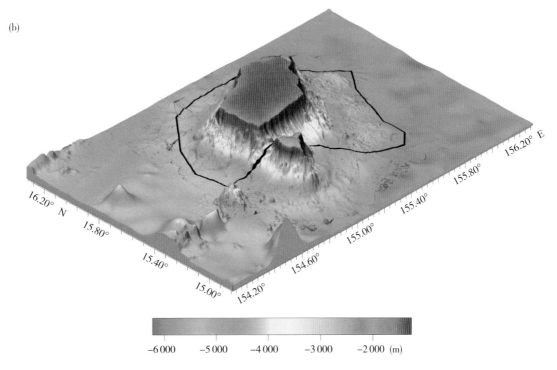

图 2-34　采薇平顶海山

(a) 地形图（等深线间隔 200 m）；(b) 三维图

Fig.2-34　Caiwei Guyot

(a) Bathymetric map (the contour interval is 200 m); (b) 3-D topographic map

2.3.14 采菽海山
Caishu Seamount

中文名称 Chinese Name	采菽海山 Caishu Haishan	英文名称 English Name	Caishu Seamount	所在大洋 Ocean or Sea	西太平洋 West Pacific Ocean
发现情况 Discovery Facts	此海山于 1997 年 6 月由中国科考船"海洋四号"在执行 DY95-7 航次时调查发现。 This seamount was discovered by the Chinese R/V *Haiyang Sihao* during the DY95-7 cruise in June, 1997.				
命名历史 Name History	在我国大洋航次调查报告和 GEBCO 地名辞典中未命名。 This seamount has not been named in Chinese cruise reports or GEBCO gazetteer.				
特征点坐标 Coordinates	15°21.07′N，154°36.05′E			长(km)×宽(km) Length (km)×Width (km)	30×30
最大水深（m） Max Depth（m）	5 560	最小水深（m） Min Depth（m）	3 227	高差（m） Total Relief（m）	2 333
地形特征 Feature Description	此海山位于采薇平顶海山群西部，发育两个山峰，规模较小，基座直径 30 km。海山最高峰顶水深 3 227 m，山麓水深 5 560 m，高差 2 333 m（图 2–35）。 This seamount is located in the west of Caiwei Guyots which develops two summits. It is small with a base diameter of 30 km. The depth of the highest summit is 3 227 m, while the piedmont depth is 5 560 m, which makes the total relief being 2 333 m (Fig.2–35).				
命名释义 Reason for Choice of Name	"采菽"出自《诗经·小雅·采菽》"采菽采菽，筐之筥之"，意为采集大豆。此句意指采集的大豆，装满筐筥的场景。大豆是周朝天子为奖励诸侯因安邦所作贡献而准备的丰厚奖赏礼物。 "Caishu" comes from a poem named *Caishu* in *Shijing · Xiaoya*. *Shijing* is a collection of ancient Chinese poems from 11th century B.C. to 6th century B.C. "They gather beans o'er there, o'er there; they gather in baskets around and square." "Caishu" means collecting legume crops. The poem describes the scene of collecting legume and filling the baskets which implies that the emperor prepared abundant gifts and prizes to award the principalities for their contributions to the dynasty.				

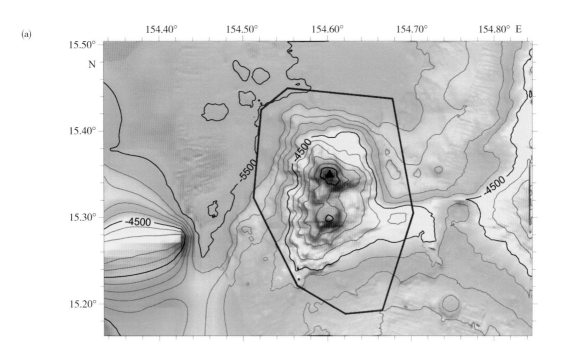

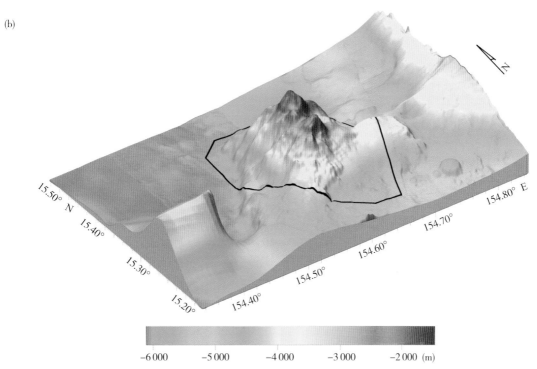

图 2-35　采菽海山

(a) 地形图（等深线间隔 200 m）；(b) 三维图

Fig.2-35　Caishu Seamount

(a) Bathymetric map (the contour interval is 200 m); (b) 3-D topographic map

2.3.15 采杞平顶海山
Caiqi Guyot

中文名称 Chinese Name	采杞平顶海山 Caiqi Pingdinghaishan	英文名称 English Name	Caiqi Guyot	所在大洋 Ocean or Sea	西太平洋 West Pacific Ocean
发现情况 Discovery Facts	此平顶海山于 1997 年 6 月由中国科考船"海洋四号"在执行 DY95-7 航次时调查发现。 This guyot was discovered by the Chinese R/V *Haiyang Sihao* during the DY95-7 cruise in June, 1997.				
命名历史 Name History	该平顶海山位于采薇平顶海山群南部，在我国大洋航次调查报告中曾暂以代号命名，2013 年由中国大洋协会命名为采杞平顶海山。 This guyot is located in the south of Caiwei Seamounts. A code was used to name it in Chinese cruise reports and it was named Caiqi Guyot by COMRA in 2013.				
特征点坐标 Coordinates	15°18.00′N，155°00.00′E			长(km)×宽(km) Length (km)× Width (km)	80×54
最大水深（m） Max Depth（m）	5 100	最小水深（m） Min Depth（m）	1 629	高差（m） Total Relief（m）	3 471
地形特征 Feature Description	此平顶海山规模较小，由地形平缓的山顶平台和陡峭的山坡组成，俯视平面形态呈椭圆形，近 EW 走向，长和宽分别为 80 km 和 54 km。山顶平台水深约 1 800 m，最浅处 1 629 m，山麓水深 5 100 m，高差 3 471 m（图 2–36）。 This guyot, running approximately E to W, is relatively small, consisting of a flat top platform and steep slopes. It has an elliptic overlook plane shape, its length and width are 80 km and 54 km respectively. The top platform depth is about 1 800 m, with the minimum depth of 1 629 m. The piedmont depth is 5 100 m, which makes the total relief being 3 471 m (Fig.2–36).				
命名释义 Reason for Choice of Name	"采杞"出自《诗经·小雅·北山》"陟彼北山，言采其杞"，是指上山采集枸杞。此诗歌是周朝时底层劳苦民众受统治阶级驱使、辛勤劳作时所唱，表达心中哀怨与不满之情。 "Caiqi" comes from a poem named *Beishan* in *Shijing · Xiaoya*. *Shijing* is a collection of ancient Chinese poems from 11th century B.C. to 6th century B.C. "The day I climb the Northern Hill, I gather wolfberries at will." "Caiqi" means collecting medlars in the mountain. The poem was a song of the toiling masses of Zhou Dynasty when they were forced to work hard, expressing their sorrow and discontent.				

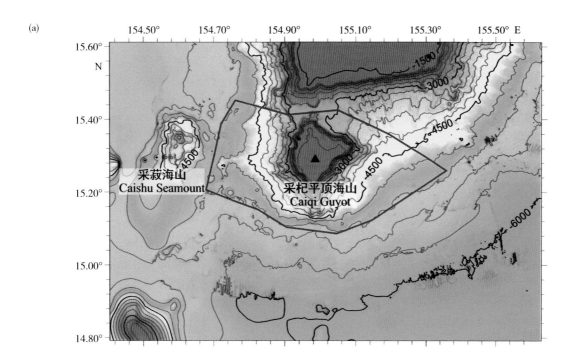

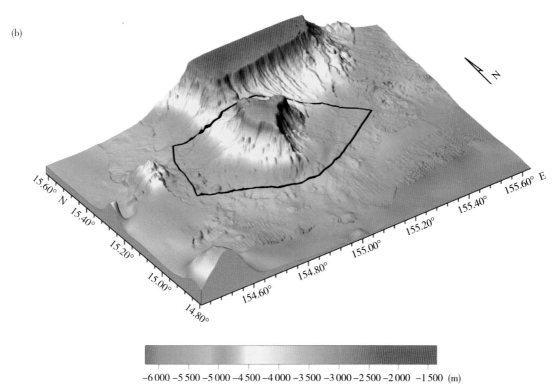

图 2-36　采杞平顶海山

(a) 地形图（等深线间隔 300 m）；(b) 三维图

Fig.2-36　Caiqi Guyot

(a) Bathymetric map (the contour interval is 300 m); (b) 3-D topographic map

2.3.16 嘉偕平顶海山群
Jiaxie Guyots

中文名称 Chinese Name	嘉偕平顶海山群 Jiaxie Pingdinghaishanqun	英文名称 English Name	Jiaxie Guyots	所在大洋 Ocean or Sea	西太平洋 West Pacific Ocean
发现情况 Discovery Facts	此平顶海山群于 1997 年 5—6 月由中国科考船"海洋四号"在执行 DY95-7 航次时调查发现。 The guyots were discovered by the Chinese R/V *Haiyang Sihao* during the DY95-7 cruise from May to June, 1997.				
命名历史 Name History	该平顶海山群由维嘉、维偕、维祯 3 个平顶海山组成,在我国大洋航次调查报告中曾暂以代号命名。维嘉平顶海山在 GEBCO 地名辞典中称为"Ita Mai Tai Guyot",维偕平顶海山在 GEBCO 地名辞典中称为"Gelendzhik Guyot",均由俄罗斯提名。 The guyots consist of 3 guyots, Weijia Guyot, Weixie Guyot and Weizhen Guyot. A code was used to name it in Chinese cruise reports. Weijia Guyot and Weixie Guyot are named Ita Mai Tai Guyot and Gelendzhik Guyot in GEBCO gazetteer respectively, both of them were proposed by Russia.				
特征点坐标 Coordinates	12°52.20′N,156°48.60′E(维嘉平顶海山 / Weijia Guyot) 12°13.20′N,156°22.80′E(维偕平顶海山 / Weixie Guyot) 12°31.80′N,156°20.63′E(维祯平顶海山 / Weizhen Guyot)			长(km)× 宽(km) Length (km) × Width (km)	150 × 75
最大水深(m) Max Depth(m)	6 120	最小水深(m) Min Depth(m)	1 209	高差(m) Total Relief(m)	4 911
地形特征 Feature Description	此平顶海山群由 3 个平顶海山组成,总体 SW—NE 走向,长和宽分别为 150 km 和 75 km,山麓水深最大 6 120 m。最高处位于南部的维偕平顶海山,水深为 1 209 m(图 2–37)。 The guyots consist of 3 guyots, and runs NE to SW overall. The length and width are 150 km and 75 km respectively. The piedmont depth is 6 120 m. The highest place is Weixie Guyot in the south, with the depth of 1 209 m (Fig.2–37).				
命名释义 Reason for Choice of Name	取词于维嘉平顶海山和维偕平顶海山。此海山群的 3 个海山采用群组化命名方法,分别命名维嘉、维偕和维祯,均取词于《诗经》。 Jiaxie Guyots consist of the last characters of Weijia and Weixie in Chinese respectively. There are three seamounts in this group, which are named using group method as Weijia, Weixie and Weizhen, all of them come from *Shijing*. *Shijing* is a collection of ancient Chinese poems from 11th century B.C. to 6th century B.C.				

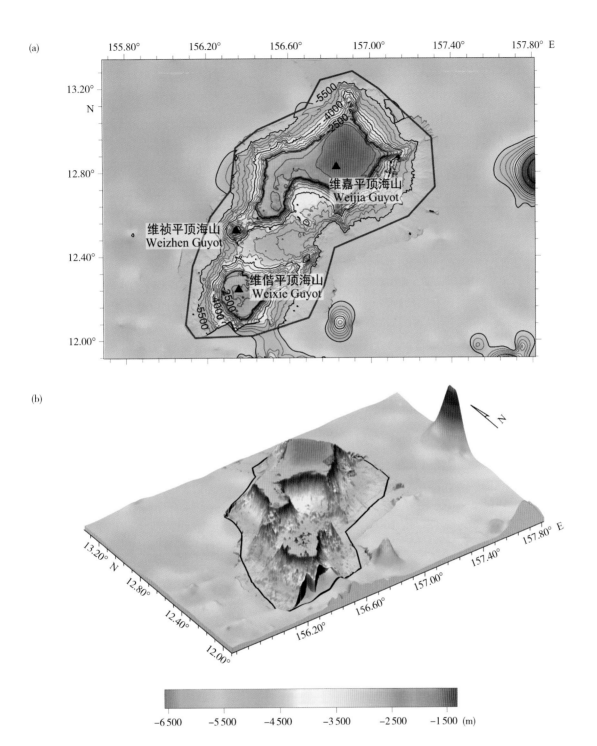

图 2-37 嘉偕平顶海山群

(a) 地形图 (等深线间隔 300 m)；(b) 三维图

Fig.2-37 Jiaxie Guyots

(a) Bathymetric map (the contour interval is 300 m); (b) 3-D topographic map

2.3.17 维嘉平顶海山
Weijia Guyot

中文名称 Chinese Name	维嘉平顶海山 Weijia Pingdinghaishan	英文名称 English Name	Weijia Guyot (Ita Mai Tai Guyot)	所在大洋 Ocean or Sea	西太平洋 West Pacific Ocean
发现情况 Discovery Facts	此平顶海山于1997年5—6月由中国科考船"海洋四号"在执行DY95-7航次时调查发现。 This guyot was discovered by the Chinese R/V *Haiyang Sihao* during the DY95-7 cruise from May to June, 1997.				
命名历史 Name History	该平顶海山在GEBCO地名辞典中称为Ita Mai Tai Guyot，2010年由俄罗斯提名，在我国大洋航次调查报告中曾暂以代号命名，2013年由中国大洋协会命名为维嘉平顶海山。 This guyot is named Ita Mai Tai Guyot in GEBCO gazetteer, proposed by Russia in 2010. A code was used to name it in Chinese cruise reports and was named Weijia Guyot by COMRA in 2013.				
特征点坐标 Coordinates	12°52.20′N，156°48.60′E			长(km)×宽(km) Length (km)× Width (km)	100×75
最大水深（m） Max Depth（m）	6 120	最小水深（m） Min Depth（m）	1 405	高差（m） Total Relief（m）	4 715
地形特征 Feature Description	此平顶海山由地形平缓的山顶平台和陡峭的山坡组成，长和宽分别为100 km和75 km。山顶平台水深约1 600 m，最浅处1 405 m，山麓水深6 120 m，高差4 715 m（图2–38）。 This guyot consists of a flat top platform and steep slopes. The length is 100 km and the width is 75 km. The top flatform depth is about 1 600 m, with the minimum depth of 1 405 m. The piedmont depth is 6 120 m, which makes the total relief being 4 715 m (Fig.2–38).				
命名释义 Reason for Choice of Name	"维嘉"出自《诗经·小雅·鱼丽》"物其多矣，维其嘉矣。物其旨矣，维其偕矣"，意为招待客人的食物丰富充足。此篇诗歌是周朝招待宾客通用之乐歌，赞美年丰物阜，表现宾主尽享美酒佳肴的场景，显示欢乐的气氛。 "Weijia" comes from a poem named *Yuli* in *Shijing · Xiaoya*. *Shijing* is a collection of ancient Chinese poems from 11th century B.C. to 6th century B.C. "Plenty on which to dine and wine, the quality is fine. Tasteful on which to dine and wine, the taste is fine." It means the host prepared abundant food to serve the guests. This poem is the song describing the scene of entertaining guests in Zhou Dynasty praising the harvest and showing the atmosphere of joy.				

(a)

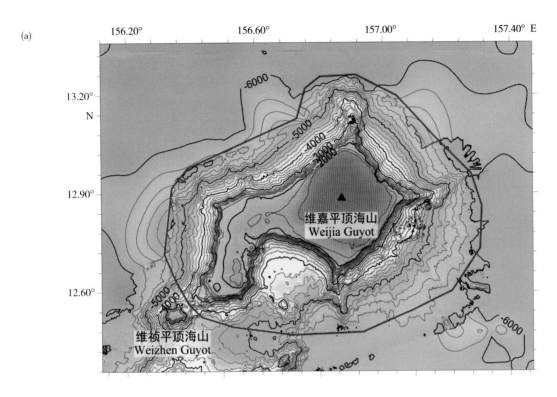

(b)

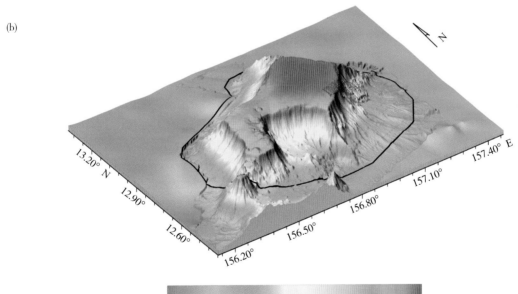

$-6\,000\ -5\,500\ -5\,000\ -4\,500\ -4\,000\ -3\,500\ -3\,000\ -2\,500\ -2\,000$ (m)

图 2-38 维嘉平顶海山

(a) 地形图 （等深线间隔 200 m）；(b) 三维图

Fig.2-38 Weijia Guyot

(a) Bathymetric map (the contour interval is 200 m); (b) 3-D topographic map

2.3.18 维祯平顶海山
Weizhen Guyot

中文名称 Chinese Name	维祯平顶海山 Weizhen Pingdinghaishan	英文名称 English Name	Weizhen Guyot	所在大洋 Ocean or Sea	西太平洋 West Pacific Ocean
发现情况 Discovery Facts	此平顶海山于 1997 年 5—6 月由中国科考船"海洋四号"在执行 DY95-7 航次时调查发现。 This guyot was discovered by the Chinese R/V *Haiyang Sihao* during the DY95-7 cruise from May to June, 1997.				
命名历史 Name History	在我国大洋航次调查报告和 GEBCO 地名辞典中未命名。 This guyot has not been named in Chinese cruise reports or GEBCO gazetteer.				
特征点坐标 Coordinates	12°31.80′N，156°20.63′E			长(km)×宽(km) Length (km)× Width (km)	20×20
最大水深（m） Max Depth（m）	5 500	最小水深（m） Min Depth（m）	2 417	高差（m） Total Relief（m）	3 083
地形特征 Feature Description	此平顶海山规模较小，由地形平缓的山顶平台和陡峭的山坡组成，基座直径 20 km。山顶平台水深约 2 550 m，最浅处 2 417 m，坡麓水深 5 500 m，高差 3 083 m（图 2-39）。 This guyot is relatively small, consisting of a flat top platform and steep slopes. Its base diameter is 20 km. The top platform depth is about 2 550 m, with the minimum depth of 2 417 m. The piedmont depth is 5 500 m, which makes the total relief being 3 083 m (Fig.2-39).				
命名释义 Reason for Choice of Name	"维祯"出自《诗经·周颂·维清》"维清缉熙，文王法典。肇禋，迄用有成，维周之祯"。本篇是周公制礼作乐时祭祀文王的宗庙乐歌，赞美周文王征战讨伐，战功赫赫，建立一个稳定和平的国家。维祯指维护国家的和谐与稳定。 "Weizhen" comes from a poem named *Weiqing* in *Shijing · Zhousong*. *Shijing* is a collection of ancient Chinese poems from 11th century B.C. to 6th century B.C. "Clear, and always reverent, are the statues of Lord Wen. Since first offerings are laid, great achievements have been made, the fame of Zhou will never fade." This poem was a temple ode, sung by the Duke of Zhou to offer a sacrifice to Lord Wen. "Weizhen" means maintaining harmony and peace of the nation.				

(a)

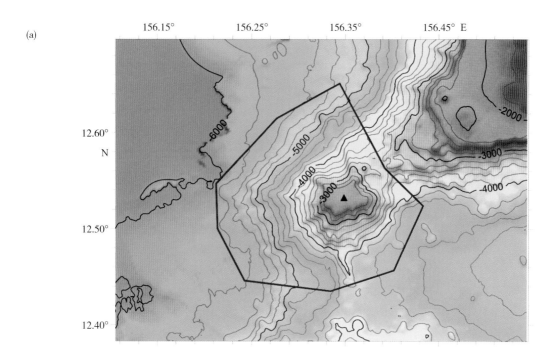

(b)

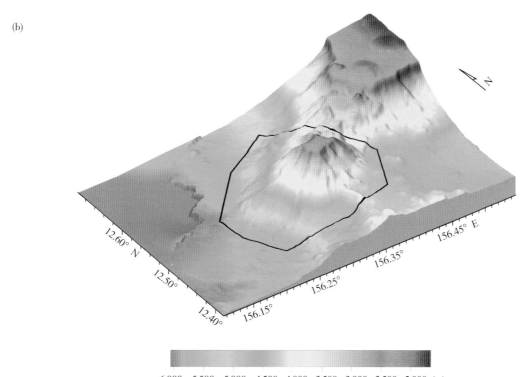

图 2-39　维祯平顶海山

(a) 地形图（等深线间隔 200 m）；(b) 三维图

Fig.2-39　Weizhen Guyot

(a) Bathymetric map (the contour interval is 200 m); (b) 3-D topographic map

2.3.19 维偕平顶海山
Weixie Guyot

中文名称 Chinese Name	维偕平顶海山 Weixie Pingdinghaishan		英文名称 English Name	Weixie Guyot (Gelendzhik Guyot)	所在大洋 Ocean or Sea	西太平洋 West Pacific Ocean
发现情况 Discovery Facts	此平顶海山于1997年5—6月由中国科考船"海洋四号"在执行DY95-7航次时调查发现。 This guyot was discovered by the Chinese R/V *Haiyang Sihao* during the DY95-7 cruise from May to June, 1997.					
命名历史 Name History	该平顶海山在GEBCO地名辞典中称为Gelendzhik Guyot，2004年由俄罗斯提名。该平顶海山在我国大洋航次调查报告中曾暂以代号命名，2013年由中国大洋协会暂命名为维偕平顶海山。 This guyot is named Gelendzhik Guyot in GEBCO gazetteer, proposed by Russia in 2004. A code was used to name it in Chinese cruise reports and was temporarily named Weixie Guyot by COMRA in 2013.					
特征点坐标 Coordinates	12°13.20′N，156°22.80′E				长(km)×宽(km) Length (km)× Width (km)	80×45
最大水深（m） Max Depth（m）	6 030	最小水深（m） Min Depth（m）		1 299	高差（m） Total Relief（m）	4 731
地形特征 Feature Description	此平顶海山由地形平缓的山顶平台和陡峭的山坡组成，整体WSW—ENE走向，长宽分别为80 km和45 km。山顶平台水深约1 500 m，最浅处1 299 m，山麓水深6 030 m，高差4 731 m（图2–40）。 This guyot consists of a flat top platform and steep slopes, running NEE to SWW overall. The length and width are 80 km and 45 km respectively. The top platform depth is about 1 500 m, with the minimum depth of 1 299 m. The piedmont depth is 6 030 m, which makes the total relief being 4 731 m (Fig.2–40).					
命名释义 Reason for Choice of Name	"维偕"出自《诗经·小雅·鱼丽》"物其多矣，维其嘉矣。物其旨矣，维其偕矣"，意为招待客人的食物美味可口。此篇是周朝招待宾客通用之乐歌，赞美年丰物阜，表现宾主尽享美酒佳肴的场景，显示欢乐的气氛。 "Gelendzhik Guyot"以俄罗斯调查船"Gelendzhik"号的名字命名。 "Weixie" comes from a poem named *Yuli* in *Shijing · Xiaoya*. *Shijing* is a collection of ancient Chinese poems from 11th century B.C. to 6th century B.C. "Plenty on which to dine and wine, the quality is fine. Tasteful on which to dine and wine, the taste is fine." It means the host prepared delicious food to serve the guests. This poem is the song describing the scene of entertaining guests in Zhou Dynasty praising the harvest and showing the atmosphere of joy. Gelendzhik Guyot comes from the Russian research vessel named *Gelendzhik*.					

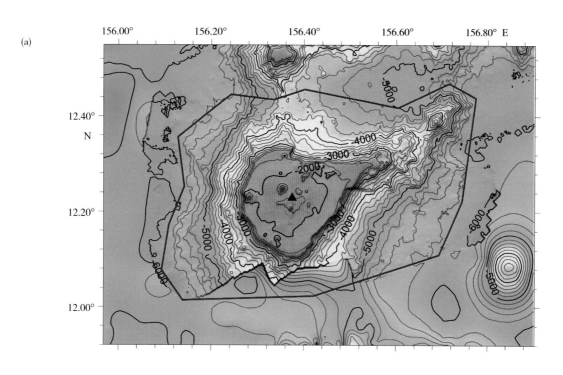

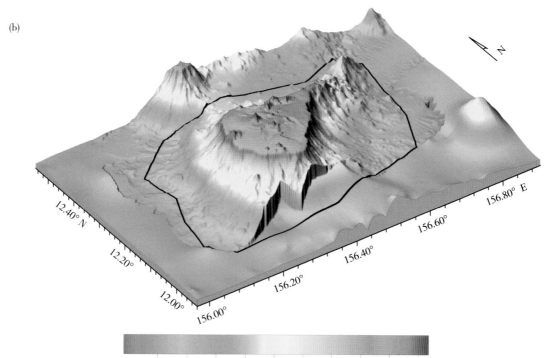

图 2-40　维偕平顶海山

(a) 地形图（等深线间隔 200 m）；(b) 三维图

Fig.2-40　Weixie guyot

(a) Bathymetric map (the contour interval is 200 m); (b) 3-D topographic map

2.3.20 布塔科夫平顶海山
Butakov Guyot

中文名称 Chinese Name	布塔科夫平顶海山 Butakefu Pingdinghaishan	英文名称 English Name	Butakov Guyot	所在大洋 Ocean or Sea	西太平洋 West Pacific Ocean
发现情况 Discovery Facts	此平顶海山于 1999 年 6 月由中国科考船"大洋一号"在执行 DY95-10 航次时调查发现；由俄罗斯使用"Gelendzhik"号科考船于 2004 年调查。 This guyot was discovered by the Chinese R/V *Dayang Yihao* during the DY95-10 cruise in June, 1999.				
命名历史 Name History	该平顶海山在 2006 年由俄罗斯提交 SCUFN 审议通过，命名为"Butakov Guyot"，中文译名为"布塔科夫平顶海山"。该平顶海山在我国大洋航次调查报告中曾暂以代号命名。 This guyot was proposed by Russia and approved by SCUFN in 2006. It is named Butakov Guyot and the translation is Butakefu Pingdinghaishan in Chinese. A code name was used temporarily in Chinese cruise reports.				
特征点坐标 Coordinates	11°26.99′N，156°35.94′E			长(km)×宽(km) Length (km)× Width (km)	150×60
最大水深（m） Max Depth（m）	6 040	最小水深（m） Min Depth（m）	1 289	高差（m） Total Relief（m）	4 751
地形特征 Feature Description	此平顶海山由地形平缓的山顶平台和陡峭的山坡组成，俯视平面形态狭长，SSE—NNW 走向，长宽分别为 150 km 和 60 km。山顶平台水深约 2 000 m，最浅处 1 289 m，山麓水深 6 040 m，高差 4 751 m（图 2–41）。 This guyot consists of a flat top platform and steep slopes. It has an elongated overlook plane shape, running NNW to SSE. The length is 150 km and the width is 60 km. The top platform depth is about 2 000 m and the minimum depth is 1 289 m. The piedmont depth is 6 040 m, which makes the total relief being 4 157 m (Fig.2–41).				
命名释义 Reason for Choice of Name	以俄罗斯海军上将 I. I. Butakov（1788—1846）的名字命名。他分别在 1852 年和 1853 年参加了"Pallada"号和"Diana"号护卫舰的环球航行。 It is named after the Russian Admiral I. I. Butakov (1788–1846), who participated in a round-the-world expedition on the frigates *Pallada* in 1852 and *Diana* in 1853.				

(a)

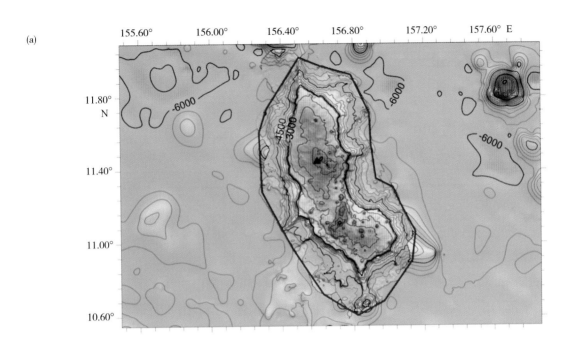

(b)

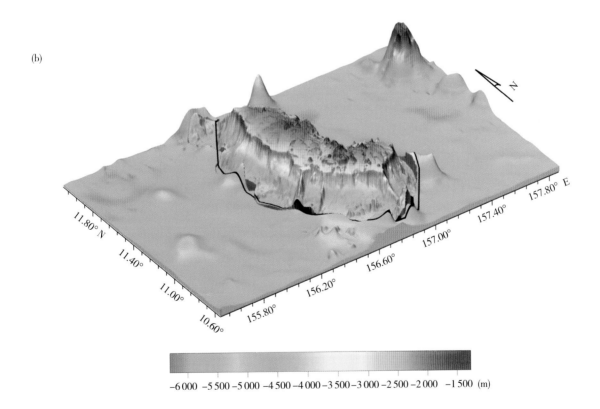

图 2-41 布塔科夫平顶海山

(a) 地形图（等深线间隔 300 m）；(b) 三维图

Fig.2-41 Butakov Guyot

(a) Bathymetric map (the contour interval is 300 m); (b) 3-D topographic map

第4节 马尔库斯-威克海山区地理实体

马尔库斯-威克海山区由西太平洋马尔库斯岛和威克岛一线的群集海山组成，西临伊豆-小笠原海沟和马里亚纳海沟，东接中太平洋海山区，南部紧邻皮嘉费他海盆，东西向跨度约2 500 km。

区内海山最浅水深约 400～500 m，临近深海盆地水深近 6 000 m。区域内的大洋基底在 1.65 亿年之前已经形成，海山在中白垩纪时由赤道以南形成后经板块构造运动漂移至现今位置，为多热点成因的幔源岩浆由板内火山喷发所致，不少海山曾经在早期出露水面 (Sager et al., 1993; Kaneda el al., 2010)。

在此海域我国命名地理实体 22 个，翻译地名 4 个。其中海山群 4 个，包括南台平顶海山群、徐福海山群、大成海山群和湛露海山群；平顶海山 15 个，包括南台平顶海山、巴蒂泽平顶海山、邦基平顶海山、德茂平顶海山、阿诺德平顶海山、拉蒙特平顶海山、徐福平顶海山、方丈平顶海山、百合平顶海山、大成平顶海山、织女平顶海山、牛郎平顶海山、湛露平顶海山、麦克唐奈平顶海山和恺悌平顶海山；海山 7 个，包括瀛洲海山、蓬莱海山、蝴蝶海山、显允海山、令德海山、令仪海山和凤鸣海山 (图 2-42)。

Section 4   Undersea Features in the Marcus-Wake Seamount Area

The Marcus-Wake Seamount Area consists of the clustered seamounts along Marcus Island-Wake Island line in Western Pacific. It is adjacent to the Izu-Bonin Trench and Mariana Trench in the west, reaches the Central Pacific Seamounts Area in the east and adjoins to the Pigafeta Basin in the south. The whole seamount area stretches across about 2 500 km.

The shallowest water depth of seamounts in this area is about 400–500 m, while the water depth is about 6 000 m near the deep-sea basins. The oceanic basement in the area had been formed 165 Ma. The seamounts were formed in the Middle Cretaceous south to the equator, and then drifted to current position through plate tectonic movements. They were caused by multi-hotspot-origin mantle magma volcanic eruptions inside the plate. Many seamounts once have reached out of the sea surface in the early time (Sager et al., 1993; Kaneda el al., 2010).

In total, 22 undersea features have been named and 4 others have been translated by China in the region of the Marcus-Wake Seamount Area. Among them, there are 4 seamounts, including Nantai Guyots, Xufu Seamounts, Dacheng Seamounts and Zhanlu Seamounts; 15 guyots, including Nantai Guyot, Batiza Guyot, Bangji Guyot, Demao Guyot, Arnold Guyot, Lamont Guyot, Xufu Guyot, Fangzhang Guyot, Baihe Guyot, Dacheng Guyot, Zhinyu Guyot, Niulang Guyot, Zhanlu Guyot, McDonnell Guyot and Kaiti Guyot; and 7 seamounts, including Yingzhou Seamount, Penglai Seamount, Hudie Seamount, Xianyun Seamount, Lingde Seamount, Lingyi Seamount and Fengming Seamount (Fig. 2–42).

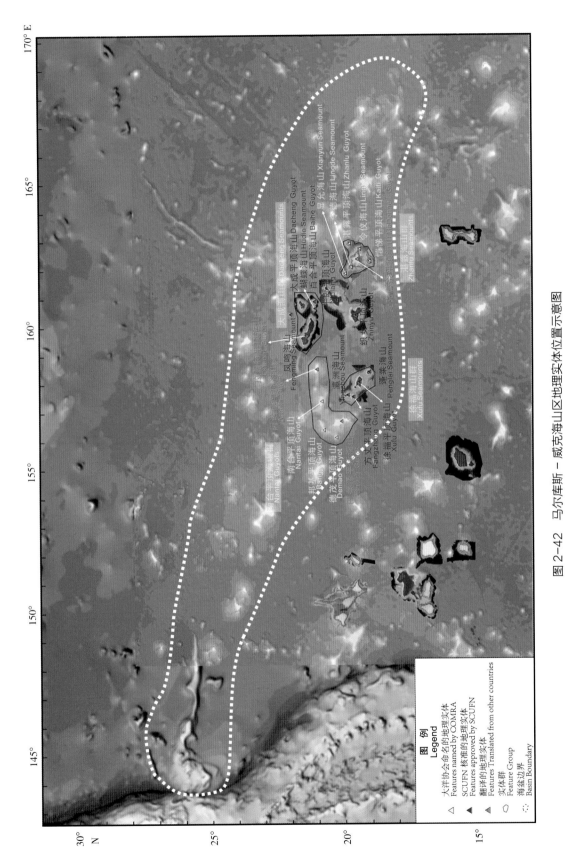

图 2-42　马尔库斯－威克海山区地理实体位置示意图

Fig.2-42　Locations of the undersea features in the Marcus-Wake Seamount Area

2.4.1　南台平顶海山群
Nantai Guyots

中文名称 Chinese Name	南台平顶海山群 Nantai Pingdinghaishanqun	英文名称 English Name	Nantai Guyots	所在大洋 Ocean or Sea	西太平洋 West Pacific Ocean
发现情况 Discovery Facts	此平顶海群山于 2011 年 8—10 月由中国科考船"海洋六号"在执行 DY125-23 航次时调查发现。 The guyots were discovered by the Chinese R/V *Haiyang Liuhao* during the DY125-23 cruise from August to October, 2011.				
命名历史 Name History	该平顶海山群包含 5 个平顶海山，在我国大洋航次调查报告中暂以代号命名。 The guyots consist of 5 guyots. A code name was used temporarily in Chinese cruise reports.				
特征点坐标 Coordinates	20°00.79′N，156°34.80′E（巴蒂泽平顶海山 / Badiza Guyot） 20°23.32′N，155°58.62′E（德茂平顶海山 / Demao Guyot） 20°53.83′N，156°14.34′E（邦基平顶海山 / Bangji Guyot） 20°55.82′N，157°09.54′E（南台平顶海山 / Nantai Guyot） 21°05.60′N，158°28.92′E（阿诺德平顶海山 / Arnold Guyot）		长(km)×宽(km) Length (km)× Width (km)		323×85
最大水深（m） Max Depth（m）	5 434	最小水深（m） Min Depth（m）	1 335	高差（m） Total Relief（m）	4 099
地形特征 Feature Description	此平顶海山群主要由 5 座规模较大的平顶海山组成，总体 SW—NE 走向，长宽分别为 323 km 和 85 km。最高处位于东部的阿诺德平顶海山，水深为 1 335 m（图 2–43）。 The guyots mainly consist of five guyots, running NE to SW overall. The length is 323 km and the width is 85 km. The highest point is in the Arnold Guyot of the eastern area, with the depth of 1 335 m (Fig.2–43).				
命名释义 Reason for Choice of Name	"南台"来源于《诗经·小雅·南山有台》。此海山群包括 5 座海山，除巴蒂泽平顶海山和阿诺德平顶海山外，其他 3 座海山采用群组化命名方法，分别命名为德茂平顶海山、邦基平顶海山和南台平顶海山，均取词于《诗经·小雅·南山有台》。 "Nantai" comes from a poem named *Nanshanyoutai* in *Shijing · Xiaoya*. *Shijing* is a collection of ancient Chinese poems from 11th century B.C. to 6th century B.C. There are five guyots in this group and three of them are named using group method as Demao Guyot, Bangji Guyot and Nantai Guyot, which come from the same poem. The other two Guyots are Batiza Guyot and Arnold Guyot.				

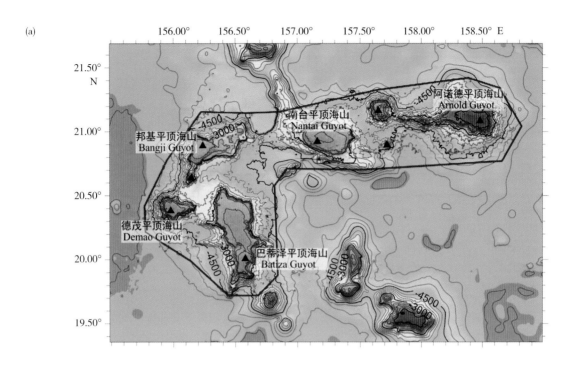

(a)

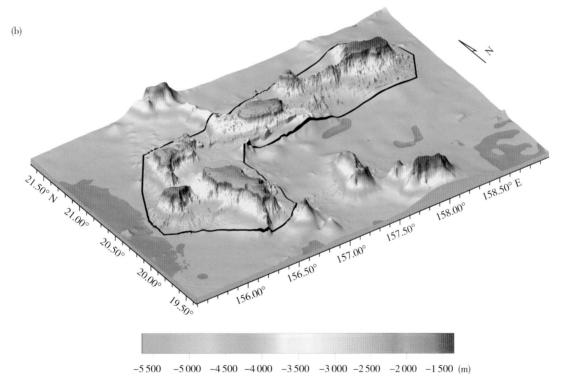

(b)

图 2-43　南台平顶海山群

(a) 地形图（等深线间隔 300 m）；(b) 三维图

Fig.2-43　Nantai Guyots

(a) Bathymetric map (the contour interval is 300 m); (b) 3-D topographic map

2.4.2 邦基平顶海山
Bangji Guyot

中文名称 Chinese Name	邦基平顶海山 Bangji Pingdinghaishan	英文名称 English Name	Bangji Guyot	所在大洋 Ocean or Sea	西太平洋 West Pacific Ocean
发现情况 Discovery Facts	此平顶海山于 2011 年 8—10 月由中国科考船"海洋六号"在执行 DY125-23 航次时调查发现。 This guyot was discovered by the Chinese R/V *Haiyang Liuhao* during the DY125-23 cruise from August to October, 2011.				
命名历史 Name History	在我国大洋航次调查报告中曾暂以代号命名。 A code name was used temporarily in Chinese cruise reports.				
特征点坐标 Coordinates	20°53.83′N，156°14.34′E（最高峰 /peak） 20°38.71′N，156°08.64′E			长(km)× 宽(km) Length (km) × Width (km)	88 × 51
最大水深（m） Max Depth（m）	5 097	最小水深（m） Min Depth（m）	1 875	高差（m） Total Relief（m）	3 222
地形特征 Feature Description	邦基平顶海山位于南台平顶海山群的西部，由地形平缓的山顶平台和陡峭的山坡组成，NE—SW 走向，长宽分别为 88 km 和 51 km。山顶平台水深约 2 000 m，最浅处 1 875 m，山麓水深 5 097 m，高差 3 222 m (图 2–44)。 Bangji Guyot is located in the west of Nantai Guyots, consisting of a flat top platform and steep slopes, running NE to SW. The length and width are 88 km and 51 km respectively. The top platform depth is about 2 000 m, with the minimum depth of 1 875 m. The piedmont depth is 5 097 m, which makes the total relief being 3 222 m (Fig.2–44).				
命名释义 Reason for Choice of Name	"邦基"出自《诗经·小雅·南山有台》"乐只君子，邦家之基"，意为君子真正的快乐在于为国家建立根基。 "Bangji" comes from a poem named *Nanshanyoutai* in *Shijing · Xiaoya*. *Shijing* is a collection of ancient Chinese poems from 11th century B.C. to 6th century B.C. "Our dear lord we congratulate; he is the mainstay of the state." This poem means that the true happiness of a gentleman derives from building foundations for the country.				

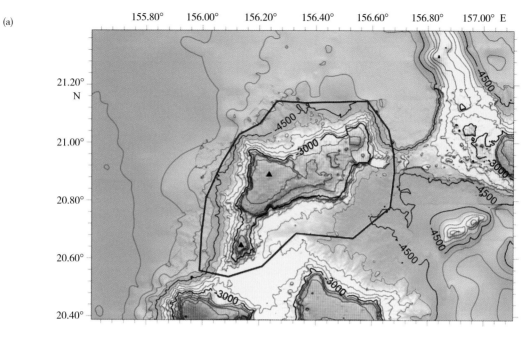

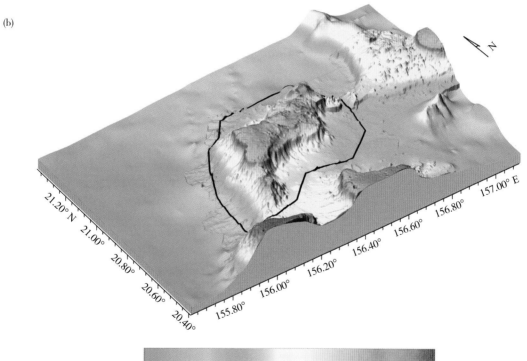

图 2-44　邦基平顶海山

(a) 地形图 （等深线间隔 300 m）；(b) 三维图

Fig.2-44　Bangji Guyot

(a) Bathymetric map (the contour interval is 300 m); (b) 3-D topographic map

2.4.3　德茂平顶海山
Demao Guyot

中文名称 Chinese Name	德茂平顶海山 Demao Pingdinghaishan	英文名称 English Name	Demao Guyot	所在大洋 Ocean or Sea	西太平洋 West Pacific Ocean
发现情况 Discovery Facts	此平顶海山于 2011 年 8—10 月由中国科考船"海洋六号"在执行 DY125-23 航次时调查发现。 This guyot was discovered by the Chinese R/V *Haiyang Liuhao* during the DY125-23 cruise from August to October, 2011.				
命名历史 Name History	在我国大洋航次调查报告中曾暂以代号命名。 A code name was used temporarily in Chinese cruise reports.				
特征点坐标 Coordinates	20°23.32′N，155°58.62′E			长(km) × 宽(km) Length (km) × Width (km)	49 × 44
最大水深（m） Max Depth（m）	5 332	最小水深（m） Min Depth（m）	1 585	高差（m） Total Relief（m）	3 747
地形特征 Feature Description	德茂平顶海山位于南台平顶海山群的西部，由地形平缓的山顶平台和陡峭的山坡组成，EW 走向，长宽分别为 49 km 和 44 km。山顶平台水深约 1 800 m，最浅处 1 585 m，山麓水深 5 332 m，高差 3 747 m（图 2–45）。 Demao Guyot is located in the west of Nantai Guyots, consisting of a flat top platform and steep slopes, running E to W. The length and width are 49 km and 44 km respectively. The top platform depth is about 1 800 m, with the minimum depth of 1 585 m. The piedmont depth is 5 332 m, which makes the total relief being 3 747 m (Fig.2–45).				
命名释义 Reason for Choice of Name	"德茂"出自《诗经·小雅·南山有台》"乐只君子，德音是茂"，意为君子真正的快乐在于养成盛大的美德。 "Demao" comes from a poem named *Nanshanyoutai* in *Shijing · Xiaoya*. *Shijing* is a collection of ancient Chinese poems from 11th century B.C. to 6th century B.C. "For our lord's sake we all cheer : his fame will last many a year." This poem means that the true happiness of a gentleman derives from his virtue.				

(a)

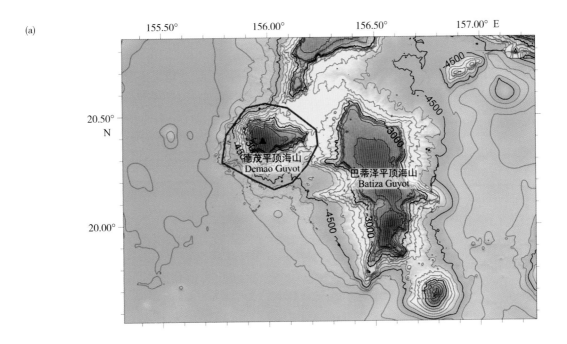

(b)

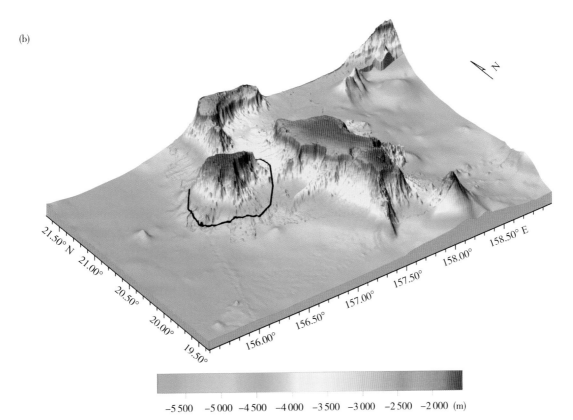

图 2-45　德茂平顶海山

(a) 地形图（等深线间隔 300 m）；(b) 三维图

Fig.2-45　Demao Guyot

(a) Bathymetric map (the contour interval is 300 m); (b) 3-D topographic map

2.4.4 巴蒂泽平顶海山
Batiza Guyot

中文名称 Chinese Name	巴蒂泽平顶海山 Badize Pingdinghaishan	英文名称 English Name	Batiza Guyot	所在大洋 Ocean or Sea	西太平洋 West Pacific Ocean
发现情况 Discovery Facts	此平顶海山于 2011 年 8—10 月由中国科考船"海洋六号"在执行 DY125-23 航次时调查发现。 This guyot was discovered by the Chinese R/V *Haiyang Liuhao* during the DY125-23 cruise from August to October, 2011.				
命名历史 Name History	该平顶海山在 GEBCO 地名辞典中称为"Batiza Guyot",中文译名为"巴蒂泽平顶海山",提名国和提名时间不详。在我国大洋航次调查报告中曾暂以代号命名。 This guyot is named Batiza Guyot in GEBCO gazetteer. Its translation is Badize Pingdinghaishan in Chinese. The proposal country and time are remained unknown. A code name was used temporarily in Chinese cruise reports.				
特征点坐标 Coordinates	20°00.79′N，156°34.80′E			长(km)×宽(km) Length (km)× Width (km)	106×76
最大水深（m） Max Depth（m）	5 130	最小水深（m） Min Depth（m）	1 570	高差（m） Total Relief（m）	3 560
地形特征 Feature Description	巴蒂泽平顶海山位于南台平顶海山群的西南部,由地形平缓的山顶平台和陡峭的山坡组成,SW—NE 走向,长宽分别为 106 km 和 76 km。山顶平台水深约 1 800 m,最浅处 1 570 m,山麓水深 5 130 m,高差 3 560 m（图 2–46）。 The Batiza Guyot is located in the southwest of Nantai Guyots. It consists of a flat top platform and steep slopes, running NE to SW. The length is 106 km and the width is 76 km. The top platform depth is about 1 800 m, with the minimum depth of 1 570 m. The piedmont depth is 5 130 m, which makes the total relief being 3 560 m (Fig.2–46).				
命名释义 Reason for Choice of Name	以海洋地质学家 Rodey Batiza（生卒年月不详）的名字命名。 The guyot is named after a marine geologist Rodey Batiza (birth and death date unknown).				

(a)

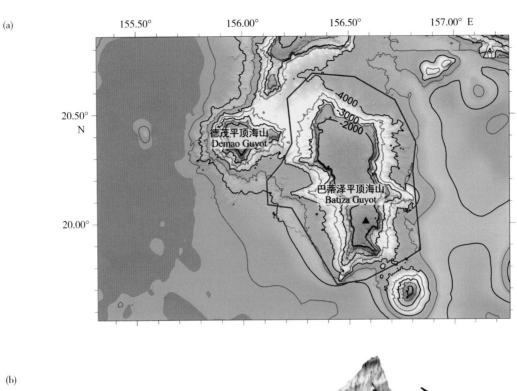

(b)

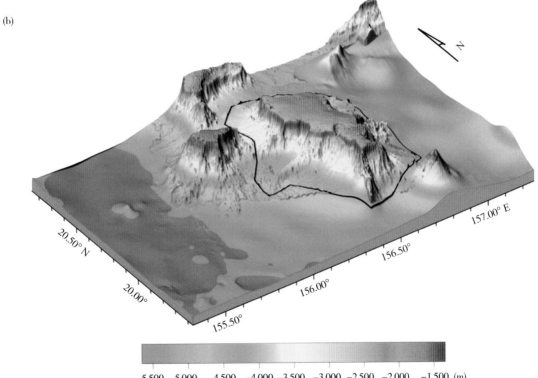

图 2-46　巴蒂泽平顶海山

(a) 地形图（等深线间隔 500 m）；(b) 三维图

Fig.2-46　Batiza Guyot

(a) Bathymetric map (the contour interval is 500 m); (b) 3-D topographic map

2.4.5 南台平顶海山
Nantai Guyot

中文名称 Chinese Name	南台平顶海山 Nantai Pingdinghaishan	英文名称 English Name	Nantai Guyot	所在大洋 Ocean or Sea	西太平洋 West Pacific Ocean
发现情况 Discovery Facts	此平顶海山于 2011 年 8—10 月由中国科考船"海洋六号"在执行 DY125-23 航次时调查发现。 This guyot was discovered by the Chinese R/V *Haiyang Liuhao* during the DY125-23 cruise from August to October, 2011.				
命名历史 Name History	在我国大洋航次调查报告中曾暂以代号命名。 A code name was used temporarily in Chinese cruise reports.				
特征点坐标 Coordinates	20°55.82′N，157°09.54′E			长 (km) × 宽 (km) Length (km) × Width (km)	105 × 56
最大水深（m） Max Depth（m）	5 073	最小水深（m） Min Depth（m）	1 720	高差（m） Total Relief（m）	3 353
地形特征 Feature Description	南台平顶海山位于南台平顶海山群的中部，由地形平缓的山顶平台和陡峭的山坡组成，EW 走向，长宽分别为 105 km 和 56 km。山顶平台水深约 2 000 m，最浅处 1 720 m，山麓水深 5 073 m，高差 3 353 m（图 2-47）。 Nantai Guyot is located in the middle of Nantai Guyots, consisting of a flat top platform and steep slopes, running E to W. The length and width are 105 km and 56 km respectively. The top platform depth is about 2 000 m, with the minimum depth of 1 720 m. The piedmont depth is 5 073 m, which makes the total relief being 3 353 m (Fig.2–47).				
命名释义 Reason for Choice of Name	"南台"出自《诗经·小雅·南山有台》"南山有台，北山有莱。乐只君子，邦家之光。""台"是一种草，"南台"意为南山上草木繁盛，此诗寓意国家拥有具备各种美德的君子贤人。 "Nantai" comes from a poem named *Nanshanyoutai Shijing · Xiaoya*. *Shijing* is a collection of ancient Chinese poems from 11th century B.C. to 6th century B.C. "The nutgrass grows on the South Hill; The goosefoot grows on the North Hill. Our lord we congratulate; He is the glory of the state." "Tai" is a kind of grass. "Nantai" describes the lush vegetation on "Nanshan". The poem means the nation needs gentleman with good virtue.				

(a)

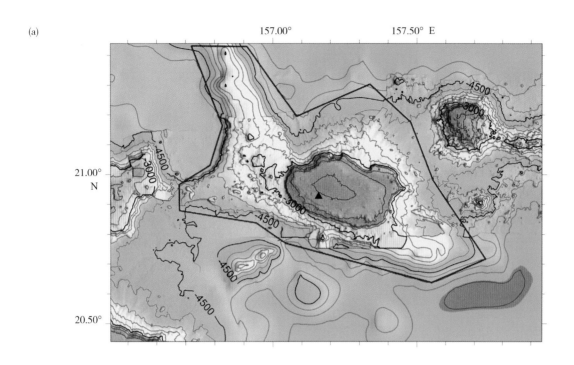

(b)

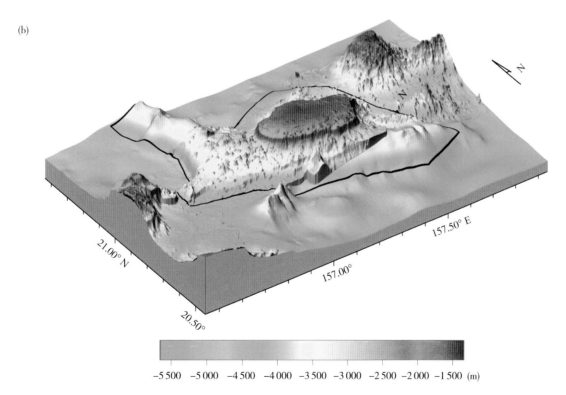

−5 500　−5 000　−4 500　−4 000　−3 500　−3 000　−2 500　−2 000　−1 500　(m)

图 2-47　南台平顶海山

(a) 地形图 （等深线间隔 300 m）；(b) 三维图

Fig.2-47　Nantai Guyot

(a) Bathymetric map (the contour interval is 300 m); (b) 3-D topographic map

2.4.6　阿诺德平顶海山
Arnold Guyot

中文名称 Chinese Name	阿诺德平顶海山 Anuode Pingdinghaishan	英文名称 English Name	Arnold Guyot	所在大洋 Ocean or Sea	西太平洋 West Pacific Ocean
发现情况 Discovery Facts	此平顶海山于 2011 年 8—10 月由中国科考船"海洋六号"在执行 DY125-23 航次时调查发现。 This guyot was discovered by the Chinese R/V *Haiyang Liuhao* during the DY125-23 cruise from August to October, 2011.				
命名历史 Name History	该平顶海山在 GEBCO 地名辞典中称为 Arnold Guyot，中文译名为阿诺德平顶海山，提名国和提名时间不详。在我国大洋航次调查报告中曾暂以代号命名。 This guyot is named Arnold Guyot in the GEBCO gazetteer and translated into Chinese as Anuode Pingdinghaishan. The proposal country and time are remained unknown. A code name was used temporarily in Chinese cruise reports.				
特征点坐标 Coordinates	21°10.30′N，157°38.94′E 21°05.60′N，158°28.92′E（最高峰 / peak）			长(km)× 宽(km) Length (km) × Width (km)	134×71
最大水深（m） Max Depth（m）	5 100	最小水深（m） Min Depth（m）	1 335	高差（m） Total Relief（m）	3 765
地形特征 Feature Description	阿诺德平顶海山位于南台平顶海山群的东部，南台平顶海山以东 20 km，由地形平缓的山顶平台和陡峭的山坡组成，EW 走向，由两个海山平台组成，中间以山脊相连，长宽分别为 134 km 和 71 km。山顶平台水深约 2 000 m，最浅处 1 335 m，山麓水深 5 100 m，高差 3 765 m（图 2-48）。 Arnold Guyot is located in the east of Nantai Guyots, 20 km east to Nantai Guyot. It consists of a flat top platform and steep slopes, running E to W. It includes two seamount platforms which are connected by a ridge. The length is 134 km and the width is 71 km. The top platform depth is about 2 000 m, and the minimum depth is 1 335 m. The piedmont depth is 5 100 m, which makes the total relief being 3 765 m (Fig.2–48).				
命名释义 Reason for Choice of Name	命名释义不详。 The naming reason is remained unknown.				

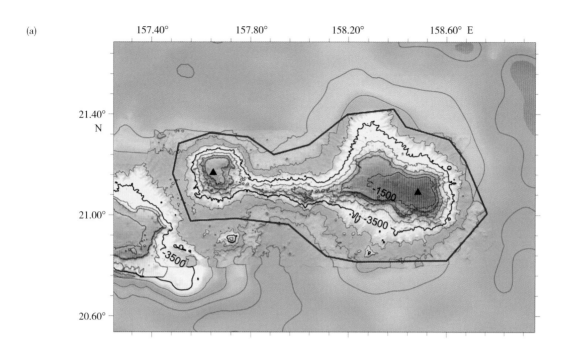

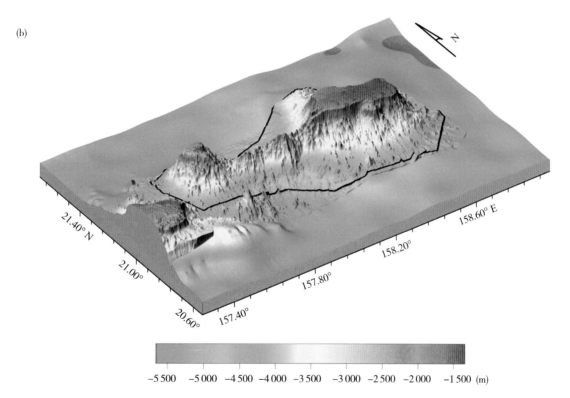

图 2-48　阿诺德平顶海山

(a) 地形图（等深线间隔 500 m）；(b) 三维图

Fig.2-48　Arnold Guyot

(a) Bathymetric map (the contour interval is 500 m); (b) 3-D topographic map

2.4.7 徐福海山群
Xufu Seamounts

中文名称 Chinese Name	徐福海山群 Xufu Haishanqun	英文名称 English Name	Xufu Seamounts	所在大洋 Ocean or Sea	西太平洋 West Pacific Ocean
发现情况 Discovery Facts	此海山群于 2004 年 8 月由中国科考船"大洋一号"在执行 DY105-16A 航次时调查发现。 The seamounts were discovered by the Chinese R/V *Dayang Yihao* during the DY105-16A cruise in August, 2004.				
命名历史 Name History	该海山群由 4 个海山组成，分别为瀛洲海山、方丈平顶海山、徐福平顶海山和蓬莱海山，在我国大洋航次调查报告中曾暂以代号命名。 The seamounts consist of four seamounts, namely Yingzhou Seamount, Fangzhang Guyot, Xufu Guyot and Penglai Seamount. A code name was used temporarily in Chinese cruise reports.				
特征点坐标 Coordinates	19°57.80′N，157°27.30′E（瀛洲海山 / Yingzhou Seamount） 19°46.30′N，157°22.80′E（方丈平顶海山 / Fangzhang Guyot） 19°32.30′N，157°56.00′E（徐福平顶海山 / Xufu Guyot） 19°12.30′N，158°14.00′E（蓬莱海山 / Penglai Seamount）			长(km)×宽(km) Length (km)× Width (km)	155×70
最大水深（m） Max Depth（m）	5 200	最小水深（m） Min Depth（m）	1 200	高差（m） Total Relief（m）	4 000
地形特征 Feature Description	海山群整体 SE—NW 走向，长宽分别为 155 km 和 70 km。此海山群的最高峰为中部的徐福平顶海山，水深约 1 200 m，坡麓水深 5 200 m，高差 4 000 m（图 2–49）。 The seamounts run NW to SE overall. The length is 155 km and the width is 70 km. The highest one among the seamounts is Xufu Guyot in the middle, its depth is about 1 200 m. The piedmont depth is 5 200 m, which makes the total relief being 4 000 m (Fig.2–49).				
命名释义 Reason for Choice of Name	以徐福平顶海山命名，徐福是秦代著名道士，秦始皇曾派遣其去海中寻找长生不老之药，瀛洲、方丈和蓬莱是传说中的目的地。此海山群 4 座海山采用群组化命名，分别选用徐福和其传说中的 3 个目的地名称。 The seamounts are named Xufu after the largest guyot among the four seamounts. "Xufu" is a famous Taoist in the Qin Dynasty. He was very erudite and had good knowledge of medicine, astronomy, navigation, etc. It is said that Xufu was sent by the first Emperor of Qin to lead thousands of people out to sea, looking for the elixir of life for the emperor, and never returned. Yingzhou, Fangzhang and Penglai are the destinations thus four seamounts in this group are named using group method as Xufu and the three destinations respectively.				

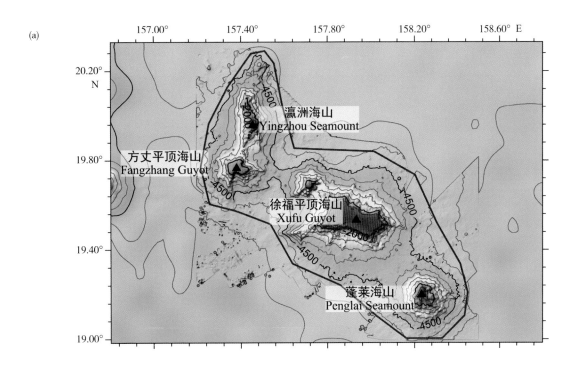

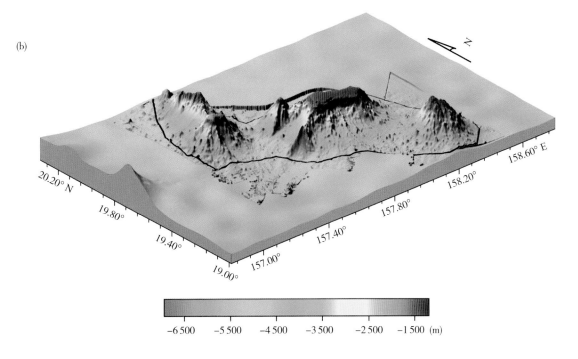

图 2-49　徐福海山群

(a) 地形图（等深线间隔 500 m）；(b) 三维图

Fig.2-49　Xufu Seamounts

(a) Bathymetric map (the contour interval is 500 m); (b) 3-D topographic map

2.4.8 瀛洲海山
Yingzhou Seamount

中文名称 Chinese Name	瀛洲海山 Yingzhou Haishan	英文名称 English Name	Yingzhou Seamount	所在大洋 Ocean or Sea	西太平洋 West Pacific Ocean
发现情况 Discovery Facts	此海山于 2004 年 8 月由中国科考船"大洋一号"在执行 DY105-16A 航次时调查发现。 This seamount was discovered by the Chinese R/V *Dayang Yihao* during the DY105-16A cruise in August, 2004.				
命名历史 Name History	由我国命名为瀛洲海山，于 2011 年提交 SCUFN 审议通过。瀛洲海山为徐福海山群的一部分，徐福海山群在我国大洋航次调查报告中曾暂以代号命名。 Yingzhou Seamount was named by China and approved by SCUFN in 2011. Yingzhou Seamount is a part of Xufu Seamounts which was assigned a code name in Chinese cruise reports.				
特征点坐标 Coordinates	19°57.80′N，157°27.30′E			长(km)×宽(km) Length (km)× Width (km)	40×18
最大水深（m） Max Depth（m）	4 000	最小水深（m） Min Depth（m）	1 400	高差（m） Total Relief（m）	2 600
地形特征 Feature Description	海山俯视平面形态呈 SN 走向的长条脊状，长宽分别为 40 km 和 18 km。海山最浅处水深 1 400 m，山麓水深 4 000 m，高差 2 600 m，边坡陡峭（图 2–50）。 This seamount has an elongated ridge-like overlook plane shape, running N to S. The length and width are 40 km and 18 km respectively. The minimum depth of this seamount is 1 400 m. The piedmont depth is 4 000 m, which makes the total relief being 2 600 m. Its slopes are steep (Fig.2–50).				
命名释义 Reason for Choice of Name	来源于中华传说，据说瀛洲乃神仙居住的海中仙山之一，从那里可以向神仙求得仙药。秦始皇曾派遣徐福去海中寻找长生不老之药，瀛洲是其目的地之一。 In the Chinese legend, Yingzhou is one of the mount in the ocean where the gods live, and it is said people could get magical medicine from gods. Xu Fu was sent by the first Emperor of Qin to look for the elixir of life from the ocean. Yingzhou Seamount was one of his destinations. As this feature is nearby Xufu Guyot, it is named Yingzhou Seamount to commemorate the whole history event.				

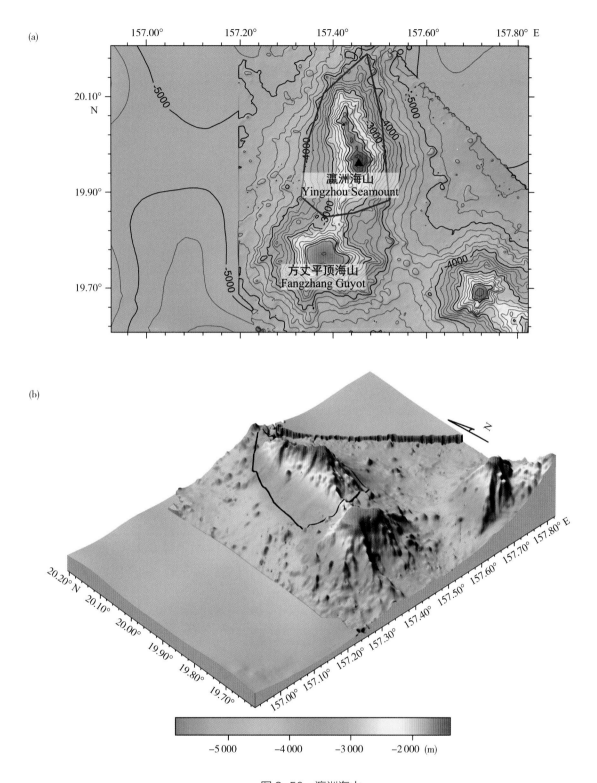

图 2-50　瀛洲海山

(a) 地形图（等深线间隔 200 m）；(b) 三维图

Fig.2-50　Yingzhou Seamount

(a) Bathymetric map (the contour interval is 200 m); (b) 3-D topographic map

2.4.9　方丈平顶海山
Fangzhang Guyot

中文名称 Chinese Name	方丈平顶海山 Fangzhang Pingdinghaishan	英文名称 English Name	Fangzhang Guyot	所在大洋 Ocean or Sea	西太平洋 West Pacific Ocean
发现情况 Discovery Facts	此海山于 2004 年 8 月由中国科考船"大洋一号"在执行 DY105-16A 航次时调查发现。 This guyot was discovered by the Chinese R/V *Dayang Yihao* during the DY105-16A cruise in August, 2004.				
命名历史 Name History	由我国命名为方丈平顶海山，于 2011 年提交 SCUFN 审议通过。方丈平顶海山为徐福海山群的一部分，徐福海山群在我国大洋航次调查报告中曾暂以代号命名。 Fangzhang Guyot was named by China and approved by SCUFN in 2011. Fangzhang Guyot is a part of Xufu Seamounts which were assigned a code name in Chinese cruise reports.				
特征点坐标 Coordinates	19°46.30′N，157°22.80′E			长(km) × 宽(km) Length (km) × Width (km)	28 × 20
最大水深（m） Max Depth（m）	4 000	最小水深（m） Min Depth（m）	1 600	高差（m） Total Relief（m）	2 400
地形特征 Feature Description	方丈平顶海山位于马尔库斯－威克海山区的徐福海山群中，与北部的瀛洲海山由鞍部相连，俯视平面形态呈近东西向的不规则状，长宽分别为 28 km 和 20 km。山顶平台水深约 1 800 m，最浅处水深 1 600 m，山麓水深 4 000 m，高差 2 400 m，顶部平坦，边坡陡峭（图 2–51）。 Connected with Yingzhou Seamount by a saddle, Fangzhang Guyot is located in the Xufu Seamounts of the Marcus-Wake Seamount Area, which has an irregular overlook plane shape, running E to W. The length and width are 28 km and 20 km respectively. The top platform depth is 1 800 m and the minimum depth of this seamount is 1 600 m. The piedmont depth is 4 000 m, which makes the total relief being 2 400 m. Its slopes are steep, and its top platform is flat (Fig.2–51).				
命名释义 Reason for Choice of Name	在中国古代传说中，方丈为海中仙山之一，从那里可以向神仙求得仙药。秦始皇曾派遣徐福去海中寻找长生不老之药，方丈就是其目的地之一。 According to a Chinese legend, Fangzhang is one of the seamounts in the ocean where the gods live. It is further said that people could get magical medicine from gods. Xu Fu was sent by the first Emperor of Qin dynasty to look for the elixir of life from the ocean. Fangzhang was one of his destinations. As this feature is located nearby Xufu Guyot, it is named Fangzhang to memorize the whole history event.				

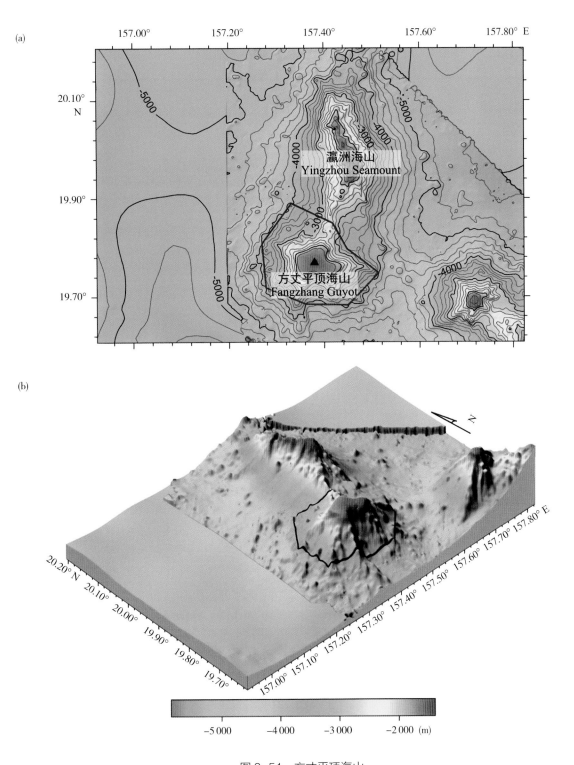

图 2-51　方丈平顶海山

(a) 地形图（等深线间隔 200 m）；(b) 三维图

Fig.2-51　Fangzhang Guyot

(a) Bathymetric map (the contour interval is 200 m); (b) 3-D topographic map

2.4.10 徐福平顶海山
Xufu Guyot

中文名称 Chinese Name	徐福平顶海山 Xufu Pingdinghaishan	英文名称 English Name	Xufu Guyot	所在大洋 Ocean or Sea	西太平洋 West Pacific Ocean
发现情况 Discovery Facts	此海山于 2004 年 8 月由中国科考船"大洋一号"在执行 DY105-16A 航次时调查发现。 This guyot was discovered by the Chinese R/V *Dayang Yihao* during the DY105-16A cruise in August, 2004.				
命名历史 Name History	由我国命名为徐福平顶海山,于 2011 年提交 SCUFN 审议通过。徐福平顶海山为徐福海山群的一部分,徐福海山群在我国大洋航次调查报告中曾暂以代号命名。 Xufu Guyot was named by China and approved by SCUFN in 2011. Xufu Guyot is a part of Xufu Seamounts which were assigned a code name in Chinese cruise reports.				
特征点坐标 Coordinates	19°32.30′N,157°56.00′E			长(km)×宽(km) Length (km) × Width (km)	62×50
最大水深(m) Max Depth(m)	4 000	最小水深(m) Min Depth(m)	1 200	高差(m) Total Relief(m)	2 800
地形特征 Feature Description	此平顶海山俯视平面形态不规则,长宽分别为 62 km 和 50 km。山顶平台水深约 1 400 m,最浅处水深 1 200 m,山麓水深 4 000 m,高差 2 800 m,顶部平坦,边坡陡峭。海山西北部发育一小型附属海山,西南部发育一小型山脊(图 2–52)。 This guyot has an irregular overlook plane shape. The length and width are 62 km and 50 km respectively. The top platform depth is 1 400 m and the minimum depth of this seamount is 1 200 m. The piedmont depth is 4 000 m, which makes the total relief being 2 800 m. Its slopes are steep, and its platform is flat. The northwest of the guyot develops a small-scale seamount, while the southwest of it develops a small ridge (Fig.2–52).				
命名释义 Reason for Choice of Name	徐福平顶海山专名取自秦代著名道士徐福。徐福,秦朝人(生卒年月不详),博学多识,精通医药、天文和航海,据传秦始皇曾派遣徐福带领数千人去海中寻求长生不老之药,经年而未归。徐福在中国沿海居民中有很高声望,许多村镇和庙宇以其命名。 The name of Xufu Guyot comes from a famous Taoist called Xu Fu in the Qin Dynasty. He was very erudite and had good knowledge of medicine, astronomy, navigation, etc. It is said that Xufu was sent by the first Emperor of Qin to lead thousands of people out to sea, looking for the elixir of life for the emperor, and never returned. Meanwhile, his reputation was fairly high as a doctor among the people in coastal areas of ancient China. In memory of Xu Fu, people named their villages and temples after him. This feature is named after Xufu to commemorate that, as early as 210 years BC, the Chinese people began to launch navigation activities in an organized manner.				

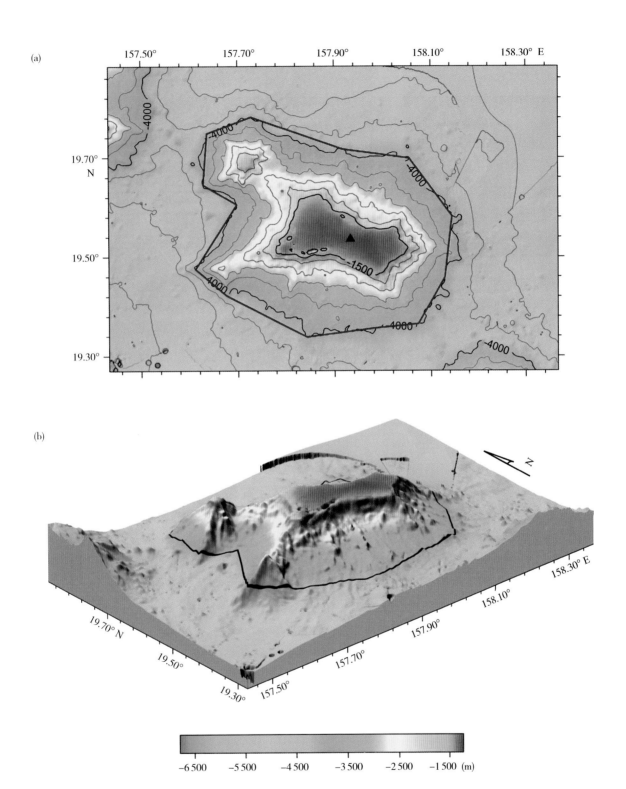

图 2-52　徐福平顶海山

(a) 地形图（等深线间隔 500 m）；(b) 三维图

Fig.2-52　Xufu Guyot

(a) Bathymetric map (the contour interval is 500 m); (b) 3-D topographic map

2.4.11 蓬莱海山
Penglai Seamount

中文名称 Chinese Name	蓬莱海山 Penglai Haishan	英文名称 English Name	Penglai Seamount	所在大洋 Ocean or Sea	西太平洋 West Pacific Ocean
发现情况 Discovery Facts	此海山于 2004 年 8 月由中国科考船"大洋一号"在执行 DY105-16A 航次时调查发现。 This seamount was discovered by the Chinese R/V *Dayang Yihao* during the DY105-16A cruise in August, 2004.				
命名历史 Name History	由我国命名为蓬莱海山，于 2011 年提交 SCUFN 审议通过。蓬莱海山为徐福海山群的一部分，徐福海山群在我国大洋航次调查报告中曾暂以代号命名。 Penglai Seamount was named by China and approved by SCUFN in 2011. Penglai Seamount is a part of Xufu Seamounts which were assigned a code name in Chinese cruise reports.				
特征点坐标 Coordinates	19°12.30′N，158°14.00′E			长(km) × 宽(km) Length (km) × Width (km)	28 × 26
最大水深（m） Max Depth（m）	4 000	最小水深（m） Min Depth（m）	1 200	高差（m） Total Relief（m）	2 800
地形特征 Feature Description	此海山长宽分别为 28 km 和 26 km。海山最浅处水深 1 200 m，山麓水深 4 000 m，高差 2 800 m，边坡陡峭（图 2-53）。 The length and width of this seamount are 28 km and 26 km respectively. The minimum depth of this seamount is 1 200 m, and the piedmont depth is 4 000 m, which makes the total relief being 2 800 m. Its slopes are steep (Fig.2–53).				
命名释义 Reason for Choice of Name	蓬莱海山的专名取自中华传说。据说蓬莱乃神仙居住的海中仙山之一，在此人们可以从神仙那里求得仙药。秦始皇曾派遣徐福去海中寻找长生不老之药，蓬莱就是其目的地之一。 According to a Chinese legend, Penglai is one of the mounts in the ocean where the gods live. It is further said that people could get magical medicine from gods. Xu Fu was sent by the first Emperor of Qin dynasty to look for the elixir of life from the ocean. Penglai was one of his destinations. As this feature is located nearby Xufu Guyot, it has been named Penglai to memorize the whole history event.				

(a)

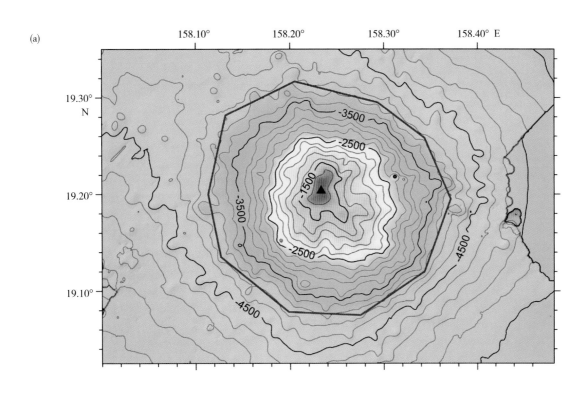

(b)

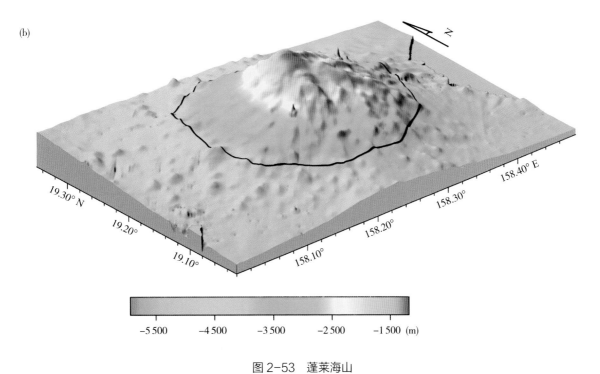

图 2-53　蓬莱海山

(a) 地形图（等深线间隔 200 m）；(b) 三维图

Fig.2-53　Penglai Seamount

(a) Bathymetric map (the contour interval is 200 m); (b) 3-D topographic map

2.4.12 大成海山群
Dacheng Seamounts

中文名称 Chinese Name	大成海山群 Dacheng Haishanqun	英文名称 English Name	Dacheng Seamounts	所在大洋 Ocean or Sea	西太平洋 West Pacific Ocean
发现情况 Discovery Facts	此海山群于 2001 年 7 月由中国科考船"大洋一号"在执行 DY105-11 航次时调查发现。 The seamounts were discovered by the Chinese R/V *Dayang Yihao* during the DY105-11 cruise in July, 2001.				
命名历史 Name History	该海山群由 4 个海山组成，分别为拉蒙特平顶海山、大成平顶海山、蝴蝶海山和百合平顶海山，在我国大洋航次调查报告中曾暂以代号命名。 The seamounts consist of 4 seamounts, namely Lamont Guyot, Dacheng Guyot, Hudie Seamount and Baihe Guyot. A code name was used temporarily in Chinese cruise reports.				
特征点坐标 Coordinates	21°32.00′N，159°32.00′E（拉蒙特平顶海山 /Lamont Guyot） 21°41.80′N，160°40.30′E（大成平顶海山 / Dacheng Guyot） 21°10.94′N，160°29.37′E（蝴蝶海山 / Hudie Seamount） 21°12.15′N，160°43.41′E（百合平顶海山 / Baihe Guyot）			长(km)×宽(km) Length (km)× Width (km)	210×59
最大水深（m） Max Depth（m）	5 500	最小水深（m） Min Depth（m）	1 213	高差（m） Total Relief（m）	4 287
地形特征 Feature Description	大成海山群位于马尔库斯－威克海山区的东部，整体长宽分别为 210 km 和 59 km。此海山群规模最大、水深最浅的海山均为拉蒙特平顶海山，水深最浅处 1 213 m，山麓水深 5 500 m，高差 4 287 m（图 2-54）。 Dacheng Seamounts are located in the east of Marcus-Wake Seamount Area and the overall length and width are 210 km and 59 km respectively. The largest and shallowest seamount among this seamounts is Lamont Guyot. The minimum depth is 1 213 m and the piedmont depth is 5 500 m, which makes the total relief being 4 287 m (Fig.2-54).				
命名释义 Reason for Choice of Name	采用此海山群中的大成平顶海山之名。 The seamounts are named after the largest seamount among it, Dacheng Guyot.				

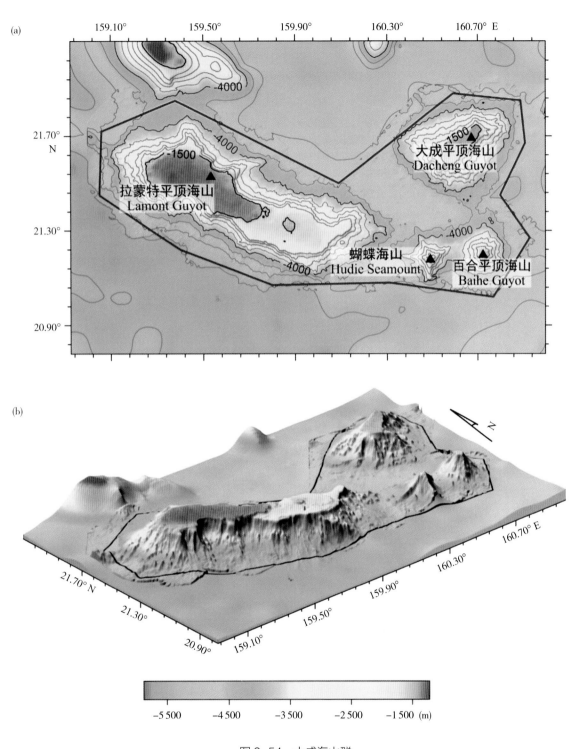

图 2-54　大成海山群

(a) 地形图（等深线间隔 500 m）；(b) 三维图

Fig.2-54　Dacheng Seamounts

(a) Bathymetric map (the contour interval is 500 m); (b) 3-D topographic map

2.4.13 拉蒙特平顶海山
Lamont Guyot

中文名称 Chinese Name	拉蒙特平顶海山 Lamengte Pingdinghaishan	英文名称 English Name	Lamont Guyot	所在大洋 Ocean or Sea	西太平洋 West Pacific Ocean
发现情况 Discovery Facts		此平顶海山于 2001 年 7 月由中国科考船"大洋一号"在执行 DY105-11 航次时调查发现。 This guyot was discovered by the Chinese R/V *Dayang Yihao* during the DY105-11 cruise in July, 2001.			
命名历史 Name History	收录于 GEBCO 地名辞典，1985 年由美国提名，中文译名为拉蒙特平顶海山。此海山在我国大洋航次调查报告中曾暂以代号命名。 The name Lamont Guyot is listed in GEBCO gazetteer. It was proposed by the U. S. in 1985 and the translation is Lamengte Pingdinghaishan in Chinese. A code was used temporarily in Chinese cruise reports.				
特征点坐标 Coordinates	21°32.00′N，159°32.00′E			长(km)×宽(km) Length (km) × Width (km)	144×57
最大水深（m） Max Depth（m）	5 000	最小水深（m） Min Depth（m）	1 213	高差（m） Total Relief（m）	3 787
地形特征 Feature Description	此平顶海山俯视平面形态呈 NW—SE 走向的长条状，长宽分别为 144 km 和 57 km，山顶分东西两部分，其中西部较大较浅，东部略深，面积较小。山顶平台水深约 1 400 m，最浅处水深 1 213 m，山麓水深 5 000 m，高差 3 787 m，顶部平坦，边坡陡峭（图 2-55）。 The guyot has an elongated overlook plane shape, running NW to SE. The length and width are 144 km and 57 km respectively. The top platform is divided into two parts. The western part is large and shallow, while the eastern part is deep and small. The top platform depth is about 1 400 m and the minimum depth is 1 213 m. The piedmont depth is 5 000 m, which makes the total relief being 3 787 m. It has a flat top and steep slopes (Fig.2-55).				
命名释义 Reason for Choice of Name	来源于 GEBCO 地名辞典，释义不详。 The name comes from GEBCO gazetteer but its interpretation is unknown.				

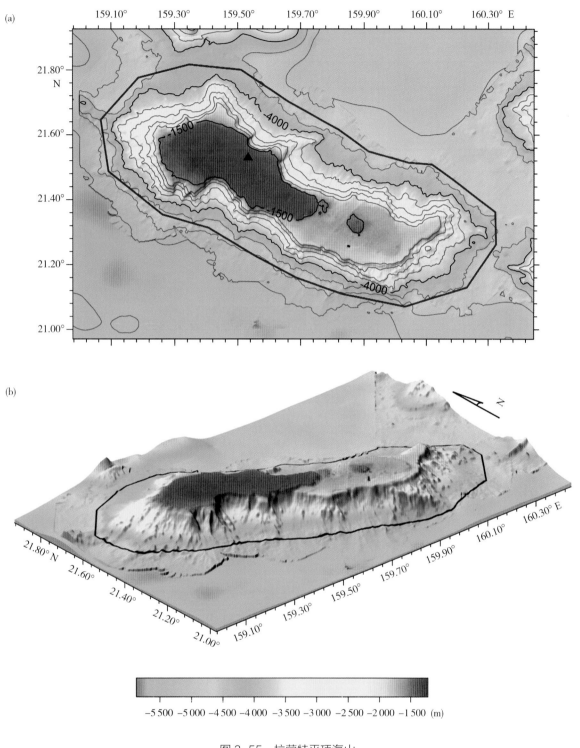

图 2-55　拉蒙特平顶海山

(a) 地形图（等深线间隔 500 m）；(b) 三维图

Fig.2-55　Lamont Guyot

(a) Bathymetric map (the contour interval is 500 m); (b) 3-D topographic map

2.4.14　大成平顶海山
Dacheng Guyot

中文名称 Chinese Name	大成平顶海山 Dacheng Pingdinghaishan	英文名称 English Name	Dacheng Guyot	所在大洋 Ocean or Sea	西太平洋 West Pacific Ocean
发现情况 Discovery Facts	此海山于 2001 年 7 月由中国科考船"大洋一号"在执行 DY105-11 航次时调查发现。 This guyot was discovered by the Chinese R/V *Dayang Yihao* during the DY105-11 cruise in July, 2001.				
命名历史 Name History	由我国命名为大成平顶海山，于 2013 年提交 SCUFN 审议通过。 This feature was named Dacheng Guyot by China and approved by SCUFN in 2013.				
特征点坐标 Coordinates	21°41.80′N，160°40.30′E			长 (km) × 宽 (km) Length (km) × Width (km)	57 × 39
最大水深（m） Max Depth（m）	4 500	最小水深（m） Min Depth（m）	1 330	高差（m） Total Relief（m）	3 170
地形特征 Feature Description	海山俯视平面形态不规则，北东部地势较高，长宽分别为 57 km 和 39 km。山顶平台水深约 1 500 m，最浅处水深 1 330 m，山麓水深 4 500 m，高差 3 170 m，顶部平坦，边坡陡峭（图 2-56）。 This guyot has an irregular overlook plane shape. The northeastern part is higher in topography. The length is 57 km and the width is 39 km. The top platform depth is about 1 500 m. The minimum depth is 1 330 m. The piedmont depth is 4 500 m, which makes the total relief being 3 170 m. The top plane is flat while the slopes are steep (Fig.2–56).				
命名释义 Reason for Choice of Name	"大成"取自《诗经·小雅·南有嘉鱼之什》"允矣君子，展也大成"，用以赞颂周宣王勇武果敢有实力，必可成功。此海山命名为大成海山，赞扬中国大洋科考队员不畏困难，勇于探索的精神和气魄。 Dacheng comes from a poem named *Nanyoujiayuzhishi* in *Shijing · Xiaoya*. *Shijing* is a collection of ancient Chinese poems from 11th century B.C. to 6th century B.C. "He is a man of true grace; His hunt is a great success." This poem is written to eulogize the Lord Xuan for his valiancy, resolution and strength. This seamount is named Dacheng to praise the China ocean expedition team for their spirit to overcome difficulties and courage to explore the ocean.				

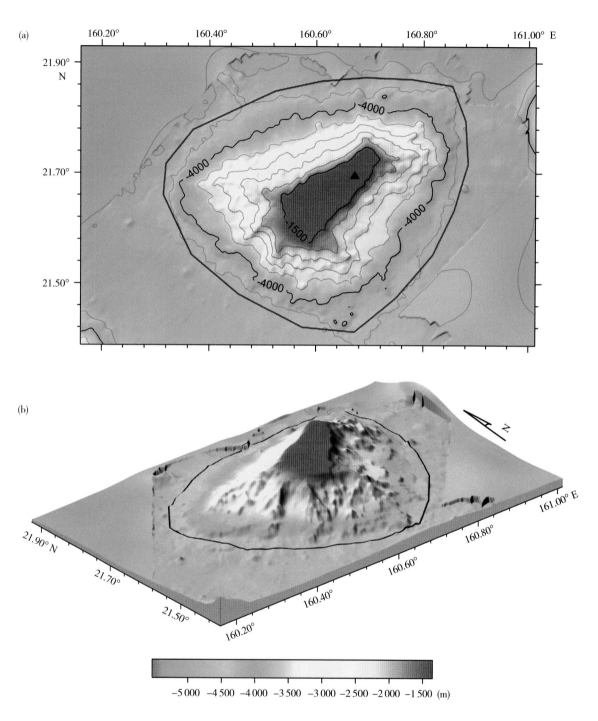

图 2-56　大成平顶海山

(a) 地形图（等深线间隔 500 m）；(b) 三维图

Fig.2-56　Dacheng Guyot

(a) Bathymetric map (the contour interval is 500 m); (b) 3-D topographic map

2.4.15 蝴蝶海山
Hudie Seamount

中文名称 Chinese Name	蝴蝶海山 Hudie Haishan	英文名称 English Name	Hudie Seamount	所在大洋 Ocean or Sea	西太平洋 West Pacific Ocean
发现情况 Discovery Facts	此海山于 2001 年 7 月由中国科考船"大洋一号"在执行 DY105-11 航次时调查发现。 This seamount was discovered by the Chinese R/V *Dayang Yihao* during the DY105-11 cruise in July, 2001.				
命名历史 Name History	由我国命名为蝴蝶海山,于 2016 年提交 SCUFN 审议通过。 Hudie Seamount was named by China and approved by SCUFN in 2016.				
特征点坐标 Coordinates	21°10.94′N,160°29.37′E			长(km)×宽(km) Length (km)× Width (km)	26 × 24
最大水深(m) Max Depth(m)	4 600	最小水深(m) Min Depth(m)	1 790	高差(m) Total Relief(m)	2 810
地形特征 Feature Description	此海山发育 4 条山脊,山顶呈尖峰状,长宽分别为 26 km 和 24 km。海山最浅处水深 1 790 m,山麓水深 4 600 m,高差 2 810 m,边坡陡峭(图 2–57)。 This seamount develops four ridges and the summit is spike shaped. The length is 26 km and the width is 24 km. The minimum depth of the seamount is 1 790 m while the piedmont depth is 4 600 m, which makes the total relief being 2 810 m. Its slopes are steep (Fig.2–57).				
命名释义 Reason for Choice of Name	该海山俯视平面形态似翩翩飞舞的"蝴蝶"。蝴蝶色彩鲜艳,形态各异,在自然界种类繁多,是植物授粉的重要媒介。 This seamount is named Hudie, which means butterflies in Chinese, because its shape looks like a flying butterfly. Butterfly is a kind of insect with colorful appearances, different shapes, a wide variety of natural species. It is an important medium of plant pollination.				

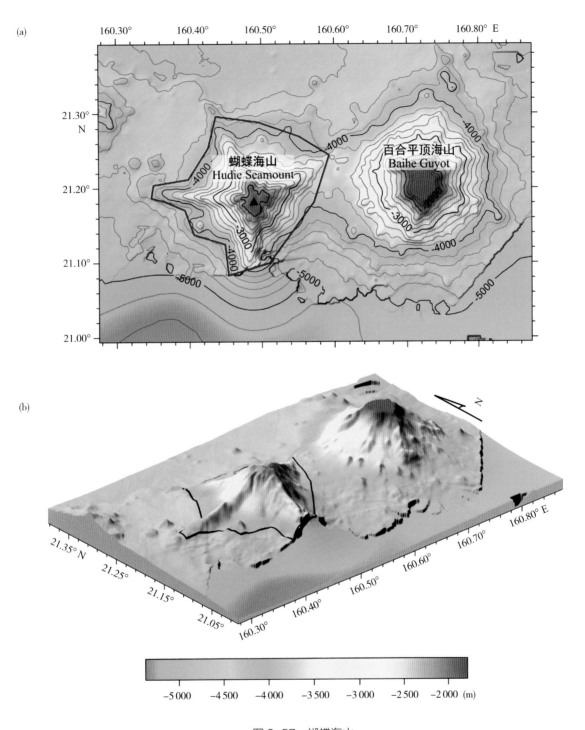

图 2-57　蝴蝶海山

(a) 地形图（等深线间隔 200 m）；(b) 三维图

Fig.2-57　Hudie Seamount

(a) Bathymetric map (the contour interval is 200 m); (b) 3-D topographic map

2.4.16　百合平顶海山
Baihe Guyot

中文名称 Chinese Name	百合平顶海山 Baihe Pingdinghaishan	英文名称 English Name	Baihe Guyot	所在大洋 Ocean or Sea	西太平洋 West Pacific Ocean
发现情况 Discovery Facts	此平顶海山于 2001 年 7 月由中国科考船"大洋一号"在执行 DY105-11 航次时调查发现。 This guyot was discovered by the Chinese R/V *Dayang Yihao* during the DY105-11 cruise in July, 2001.				
命名历史 Name History	由我国命名为百合平顶海山，于 2016 年提交 SCUFN 审议通过。 Baihe Guyot was named by China and approved by SCUFN in 2016.				
特征点坐标 Coordinates	21°12.15′N，160°43.41′E			长(km)×宽(km) Length (km)× Width (km)	31 × 26
最大水深（m） Max Depth（m）	4 600	最小水深（m） Min Depth（m）	1 525	高差（m） Total Relief（m）	3 075
地形特征 Feature Description	此平顶海山长宽分别为 31 km 和 26 km。山顶平台水深约 1 700 m，最浅处水深 1 525 m，山麓水深 4 600 m，高差 3 075 m，顶部平坦，边坡陡峭（图 2–58）。 The length and width of this guyot are 31 km and 26 km respectively. The top platform depth is about 1 700 m and the minimum depth is 1 525 m. The piedmont depth is 4 600 m, which makes the total relief being 3 075 m. The guyot has a flat top and steep slopes (Fig.2–58).				
命名释义 Reason for Choice of Name	该平顶海山发育 6 条山脊，平面俯视形态与百合发育 6 条花瓣类似，故以此命名。百合为中国原产多年生草本植物，大气含蓄，雅致美丽，在中华文化中常用于对美好婚姻生活的祝福。 This guyot develops six ridges and its overlook plane shape is similar to flower Lily, which bears six petals. Baihe is a Chinese word for Lily. Lily is a perennial herb of Chinese origin, elegant and beautiful, in the Chinese culture is commonly used to bless a good marriage life.				

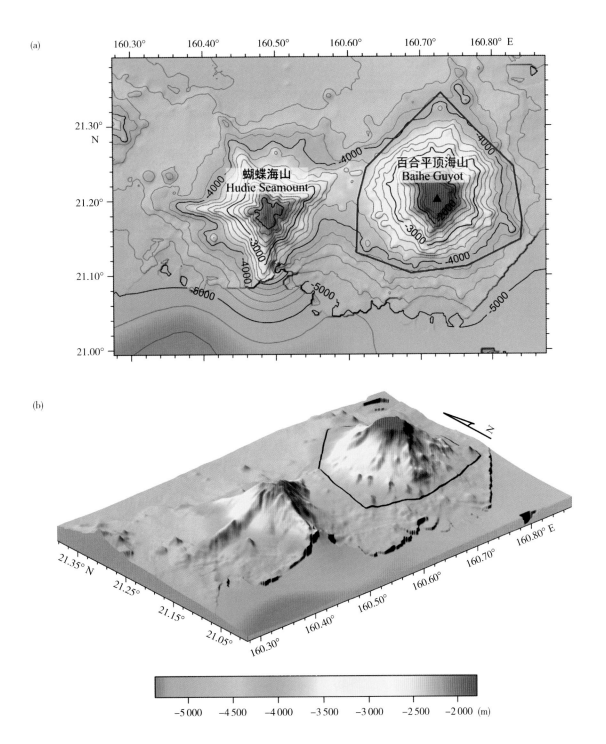

图 2-58　百合平顶海山

(a) 地形图（等深线间隔 200 m）；(b) 三维图

Fig.2-58　Baihe Guyot

(a) Bathymetric map (the contour interval is 200 m); (b) 3-D topographic map

2.4.17　牛郎平顶海山
Niulang Guyot

中文名称 Chinese Name	牛郎平顶海山 Niulang Pingdinghaishan	英文名称 English Name	Niulang Guyot	所在大洋 Ocean or Sea	西太平洋 West Pacific Ocean
发现情况 Discovery Facts	此平顶海山于2003年4月由中国科考船"大洋一号"在执行DY105-12/14航次时调查发现。 This guyot was discovered by the Chinese R/V *Dayang Yihao* during the DY105-12/14 cruise in April, 2003.				
命名历史 Name History	由我国命名为牛郎平顶海山，于2012年提交SCUFN审议通过。 Niulang Guyot was named by China and approved by SCUFN in 2012.				
特征点坐标 Coordinates	20°22.80′N，160°45.40′E 20°36.40′N，161°01.66′E 20°44.08′N，160°11.68′E		长(km) × 宽(km) Length (km) × Width (km)		135 × 85
最大水深（m） Max Depth（m）	4 700	最小水深（m） Min Depth（m）	1 600	高差（m） Total Relief（m）	3 100
地形特征 Feature Description	海山由3个小型海山组成，整体长宽分别为135 km和85 km。山顶平台水深约1 800 m，最浅处水深1 600 m，山麓水深4 700 m，高差3 100 m，顶部平坦，边坡陡峭（图2-59）。 This guyot consists of three small seamounts. The overall length and width are 135 km and 85 km respectively. The top platform depth is about 1 800 m. The minimum depth of this seamount is 1 600 m, and the piedmont depth is 4 700 m, which makes the total relief being 3 100 m. Its slopes are steep, while the top is flat (Fig.2-59).				
命名释义 Reason for Choice of Name	"牛郎"出自《诗经·小雅·大东》"跂彼织女，终日七襄。虽则七襄，不成报章。睆彼牵牛，不以服箱。"织女和牛郎的爱情故事在中国民间广泛流传，传说两人为银河所隔，唯每年的农历七月初七方能通过跨河的鹊桥得以相见。两个平顶海山距离较近，以鞍部相隔，寓意牛郎织女两地分隔，分别命名织女平顶海山和牛郎平顶海山。 "Niulang" comes from a poem named *Dadong* in *Shijing · Xiaoya*. *Shijing* is a collection of ancient Chinese poems from 11th century B.C. to 6th century B.C. "The Weaver Maiden Stars there stay and travel seven times a day. They travel seven times a day, but weaver's role they do not play. There also shine the Cowherd Stars, but never draw the travellers' cars." The poem is about a love story between a couple named Niulang and Zhinyu. In the story, Niulang is transformed into a star called "Niulang Star", while Zhinyu is transformed into a star called "Zhinyu Star", and they are separated by the Milky Way. Similarly, Zhinyu Guyot and Niulang Guyot are separated by a saddle valley.				

(a)

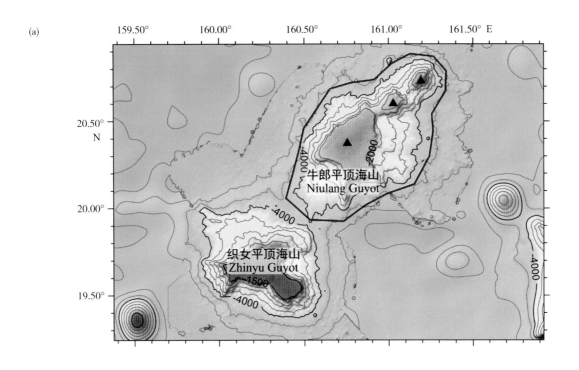

(b)

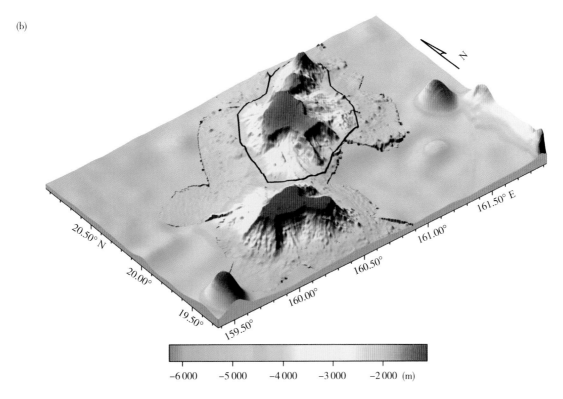

图 2-59　牛郎平顶海山

(a) 地形图（等深线间隔 500 m）；(b) 三维图

Fig.2-59　Niulang Guyot

(a) Bathymetric map (the contour interval is 500 m); (b) 3-D topographic map

2.4.18 织女平顶海山
Zhinyu Guyot

中文名称 Chinese Name	织女平顶海山 Zhinyu Pingdinghaishan	英文名称 English Name	Zhinyu Guyot	所在大洋 Ocean or Sea	西太平洋 West Pacific Ocean
发现情况 Discovery Facts	此平顶海山于2003年4月由中国科考船"大洋一号"在执行DY105-12/14航次时调查发现。 This guyot was discovered by the Chinese R/V *Dayang Yihao* during the DY105-12/14 cruise in April, 2003.				
命名历史 Name History	由我国命名为织女平顶海山，于2012年提交 SCUFN 审议通过。 Zhinyu Guyot was named by China and approved by SCUFN in 2012.				
特征点坐标 Coordinates	19°39.20′N，160°09.40′E			长(km)× 宽(km) Length (km) × Width (km)	90 × 75
最大水深（m） Max Depth（m）	4 700	最小水深（m） Min Depth（m）	1 200	高差（m） Total Relief（m）	3 500
地形特征 Feature Description	海山形状不规则，长宽分别为90 km和75 km。山顶平台水深约1 400 m，最浅处水深1 200 m，山麓水深4 700 m，高差3 500 m，顶部平坦，边坡陡峭（图2-60）。 This guyot has an irregular shape. The length and width are 90 km and 75 km respectively. The top platform depth is 1 400 m. The minimum depth of this seamount is 1 200 m, and the piedmont depth is 4 700 m, which makes the total relief being 3 500 m. Its slopes are steep, while the top is flat (Fig.2-60).				
命名释义 Reason for Choice of Name	"织女"出自《诗经·小雅·大东》"跂彼织女，终日七襄。虽则七襄，不成报章。睆彼牵牛，不以服箱。"织女和牛郎的爱情故事在中国民间广泛流传，传说两人为银河所隔，唯每年的农历七月初七方能通过跨河的鹊桥得以相见。两个平顶海山距离较近，以鞍部相隔，寓意牛郎织女两地分隔，分别命名织女平顶海山和牛郎平顶海山。 Zhinyu comes from a poem named *Dadong* in *Shijing · Xiaoya*. *Shijing* is a collection of ancient Chinese poems from 11th century B.C. to 6th century B.C. "The Weaver Maiden Stars there stay and travel seven times a day. They travel seven times a day, but weaver's role they do not play. There also shine the Cowherd Stars, but never draw the travellers' cars." The poem is about a love story between a couple named Niulang and Zhinyu. In the story, Niulang is transformed into a star called "Niulang Star", while Zhinyu is transformed into a star called "Zhinyu Star", and they are separated by the Milky Way. Similarly, Zhinyu Guyot and Niulang Guyot are separated by a saddle valley.				

(a)

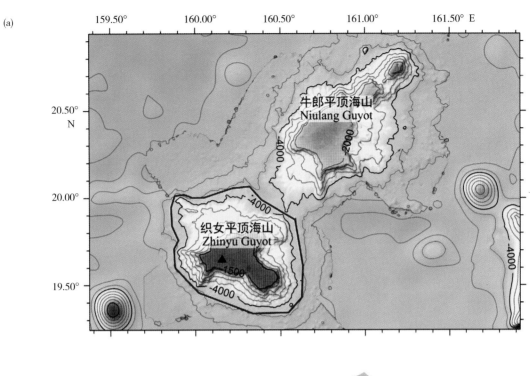

(b)

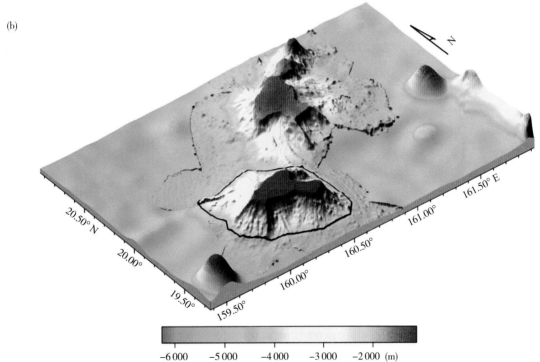

图 2-60　织女平顶海山

(a) 地形图（等深线间隔 500 m）；(b) 三维图

Fig.2-60　Zhinyu Guyot

(a) Bathymetric map (the contour interval is 500 m); (b) 3-D topographic map

2.4.19 湛露海山群
Zhanlu Seamounts

中文名称 Chinese Name	湛露海山群 Zhanlu Haishanqun	英文名称 English Name	Zhanlu Seamounts	所在大洋 Ocean or Sea	西太平洋 West Pacific Ocean
发现情况 Discovery Facts					

发现情况 Discovery Facts	此海山群于 2003 年 7—8 月由中国科考船 "海洋四号" 在执行 DY105-15 航次时调查发现。 The seamounts were discovered by the Chinese R/V *Haiyang Sihao* during the DY105-15 cruise from July to August, 2003.
命名历史 Name History	该海山群由 6 座海山组成，中部规模最大的平顶海山在 ACUF 地名辞典中称为 "McDonnell Guyot"，中文译名为 "麦克唐奈平顶海山"。该海山群在我国大洋航次调查报告中曾暂以代号命名。 The seamounts consist of 6 guyots or seamounts, among which the largest one is located in the middle of the guyots, which was named McDonnell Guyot in ACUF gazetteer. Its translation is Maiketangnai Pingdinghaishan in Chinese. A code was used to name Zhanlu Seamounts temporarily in Chinese cruise reports.
特征点坐标 Coordinates	20°01.92′N，161°41.99′E（显允海山 / Xianyun Seamount） 19°51.23′N，161°56.03′E（令德海山 / Lingde Seamount） 19°53.01′N，162°45.97′E（湛露平顶海山 / Zhanlu Guyot） 19°32.52′N，162°34.80′E（令仪海山 / Lingyi Seamount） 19°08.25′N，162°07.58′E（恺悌平顶海山 / Kaiti Guyot） 19°37.09′N，162°13.74′E（麦克唐奈平顶海山 / McDonnell Guyot）	长(km)×宽(km) Length (km)× Width (km)	110×105

最大水深（m） Max Depth（m）	5 460	最小水深（m） Min Depth（m）	1 127	高差（m） Total Relief（m）	4 333

地形特征 Feature Description	湛露海山群位于马尔库斯－威克海山区，由 6 座平顶海山 / 海山组成。海山群长和宽分别为 110 km 和 105 km，最高处位于中部的麦克唐奈平顶海山，水深为 1 127 m（图 2-61）。 Zhanlu seamounts are located in the Marcus-Wake Seamount Area, consisting of 6 guyots or seamounts. The length and the width of the seamounts are 110 km and 105 km respectively. Among the seamounts, the highest one is McDonnell Guyot, with the depth of 1 127 m (Fig.2-61).
命名释义 Reason for Choice of Name	"湛露" 出自《诗经·小雅·湛露》。此诗歌为周朝君王与贵族夜宴时所唱，表现君臣和谐共欢的场景。此海山群包含 6 个海山，除麦克唐奈海山外，显允、令德、湛露、令仪、恺悌均取词于《诗经·小雅·湛露》，以群组化方法分别命名。 "Zhanlu" comes from a poem named *Zhanlu* in *Shijing · Xiaoya*. *Shijing* is a collection of ancient Chinese poems from 11th century B.C. to 6th century B.C. This poem is the song when the emperor and nobles' banquet showing the happiness of them. To name these seamounts using group method, all names are selected from the same poem except the McDonnell Guyot.

(a)

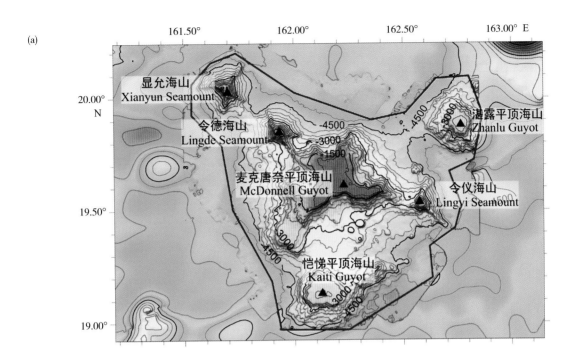

(b)

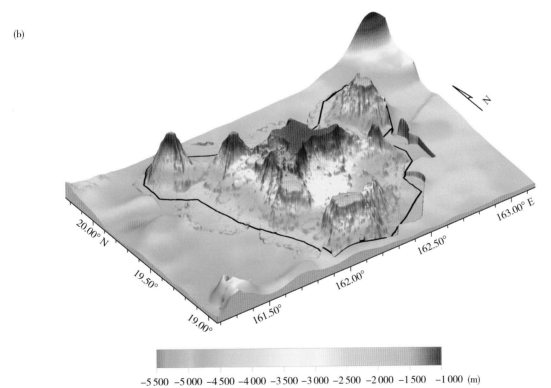

−5 500　−5 000　−4 500　−4 000　−3 500　−3 000　−2 500　−2 000　−1 500　−1 000　(m)

图 2-61　湛露海山群

(a) 地形图（等深线间隔 300 m）；(b) 三维图

Fig.2-61　Zhanlu Seamounts

(a) Bathymetric map (the contour interval is 300 m); (b) 3-D topographic map

2.4.20 显允海山
Xianyun Seamount

中文名称 Chinese Name	显允海山 Xianyun Haishan	英文名称 English Name	Xianyun Seamount	所在大洋 Ocean or Sea	西太平洋 West Pacific Ocean
发现情况 Discovery Facts	此海山于 2003 年 7—8 月由中国科考船"海洋四号"在执行 DY105-15 航次时调查发现。 This seamount was discovered by the Chinese R/V *Haiyang Sihao* during the DY105-15 cruise from July to August, 2003.				
命名历史 Name History	此海山为湛露海山群的一部分，湛露海山群在我国大洋航次调查报告中曾暂以代号命名。 This seamount is a part of Zhanlu Seamounts. A code was used to name Zhanlu Seamounts temporarily in Chinese cruise reports.				
特征点坐标 Coordinates	20°01.92′N，161°41.99′E			长(km)×宽(km) Length (km) × Width (km)	35×35
最大水深（m） Max Depth（m）	5 130	最小水深（m） Min Depth（m）	1 460	高差（m） Total Relief（m）	3 670
地形特征 Feature Description	显允海山位于湛露海山群西北部，海山平面形态近圆形，基座直径 35 km，峰顶水深 1 460 m，山麓水深 5 130 m，高差 3 670 m，山坡陡峭。此海山与令德海山之间以鞍部分隔，鞍部水深 4 050 m（图 2-62）。 Xianyun Seamount is located in the northwest of Zhanlu Seamounts. It has a nearly round overlook plane shape and the base diameter is 35 km. The depth of the summit is 1 460 m, while the piedmont depth is 5 130 m, which makes the total relief being 3 670 m. The slopes of the seamount are steep. This seamount is separated from Lingde Seamount by a saddle where the depth is 4 050 m (Fig.2-62).				
命名释义 Reason for Choice of Name	"显允"出自《诗经·小雅·湛露》"显允君子，莫不令德"，意为"坦荡诚信的君子，无不具有美善的情操"，显允意指坦荡诚信。该篇描述了贵族们在举行宴会，尽情饮乐，互相赞美的情景。 "Xianyun" comes from a poem named *Zhanlu* in *Shijing · Xiaoya*. *Shijing* is a collection of ancient Chinese poems from 11th century B.C. to 6th century B.C. "You are my distinguished guests; Your virtue I can verify." "Xianyun" means the candid and ingenuous qualities. The article describes the nobility in a party, enjoying the music and praising each other.				

(a)

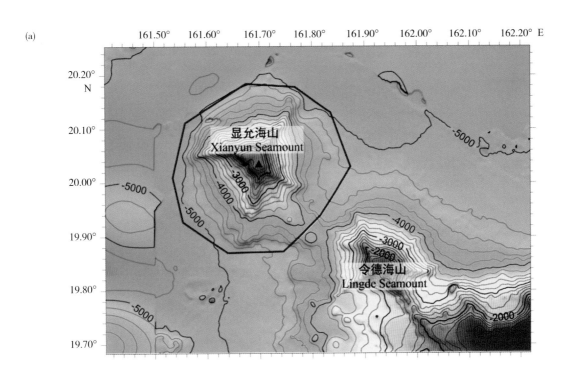

(b)

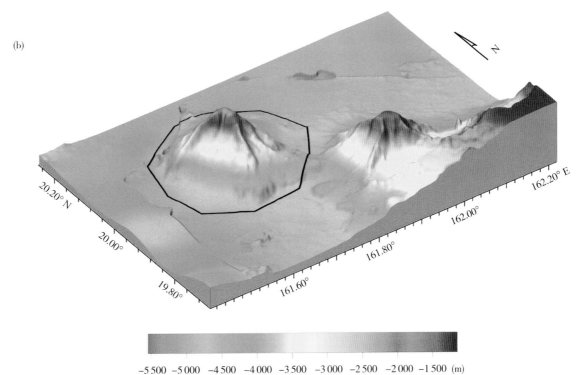

图 2-62　显允海山

(a) 地形图（等深线间隔 200 m）；(b) 三维图

Fig.2-62　Xianyun Seamount

(a) Bathymetric map (the contour interval is 200 m); (b) 3-D topographic map

2.4.21　令德海山
Lingde Seamount

中文名称 Chinese Name	令德海山 Lingde Haishan	英文名称 English Name	Lingde Seamount	所在大洋 Ocean or Sea	西太平洋 West Pacific Ocean
发现情况 Discovery Facts	此海山于2003年7—8月由中国科考船"海洋四号"在执行DY105-15航次时调查发现。 This seamount was discovered by the Chinese R/V *Haiyang Sihao* during the DY105-15 cruise from July to August, 2003.				
命名历史 Name History	此海山为湛露海山群的一部分，湛露海山群在我国大洋航次调查报告中曾暂以代号命名。 This seamount is a part of Zhanlu Seamounts. A code was used to name Zhanlu Seamounts temporarily in Chinese cruise reports.				
特征点坐标 Coordinates	19°51.23′N，161°56.03′E			长(km)×宽(km) Length (km)× Width (km)	30×30
最大水深（m） Max Depth（m）	4 880	最小水深（m） Min Depth（m）	1 286	高差（m） Total Relief（m）	3 594
地形特征 Feature Description	令德海山位于湛露海山群中部，显允海山和麦克唐奈平顶海山之间。海山峰顶水深1 286 m，山麓水深4 880 m，高差3 594 m（图2–63）。 Lingde Seamount is located in the middle of Zhanlu Seamounts, between Xianyun Seamount and McDonnell Guyot. The depth of the seamount summit is 1 286 m, while the piedmont depth is 4 880 m, which makes the total relief being 3 594 m (Fig.2–63).				
命名释义 Reason for Choice of Name	"令德"出自《诗经·小雅·湛露》"显允君子，莫不令德"，意为"坦荡诚信的君子，无不具有美善的情操"，令德指美善的道德情操。该篇描述了贵族们在举行宴会，尽情饮乐，互相赞美的情景。 "Lingde" comes from a poem named *Zhanlu* in *Shijing · Xiaoya*. *Shijing* is a collection of ancient Chinese poems from 11th century B.C. to 6th century B.C. "You are my distinguished guests; Your virtue I can verify." This poem means that a candid and ingenuous gentleman has good characters. "Lingde" means good character. The poem describes the nobility in a party, enjoying the music and praising each other.				

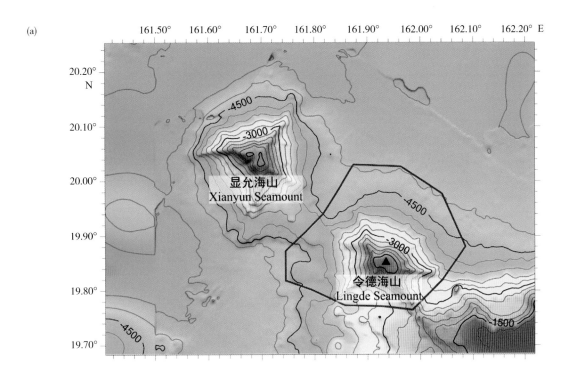

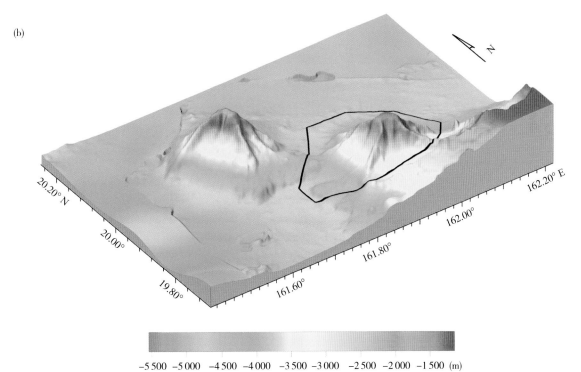

图 2-63　令德海山

(a) 地形图（等深线间隔 300 m）；(b) 三维图

Fig.2-63　Lingde Seamount

(a) Bathymetric map (the contour interval is 300 m); (b) 3-D topographic map

2.4.22 麦克唐奈平顶海山
McDonnell Guyot

中文名称 Chinese Name	麦克唐奈平顶海山 Maiketangnai Pingdinghaishan	英文名称 English Name	McDonnell Guyot	所在大洋 Ocean or Sea	西太平洋 West Pacific Ocean
发现情况 Discovery Facts	此平顶海山于 2003 年 7—8 月由中国科考船"海洋四号"在执行 DY105-15 航次时调查发现。 This guyot was discovered by the Chinese R/V *Haiyang Sihao* during the DY105-15 cruise from July to August, 2003.				
命名历史 Name History	该平顶海山在 ACUF 地名辞典中称为 McDonnell Guyot，中文译名为麦克唐奈平顶海山。此平顶海山为湛露海山群中规模最大的海山，湛露海山群在我国大洋航次调查报告中曾暂以代号命名。 This guyot is named McDonnell Guyot in ACUF gazetteer. Its translation is Maiketangnai Pingdinghaishan in Chinese. This guyot is the largest seamount among Zhanlu Seamounts. A code was used to name Zhanlu Seamounts temporarily in Chinese cruise reports.				
特征点坐标 Coordinates	19°37.09′N，162°13.74′E			长(km)×宽(km) Length (km)× Width (km)	80×70
最大水深（m） Max Depth（m）	4 880	最小水深（m） Min Depth（m）	1 127	高差（m） Total Relief（m）	3 753
地形特征 Feature Description	麦克唐奈平顶海山位于湛露海山群中部，由地形平缓的山顶平台和陡峭的山坡组成。海山俯视平面形态呈不规则状，长宽分别为 80 km 和 70 km，山顶平台水深约 1 400 m，最浅处 1 127 m，山麓水深 4 880 m，高差 3 753 m。此海山与令德海山、令仪海山和恺悌平顶海山之间以鞍部分隔（图 2-64）。 McDonnell Guyot is located in the middle of Zhanlu Seamounts. It consists of a flat top platform and steep slopes. The guyot has an irregular overlook plane shape. The length is 80 km and the width is 70 km. The top platform depth is about 1 400 m, and the minimum depth is 1 127 m, while the piedmont depth is 4 880 m, which makes the total relief being 3 753 m. The guyot is separated from Lingde Seamount, Lingyi Seamount and Kaiti Guyot by saddles (Fig.2-64).				
命名释义 Reason for Choice of Name	以海军科考船"John McDonnell"号的名字命名。 This guyot is named after *John McDonnell*, a naval research vessel.				

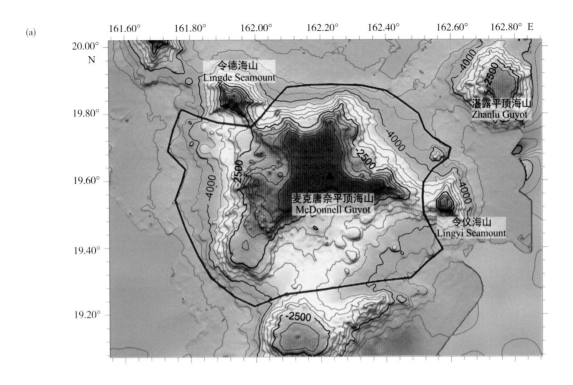

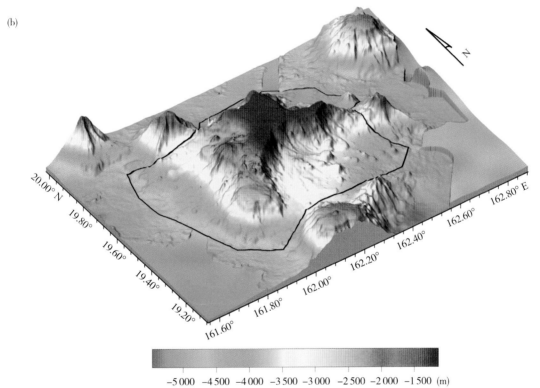

图 2-64　麦克唐奈平顶海山

(a) 地形图（等深线间隔 300 m）；(b) 三维图

Fig.2-64　McDonnell Guyot

(a) Bathymetric map (the contour interval is 300 m); (b) 3-D topographic map

2.4.23 湛露平顶海山
Zhanlu Guyot

中文名称 Chinese Name	湛露平顶海山 Zhanlu Pingdinghaishan	英文名称 English Name	Zhanlu Guyot	所在大洋 Ocean or Sea	西太平洋 West Pacific Ocean
发现情况 Discovery Facts	此平顶海山于 2003 年 7—8 月由中国科考船"海洋四号"在执行 DY105-15 航次时调查发现。 This guyot was discovered by the Chinese R/V *Haiyang Sihao* during the DY105-15 cruise from July to August, 2003.				
命名历史 Name History	此平顶海山为湛露海山群的一部分，湛露海山群在我国大洋航次调查报告中曾暂以代号命名。 This guyot is a part of Zhanlu Seamounts. A code was used to name Zhanlu Seamounts temporarily in Chinese cruise reports.				
特征点坐标 Coordinates	19°53.01′N，162°45.97′E			长(km) × 宽(km) Length (km) × Width (km)	30 × 30
最大水深（m） Max Depth（m）	5 450	最小水深（m） Min Depth（m）	2 005	高差（m） Total Relief（m）	3 445
地形特征 Feature Description	湛露平顶海山位于湛露海山群东部，由地形平缓的山顶平台和陡峭的山坡组成，俯视平面形态近圆形，基座直径 30 km，山顶平台规模较小，水深约 2 200 m，最浅处 2 005 m，山麓水深 5 450 m，高差 3 445 m，海山山坡陡峭（图 2–65）。 Zhanlu Guyot is located in the east of Zhanlu Seamounts, consisting of a flat top platform and steep slopes. It has nearly a round overlook plane shape with a base diameter of 30 km. The scale of platform is small. The top platform depth is about 2 200 m, with the minimum depth of 2 005 m, while the piedmont depth is 5 450 m, which makes the total relief being 3 435 m (Fig.2–65).				
命名释义 Reason for Choice of Name	"湛露"出自《诗经·小雅·湛露》"湛湛露斯，在彼丰草"，意为浓浓的露水沾满在茂盛的青草上，湛露为浓重的露水。该篇描述了贵族们在举行宴会，尽情饮乐，互相赞美的情景。 "Zhanlu" comes from a poem named *Zhanlu* in *Shijing · Xiaoya*. *Shijing* is a collection of ancient Chinese poems from 11th century B.C. to 6th century B.C. "Heavy, heavy the dews lie; No grasses in the field are dry." This poem means that the heavy dew is lying on the lush grass. "Zhanlu" means the heavy dew. The poem describes the nobility in a party, enjoying the music and praise each other.				

(a)

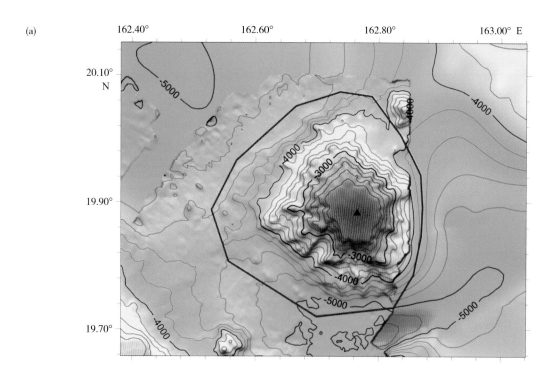

(b)

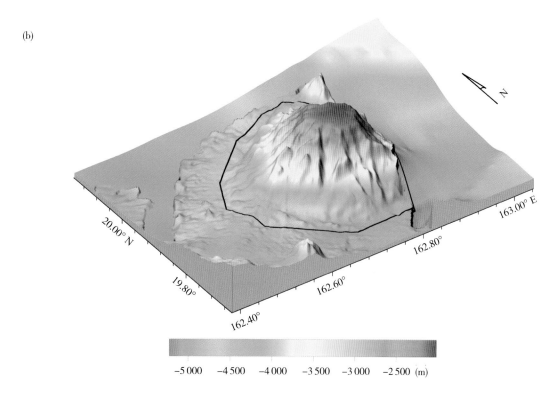

图 2-65　湛露平顶海山

(a) 地形图（等深线间隔 200 m）；(b) 三维图

Fig.2-65　Zhanlu Guyot

(a) Bathymetric map (the contour interval is 200 m); (b) 3-D topographic map

2.4.24　令仪海山
Lingyi Seamount

中文名称 Chinese Name	令仪海山 Lingyi Haishan		英文名称 English Name	Lingyi Seamount	所在大洋 Ocean or Sea	西太平洋 West Pacific Ocean
发现情况 Discovery Facts	此海山于 2003 年 7—8 月由中国科考船 "海洋四号" 在执行 DY105-15 航次时调查发现。 This seamount was discovered by the Chinese R/V *Haiyang Sihao* during the DY105-15 cruise from July to August, 2003.					
命名历史 Name History	此海山为湛露海山群的一部分，湛露海山群在我国大洋航次调查报告中曾暂以代号命名。 This seamount is a part of Zhanlu Seamounts. A code was used to name Zhanlu Seamounts temporarily in Chinese cruise reports.					
特征点坐标 Coordinates	19°32.52′N，162°34.80′E				长(km)×宽(km) Length (km)× Width (km)	25×20
最大水深（m） Max Depth（m）	5 000	最小水深（m） Min Depth（m）	1 682		高差（m） Total Relief（m）	3 318
地形特征 Feature Description	令仪海山位于湛露海山群东部，海山近圆锥形，基座直径 20～25 km，峰顶水深 1 682 m，山麓水深 5 000 m，高差 3 318 m。海山山坡陡峭，与麦克唐奈平顶海山之间以鞍部分隔，鞍部水深 2 900 m（图 2–66）。 Lingyi Seamount is located in the east of Zhanlu Seamounts. Its shape is nearly a cone with the base diameter of 20–25 km. The depth of the summit is 1 682 m, while the piedmont depth is 5 000 m, which makes the total relief being 3 318 m. The slopes of the seamount are steep. This seamount is separated from McDonnell Guyot by a saddle where the depth is 2 900 m (Fig.2–66).					
命名释义 Reason for Choice of Name	"令仪" 出自《诗经·小雅·湛露》"岂弟君子，莫不令仪"，意为 "和悦平易的君子，看上去风度翩翩"，令仪指人的仪态风度翩翩。该篇描述了贵族们在举行宴会，尽情饮乐，互相赞美的情景。 "Lingyi" comes from a poem named *Zhanlu* in *Shijing · Xiaoya*. *Shijing* is a collection of ancient Chinese poems from 11th century B.C. to 6th century B.C. "You are my gentle guests; You have good manners in the spree." This poem means that an agreeable and pleasant gentleman is always chivalrous. "Lingyi" means to be chivalrous. The poem describes the nobility in a party, enjoying the music and praise each other.					

(a)

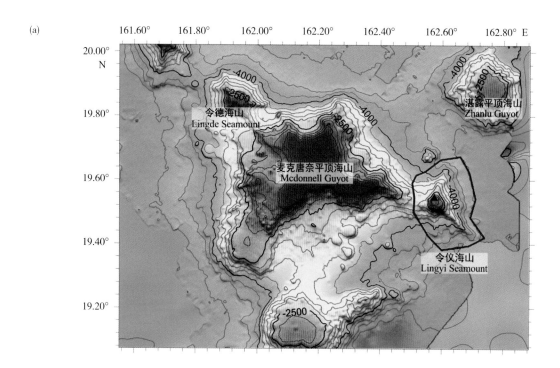

(b)

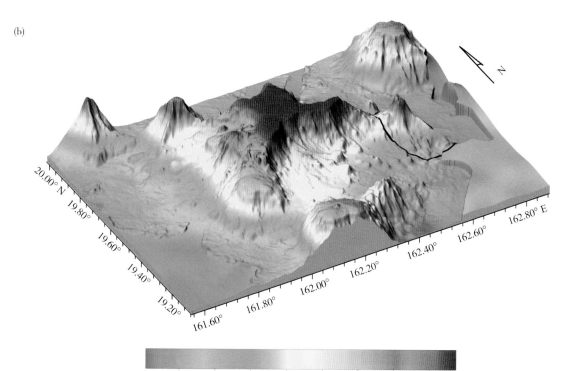

图 2-66　令仪海山

(a) 地形图（等深线间隔 300 m）；(b) 三维图

Fig.2-66　Lingyi Seamount

(a) Bathymetric map (the contour interval is 300 m); (b) 3-D topographic map

2.4.25 恺悌平顶海山
Kaiti Guyot

中文名称 Chinese Name	恺悌平顶海山 Kaiti Pingdinghaishan	英文名称 English Name	Kaiti Guyot	所在大洋 Ocean or Sea	西太平洋 West Pacific Ocean
发现情况 Discovery Facts					
命名历史 Name History					
特征点坐标 Coordinates	19°08.25′N，162°07.58′E			长(km)×宽(km) Length (km) × Width (km)	75×35
最大水深（m） Max Depth（m）	5 460	最小水深（m） Min Depth（m）	1 957	高差（m） Total Relief（m）	3 503

发现情况 / Discovery Facts:

此平顶海山于 2003 年 7—8 月由中国科考船"海洋四号"在执行 DY105-15 航次时调查发现。

This Guyot was discovered by the Chinese R/V *Haiyang Sihao* during the DY105-15 cruise from July to August, 2003.

命名历史 / Name History:

此平顶海山为湛露海山群的一部分，湛露海山群在我国大洋航次调查报告中曾暂以代号命名。

This guyot is a part of Zhanlu Seamounts. A code was used to name Zhanlu Seamounts temporarily in Chinese cruise reports.

地形特征 / Feature Description:

此平顶海山位于湛露海山群南部，由地形平缓的山顶平台和陡峭的山坡组成。平顶海山俯视平面形态狭长，SW—NE 走向，长宽分别为 75 km 和 35 km。山顶平台水深约 2 200 m，最浅处 1 957 m，山麓水深 5 460 m，高差 3 503 m。平顶海山与麦克唐奈平顶海山之间以鞍部分隔，鞍部水深 3 375 m（图 2–67）。

Kaiti Guyot is located in the south of Zhanlu Seamounts, consisting of a flat top platform and steep slopes. It has a an elongated overlook plane shape, running NE to SW. The length is 75 km and the width is 35 km. The top platform depth is about 2 200 m, with the minimum depth of 1 957 m, while the piedmont depth is 5 460 m, which makes the total relief being 3 503 m. This guyot is separated from McDonnell Guyot by a saddle where the depth is 3 375 m (Fig.2–67).

命名释义 / Reason for Choice of Name:

"恺悌"出自《诗经·小雅·湛露》"岂弟君子，莫不令仪"，意为"和悦平易的君子，看上去风度翩翩"。该篇描述了贵族们在举行宴会，尽情饮乐，互相赞美的情景。古文中"岂弟"与"恺悌"同义，指和乐平易的态度。为了辨识方便，此处采用"恺悌"一词。

"Kaiti" comes from a poem named *Zhanlu* in *Shijing · Xiaoya*. *Shijing* is a collection of ancient Chinese poems from 11th century B.C. to 6th century B.C. "You are my gentle guests; You have good manners in the spree." This poem means that an agreeable and pleasant gentleman is always chivalrous. "Kaiti" means agreeable and pleasant. The poem describes the nobility in a party, enjoying the music and praise each other.

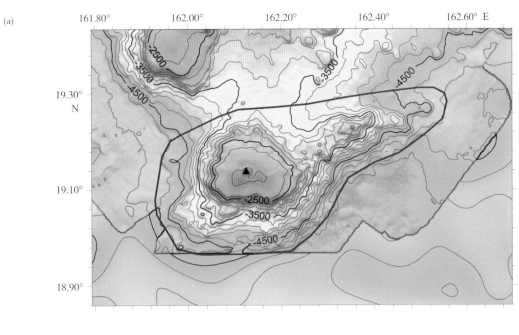

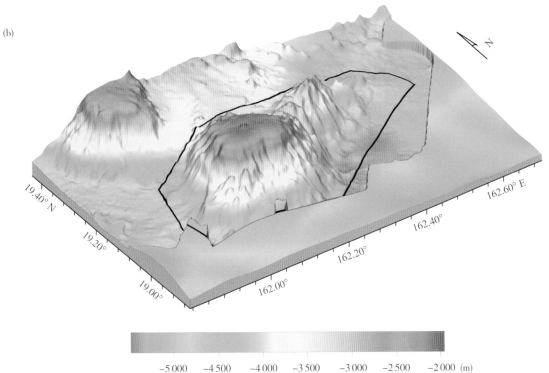

图 2-67 恺悌平顶海山

(a) 地形图（等深线间隔 200 m）；(b) 三维图

Fig.2-67 Kaiti Guyot

(a) Bathymetric map (the contour interval is 200 m); (b) 3-D topographic map

2.4.26 凤鸣海山
Fengming Seamount

中文名称 Chinese Name	凤鸣海山 Fengming Haishan	英文名称 English Name	Fengming Seamount	所在大洋 Ocean or Sea	西太平洋 West Pacific Ocean
发现情况 Discovery Facts	此海山于 2015 年由中国科考船"竺可桢号"在执行测量任务时调查发现。 This seamount was discovered by the Chinese R/V *Chukochenhao* during the survey in 2015.				
命名历史 Name History	由我国命名为凤鸣海山，于 2018 年提交 SCUFN 审议通过。 This feature was named Fengming Seamount by China and the name was approved by SCUFN in 2018.				
特征点坐标 Coordinates	22°05.5′N，160°15.6′E		长(km)×宽(km) Length (km)× Width (km)		31×28
最大水深（m） Max Depth（m）	5 460	最小水深（m） Min Depth（m）	1 561	高差（m） Total Relief（m）	3 899
地形特征 Feature Description	海山俯视形态近圆形，立体锥状，近等轴延伸，发育 7 条山脊，长宽分别为 31 km 和 28 km。最浅处水深 1 561 m，山麓水深 5 460 m，高差 3 899 m，顶端尖突，边坡陡峭（图 2–68）。 Fengming Seamount has a near round overlook plane shape and looks like a solid cone. It has a nearly equiaxial extension and develops 7 ridges. The length is 31 km and the width is 28 km. The minimum water depth is 1561 m. The piedmont depth is 5460 m, which makes the total relief being 3899 m. It has a spiky top and steep slopes (Fig.2–68).				
命名释义 Reason for Choice of Name	此海山形状似展翅而鸣的凤凰，故名。 This seamount is named after Fengming because it looks like a phoenix which spreads its wings and singing.				

(a)

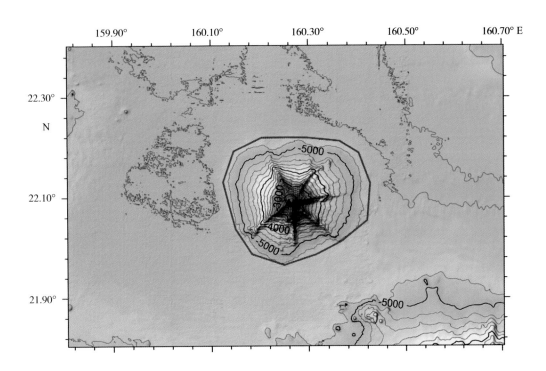

(b)

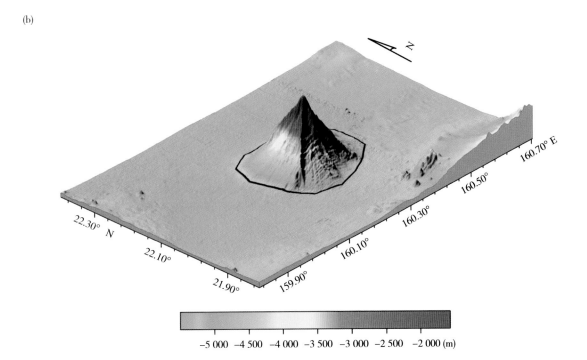

图 2-68　凤鸣海山

(a) 地形图（等深线间隔 200 m）；(b) 三维图

Fig.2-68　Fengming Seamount

(a) Bathymetric map (the contour interval is 200 m); (b) 3-D topographic map

第5节　马绍尔海山区地理实体

　　马绍尔海山区位于麦哲伦海山区以东、马尔库斯－威克海山区以南，中太平洋海盆西侧，由数十个海山构成。海山区平面形状不规则，大致上呈NW—SE向延伸，长度超过2 000 km，宽度达1 000 km。此海山区内的地理实体以大型平顶海山为主，平顶海山由地形平坦的山顶平台和陡峭的山坡构成，平台水深1 600～2 300 m，山麓水深5 200～6 000 m。

　　马绍尔海山区所在位置是一条规模宏大的北西向断裂带，表明岩浆沿断裂带喷发或溢出形成众多的海山。区内两个或多个海山相对集中构成海山群，推测是同一区域内发生多期火山活动所致。海山的平顶地貌因其曾出露海平面遭受风化剥蚀。此区域内部分海山出露海面，构成环形珊瑚礁，即为马绍尔群岛。

　　在此海域我国翻译地名1个，为纳济莫夫平顶海山群（图2-69）。

Section 5　Undersea Features in the Marshall Seamount Area

The Marshall Seamount Area is located in the east to the Magellan Seamount Area, south to the Marcus-Wake Seamount Area, west to the Central Pacific Basin, consisting of dozens of seamounts. This seamount area has an irregular plane shape, generally running NW to SE with the length over 2 000 km and width over 1 000 km. The undersea features in this area are mainly large guyots as well, with flat top platforms and steep slopes. The water depth of the platform is 1 600–2 300 m, while the piedmont depth is 5 200–6 000 m.

The location of the Marshall Seamount Area is a large-scale NW trend fracture zone, which indicates that numerous seamounts were created by magma eruption or overflow along the fracture zone. Two or more relatively gathering seamounts constituted seamount groups in this area. It is presumed that was caused by several volcanic activities in the same area. The geomorphology of the tops is flat because they had reached out of the sea surface and suffered from weather denudation. Some seamounts in this area are still out of the sea surface, constituting atolls, namely Marshall Islands.

In total, one undersea feature has been translated by China in the region of the Marshall Seamount Area, that is, Nazimov Guyots (Fig. 2–69).

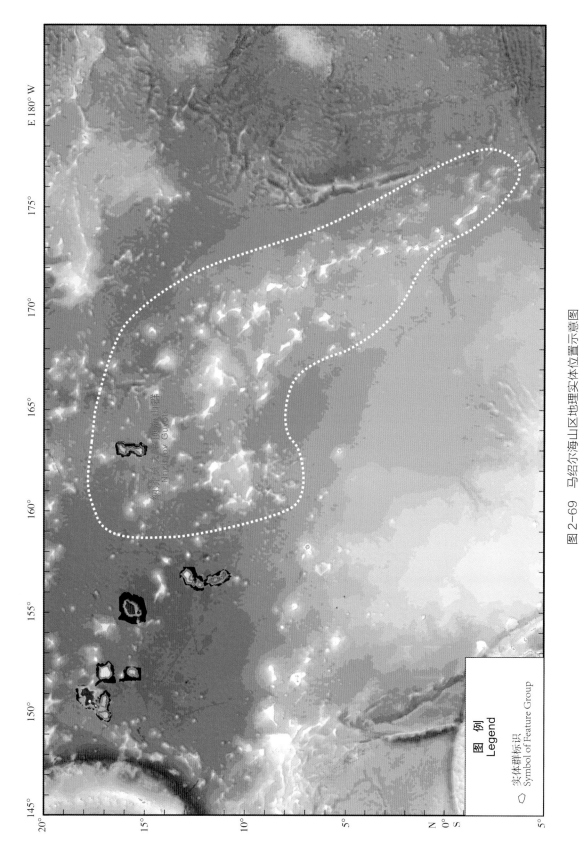

图 2-69　马绍尔海山区地理实体位置示意图

Fig.2-69　Locations of the undersea features in the Marshall Seamount Area

2.5.1 纳济莫夫平顶海山群
Nazimov Guyots

中文名称 Chinese Name	纳济莫夫平顶海山群 Najimofu Pingdinghaishanqun	英文名称 English Name	Nazimov Guyots	所在大洋 Ocean or Sea	西太平洋 West Pacific Ocean
发现情况 Discovery Facts	此平顶海山群于 1999 年 6—8 月由中国科考船"大洋一号"在执行 DY95-10 航次时调查发现。 The guyots were discovered by the Chinese R/V *Dayang Yihao* during the DY95-10 cruise from June to August, 1999.				
命名历史 Name History	该平顶海山群在 GEBCO 地名辞典中称为 Nazimov Guyots，中文译名为纳济莫夫平顶海山群，2006 年由俄罗斯提名。该平顶海山群在我国大洋航次调查报告中曾暂以代号命名。 The guyots are named Nazimov Guyots in GEBCO gazetteer. Its translation is Najimofu Pingdinghaishanqun in Chinese. It was proposed by Russia in 2006. A code was used to name the guyots temporarily in Chinese cruise reports.				
特征点坐标 Coordinates	15°09.98′N，162°52.61′E（南 /South） 16°06.04′N，162°59.88′E（北 /North）			长 (km) × 宽 (km) Length (km) × Width (km)	160 × 40
最大水深（m） Max Depth（m）	5 080	最小水深（m） Min Depth（m）	1 292	高差（m） Total Relief（m）	3 788
地形特征 Feature Description	此平顶海山群由南北两座平顶海山组成，两海山之间以狭长的山脊相连，呈 SN 走向，长宽分别 160 km 和 40 km。山顶平台水深约 1 500 m，南部海山最浅处 1 292 m，山麓水深 5 080 m，高差为 3 788 m（图 2-70）。 The guyots consist of two guyots, located in the north and south respectively, which are connected by a narrow ridge running N to S. The length and width of the guyots are 160 km and 40 km respectively. The top platform depth is about 1 500 m. The depth of shallowest place in the southern guyot is 1 292 m, while the piedmont depth is 5 080 m, which makes the total relief being 3 788 m (Fig.2-70).				
命名释义 Reason for Choice of Name	本平顶海山群以俄罗斯海军上将纳济莫夫（1829—1902）命名，他是"Nadezhd"号、"Pallada"号和"Cesarevich"号等船上的科学家和指挥官。他重新测量了马绍尔群岛的位置。 The guyots are named after Admiral P. N. Nazimov (1829–1902), a Russian researcher and commanding officer on the vessels *Nadezhd*, *Pallada* and *Cesarevich*. He remeasured the locations of islands in the Marshall Islands .				

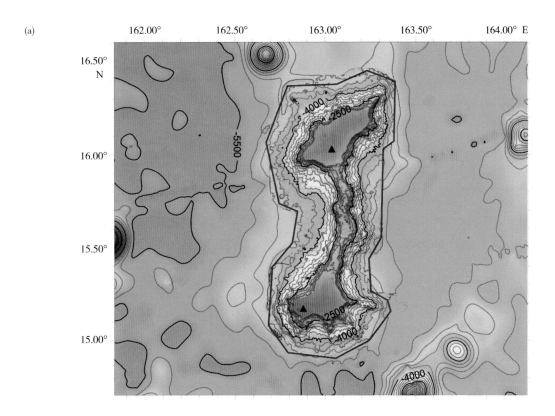

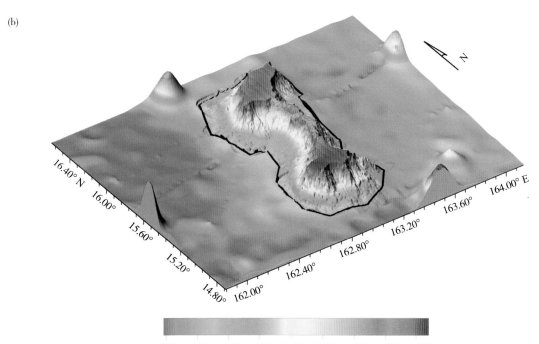

图 2-70 纳济莫夫平顶海山群

(a) 地形图（等深线间隔 300 m）；(b) 三维图

Fig.2-70 Nazimov Guyots

(a) Bathymetric map (the contour interval is 300 m); (b) 3-D topographic map

第 6 节 中太平洋海山区地理实体

中太平洋海山区群山呈簇形排列，近东西向展布于夏威夷群岛与马绍尔群岛之间，东西向延伸近 2 500 km。西北临马尔库斯－威克海山区，东南接莱恩海岭，南北分置中太平洋海盆和西北太平洋海盆。海山区内海山最浅水深约 800 ～ 1 000 m，相邻深海盆地水深近6 000 m。

海山区发育大量火山岛链、平顶海山与海底高原，海山主要形成于中白垩世和晚白垩至早第三纪，区域大洋基底在 1.3 亿年之前即已形成。海山形成与热点作用的板内火山作用关系密切，热点源主要位于赤道以南现今法属玻利尼西亚群岛区的超级地幔羽。海山形成后伴随太平洋板块运移至现今位置，不少海山曾出露水面，在沉降过程中形成巨厚的碳酸盐沉积 (Winterer and Metzler, 1984; Flood, 1999)。

在此海域我国命名地理实体 26 个，翻译地名 5 个。其中海山群 5 个，包括鹤鸣平顶海山群、鸿雁平顶海山群、斯干平顶海山群、白驹平顶海山群和无羊海山群；平顶海山 22 个，包括紫檀平顶海山、九皋平顶海山、乐园平顶海山、刘徽平顶海山、法显平顶海山、贾耽平顶海山、义净平顶海山、窦叔蒙平顶海山、如竹平顶海山、如松平顶海山、如翼平顶海山、如翚平顶海山、潜鱼平顶海山、白驹平顶海山、维鱼平顶海山、簑笠平顶海山、牧来平顶海山、年丰平顶海山、路易斯·阿加西斯平顶海山、亚历山大·阿加西斯平顶海山、艾利森平顶海山和希格平顶海山；海山 3 个，为麀胘海山、犉羊海山和尤因海山；海脊 1 个，为阿池海脊（图 2-71）。

Section 6　Undersea Features in the Central Pacific Seamount Area

Seamounts in the Central Pacific Seamount Area are arranged in clustered shape, distributing between the Hawaiian Islands and the Marshall Islands nearly EW trending with the extension of about 2 500 km. The Marcus-Wake Seamount Area is near the northwest of it, with the Line Islands Chain being at the southeast of it, the Central Pacific Basin being at the south of it and the North-West Pacific Basin being at the north of it. The shallowest water depth of seamounts in this area is about 800–1 000 m, while the water depth of adjacent basins is about 6 000 m.

The geomorphology and topography of this seamount area are characterized by a large number of volcanic island chains, guyots and plateaus. The seamounts were mainly formed in the Middle Cretaceous and Late Cretaceous to Early Tertiary, and the regional oceanic basement had already been formed 130 million years ago. The formation of regional seamounts is in a close relationship with inner plate volcanism caused by hotspot effect. The sources of hotspots are mainly located in the super mantle plume in the region of present French Polynesian Islands south to the equator. After formation, the seamounts migrated to current position with the Pacific Plate. Many seamounts once had reached out of the sea surface and formed very thick carbonate deposits during the process of settlement (Winterer and Metzler et al., 1984; Flood, 1999).

In total, 26 undersea features have been named and 5 others have been translated by China in the Central Pacific Seamount Area. Among them, there are 5 seamounts, including Heming Guyots, Hongyan Guyots, Sigan Guyots, Baiju Guyots and Wuyang Seamounts; 22 guyots, including Zitan Guyot, Jiugao Guyot, Leyuan Guyot, Liuhui Guyot, Faxian Guyot, Jiadan Guyot, Yijing Guyot, Doushumeng Guyot, Ruzhu Guyot, Rusong Guyot, Ruyi Guyot, Ruhui Guyot, Qianyu Guyot, Baiju Guyot, Weiyu Guyot, Suoli Guyot, Mulai Guyot, Nianfeng Guyot, Louis Agassiz Guyot, Alexander Agassiz Guyot, Allison Guyot and HIG Guyot; 3 seamounts, including Huigong Seamount, Chunyang Seamount and Ewing Seamount; 1 ridge, that is Echi Ridge (Fig. 2–71).

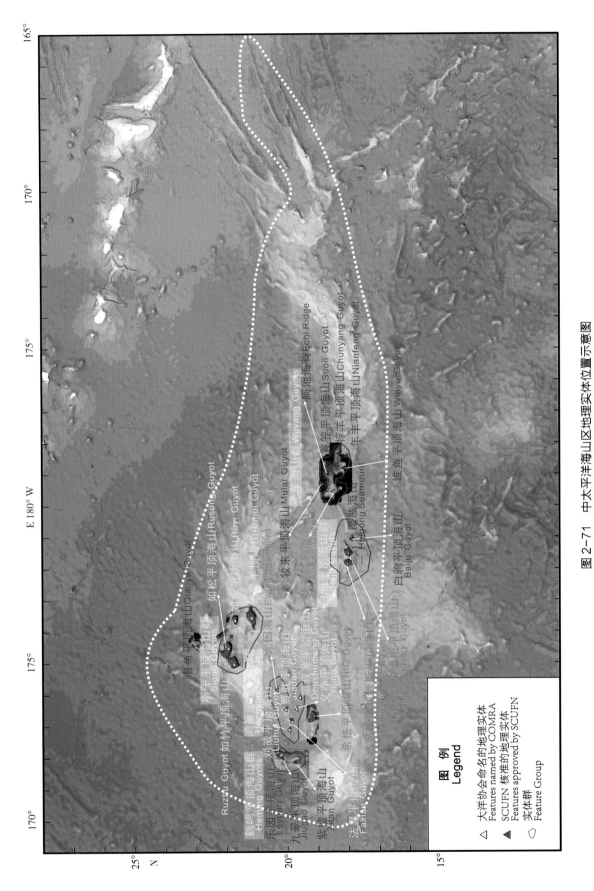

图 2-71　中大平洋海山区地理实体位置示意图

Fig.2-71　Locations of the undersea features in the Central Pacific Seamount Area

2.6.1 潜鱼平顶海山
Qianyu Guyot

中文名称 Chinese Name	潜鱼平顶海山 Qianyu Pingdinghaishan	英文名称 English Name	Qianyu Guyot	所在大洋 Ocean or Sea	中太平洋 Central Pacific Ocean
发现情况 Discovery Facts	此平顶海山于2011年8月由中国科考船"海洋六号"在执行DY125-23航次时调查发现。 This guyot was discovered by the Chinese R/V *Haiyang Liuhao* during the DY125-23 cruise in August, 2011.				
命名历史 Name History	由我国命名为潜鱼平顶海山,于2012年提交SCUFN审议通过。 Qianyu Guyot was named by China and approved by SCUFN in 2012.				
特征点坐标 Coordinates	22°58.40′N,175°38.50′E		长(km)×宽(km) Length (km) × Width (km)		47×32
最大水深（m） Max Depth（m）	5 200	最小水深（m） Min Depth（m）	1 200	高差（m） Total Relief（m）	4 000
地形特征 Feature Description	潜鱼平顶海山位于中太平洋海山区的北部,由地形平缓的山顶平台和陡峭的山坡组成,NE—SW走向,长宽分别为47 km和32 km。山顶平台水深约1 500 m,最浅处1 200 m,山麓水深5 200 m,高差4 000 m(图2-72)。 Qianyu Guyot is located in the north of the Central Pacific Seamount Area, consisting of a flat top platform and steep slopes, running NE to SW. The length is 47 km and the width is 32 km. The top platform depth is 1 500 m with the minimum depth of 1 200 m, while the piedmont depth is 5 200 m, which make the total relief being 4 000 m (Fig.2-72).				
命名释义 Reason for Choice of Name	"潜鱼"出自《诗经·小雅·鹤鸣》"鱼潜在渊,或在于渚"。"潜鱼"表示在深水中的鱼。此海山俯视平面形态貌似一条下潜的鱼,故以此命名。 "Qianyu" comes from a poem named *Heming* in *Shijing · Xiaoya*. *Shijing* is a collection of ancient Chinese poems from 11th century B.C. to 6th century B.C. "The fish may swim in the deep, or near the bar make a leap." The fish may make a leap, or remain in the deep. It means fish diving in deep water. It was named because its overlook plane shape looks like a diving fish.				

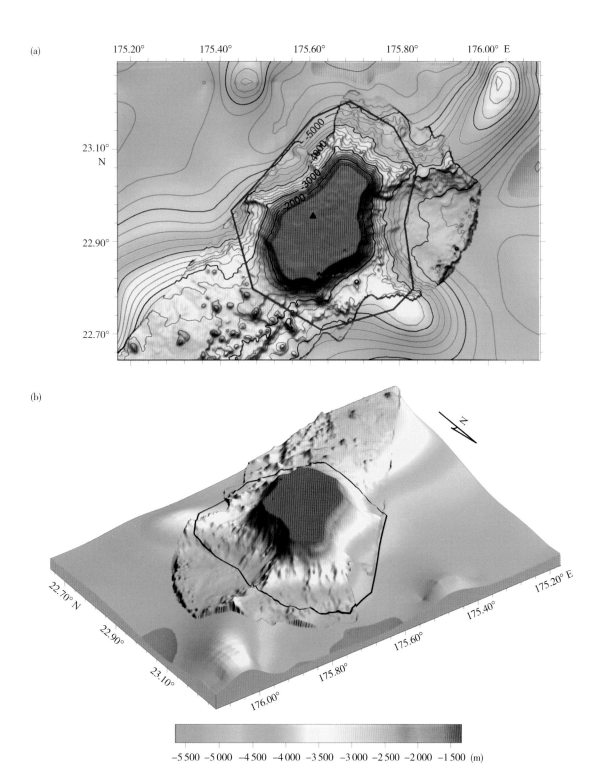

图 2-72　潜鱼平顶海山

(a) 地形图（等深线间隔 200 m）；(b) 三维图

Fig.2-72　Qianyu Guyot

(a) Bathymetric map (the contour interval is 200 m); (b) 3-D topographic map

2.6.2 斯干平顶海山群
Sigan Guyots

中文名称 Chinese Name	斯干平顶海山群 Sigan Pingdinghaishanqun	英文名称 English Name	Sigan Guyots	所在大洋 Ocean or Sea	中太平洋 Central Pacific Ocean
发现情况 Discovery Facts	此平顶海山群于 1999 年 9 月由中国科考船 "大洋一号" 在执行 DY95-10 航次时调查发现。 The guyots were discovered by the Chinese R/V *Dayang Yihao* during the DY95-10 cruise in September, 1999.				
命名历史 Name History	该平顶海山群包括 4 个平顶海山，在我国大洋航次调查报告中曾暂以代号命名。 The guyots consist of 4 guyots named Ruzhu, Rusong, Ruyi and Ruhui. A code name was used temporarily in Chinese cruise reports.				
特征点坐标 Coordinates	21°44.72′N，175°00.85′E（如竹平顶海山 / Ruzhu Guyot） 21°57.03′N，175°32.57′E（如松平顶海山 / Rusong Guyot） 21°41.40′N，176°23.31′E（如翼平顶海山 / Ruyi Guyot） 21°10.56′N，176°27.42′E（如辈平顶海山 / Ruhui Guyot）			长 (km) × 宽 (km) Length (km) × Width (km)	251 × 112
最大水深（m） Max Depth（m）	4 500	最小水深（m） Min Depth（m）	1 152	高差（m） Total Relief（m）	3 348
地形特征 Feature Description	斯干平顶海山群位于中太平洋海山区西北部，整体 SE—NW 走向，长宽分别为 251 km 和 112 km。此平顶海山群的最高峰位于如松平顶海山，最浅处水深 1 152 m。如辈平顶海山为群内规模最大的海山（图 2–73）。 Sigan Guyots are located in the northwest of Central Pacific Seamount Area, running NW to SE overall. The length and width are 251 km and 112 km respectively. The summit of these guyots is in Rusong Guyot, with the minimum depth 1 152 m. Ruhui Guyot is the largest one among these guyots (Fig.2–73).				
命名释义 Reason for Choice of Name	"斯干" 出自《诗经·小雅·斯干》。"斯干" 歌咏周宣王修建的王宫规模之宏大、壮丽以及环境之优美，表达了对生活的良好祝愿和赞美。采用群组化的方法，选取 "斯干" 篇名作为海山群名称，选取篇中的如竹、如松、如翼和如辈分别命名群中的 4 个平顶海山。 "Sigan" comes from a poem named *Sigan* in *Shijing · Xiaoya*. *Shijing* is a collection of ancient Chinese poems from 11th century B.C. to 6th century B.C. "Sigan" sang the grand, magnificent scale and beautiful environment of the palace built by the King Xuan (the Zhou Dynasty), and moreover, expressed the good wishes and praises for the life of the master. The names of the four guyots in this group, Ruzhu, Rusong, Ruyi and Rushi are all from the same poem and the title of the poem is selected as the name of guyots.				

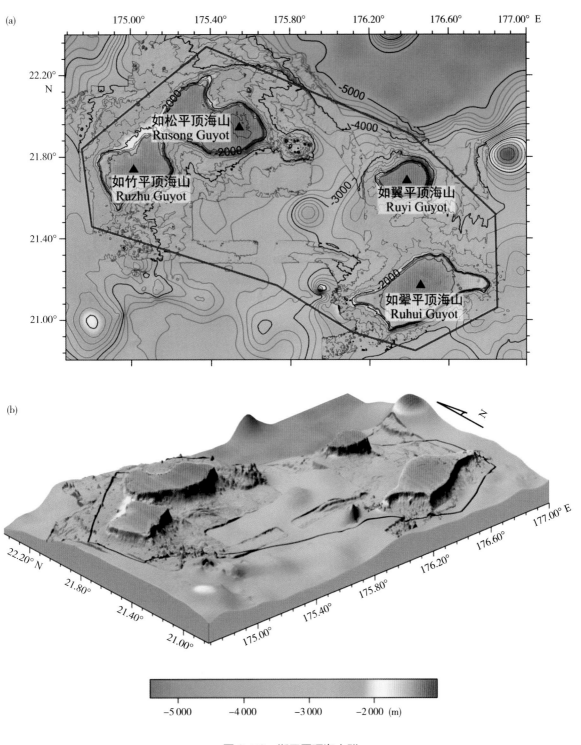

图 2-73　斯干平顶海山群

(a) 地形图（等深线间隔 200 m）；(b) 三维图

Fig.2-73　Sigan Guyots

(a) Bathymetric map (the contour interval is 200 m); (b) 3-D topographic map

2.6.3 如竹平顶海山
Ruzhu Guyot

中文名称 Chinese Name	如竹平顶海山 Ruzhu Pingdinghaishan	英文名称 English Name	Ruzhu Guyot	所在大洋 Ocean or Sea	中太平洋 Central Pacific Ocean
发现情况 Discovery Facts	此平顶海山于 1999 年 9 月由中国科考船"大洋一号"在执行 DY95-10 航次时调查发现。 This guyot was discovered by the Chinese R/V *Dayang Yihao* during the DY95-10 cruise in September, 1999.				
命名历史 Name History	如竹平顶海山为斯干平顶海山群的一部分。如竹平顶海山在我国大洋航次调查报告中曾暂以代号命名。 Ruzhu Guyot is a part of Sigan Guyots. A code was used to name Ruzhu Guyot temporarily in Chinese cruise reports.				
特征点坐标 Coordinates	21°43.90′N，175°03.59′E			长(km)×宽(km) Length (km)× Width (km)	45×26
最大水深（m） Max Depth（m）	2 700	最小水深（m） Min Depth（m）	1 231	高差（m） Total Relief（m）	1 469
地形特征 Feature Description	该平顶海山俯视平面形态近似四边形，长宽分别为 45 km 和 26 km。山顶平台水深约 1 400 m，最浅处水深 1 231 m，山麓水深 2 700 m，高差 1 469 m，顶部平坦，边坡陡峭（图 2-74）。 This guyot has nearly a quadrangle overlook plane shape. The length and width are 45 km and 26 km respectively. The top platform depth is about 1 400 m and the minimum depth is 1 231 m. The piedmont depth is 2 700 m, which makes the total relief being 1 469 m. The top platform is flat while the slopes are steep (Fig.2-74).				
命名释义 Reason for Choice of Name	如竹出自《诗经·小雅·斯干》"如竹苞矣，如松茂矣"。"斯干"歌咏周宣王修建的王宫规模之宏大、壮丽以及环境之优美。"如竹苞矣"，意指王宫周边修竹繁盛，环境优雅。 "Ruzhu" comes from a poem named *Sigan* in *Shijing · Xiaoya*. *Shijing* is a collection of ancient Chinese poems from 11th century B.C. to 6th century B.C. "Lush bamboos are growing there; Green towering pine-trees are there." The poem is in praise of the grandness, magnificence and elegance of palace. "Lush bamboos are growing there" means bamboo growing around the place makes the enviroment very elegant.				

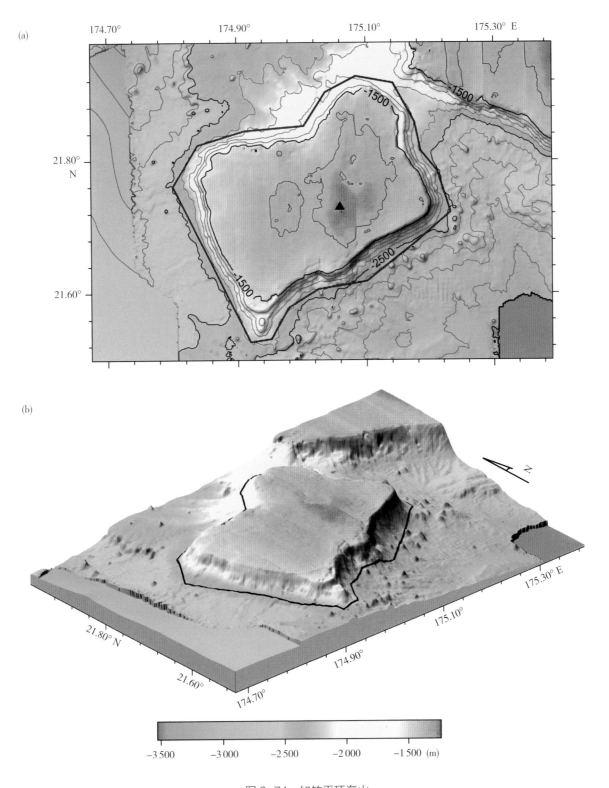

图 2-74　如竹平顶海山

(a) 地形图（等深线间隔 200 m）；(b) 三维图

Fig.2-74　Ruzhu Guyot

(a) Bathymetric map (the contour interval is 200 m); (b) 3-D topographic map

2.6.4　如松平顶海山
Rusong Guyot

中文名称 Chinese Name	如松平顶海山 Rusong Pingdinghaishan	英文名称 English Name	Rusong Guyot	所在大洋 Ocean or Sea	中太平洋 Central Pacific Ocean
发现情况 Discovery Facts	此平顶海山于1999年9月由中国科考船"大洋一号"在执行DY95-10航次时调查发现。 This guyot was discovered by the Chinese R/V *Dayang Yihao* during the DY95-10 cruise in September, 1999.				
命名历史 Name History	如松平顶海山为斯干平顶海山群的一部分。如松平顶海山在我国大洋航次调查报告中曾暂以代号命名。 Rusong Guyot is a part of Sigan Guyots. A code was used to name Rusong Guyot temporarily in Chinese cruise reports.				
特征点坐标 Coordinates	21°57.03′N，175°32.57′E			长(km)×宽(km) Length (km) × Width (km)	56 × 26
最大水深（m） Max Depth（m）	2 650	最小水深（m） Min Depth（m）	1 152	高差（m） Total Relief（m）	1 498
地形特征 Feature Description	该平顶海山形状不规则，长宽分别为56 km和26 km，北部发育一弯曲山脊。山顶平台水深约1 300 m，最浅处水深1 152 m，山麓水深2 650 m，高差1 498 m，顶部平坦，斜坡陡峭（图2-75）。 This guyot has an irregular shape. Its length and width are 56 km and 26 km respectively. A curved ridge is developed in the northern part. The top platform depth is about 1 300 m and the minimum depth is 1 152 m. The piedmont depth is 2 650 m, which makes the total relief being 1 498 m. The top platform is flat while the slopes are steep (Fig.2-75).				
命名释义 Reason for Choice of Name	如松出自《诗经·小雅·斯干》"如竹苞矣，如松茂矣"。"斯干"歌咏周宣王修建的王宫规模之宏大、壮丽以及环境之优美。"如松茂矣"，意指王宫周边松林茂密，景色宜人。 "Rusong" comes from a poem named *Sigan* in *Shijing · Xiaoya*. *Shijing* is a collection of ancient Chinese poems from 11th century B.C. to 6th century B.C. "Lush bamboos are growing there; Green towering pine-trees are there." The poem is in praise of the grandness, magnificence and elegance of palace. "Green towering pine-trees are there" means thick pine trees were near the palace.				

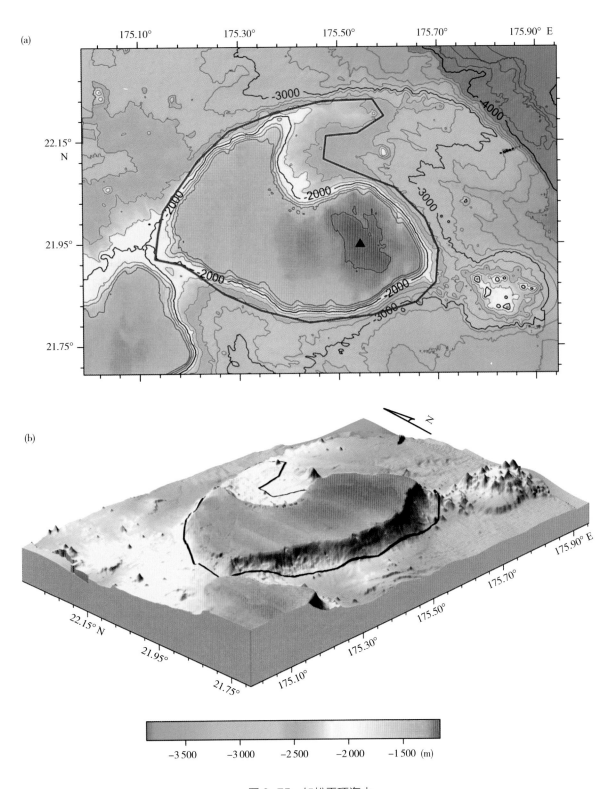

图 2-75　如松平顶海山

(a) 地形图（等深线间隔 200 m）；(b) 三维图

Fig.2-75　Rusong Guyot

(a) Bathymetric map (the contour interval is 200 m); (b) 3-D topographic map

2.6.5 如翼平顶海山
Ruyi Guyot

中文名称 Chinese Name	如翼平顶海山 Ruyi Pingdinghaishan	英文名称 English Name	Ruyi Guyot	所在大洋 Ocean or Sea	中太平洋 Central Pacific Ocean
发现情况 Discovery Facts	此平顶海山于1999年9月由中国科考船"大洋一号"在执行DY95-10航次时调查发现。 This guyot was discovered by the Chinese R/V *Dayang Yihao* during the DY95-10 cruise in September, 1999.				
命名历史 Name History	如翼平顶海山为斯干平顶海山群的一部分。如翼平顶海山在我国大洋航次调查报告中曾暂以代号命名。 Ruyi Guyot is a part of Sigan Guyots. A code was used to name Ruyi Guyot temporarily in Chinese cruise reports.				
特征点坐标 Coordinates	21°41.40′N，176°23.31′E			长(km)×宽(km) Length (km)× Width (km)	32×21
最大水深（m） Max Depth（m）	3 050	最小水深（m） Min Depth（m）	1 247	高差（m） Total Relief（m）	1 803
地形特征 Feature Description	该平顶海山形状不规则，长宽分别为32 km和21 km。山顶平台水深约1 400 m，最浅处水深1 247 m，山麓水深3 050 m，高差1 803 m，顶部平坦，边坡陡峭（图2-76）。 This guyot has an irregular shape. Its length and width are 32 km and 21 km respectively. The top platform depth is about 1 400 m and the minimum depth is 1 247 m. The piedmont depth is 3 050 m, which makes the total relief being 1 803 m. The top platform is flat while the slopes are steep (Fig.2-76).				
命名释义 Reason for Choice of Name	如翼出自《诗经·小雅·斯干》"如跂斯翼，如矢斯棘"。"斯干"歌咏周宣王修建的王宫规模之宏大、壮丽以及环境之优美。"如跂斯翼"，意指王宫殿宇端正，恰如人立。 "Ruyi" comes from a poem named *Sigan* in *Shijing · Xiaoya*. *Shijing* is a collection of ancient Chinese poems from 11th century B.C. to 6th century B.C. "As steady as a man standing, as straight as an arrow flying." The poem is in praise of the grandness, magnificence and elegance of palace. "As steady as a man standing" means the palace is upright, just like a man stands.				

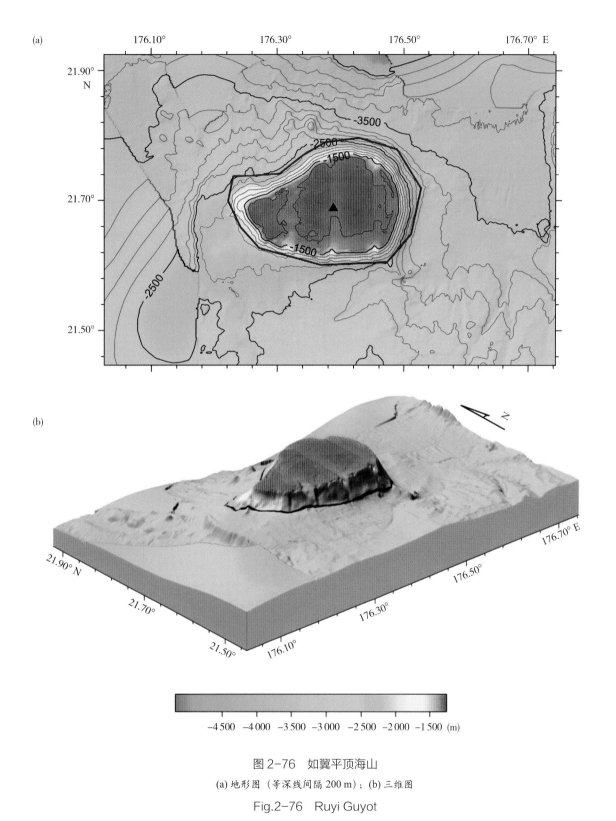

图 2-76 如翼平顶海山

(a) 地形图（等深线间隔 200 m）；(b) 三维图

Fig.2-76 Ruyi Guyot

(a) Bathymetric map (the contour interval is 200 m); (b) 3-D topographic map

2.6.6　如翚平顶海山
Ruhui Guyot

中文名称 Chinese Name	如翚平顶海山 Ruhui Pingdinghaishan	英文名称 English Name	Ruhui Guyot	所在大洋 Ocean or Sea	中太平洋 Central Pacific Ocean
发现情况 Discovery Facts	此平顶海山于 1999 年 9 月由中国科考船"大洋一号"在执行 DY95-10 航次时调查发现。 This guyot was discovered by the Chinese R/V *Dayang Yihao* during the DY95-10 cruise in September, 1999.				
命名历史 Name History	如翚平顶海山为斯干平顶海山群的一部分。如翚平顶海山在我国大洋航次调查报告中曾暂以代号命名。 Ruhui Guyot is a part of Sigan Guyots. A code was used to name Ruhui Guyot temporarily in Chinese cruise reports.				
特征点坐标 Coordinates	21°10.56′N，176°27.42′E			长 (km) × 宽 (km) Length (km) × Width (km)	75 × 34
最大水深（m） Max Depth（m）	2 800	最小水深（m） Min Depth（m）	1 236	高差（m） Total Relief（m）	1 564
地形特征 Feature Description	该平顶海山长宽分别为 75 km 和 34 km。山顶平台水深约 1 400 m，最浅处水深 1 236 m，山麓水深 2 800 m，高差 1 564 m，顶部平坦，边坡陡峭（图 2–77）。 The length and width of this guyot are 75 km and 34 km respectively. The top platform depth is about 1 400 m and the minimum depth is 1 236 m. The piedmont depth is 2 800 m, which makes the total relief being 1 564 m. The top platform is flat while the slopes are steep (Fig.2–77).				
命名释义 Reason for Choice of Name	如翚出自《诗经·小雅·斯干》"如鸟斯革，如翚斯飞"。"斯干"歌咏周宣王修建的王宫规模之宏大、壮丽以及环境之优美。"如翚斯飞"，意指王宫檐角飞挑，如锦鸡展翅。 "Ruhui" comes from a poem named *Sigan* in *Shijing · Xiaoya*. *Shijing* is a collection of ancient Chinese poems from 11th century B.C. to 6th century B.C. "As balanced as a bird hovering, as gaudy as a pheasant displaying." The poem is in praise of the grandness, magnificence and elegance of palace. "As gaudy as a pheasant displaying" means that cornices of the palaces have such dynamic beauty like flying caragana.				

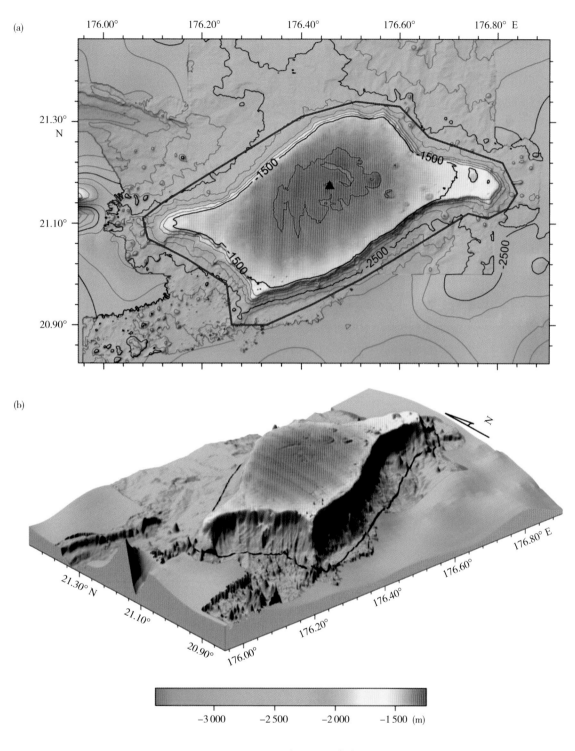

图 2-77　如羿平顶海山

(a) 地形图（等深线间隔 200 m）；(b) 三维图

Fig.2-77　Ruhui Guyot

(a) Bathymetric map (the contour interval is 200 m); (b) 3-D topographic map

2.6.7　鹤鸣平顶海山群
Heming Guyots

中文名称 Chinese Name	鹤鸣平顶海山群 Heming Pingdinghaishanqun	英文名称 English Name	Heming Guyots	所在大洋 Ocean or Sea	中太平洋 Central Pacific Ocean
发现情况 Discovery Facts	此平顶海山群于2001年8月由中国科考船"大洋一号"在执行DY105-11航次时调查发现。 The guyots were discovered by the Chinese R/V *Dayang Yihao* during the DY105-11 cruise in July, 2001.				
命名历史 Name History	该平顶海山群由3个平顶海山组成,分别为紫檀平顶海山、九皋平顶海山和乐园平顶海山,在我国大洋航次调查报告中曾暂以代号命名。 The guyots consist of 3 guyots, namely Zitan Guyot, Jiugao Guyot and Leyuan Guyot. A code name was used temporarily in Chinese cruise reports.				
特征点坐标 Coordinates	19°44.88′N,171°54.88′E（紫檀平顶海山/Zitan Guyot） 20°09.14′N,172°03.04′E（九皋平顶海山/Jiugao Guyot） 20°16.27′N,172°08.64′E（乐园平顶海山/Leyuan Guyot）			长(km)×宽(km) Length (km)× Width (km)	128×58
最大水深（m） Max Depth（m）	4 800	最小水深（m） Min Depth（m）	1 166	高差（m） Total Relief（m）	3 634
地形特征 Feature Description	鹤鸣平顶海山群长宽分别约128 km和58 km,最浅处水深1 166 m,山麓水深4 800 m,高差3 634 m。此海山群的最高峰位于南部的紫檀平顶海山,水深1 166 m,紫檀平顶海山也是群内规模最大的海山（图2-78）。 The length and width of Heming Guyots are 128 km and 58 km respectively. The minimum depth is 1 166 m and the piedmont depth is 4 800 m, which makes the total relief being 3 634 m. The highest summit where the depth is 1 166 m is in Zitan Guyot, which is located in the south of the guyots. Zitan Guyot is also the largest one in the guyots (Fig.2-78).				
命名释义 Reason for Choice of Name	"鹤鸣"出自《诗经·小雅·鹤鸣》。"鹤鸣"是一首即景抒情诗歌,描绘了一幅由青山、深泽、潜鱼、飞鹤、苍树等组合而成的园林图画。此平顶海山群命名为"鹤鸣",展示园林之中鹤鸣之声高入云霄,震动四野;选用该篇中的紫檀、九皋和乐园分别命名群中的3个平顶海山。 "Heming" comes from a poem named *Heming* in *Shijing · Xiaoya*. *Shijing* is a collection of ancient Chinese poems from 11th century B.C. to 6th century B.C. "Heming" is a scenery lyrics depicting a landscape picture including green mountain, deep swamps, flying cranes, verdant trees and so on. The guyots are named Heming to show that the singing of the flying crane penetrates the clouds and shakes the fields. "Zitan", "Jiugao" and "Leyuan" from this poem are selected as the name of the three guyots in this guyots.				

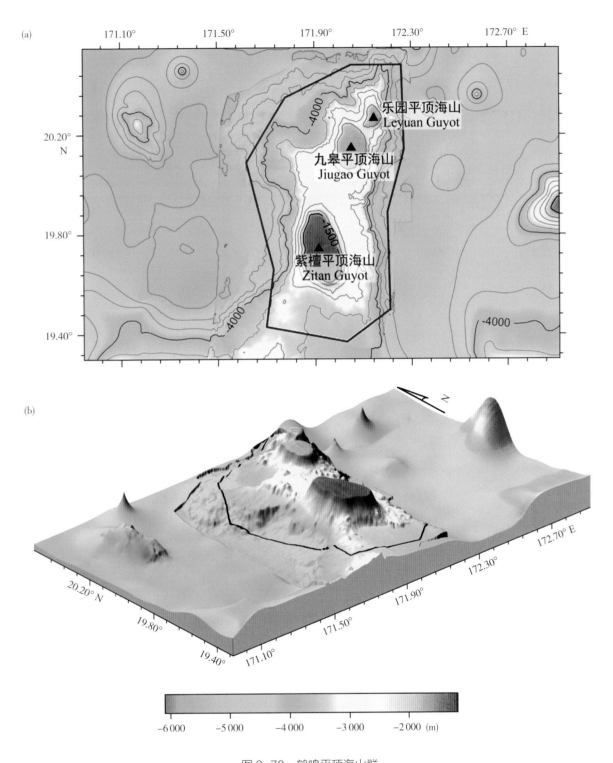

图 2-78　鹤鸣平顶海山群

(a) 地形图（等深线间隔 500 m）；(b) 三维图

Fig.2-78　Heming Guyots

(a) Bathymetric map (the contour interval is 500 m); (b) 3-D topographic map

2.6.8 乐园平顶海山
Leyuan Guyot

中文名称 Chinese Name	乐园平顶海山 Leyuan Pingdinghaishan	英文名称 English Name	Leyuan Guyot	所在大洋 Ocean or Sea	中太平洋 Central Pacific Ocean
发现情况 Discovery Facts	此平顶海山于2001年8月由中国科考船"大洋一号"在执行DY105-11航次时调查发现。 This guyot was discovered by the Chinese R/V *Dayang Yihao* during the DY105-11 cruise in August, 2001.				
命名历史 Name History	由我国命名为乐园平顶海山，于2017年提交SCUFN审议通过。 Leyuan Guyot was named by China and approved by SCUFN in 2017.				
特征点坐标 Coordinates	20°16.27′N，172°08.64′E			长(km)×宽(km) Length (km)× Width (km)	19×11
最大水深（m） Max Depth（m）	3 200	最小水深（m） Min Depth（m）	1 508	高差（m） Total Relief（m）	1 692
地形特征 Feature Description	乐园平顶海山俯视平面形态不规则，发育西北向、东北向和东西向3条山脊，长宽分别为19 km和11 km。山顶平台水深约1 700 m，最浅处水深1 508 m，山麓水深3 200 m，高差1 692 m，顶部平坦，边坡陡峭（图2-79）。 Leyuan Guyot has an irregular overlook plane shape and develops three ridges running NW to SE, NE to SW, and E to W respectively. The length and width are 19 km and 11 km respectively. The top platform depth is about 1 700 m and the minimum depth is 1 508 m. The piedmont depth is 3 200 m, which makes the total relief being 1 692 m. The top platform is flat while the slopes are steep (Fig.2-79).				
命名释义 Reason for Choice of Name	"乐园"出自《诗经·小雅·鹤鸣》。"鹤鸣"是一首即景抒情小诗，全诗描绘了一幅由青山、深泽、潜鱼、飞鹤、苍树等组合而成的园林图画。此平顶海山命名为"乐园"，"乐园"意指如诗如画、风光秀丽的园林。 "Leyuan" comes from a poem named *Heming* in *Shijing · Xiaoya*. *Shijing* is a collection of ancient Chinese poems from 11th century B.C. to 6th century B.C. Heming is a scenic and lyric poetry, the poem depicts a landscape painting by a combination of green mountain, deep swamps, flying cranes, verdant trees and so on. This guyot is named "Leyuan", meaning the picturesque scenery, beautiful gardens.				

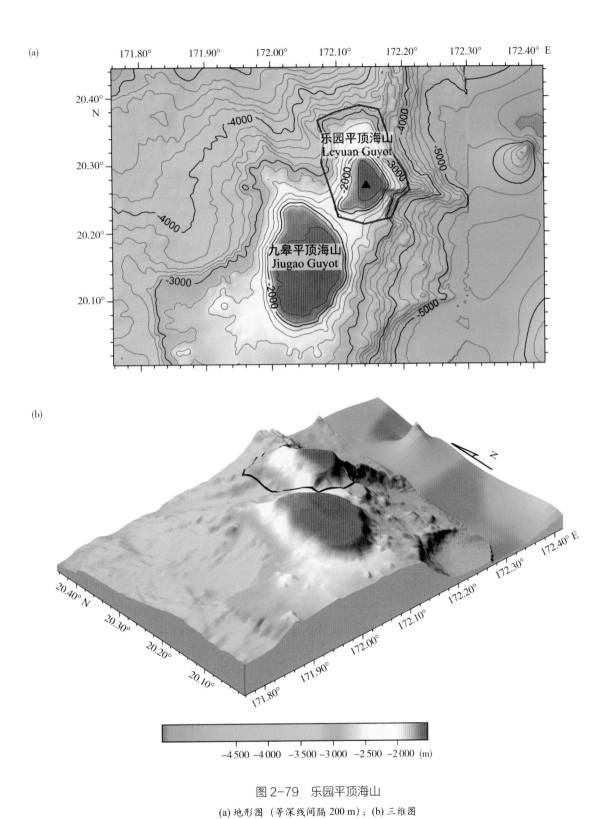

图 2-79　乐园平顶海山

(a) 地形图（等深线间隔 200 m）；(b) 三维图

Fig.2-79　Leyuan Guyot

(a) Bathymetric map (the contour interval is 200 m); (b) 3-D topographic map

2.6.9　九皋平顶海山
Jiugao Guyot

中文名称 Chinese Name	九皋平顶海山 Jiugao Pingdinghaishan	英文名称 English Name	Jiugao Guyot	所在大洋 Ocean or Sea	中太平洋 Central Pacific Ocean
发现情况 Discovery Facts	此平顶海山于 2001 年 8 月由中国科考船"大洋一号"在执行 DY105-11 航次时调查发现。 This guyot was discovered by the Chinese R/V *Dayang Yihao* during the DY105-11 cruise in August, 2001.				
命名历史 Name History	由我国命名为九皋平顶海山，于 2017 年提交 SCUFN 审议通过。 Jiugao Guyot was named by China and approved by SCUFN in 2017.				
特征点坐标 Coordinates	20°09.14′N，172°03.04′E			长(km) × 宽(km) Length (km) × Width (km)	27 × 15
最大水深（m） Max Depth（m）	2 600	最小水深（m） Min Depth（m）	1 514	高差（m） Total Relief（m）	1 086
地形特征 Feature Description	九皋平顶海山俯视平面形态呈南北向延伸，长宽分别为 27 km 和 15 km。山顶平台水深约 1 700 m，最浅处水深 1 514 m，山麓水深 2 600 m，高差 1 086 m，顶部平坦，边坡陡峭（图 2–80）。 Jiugao Guyot runs from N to S. The length and width are 27 km and 15 km respectively. The top platform depth is about 1 700 m and the minimum depth is 1 514 m. The piedmont depth is 2 600 m, which makes the total relief being 1 086 m. The top platform is flat while the slopes are steep (Fig.2–80).				
命名释义 Reason for Choice of Name	"九皋"出自《诗经·小雅·鹤鸣》。"鹤鸣"是一首即景抒情小诗，全诗描绘了一幅由青山、深泽、潜鱼、飞鹤、苍树等组合而成的园林图画。此平顶海山命名为"九皋"，"九皋"意指园林之中曲折深远的沼泽。 "Jiugao" comes from a poem named *Heming* in *Shijing · Xiaoya*. *Shijing* is a collection of ancient Chinese poems from 11th century B.C. to 6th century B.C. *Heming* is a scenic and lyric poetry, the poem depicts a landscape painting by combination of green mountains, deep swamps, flying cranes, verdant trees and so on. This guyot is named "Jiugao", meaning the tortuous and far-reaching swamp in the garden.				

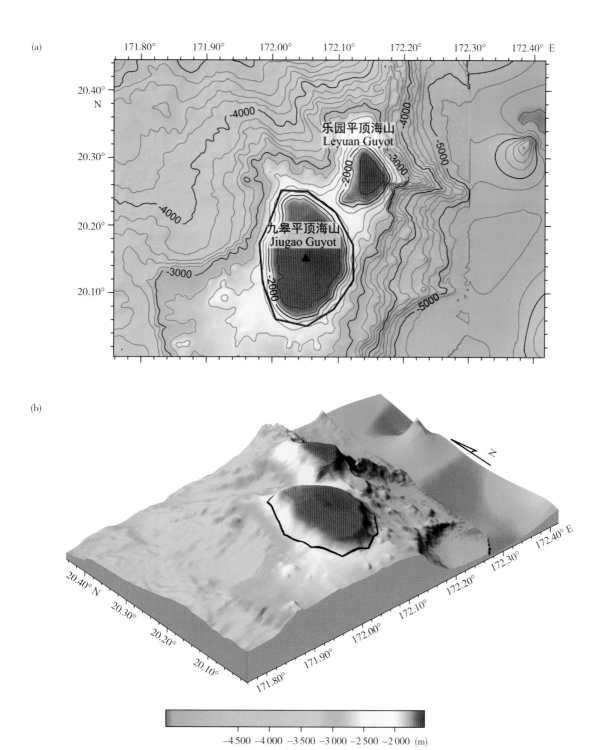

图 2-80　九皋平顶海山

(a) 地形图（等深线间隔 200 m）；(b) 三维图

Fig.2-80　Jiugao Guyot

(a) Bathymetric map (the contour interval is 200 m); (b) 3-D topographic map

2.6.10 紫檀平顶海山
Zitan Guyot

中文名称 Chinese Name	紫檀平顶海山 Zitan Pingdinghaishan	英文名称 English Name	Zitan Guyot	所在大洋 Ocean or Sea	中太平洋 Central Pacific Ocean
发现情况 Discovery Facts	此平顶海山于 2001 年 8 月由中国科考船"大洋一号"在执行 DY105-11 航次时调查发现。 This guyot was discovered by the Chinese R/V *Dayang Yihao* during the DY105-11 cruise in August, 2001.				
命名历史 Name History	由我国命名为紫檀平顶海山,于 2017 年提交 SCUFN 审议通过。 Zitan Guyot was named by China and approved by SCUFN in 2017.				
特征点坐标 Coordinates	19°44.88′N,171°54.88′E			长(km)×宽(km) Length (km) × Width (km)	43 × 26
最大水深（m） Max Depth（m）	3 900	最小水深（m） Min Depth（m）	1 166	高差（m） Total Relief（m）	2 734
地形特征 Feature Description	该平顶海山俯视平面形态近椭圆形,南北向延伸,长宽分别为 43 km 和 26 km。山顶平台水深约 1 300 m,最浅处水深 1 166 m,山麓水深 3 900 m,高差 2 734 m,顶部平坦,边坡陡峭(图 2–81)。 Zitan Guyot has a nearly elliptic shape, running N to S. The length and width are 43 km and 26 km respectively. The top platform depth is about 1 300 m and the minimum depth is 1 166 m, while the piedmont depth is 3 900 m, which makes the total relief being 2 734 m. The top platform is flat while the slopes are steep (Fig.2–81).				
命名释义 Reason for Choice of Name	"紫檀"出自《诗经·小雅·鹤鸣》。"鹤鸣"是一首即景抒情诗歌,全诗描绘了一幅由青山、深泽、潜鱼、飞鹤、苍树等组合而成的园林图画。此平顶海山命名为"紫檀","紫檀"意指园林之中高大繁盛的紫檀。 "Zitan" comes from a poem named *Heming* in *Shijing · Xiaoya*. *Shijing* is a collection of ancient Chinese poems from 11th century B.C. to 6th century B.C. Heming is a scenic and lyric poetry, the poem depicts a landscape painting by combination of green mountain, deep swamps, flying cranes, verdant trees and so on. This guyot is named "Zitan", meaning the tall prosperous red sandalwood in the garden.				

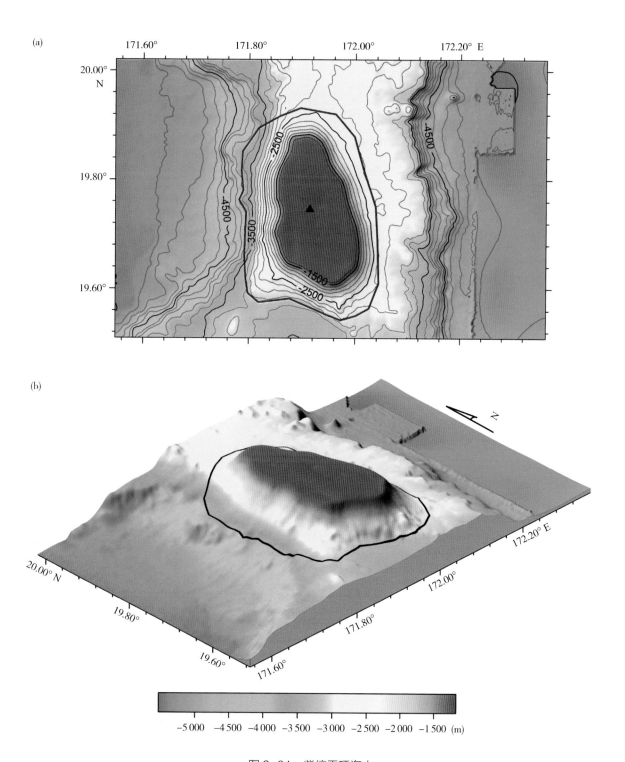

图 2-81　紫檀平顶海山

(a) 地形图（等深线间隔 200 m）；(b) 三维图

Fig.2-81　Zitan Guyot

(a) Bathymetric map (the contour interval is 200 m); (b) 3-D topographic map

2.6.11 鸿雁平顶海山群
Hongyan Guyots

中文名称 Chinese Name	鸿雁平顶海山群 Hongyan Pingdinghaishanqun	英文名称 English Name	Hongyan Guyots	所在大洋 Ocean or Sea	中太平洋 Central Pacific Ocean
发现情况 Discovery Facts	此平顶海山群于2001年9月由中国科考船"大洋一号"在执行DY105-11航次时所发现。 The guyots were discovered by the Chinese R/V *Dayang Yihao* during the DY105-11 cruise in September, 2001.				
命名历史 Name History	该平顶海山群由6个平顶海山组成，在我国大洋航次调查报告中曾暂以代号命名。 The guyots consist of 6 guyots. A code was used temporarily in Chinese cruise reports.				
特征点坐标 Coordinates	19°54.41′N，172°55.21′E（刘徽平顶海山 / Liuhui Guyot） 19°51.13′N，173°10.58′E（法显平顶海山 / Faxian Guyot） 20°01.98′N，173°33.86′E（贾耽平顶海山 / Jiadan Guyot） 19°40.52′N，173°35.40′E（义净平顶海山 / Yijing Guyot） 19°34.21′N，173°57.53′E（窦叔蒙平顶海山 / Doushumeng Guyot） 20°20.00′N，174°10.00′E（尤因海山 / Ewing Seamount）		长(km)×宽(km) Length (km) × Width (km)		197 × 83
最大水深（m） Max Depth（m）	4 500	最小水深（m） Min Depth（m）	1 287	高差（m） Total Relief（m）	3 213
地形特征 Feature Description	鸿雁平顶海山群位于中太平洋海山区西部，整体长宽分别为197 km和83 km。此平顶海山群的最高峰位于法显平顶海山，山顶平台水深约1 500 m，最浅处水深1 287 m。尤因海山为群内最大海山（图2–82）。 Hongyan Guyots are located in the west of Central Pacific Seamount Area. The overall length and width are 197 km and 83 km respectively. The highest summit of these guyots is in Faxian Guyot. The top platform water depth is about 1 500 m and the minimum depth is 1 287 m. The largest one among these guyots is Ewing Seamount (Fig.2–82).				
命名释义 Reason for Choice of Name	此平顶海山群的俯视平面形态从西向东看酷似"人"字形排列高翔的雁群，故选取"鸿雁"命名此海山群。雁群每年要进行两次长距离迁徙，路遥万险，但其坚韧不拔和团队精神为人称道。本海山群的6座平顶海山选择刘徽、法显、贾耽、义净、窦叔蒙和尤因等命名，这些人在海洋岛礁测量、航海、地理、潮汐观测等海洋科学有关方面功绩卓著，都具有鸿雁般的毅力，虽历重重困难与挑战，但始终坚持不懈，终成大业，除尤因是美国命名的近代人名外，其他5位均为中国古代科学家。 This guyots are named Hongyan because the shape of them looks like a flock of swan geese. The swan geese migrate twice for long distance every year, experiencing a lot of difficulties, but their diligency and team spirit are commendable.The six guyots are named after Liuhui, Faxian, Jiadan, Yijing, Doushumeng and Ewing respectively. These people made great contributions to the island measurement, navigation, geography, tidal and other activities of marine science. They had amazing perseverance just like swan geese. Though suffering numerous difficulties and challenges, they never gave up and finally succeeded. Except Ewing is a modern American name, the other five are all ancient Chinese scientists.				

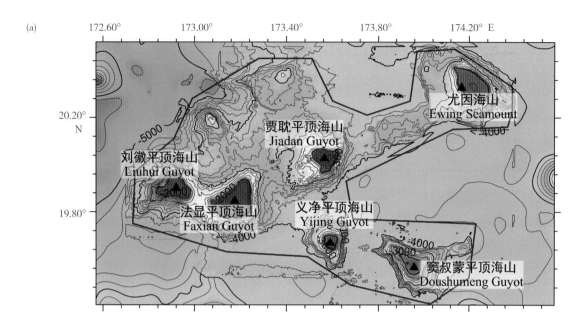

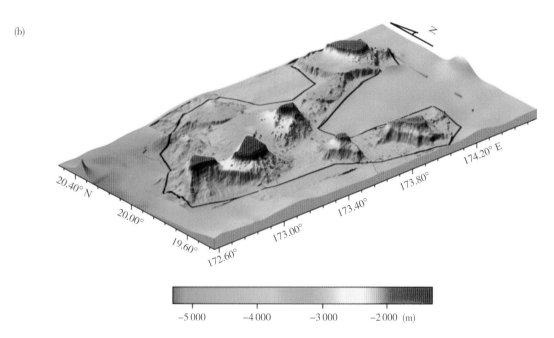

图 2-82　鸿雁平顶海山群

(a) 地形图（等深线间隔 200 m）；(b) 三维图

Fig.2-82　Hongyan Guyots

(a) Bathymetric map (the contour interval is 200 m); (b) 3-D topographic map

2.6.12 刘徽平顶海山
Liuhui Guyot

中文名称 Chinese Name	刘徽平顶海山 Liuhui Pingdinghaishan	英文名称 English Name	Liuhui Guyot	所在大洋 Ocean or Sea	中太平洋 Central Pacific Ocean
发现情况 Discovery Facts	此平顶海山于 2001 年 9 月由中国科考船"大洋一号"在执行 DY105-11 航次时调查发现。 This guyot was discovered by the Chinese R/V *Dayang Yihao* during the DY105-11 cruise in September, 2001.				
命名历史 Name History	此平顶海山为鸿雁平顶海山群的一部分,鸿雁平顶海山群在我国大洋航次调查报告中曾暂以代号命名。 This guyot is a part of Hongyan Guyots. A code was used to name Hongyan Guyots temporarily in Chinese cruise reports.				
特征点坐标 Coordinates	19°54.41′N,172°55.21′E			长(km)×宽(km) Length (km)×Width (km)	28×23
最大水深(m) Max Depth(m)	4 500	最小水深(m) Min Depth(m)	1 521	高差(m) Total Relief(m)	2 979
地形特征 Feature Description	该平顶海山俯视平面形态呈近三角形,长宽分别为 28 km 和 23 km。山顶平台水深约 1 700 m,最浅处水深 1 521 m,山麓水深 4 500 m,高差 2 979 m,顶部平坦,边坡陡峭(图 2–83)。 This guyot has nearly a triangular overlook shape. The length and width are 28 km and 23 km respectively. The top platform depth is about 1 700 m and the minimum depth is 1 521 m. The piedmont depth is 4 500 m, which makes the total relief being 2 979 m. The top platform is flat while the slopes are steep (Fig.2–83).				
命名释义 Reason for Choice of Name	刘徽(约公元 225—295 年),三国时代魏国人,撰《海岛算经》一书,通过测量法来计算海岛的高与远,是我国最早的测量学著作。此平顶海山命名为"刘徽",以纪念刘徽在我国岛礁测量方面的重要贡献。 Liu Hui (about A.D.225–295), who lived in the State of Wei during the Three Kingdoms period, wrote a famous book, which calculated the island's height and distance by measuring methods and it is China's earliest surveying activities. This guyot is named after "Liuhui", to commemorate his important contributions in the field of islands measurement of China.				

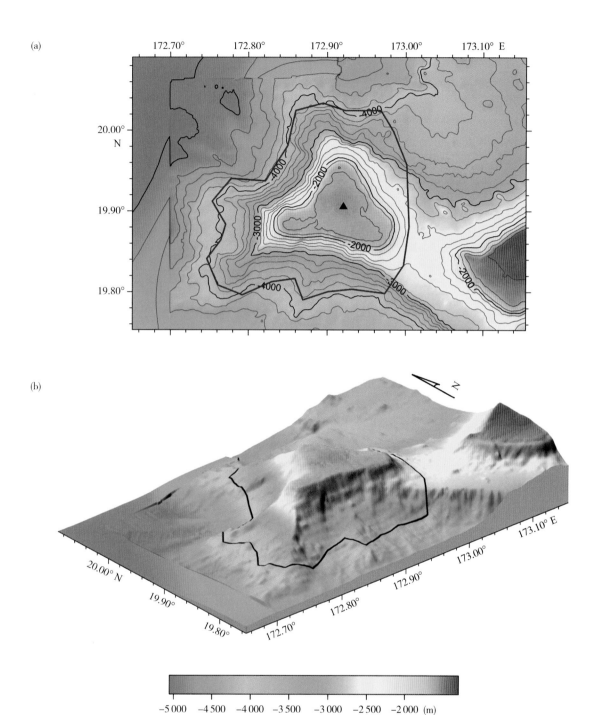

图 2-83 刘徽平顶海山

(a) 地形图（等深线间隔 200 m）；(b) 三维图

Fig.2-83 Liuhui Guyot

(a) Bathymetric map (the contour interval is 200 m); (b) 3-D topographic map

2.6.13 法显平顶海山
Faxian Guyot

中文名称 Chinese Name	法显平顶海山 Faxian Pingdinghaishan	英文名称 English Name	Faxian Guyot	所在大洋 Ocean or Sea	中太平洋 Central Pacific Ocean
发现情况 Discovery Facts	此平顶海山于 2001 年 9 月由中国科考船 "大洋一号" 在执行 DY105-11 航次时调查发现。 This guyot was discovered by the Chinese R/V *Dayang Yihao* during the DY105-11 cruise in September, 2001.				
命名历史 Name History	此平顶海山为鸿雁平顶海山群的一部分，鸿雁平顶海山群在我国大洋航次调查报告中曾暂以代号命名。 This guyot is a part of Hongyan Guyots. A code was used to name Hongyan Guyots temporarily in Chinese cruise reports.				
特征点坐标 Coordinates	19°51.13′N，173°10.58′E			长(km)×宽(km) Length (km)× Width (km)	35×16
最大水深（m） Max Depth（m）	3 600	最小水深（m） Min Depth（m）	1 287	高差（m） Total Relief（m）	2 313
地形特征 Feature Description	该平顶海山长宽分别为 35 km 和 16 km。山顶平台水深约 1 500 m，最浅处水深 1 287 m，山麓水深 3 600 m，高差 2 313 m，顶部平坦，边坡陡峭（图 2–84）。 The length and width of this guyot are 35 km and 16 km respectively. The top platform water depth is about 1 500 m and the minimum depth is 1 287 m. The piedmont depth is 3 600 m, which makes the total relief being 2 313 m. The top platform is flat while the slopes are steep (Fig.2–84).				
命名释义 Reason for Choice of Name	法显（约公元342—423 年），山西临汾人。公元399 年，年近花甲的法显率领僧徒西行求法，至天竺，获得大量佛教典籍。公元 409 年乘坐商船自海路返回，历尽艰险。他根据亲身经历撰写了名著《法显传》（又名《佛国记》），记载了中亚、南亚、东南亚各国的风土人情，是中国历史上第一部关于远洋航行的纪实性文献。此平顶海山命名为 "法显"，以纪念法显在航海和文化传播方面的重要贡献。 Fa Xian (about A.D.342–423), who lived in Linfen, Shanxi Province, China. In A.D. 399, Fa Xian, nearly 60 years old, led several Buddhist monks to Tianzhu, west of China, for learning buddhist scriptures, and got a large number of Buddhist books. In A.D. 409, he returned by merchant ship difficultly. According to personal experience he wrote a famous book, which recorded the local conditions and customs of Central Asia, South Asia and Southeast Asia. It is the first documentary literature on ocean-going voyage in Chinese history. The guyot is named after "Faxian", to commemorate his significant contributions in the fields of navigation and cultural communication.				

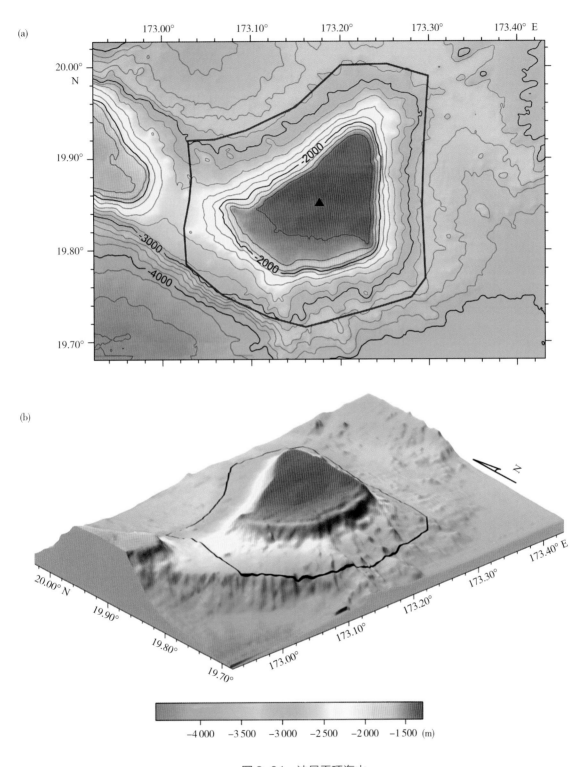

图 2-84　法显平顶海山

(a) 地形图（等深线间隔 200 m）；(b) 三维图

Fig.2-84　Faxian Guyot

(a) Bathymetric map (the contour interval is 200 m); (b) 3-D topographic map

2.6.14　贾耽平顶海山
Jiadan Guyot

中文名称 Chinese Name	贾耽平顶海山 Jiadan Pingdinghaishan	英文名称 English Name	Jiadan Guyot	所在大洋 Ocean or Sea	中太平洋 Central Pacific Ocean
发现情况 Discovery Facts	此平顶海山于2001年9月由中国科考船"大洋一号"在执行DY105-11航次时调查发现。 This guyot was discovered by the Chinese R/V *Dayang Yihao* during the DY105-11 cruise in September, 2001.				
命名历史 Name History	此平顶海山为鸿雁平顶海山群的一部分，鸿雁平顶海山群在我国大洋航次调查报告中曾暂以代号命名。 This guyot is a part of Hongyan Guyots. A code was used to name Hongyan Guyots temporarily in Chinese cruise reports.				
特征点坐标 Coordinates	20°1.98′N，173°33.86′E		长(km)×宽(km) Length (km)× Width (km)		33×21
最大水深（m） Max Depth（m）	3 200	最小水深（m） Min Depth（m）	1 338	高差（m） Total Relief（m）	1 862
地形特征 Feature Description	该平顶海山发育4条山脊，长宽分别为33 km和21 km。山顶平台水深约1 500 m，最浅处水深1 338 m，山麓水深3 200 m，高差1 862 m，顶部平坦，边坡陡峭（图2–85）。 This guyot develops four ridges. The length and width are 33 km and 21 km respectively. The top platform depth is about 1 500 m and the minimum depth is 1 338 m. The piedmont depth is 3 200 m, which makes the total relief being 1 862 m. The top platform is flat while the slopes are steep (Fig.2–85).				
命名释义 Reason for Choice of Name	贾耽（公元730—805年），字敦诗，河北南皮人。唐代著名政治家、地理学家。曾花17年绘出《海内华夷图》。他又撰《广州通海夷道》一文，详细记载了唐代中国与南洋、西亚、东非地区的远洋航路。此平顶海山命名为"贾耽"，以纪念贾耽在航海方面的重要贡献。 Jiadan (about A.D.730–805), whose another name taken at the age of twenty was Dunshi, lived in Nanpi, Hebei Province, China. He was a famous politician and geographer in Tang Dynasty, who spent seventeen years drawing a map of worldwide countries. He wrote a famous article, which recorded in detail ocean routes among the Tang Dynasty and Southeast Asia, West Asia, East African et al. The guyot is named after "Jiadan" to commemorate his great contributions in the field of navigation.				

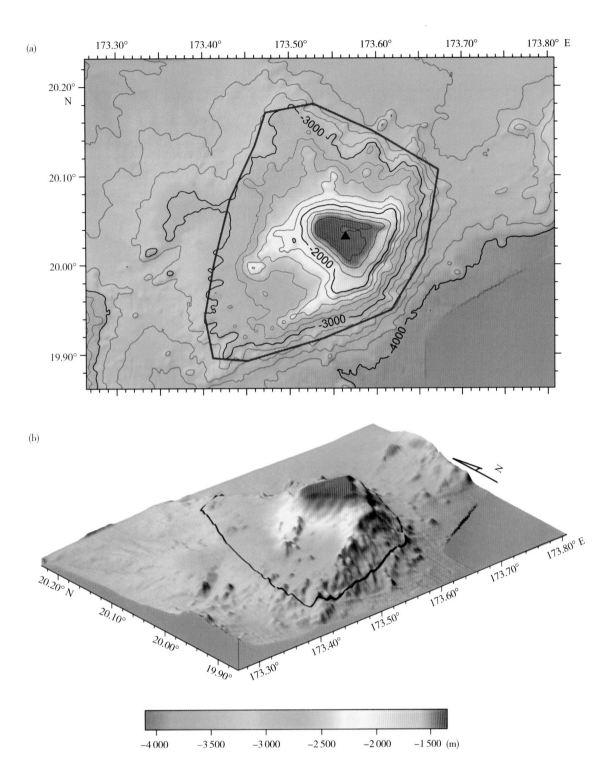

图 2-85　贾耽平顶海山

(a) 地形图（等深线间隔 200 m）；(b) 三维图

Fig.2-85　Jiadan Guyot

(a) Bathymetric map (the contour interval is 200 m); (b) 3-D topographic map

2.6.15　义净平顶海山
Yijing Guyot

中文名称 Chinese Name	义净平顶海山 Yijing Pingdinghaishan	英文名称 English Name	Yijing Guyot	所在大洋 Ocean or Sea	中太平洋 Central Pacific Ocean
发现情况 Discovery Facts	此平顶海山于 2001 年 9 月由中国科考船"大洋一号"在执行 DY105-11 航次时调查发现。 This guyot was discovered by the Chinese R/V *Dayang Yihao* during the DY105-11 cruise in September, 2001.				
命名历史 Name History	此平顶海山为鸿雁平顶海山群的一部分，鸿雁平顶海山群在我国大洋航次调查报告中曾暂以代号命名。 This guyot is a part of Hongyan Guyots. A code was used to name Hongyan Guyots temporarily in Chinese cruise reports.				
特征点坐标 Coordinates	19°40.52′N，173°35.40′E			长(km)×宽(km) Length (km)× Width (km)	29×19
最大水深（m） Max Depth（m）	4 100	最小水深（m） Min Depth（m）	1 817	高差（m） Total Relief（m）	2 283
地形特征 Feature Description	该平顶海山长宽分别为 29 km 和 19 km。山顶平台水深约 2 000 m，最浅处水深 1 817 m，山麓水深 4 100 m，高差 2 283 m，顶部平坦，边坡陡峭 (图 2–86)。 The length and width of this guyot are 29 km and 19 km respectively. The top platform depth is about 2 000 m and the minimum depth is 1 817 m. The piedmont depth is 4 100 m, which makes the total relief being 2 283 m. The top platform is flat while the slopes are steep (Fig.2–86).				
命名释义 Reason for Choice of Name	义净（公元 635—713 年），山东历城人，初唐时期著名僧人。咸亨二年（671 年）从广州乘船前往印度，求经 20 余载，695 年航海回国。以其亲身经历撰写著名的《南海寄归内法传》与《大唐西域求法高僧传》，是研究唐代航海史的重要史料。此平顶海山命名为"义净"，以纪念义净在航海和文化传播方面的重要贡献。 Yi Jing (about A.D. 635–713), who lived in Licheng, Shandong Province, China, was a famous monk in the early period of Tang Dynasty. In Xianheng 2nd year (A.D. 671), he went to India from Guangzhou by ship for Buddhism over 20 years. He returned by sea in A.D. 695 and wrote two famous books about navigation according to personal experience. It is the important historical data to study the history of navigation of the Tang Dynasty. The guyot is named after "Yijing", to commemorate his significant contributions in the fields of navigation and cultural communication.				

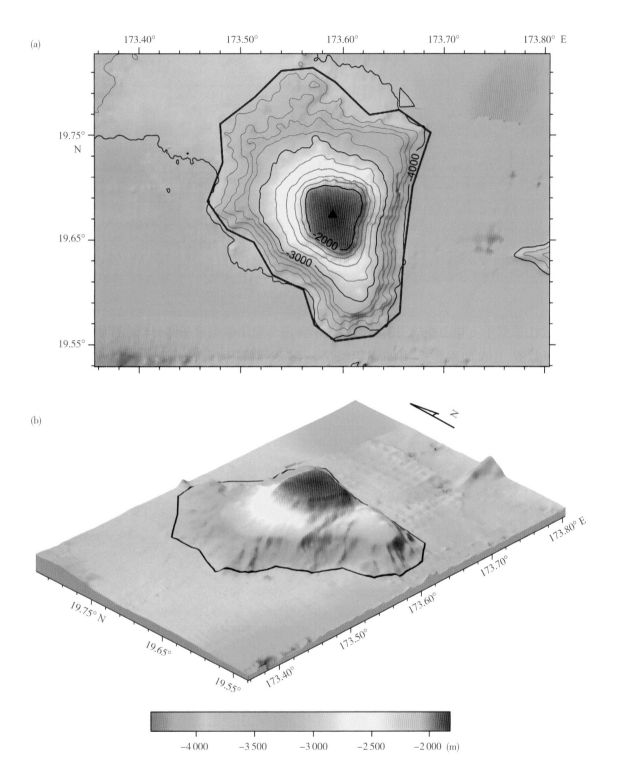

图 2-86 义净平顶海山

(a) 地形图（等深线间隔 200 m）；(b) 三维图

Fig.2-86 Yijing Guyot

(a) Bathymetric map (the contour interval is 200 m); (b) 3-D topographic map

2.6.16 尤因海山
Ewing Seamount

中文名称 Chinese Name	尤因海山 Youyin Haishan	英文名称 English Name	Ewing Seamount	所在大洋 Ocean or Sea	中太平洋 Central Pacific Ocean
发现情况 Discovery Facts	此海山于 2001 年 9 月由中国科考船"大洋一号"在执行 DY105-11 航次时调查发现(美国于 1960 年由"VEMA"号发现)。 This seamount was discovered by the Chinese R/V *Dayang Yihao* during the DY105-11 cruise in September, 2001 (It was discovered by *VEMA* vessel of the U. S. in 1960).				
命名历史 Name History	收录于 GEBCO 地名辞典,1960 年由美国提名,中文译名为"尤因海山"。此海山为鸿雁平顶海山群的一部分,鸿雁平顶海山群在我国大洋航次调查报告中曾暂以代号命名。 The name Ewing Seamount is listed in GEBCO gazetteer. It was proposed by the U. S. in 1960. Its translation is Youyin Haishan in Chinese. The seamount is a part of Hongyan Guyots. A code was used to name Hongyan Guyots temporarily in Chinese cruise reports.				
特征点坐标 Coordinates	20°19.80′N,174°13.10′E			长(km)× 宽(km) Length (km)× Width (km)	44×37
最大水深(m) Max Depth(m)	4 250	最小水深(m) Min Depth(m)	1 775	高差(m) Total Relief(m)	2 475
地形特征 Feature Description	海山俯视平面形态呈 NW—SE 走向的"梨"形,西北端狭小,东南端宽大,长宽分别为 44 km 和 37 km。海山最浅处水深 1 775 m,山麓水深 4 250 m,高差 2 475 m,顶部平坦,边坡陡峭(图 2–87)。 The seamount has a pear-like overlook shape, running NW to SE, narrow end in the northwest and broad end in the southeast. The length is 44 km and width is 37 km. The minimum depth of the seamount is 1 775 m, while the piedmont depth is 4 250 m, which makes the total relief being 2 475 m. The top platform is flat and the slopes are steep (Fig.2–87).				
命名释义 Reason for Choice of Name	本海山是根据美国地球物理学家 Maurice Ewing(1906—1974)之名命名。 This seamount is named after Maurice Ewing (1906–1974), an American geophysicist and oceanographer.				

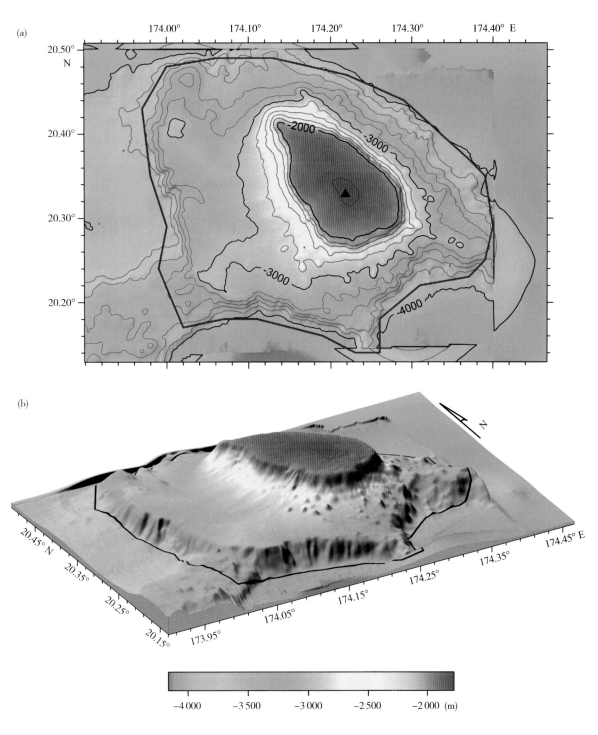

图 2-87　尤因海山

(a) 地形图（等深线间隔 200 m）；(b) 三维图

Fig.2-87　Ewing Seamount

(a) Bathymetric map (the contour interval is 200 m); (b) 3-D topographic map

2.6.17 窦叔蒙平顶海山
Doushumeng Guyot

中文名称 Chinese Name	窦叔蒙平顶海山 Doushumeng Pingdinghaishan	英文名称 English Name	Doushumeng Guyot	所在大洋 Ocean or Sea	中太平洋 Central Pacific Ocean
发现情况 Discovery Facts	此平顶海山于2005年6月由中国科考船"大洋一号"在执行DY105-17B航次时调查发现。 This guyot was discovered by the Chinese R/V *Dayang Yihao* during the DY105-17B cruise in June, 2005.				
命名历史 Name History	此平顶海山为鸿雁平顶海山群的一部分，鸿雁平顶海山群在我国大洋航次调查报告中曾暂以代号命名。 This guyot is a part of Hongyan Guyots. A code was used to name Hongyan Guyots temporarily in Chinese cruise reports.				
特征点坐标 Coordinates	19°34.21′N，173°57.53′E			长(km)×宽(km) Length (km) × Width (km)	40×30
最大水深（m） Max Depth（m）	4 100	最小水深（m） Min Depth（m）	1 893	高差（m） Total Relief（m）	2 207
地形特征 Feature Description	该平顶海山长宽分别为40 km和30 km。山顶平台水深约2 000 m，最浅处水深1 893 m，山麓水深4 100 m，高差2 207 m，顶部平坦，边坡陡峭（图2–88）。 The length and width of this guyot are 40 km and 30 km respectively. The top platform depth is about 2 000 m and the minimum depth is 1 893 m. The piedmont depth is 4 100 m, which makes the total relief being 2 207 m. The top platform is flat while the slopes are steep (Fig.2–88).				
命名释义 Reason for Choice of Name	窦叔蒙（生卒年不详），唐代著名科学家，其撰写的《海涛志》是我国现存最早的论述海洋潮汐的著作。此平顶海山命名为"窦叔蒙"，以纪念窦叔蒙在海洋学方面的重要贡献。 Doushumeng (his birth and death dates are unknown) was a famous scientist of the Tang Dynasty. He wrote a famous book, which is the first existent works discussing ocean tides in China. The guyot is named after "Doushumeng", to commemorate his great contributions in the field of oceanography.				

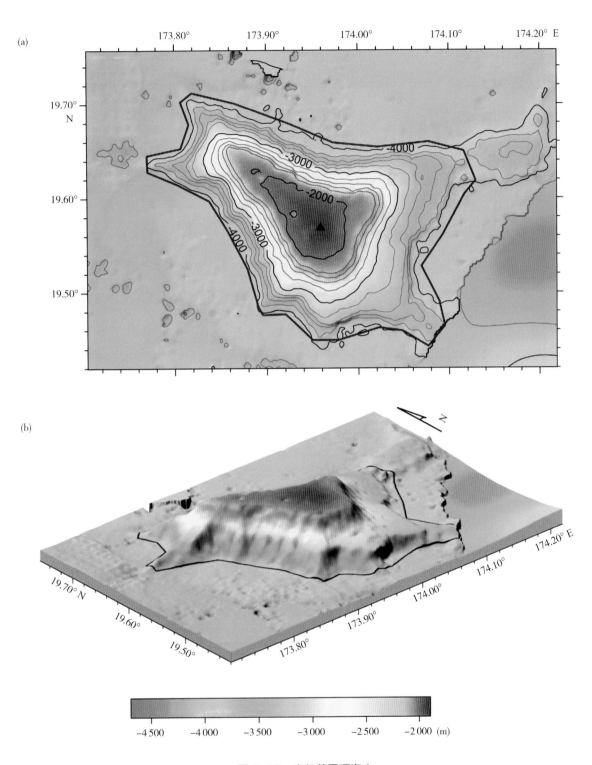

图 2-88 窦叔蒙平顶海山

(a) 地形图 (等深线间隔 200 m)；(b) 三维图

Fig.2-88 Doushumeng Guyot

(a) Bathymetric map (the contour interval is 200 m); (b) 3-D topographic map

2.6.18 希格平顶海山
HIG Guyot

中文名称 Chinese Name	希格平顶海山 Xige Pingdinghaishan	英文名称 English Name	HIG Guyot	所在大洋 Ocean or Sea	中太平洋 Central Pacific Ocean
发现情况 Discovery Facts	此平顶海山于 2011 年 8 月由中国科考船"海洋六号"在执行 DY125-23 航次时调查发现。 The guyot was discovered by the Chinese R/V *Haiyang Liuhao* during the DY125-23 cruise in August, 2011.				
命名历史 Name History	收录于 GEBCO 地名辞典，由美国夏威夷地球物理研究所 Keating 和 Kroenke 提名，1995 年由 SCUFN 接受。我国大洋航次调查报告中曾暂以代号命名。 The name HIG Guyot is listed in GEBCO gazetteer. It was proposed by Keating and Kroenke from Hawaiian Institute of Geophysics (HIG), America and approved by SCUFN in 1995. A code name was used temporarily in Chinese cruise reports.				
特征点坐标 Coordinates	19°10.00′N，173°15.00′E			长(km)×宽(km) Length (km) × Width (km)	57×37
最大水深（m） Max Depth（m）	3 900	最小水深（m） Min Depth（m）	1 294	高差（m） Total Relief（m）	2 606
地形特征 Feature Description	此平顶海山由地形宽阔的平台和陡峭山坡组成，长宽分别为 57 km 和 37 km。山顶平台水深约 1 500 m，最浅处水深 1 294 m，山麓水深 3 900 m，高差 2 606 m（图 2–89）。 This guyot consists of a broad top platform and steep slopes. The length is 57 km and the width is 37 km. The top platform depth is 1 500 m and the minimum depth is 1 294 m, while the piedmont depth is 3 900 m, which makes the total relief being 2 606 m (Fig.2–89).				
命名释义 Reason for Choice of Name	本平顶海山是根据美国夏威夷地球物理研究所命名，该研究所在 1981 年对一个非常大的平顶海山进行了地球物理调查研究。 This guyot is named from the Hawaiian Institute of Geophysics (HIG), which conducted a geographical investigation focusing on an exceptionally large guyot in 1981 (Nemeto & Kroenke, 1985).				

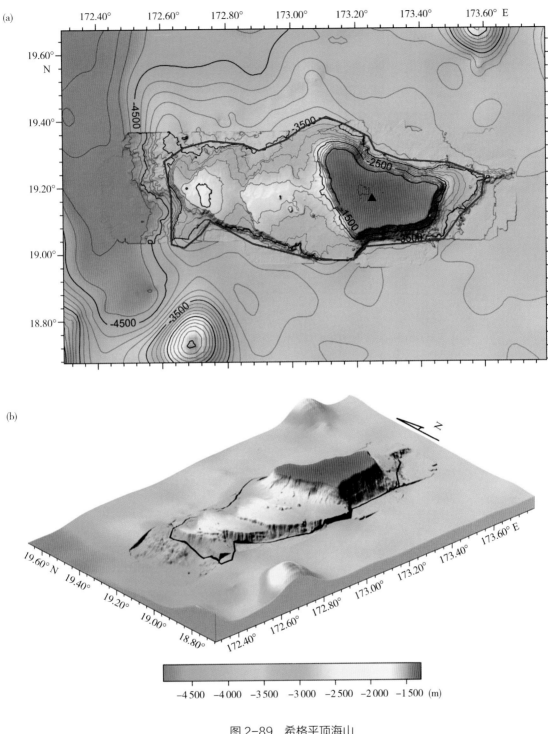

图 2-89　希格平顶海山

(a) 地形图（等深线间隔 200 m）；(b) 三维图

Fig.2-89　HIG Guyot

(a) Bathymetric map (the contour interval is 200 m); (b) 3-D topographic map

2.6.19　白驹平顶海山群
Baiju Guyots

中文名称 Chinese Name	白驹平顶海山群 Baiju Pingdinghaishanqun	英文名称 English Name	Baiju Guyots	所在大洋 Ocean or Sea	中太平洋 Central Pacific Ocean
发现情况 Discovery Facts	此平顶海山群于 1998 年 11 月由中国科考船 "大洋一号" 在执行 DY95-8 航次时调查发现。 The guyots were discovered by the Chinese R/V *Dayang Yihao* during the DY95-8 cruise in November, 1998.				
命名历史 Name History	该平顶海山群由 3 个平顶海山组成，分别为西部的路易斯·阿加西斯平顶海山、中部的亚历山大·阿加西斯平顶海山和东部的白驹平顶海山，在我国大洋航次调查报告中曾暂以代号命名。 The guyots consist of 3 guyots, which are the western Louis Agassiz Guyot, central Alexander Agassiz Guyot and eastern Baiju Guyot. A code name was used temporarily in Chinese cruise reports.				
特征点坐标 Coordinates	17°52.00′N，178°12.00′E*（路易斯·阿加西斯平顶海山 / Louis Agassiz Guyot） 17°54.00′N，178°33.00′E*（亚历山大·阿加西斯平顶海山 / Alexander Agassiz Guyot） 17°53.90′N，178°58.70′E（白驹平顶海山 / Baiju Guyot）			长(km) × 宽(km) Length (km) × Width (km)	140 × 80
最大水深（m） Max Depth（m）	4 600	最小水深（m） Min Depth（m）	1 544	高差（m） Total Relief（m）	3 056
地形特征 Feature Description	白驹平顶海山群位于中太平洋海山区，东西走向，整体长宽分别约为 140 km 和 80 km。海山群最浅处水深 1 544 m，山麓水深 4 600 m，高差 3 056 m。此平顶海山群的最高峰位于路易斯·阿加西斯平顶海山，最浅处水深 1 544 m，亚历山大·阿加西斯平顶海山为群内最大海山（图 2–90）。 Baiju Guyots are located in the Central Pacific Seamount Area, running E to W. The overall length and width are 140 km and 80 km respectively. The minimum water depth is 1 544 m. The piedmont depth is 4 600 m, which makes the total relief being 3 056 m. The highest summit of these guyots is in Louis Agassiz Guyot with a minimum depth of 1 544 m. Alexander Agassiz Guyot is the largest among these guyots (Fig.2–90).				
命名释义 Reason for Choice of Name	采用平顶海山群中的白驹平顶海山之名。 The guyots are named after the name of Baiju Guyot among these guyots.				

　　* 表中路易斯·阿加西斯平顶海山和亚历山大·阿加西斯平顶海山的特征点坐标为 ACUF 地名辞典给出的数值。我国根据自己的调查数据在 SCUFN 第 24 次会议上对这两座海山的特征点坐标提出纠正，纠正后路易斯.阿加西斯平顶海山的特征点坐标为 17°59.70′N，178°10.90′E，亚历山大·阿加西斯平顶海山的特征点坐标为 17°59.70′N，178°30.40′E。为了尊重原提交者，表格中保留了 ACUF 地名辞典给出的数值，但是在图件制作中采用了我国给出的纠正后的特征点坐标。

　　In this table, the coordinates of Louis Agassiz Guyot and Alexander Agassiz Guyot are the values given by the ACUF gazetteer. China proposed revisions to the coordinates of these two guyots according to their own investigation data during the 24th conference of SCUFN. After revisions, the coordinate of Louis Agassiz Guyot is 17°59.70′N, 178°10.90′E, while that of Alexander Agassiz Guyot is 17°59.70′N, 178°30.40′E. Due to full respect to the original submitter, this table keeps values from ACUF gazetteer. However, we use the coordinates after revisions by China in producing the corresponding figures.

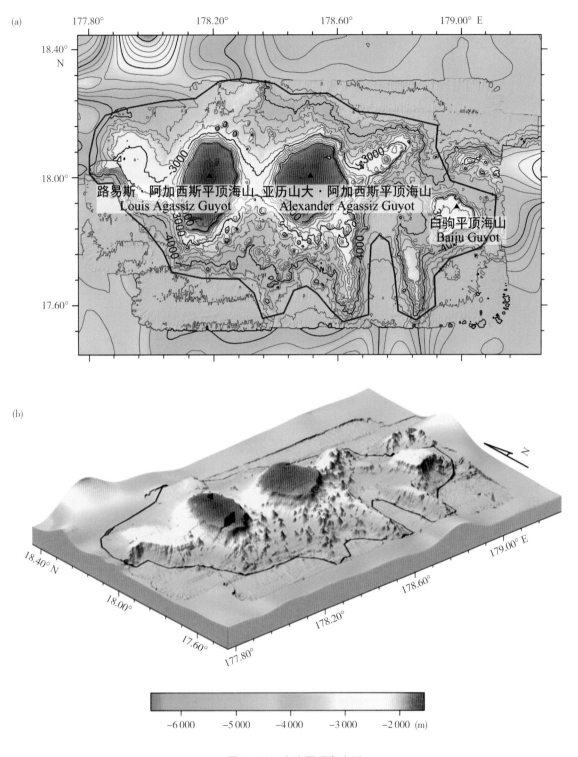

图 2-90　白驹平顶海山群

(a) 地形图（等深线间隔 200 m）；(b) 三维图

Fig.2-90　Baiju Guyots

(a) Bathymetric map (the contour interval is 200 m); (b) 3-D topographic map

2.6.20 路易斯·阿加西斯平顶海山
Louis Agassiz Guyot

中文名称 Chinese Name	路易斯·阿加西斯 平顶海山 Luyisi · Ajiaxisi Pingdinghaishan	英文名称 English Name	Louis Agassiz Guyot	所在大洋 Ocean or Sea	中太平洋 Central Pacific Ocean
发现情况 Discovery Facts	此平顶海山于 1998 年 11 月由中国科考船 "大洋一号" 在执行 DY95-8 航次时调查发现。 This guyot was discovered by the Chinese R/V *Dayang Yihao* during the DY95-8 cruise in November, 1998.				
命名历史 Name History	2011 年收录于 GEBCO 地名辞典，中文译名为路易斯·阿加西斯平顶海山。此平顶海山为白驹平顶海山群的一部分，白驹平顶海山群在我国大洋航次调查报告中曾暂以代号命名。 The name Louis Agassiz Guyot is listed in GEBCO gazetteer. The guyot is a part of Baiju Guyots. A code name was used temporarily in Chinese cruise reports.				
特征点坐标 Coordinates	17°52.00′N，178°12.00′E*			长(km)×宽(km) Length (km) × Width (km)	53 × 27
最大水深（m） Max Depth（m）	3 300	最小水深（m） Min Depth（m）	1 544	高差（m） Total Relief（m）	1 756
地形特征 Feature Description	平顶海山俯视平面形态呈南北向延伸的椭圆形，长宽分别为 53 km 和 27 km。山顶平台水深约 1 700 m，最浅处水深 1 544 m，山麓水深 3 300 m，高差 1 756 m，顶部平坦，边坡陡峭（图 2–91）。 This guyot has an elliptic overlook plane shape, running N to S. The length is 53 km and the width is 27 km. The top platform depth is about 1 700 m with the minimum depth is 1 544 m, while the piedmont depth is 3 300 m, which makes the total relief being 1 756 m. The guyot has a flat top platform and steep slopes (Fig.2–91).				
命名释义 Reason for Choice of Name	以美国科学家 Louis Agassiz (1807—1873) 的名字命名。他出生于瑞士，是一位自然学家、地质学家和教师。在冰川活动和鱼类灭绝等自然科学研究方面做出了革命性贡献。他的创新性教学方法改进了美国自然科学的教育。 This guyot is named after Louis Agassiz (1807–1873), a Swiss-born American naturalist, geologist, and teacher who made revolutionary contributions to the study of natural science with landmark work on glacier activity and extinct fishes. He achieved lasting fame through his innovative teaching methods, which altered the character of natural science education in the United States.				

* 注解见 168 页。
　See the annotation at page 168.

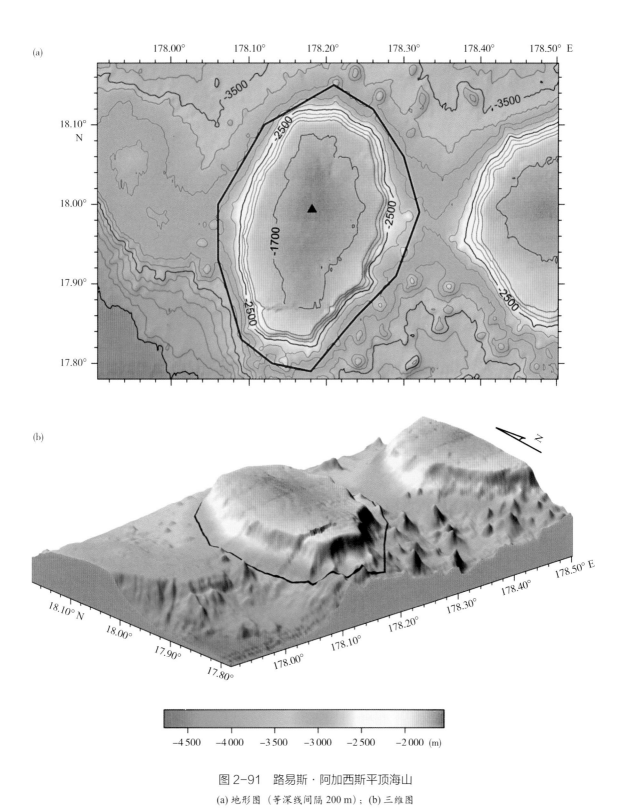

图 2-91　路易斯·阿加西斯平顶海山

(a) 地形图（等深线间隔 200 m）；(b) 三维图

Fig.2-91　Louis Agassiz Guyot

(a) Bathymetric map (the contour interval is 200 m); (b) 3-D topographic map

2.6.21 亚历山大·阿加西斯平顶海山
Alexander Agassiz Guyot

中文名称 Chinese Name	亚历山大·阿加西斯平顶海山 Yalishanda·Ajiaxisi Pingdinghaishan	英文名称 English Name	Alexander Agassiz Guyot	所在大洋 Ocean or Sea	中太平洋 Central Pacific Ocean
发现情况 Discovery Facts	此平顶海山于 1998 年 11 月由中国科考船"大洋一号"在执行 DY95-8 航次时调查发现。 The guyot was discovered by the Chinese R/V *Dayang Yihao* during the DY95-8 cruise in November, 1998.				
命名历史 Name History	2011 年收录于 GEBCO 地名辞典，中文译名为亚历山大·阿加西斯平顶海山。此平顶海山为白驹平顶海山群的一部分，白驹平顶海山群在我国大洋航次调查报告中曾暂以代号命名。 The name Alexander Agassiz Guyot is listed in GEBCO gazetteer. Its translation is Yalishanda·Ajiaxisi Pingdinghaishan in Chinese. The guyot is a part of Baiju Guyots. A code was used to name Baiju Guyots temporarily in Chinese cruise reports.				
特征点坐标 Coordinates	17°54.00′N，178°33.00′E*			长(km)× 宽(km) Length (km) × Width (km)	46 × 40
最大水深（m） Max Depth（m）	3 700	最小水深（m） Min Depth（m）	1 600	高差（m） Total Relief(m)	2 100
地形特征 Feature Description	此平顶海山俯视平面形态近圆形，东侧发育东北和南北向的两条山脊，整体长宽分别为 46 km 和 40 km。山顶平台水深约 1 800 m，最浅处水深 1 600 m，山麓水深 3 700 m，高差 2 100 m，顶部平坦，边坡陡峭（图 2–92）。 This guyot has a nearly round overlook plane shape. There are two ridges developing in the east, one running NE to SW, the other running N to S. The length and width of this guyot are 46 km and 40 km respectively. The top platform depth is about 1 800 m, and the minimum depth is 1 600 m, while the piedmont depth is 3 700 m, which makes the total relief being 2 100 m, the guyot has a flat top platform and steep slopes (Fig.2–92).				
命名释义 Reason for Choice of Name	此平顶海山以 Alexander Agassiz（1835—1910）名字命名，他出生于瑞士，是一位海洋动物学家、海洋学家和采矿工程师。他对系统的动物学研究、海底认知及一个大铜矿发现等都做出了重要贡献。 This guyot is named after Alexander Agassiz (1835–1910), a Swiss-born American marine zoologist, oceanographer, and mining engineer who made important contributions to systematic zoology, to the knowledge of ocean beds, and to the discovery of a major copper mine.				

* 注解见 168 页。
See the annotation at page 168.

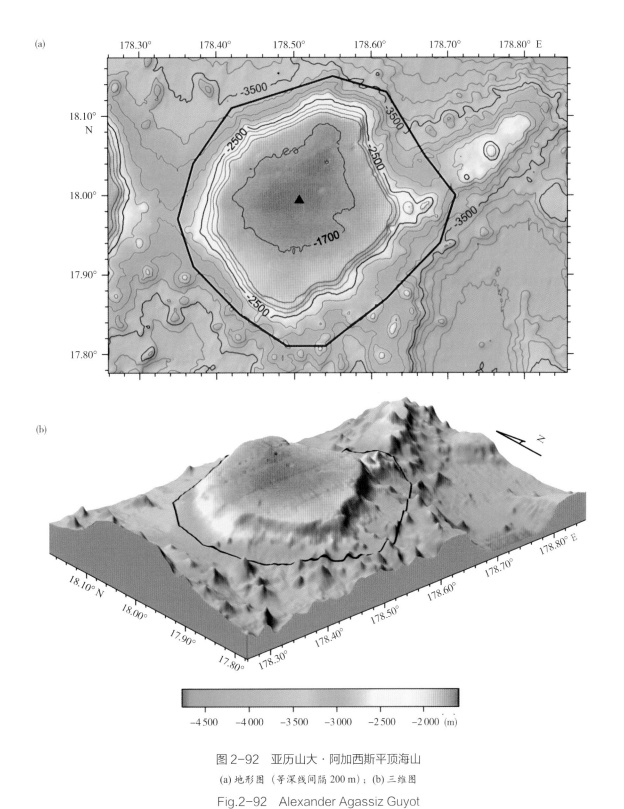

图 2-92　亚历山大·阿加西斯平顶海山

(a) 地形图（等深线间隔 200 m）；(b) 三维图

Fig.2-92　Alexander Agassiz Guyot

(a) Bathymetric map (the contour interval is 200 m); (b) 3-D topographic map

2.6.22 白驹平顶海山
Baiju Guyot

中文名称 Chinese Name	白驹平顶海山 Baiju Pingdinghaishan	英文名称 English Name	Baiju Guyot	所在大洋 Ocean or Sea	中太平洋 Central Pacific Ocean
发现情况 Discovery Facts	此平顶海山于 1998 年 11 月由中国科考船"大洋一号"在执行 DY95-8 航次时调查发现。 This guyot was discovered by the Chinese R/V *Dayang Yihao* during the DY95-8 cruise in November, 1998.				
命名历史 Name History	该平顶海山由我国命名为白驹平顶海山，于 2011 年提交 SCUFN 审议通过。此平顶海山为白驹平顶海山群的一部分，白驹平顶海山群在我国大洋航次调查报告中曾暂以代号命名。 Baiju Guyot is named by China and approved by SCUFN in 2011. Baiju Guyot is a part of Baiju Guyots. A code was used to name Baiju Guyots temporarily in Chinese cruise reports.				
特征点坐标 Coordinates	17°53.90′N，178°58.70′E			长(km) × 宽(km) Length (km) × Width (km)	19 × 17
最大水深（m） Max Depth（m）	3 700	最小水深（m） Min Depth（m）	2 520	高差（m） Total Relief（m）	1 180
地形特征 Feature Description	此平顶海山长宽分别为 19 km 和 17 km。山顶平台水深约 2700 m，最浅处水深 2 520 m，山麓水深 3 700 m，高差 1 180 m，顶部较平，边坡陡峭（图 2–93）。 The length and width of this guyot are 19 km and 17 km respectively. The top platform depth is 2 700 m and the minimum depth is 2 520 m, while the piedmont depth is 3 700 m, which makes the total relief being 1 180 m. The top platform is flat and slopes are steep (Fig.2–93).				
命名释义 Reason for Choice of Name	"白驹"出自《诗经·小雅·白驹》的篇名，白驹即为白色的骏马。该篇描写了送别朋友时依依不舍，主人试图拴马以挽留客人的场景，尽表主人的好客之情。 "Baiju" comes from a poem named *Baiju* in *Shijing · Xiaoya*. *Shijing* is a collection of ancient Chinese poems from 11th century B.C. to 6th century B.C. It means "white horse" in Chinese. This poem described that the host tried to tie down the white horse (Baiju) of the guest. Tying down the horse is for keeping the guest to stay. The poem is outpouring the warm and sincere hospitality of the host. The feature is named for its shape similar to a horse.				

(a)

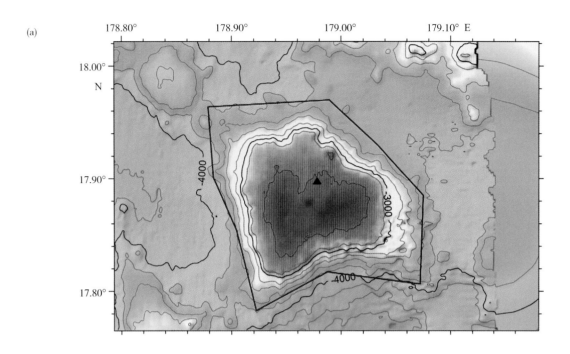

(b)

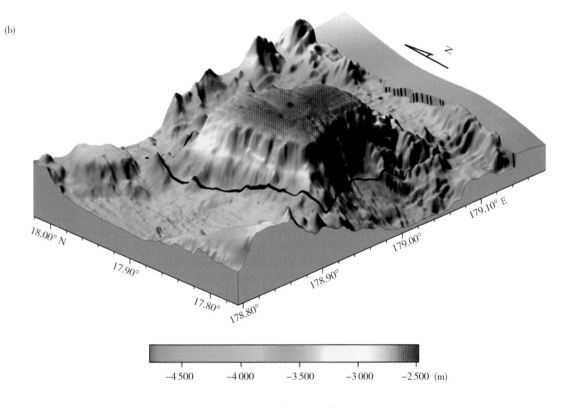

图 2-93　白驹平顶海山

(a) 地形图（等深线间隔 200 m）；(b) 三维图

Fig.2-93　Baiju Guyot

(a) Bathymetric map (the contour interval is 200 m); (b) 3-D topographic map

2.6.23 无羊海山群
Wuyang Seamounts

中文名称 Chinese Name	无羊海山群 Wuyang haishanqun	英文名称 English Name	Wuyang Seamounts	所在大洋 Ocean or Sea	中太平洋 Central Pacific Ocean
发现情况 Discovery Facts	此海山群于 1998 年 11 月由中国科考船"大洋一号"在执行 DY95-8 航次时调查发现。 The seamounts were discovered by the Chinese R/V *Dayang Yihao* during the DY95-8 cruise in November, 1998.				
命名历史 Name History	无羊海山群由 8 个海山组成，在我国大洋航次调查报告中曾暂以代号命名。 Wuyang Seamounts consist of 8 seamounts. A code name was used temporarily in Chinese cruise reports.				
特征点坐标 Coordinates	18°31.00′N，179°36.00′W（艾利森平顶海山 /Allison Guyot） 18°03.52′N，179°22.64′W（麾肱海山 / Huigong Seamount） 18°49.80′N，179°16.80′W（牧来平顶海山 / Mulai Guyot） 18°40.28′N，178°58.00′W（阿池海脊 / Echi Ridge） 18°06.50′N，178°42.50′W（维鱼平顶海山 / Weiyu Guyot） 18°42.60′N，178°42.50′W（蓑笠平顶海山 / Suoli Guyot） 18°19.70′N，178°15.90′W（犉羊海山 / Chunyang Seamount） 18°01.40′N，178°24.00′W（年丰平顶海山 / Nianfeng Guyot）			长 (km)×宽 (km) Length (km)× Width (km)	213×68
最大水深（m） Max Depth（m）	4 750	最小水深（m） Min Depth（m）	1 249	高差（m） Total Relief（m）	3 501
地形特征 Feature Description	无羊海山群长宽分别为 213 km 和 68 km，最浅处水深 1 249 m，山麓水深 4 750 m，高差 3 501 m。群内最大和最高的山峰均为艾利森平顶海山，最浅处 1 249 m（图 2-94）。 The length and width of the Wuyang Seamounts are 213 km and 68 km respectively. The minimum water depth is 1 249 m while the piedmont depth is 4 750 m, which makes the total relief being 3 501 m. The largest and highest guyot among these seamounts is Allison Guyot with a minimum depth of 1 249 m (Fig.2-94).				
命名释义 Reason for Choice of Name	"无羊"出自《诗经·小雅·无羊》的篇名，描绘了一幅牛羊繁盛、生动活泼的放牧狩猎生活画卷。此海山群包含 8 个海山，除艾利森平顶海山外，麾肱、牧来、阿池、维鱼、蓑笠、犉羊和年丰均取词于《诗经·小雅·无羊》，以群组化方法予以命名。 "Wuyang" comes from a poem named *Wuyang* in *Shijing · Xiaoya*. *Shijing* is a collection of ancient Chinese poems from 11th century B.C. to 6th century B.C. It depicts a vivid, lively picture of grazing and hunting life, in which cattle and sheep are prosperous. This seamounts consist of 8 seamounts, except Allison Guyot, all others name, i.e., Huigong, Mulai, Echi, Weiyu, Suoli, Chunyang and Nianfeng, are taken from poem named *Wuyang* in *Shijing · Xiaoya*. They are named using group method.				

(a)

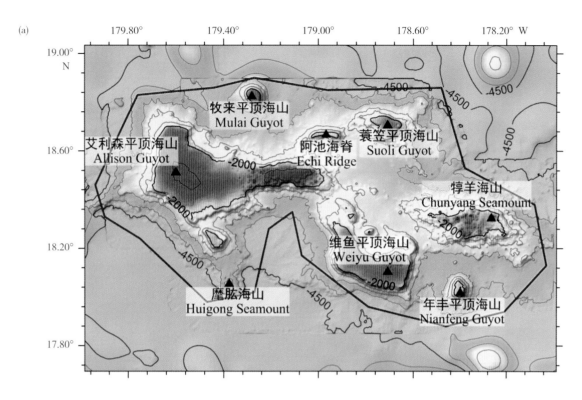

(b)

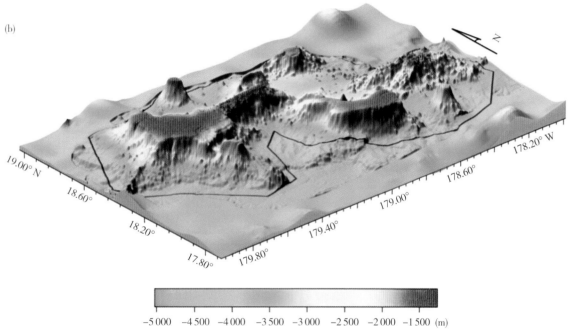

图 2-94　无羊海山群

(a) 地形图（等深线间隔 500 m）；(b) 三维图

Fig.2-94　Wuyang Seamounts

(a) Bathymetric map (the contour interval is 500 m); (b) 3-D topographic map

2.6.24　艾利森平顶海山
Allison Guyot

中文名称 Chinese Name	艾利森平顶海山 Ailisen Pingdinghaishan	英文名称 English Name	Allison Guyot	所在大洋 Ocean or Sea	中太平洋 Central Pacific Ocean
发现情况 Discovery Facts	此平顶海山于 1998 年 11 月由中国科考船"大洋一号"在执行 DY95-8 航次时调查发现。 This guyot was discovered by the Chinese R/V *Dayang Yihao* during the DY95-8 cruise in November, 1998.				
命名历史 Name History	此平顶海山收录于 ACUF 地名辞典。此平顶海山为无羊海山群的一部分，无羊海山群在我国大洋航次调查报告中曾暂以代号命名。 This guyot is listed in ACUF gazetteer. This guyot is a part of Wuyang Seamounts. A code was used to name Wuyang Seamounts temporarily in Chinese cruise reports.				
特征点坐标 Coordinates	18°31.00′N，179°36.00′W			长(km)×宽(km) Length (km) × Width (km)	110×90
最大水深（m） Max Depth（m）	4 500	最小水深（m） Min Depth（m）	1 250	高差（m） Total Relief（m）	3 250
地形特征 Feature Description	此平顶海山俯视平面形态不规则，长宽分别为 110 km 和 90 km。山顶平台水深约 1 400 m，最浅处水深 1 250 m，山麓水深 4 500 m，高差 3 250 m，西侧顶部平坦，东侧顶部起伏，边坡陡峭（图 2-95）。 This guyot has an irregular overlook plane shape. The length and width are 110 km and 90 km respectively. The top platform depth is about 1 400 m and the minimum depth is 1 250 m. The piedmont depth is 4 500 m, which makes the total relief being 3 250 m. The western top platform is flat while the eastern is fluctuant and the slopes are steep (Fig.2–95).				
命名释义 Reason for Choice of Name	来源于 ACUF 地名辞典，释义不详。 The name comes from ACUF gazetteer and the reason for choice of name remained unknown.				

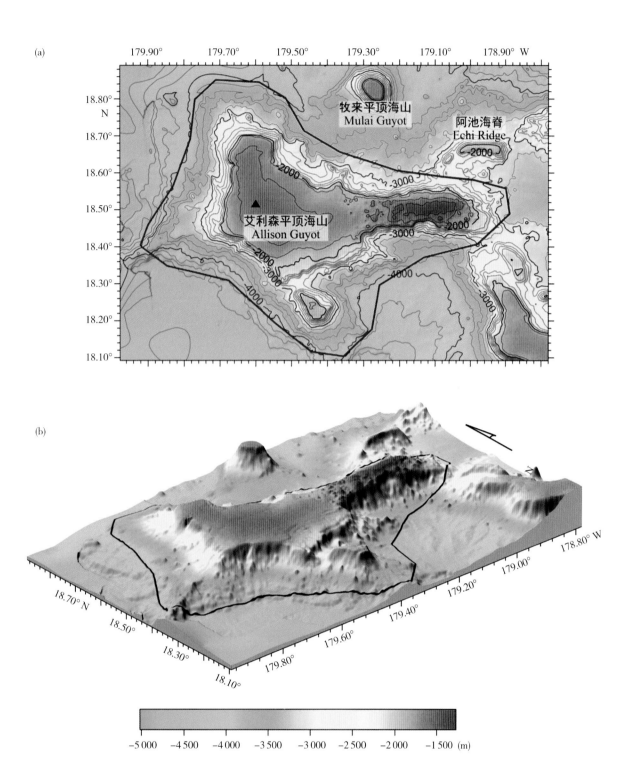

图 2-95　艾利森平顶海山

(a) 地形图（等深线间隔 200 m）；(b) 三维图

Fig.2-95　Allison Guyot

(a) Bathymetric map (the contour interval is 200 m); (b) 3-D topographic map

2.6.25　牧来平顶海山
Mulai Guyot

中文名称 Chinese Name	牧来平顶海山 Mulai Pingdinghaishan	英文名称 English Name	Mulai Guyot	所在大洋 Ocean or Sea	中太平洋 Central Pacific Ocean
发现情况 Discovery Facts	此平顶海山于 1998 年 11 月由中国科考船"大洋一号"在执行 DY95-8 航次时调查发现。 This guyot was discovered by the Chinese R/V *Dayang Yihao* during the DY95-8 cruise in November, 1998.				
命名历史 Name History	由我国命名为牧来平顶海山，于 2014 年提交 SCUFN 审议通过。 This guyot was named Mulai Guyot by China and approved by SCUFN in 2014.				
特征点坐标 Coordinates	18°49.80′N，179°16.80′W			长 (km) × 宽 (km) Length (km) × Width (km)	24 × 21
最大水深（m） Max Depth（m）	3 700	最小水深（m） Min Depth（m）	1 600	高差（m） Total Relief（m）	2 100
地形特征 Feature Description	此平顶海山俯视平面形态呈近圆形，长宽分别为 24 km 和 21 km。山顶平台水深约 1 800 m，最浅处水深 1 600 m，山麓水深 3 700 m，高差 2 100 m，顶部平坦，边坡陡峭（图 2-96）。 This guyot has a nearly round overlook plane shape. The length is 24 km and the width is 21 km. The top platform depth is about 1 800 m and the minimum depth is 1 600 m, while the piedmont depth is 3 700 m, which make the total relief being 2 100 m. The guyot has a flat top platform and steep slopes (Fig.2-96).				
命名释义 Reason for Choice of Name	"牧来"出自《诗经·小雅·无羊》。全诗描绘了一幅牛羊繁盛、生动活泼的放牧狩猎生活画卷。身着蓑衣、头戴斗笠的牧人放牧的牛羊遍布在山丘和池边，牧人轻轻挥鞭，成群的牛羊跃上坡顶。牧人梦见蝗虫都化为鱼，梦见了画着鹰隼的旗帜，太卜占梦之后，认为这个梦预示着来年五谷丰登、人畜兴旺，表现了古人对美好生活的追求与向往。此区域包含"维鱼"等多座海山，分别以"无羊"这首诗中的"蓑笠、犉羊、年丰、维鱼、阿池、麾肱、牧来"等词加以命名，"牧来"意指牧人来野外放牧。 "Mulai" comes from a poem named *Wuyang* in *Shijing · Xiaoya*. *Shijing* is a collection of ancient Chinese poems from 11th century B.C. to 6th century B.C. This poem describes the lively hunting life with prosperous cattle and sheep spreading over the hills and ponds. The shepherd wearing straw rain cape and bamboo hat gently waves his whip and the cattle and sheep jump onto the top of the hill. The shepherd dreams locusts turning into fish and flags with eagles. A seer tells him that the dream forebodes a great harvest the next year. This region has many seamounts such as the Weiyu Guyot. Six words in the poem *Wuyang* are used to name, the other six seamounts or ridges located in this region, namely "Suoli", "Chunyang", "Nianfeng", "Echi", "Huigong" and "Mulai". The dense seamounts in this region are like a flock of cattle and sheep. "Mulai" represents the shepherd grazing in the wild.				

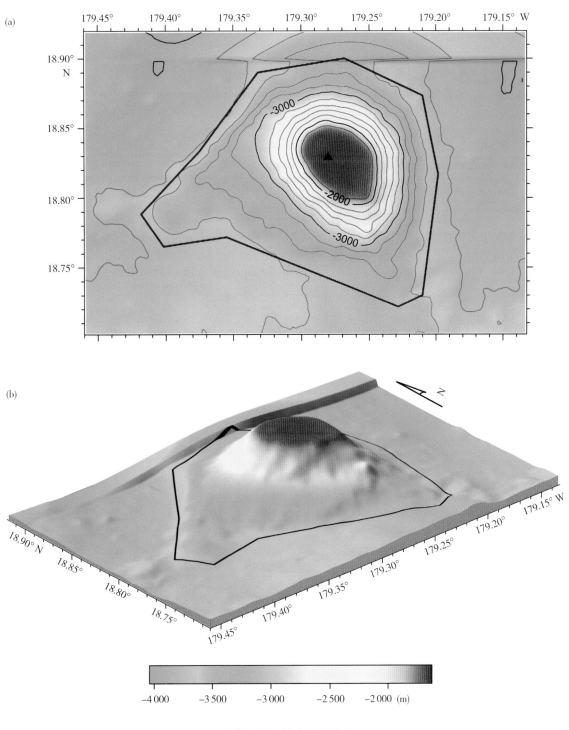

图 2-96　牧来平顶海山

(a) 地形图 （等深线间隔 200 m）；(b) 三维图

Fig.2-96　Mulai Guyot

(a) Bathymetric map (the contour interval is 200 m); (b) 3-D topographic map

2.6.26　阿池海脊
Echi Ridge

中文名称 Chinese Name	阿池海脊 Echi Haiji	英文名称 English Name	Echi Ridge	所在大洋 Ocean or Sea	中太平洋 Central Pacific Ocean
发现情况 Discovery Facts	此海脊于 1998 年 11 月由中国科考船"大洋一号"在执行 DY95-8 航次时调查发现。 This ridge was discovered by the Chinese R/V *Dayang Yihao* during the DY95-8 cruise in November, 1998.				
命名历史 Name History	由我国命名为阿池海脊，于 2015 年提交 SCUFN 审议通过。此海脊为无羊海山群的一部分，无羊海山群在我国大洋航次调查报告中曾暂以代号命名。 This ridge was named Echi Ridge by China and approved by SCUFN in 2015. It is a part of Wuyang Seamounts. A code was used to name Wuyang Seamounts temporarily in Chinese cruise reports.				
特征点坐标 Coordinates	18°40.28′N，178°58.00′W			长(km)×宽(km) Length (km)× Width (km)	22×7
最大水深（m） Max Depth（m）	3 100	最小水深（m） Min Depth（m）	1 860	高差（m） Total Relief（m）	1 240
地形特征 Feature Description	此海脊俯视平面形态呈长长条形，长宽分别为 22 km 和 7 km。山顶平台水深约 2 000 m，最浅处水深 1 860 m，山麓水深 3 100 m，高差 1 240 m，边坡陡峭（图 2–97）。 This ridge has an elongated overlook plane shape. The length is 22 km and the width is 7 km. The top platform depth is 2 000 m and the minimum depth is 1 860 m. The piedmont depth is 3 100 m, which makes the total relief being 1 240 m (Fig.2–97).				
命名释义 Reason for Choice of Name	"阿池"出自《诗经·小雅·无羊》。这首诗描绘了一幅牛羊繁盛、生动活泼的放牧狩猎生活画卷。身着蓑衣、头戴斗笠的牧人放牧的牛羊遍布在山丘和池边，牧人轻轻挥鞭，成群的牛羊跃上坡顶。牧人梦见蝗虫都化为鱼，梦见了画着鹰隼的旗帜，太卜占梦之后，认为这个梦预示着来年五谷丰登、人畜兴旺，表现了古人对美好生活的追求与向往。此区域包含"维鱼"等多座海山，分别以"无羊"这首诗中的"蓑笠、矜羊、年丰、维鱼、阿池、麾肱、牧来"等词加以命名，"阿池"意指牛羊遍布的山坡和池边。 "Echi" comes from a poem named *Wuyang* in *Shijing · Xiaoya*. *Shijing* is a collection of ancient Chinese poems from 11th century B.C. to 6th century B.C. This poem describes the lively hunting life with prosperous cattle and sheep spreading over the hills and ponds. The shepherd wearing straw rain cape and bamboo hat gently waves his whip and the cattle and sheep jump onto the top of the hill. The shepherd dreams locusts turning into fish and flags with eagles. A seer tells him that the dream forebodes a great harvest the next year. This region has many seamounts such as the Weiyu Seamount. Six words in the poem *Wuyang* are used to name, the other six seamounts or ridges located in this region, namely "Suoli", "Chunyang", "Nianfeng", "Echi", "Huigong" and "Mulai". The dense seamounts in this region are like a flock of cattle and sheep. "Echi" represents the flock of cattle and sheep all over the hillside and poolside.				

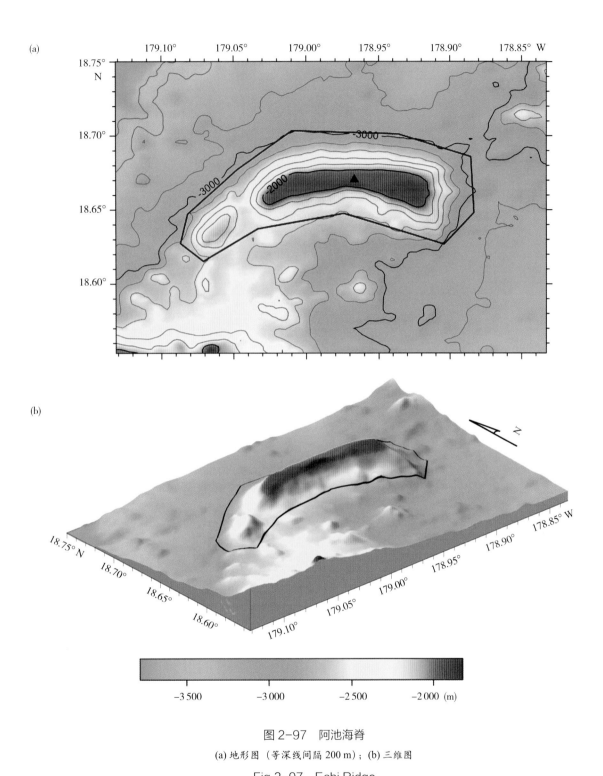

图 2-97　阿池海脊

(a) 地形图（等深线间隔 200 m）；(b) 三维图

Fig.2-97　Echi Ridge

(a) Bathymetric map (the contour interval is 200 m); (b) 3-D topographic map

2.6.27 蓑笠平顶海山
Suoli Guyot

中文名称 Chinese Name	蓑笠平顶海山 Suoli Pingdinghaishan		英文名称 English Name	Suoli Guyot	所在大洋 Ocean or Sea	中太平洋 Central Pacific Ocean
发现情况 Discovery Facts	此平顶海山于 1998 年 11 月由中国科考船"大洋一号"在执行 DY95-8 航次时调查发现。 This guyot was discovered by the Chinese R/V *Dayang Yihao* during the DY95-8 cruise in November, 1998.					
命名历史 Name History	由我国命名为蓑笠平顶海山，于 2014 年提交 SCUFN 审议通过。 This guyot was named Suoli Guyot by China and approved by SCUFN in 2014.					
特征点坐标 Coordinates	18°42.60′N，178°42.50′W				长(km)×宽(km) Length (km) × Width (km)	38×35
最大水深（m） Max Depth（m）	4 400	最小水深（m） Min Depth（m）		1 600	高差（m） Total Relief（m）	2 800
地形特征 Feature Description	该平顶海山长宽分别为 38 km 和 35 km。山顶平台水深约 1 800 m，最浅处水深 1 600 m，山麓水深 4 400 m，高差 2 800 m，顶部平坦，边坡陡峭（图 2–98）。 The length and width of this guyot are 38 km and 35 km respectively. The top platform depth is about 1 800 m. The minimum depth is 1 600 m. The piedmont depth is 4 400 m, which makes the total relief being 2 800 m. The top platform is flat while the slopes are steep (Fig.2–98).					
命名释义 Reason for Choice of Name	"蓑笠"出自《诗经·小雅·无羊》。全诗描绘了一幅牛羊繁盛、生动活泼的放牧狩猎生活画卷。身着蓑衣、头戴斗笠的牧人放牧的牛羊遍布在山丘和池边，牧人轻轻挥鞭，成群的牛羊跃上坡顶。牧人梦见蝗虫都化为鱼，梦见了画着鹰隼的旗帜，太卜占梦之后，认为这个梦预示着来年五谷丰登、人畜兴旺，表现了古人对美好生活的追求与向往。此区域包含"维鱼"等多座海山，分别以"无羊"这首诗中的"蓑笠、犉羊、年丰、维鱼、阿池、麾肱、牧来"等词加以命名，"蓑笠"指牧人雨中放牧时穿着的草棕雨衣和竹编帽子。 "Suoli" comes from a poem named *Wuyang* in *Shijing · Xiaoya*. *Shijing* is a collection of ancient Chinese poems from 11th century B.C. to 6th century B.C. This poem describes the lively hunting life with prosperous cattle and sheep spreading over the hills and ponds. The shepherd wearing straw rain cape and bamboo hat gently waves his whip and the cattle and sheep jump onto the top of the hill. The shepherd dreams locusts turning into fish and flags with eagles. A seer tells him that the dream forebodes a great harvest the next year. This region has many seamounts such as the Weiyu Seamount. Six words in the poem *Wuyang* are used to name, the other six seamounts or ridges located in this region, namely "Suoli", "Chunyang", "Nianfeng", "Echi", "Huigong" and "Mulai". The dense seamounts in this region are like a flock of cattle and sheep. "Suoli" represents the shepherd wears grass raincoat and bamboo hat.					

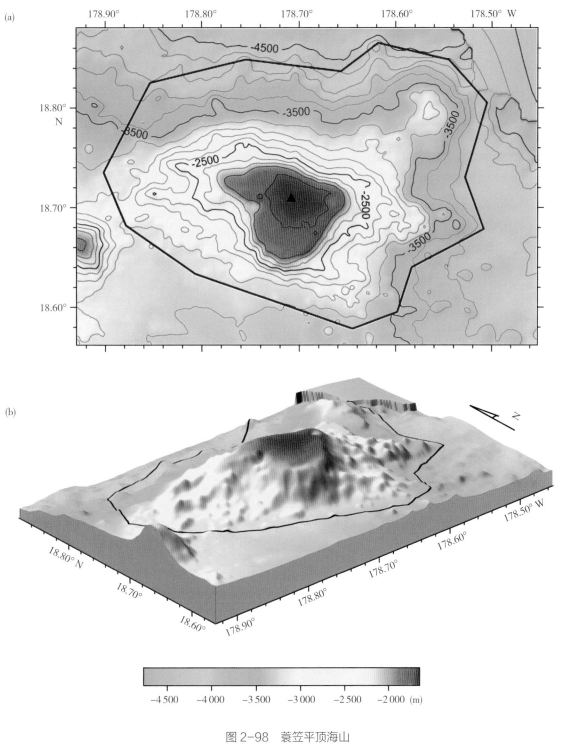

图 2-98　蓑笠平顶海山

(a) 地形图（等深线间隔 200 m）；(b) 三维图

Fig.2-98　Suoli Guyot

(a) Bathymetric map (the contour interval is 200 m); (b) 3-D topographic map

2.6.28 犉羊海山
Chunyang Seamount

中文名称 Chinese Name	犉羊海山 Chunyang Haishan		英文名称 English Name	Chunyang Seamount	所在大洋 Ocean or Sea	中太平洋 Central Pacific Ocean
发现情况 Discovery Facts	此海山于 1998 年 11 月由中国科考船"大洋一号"在执行 DY95-8 航次时调查发现。 This seamount was discovered by the Chinese R/V *Dayang Yihao* during the DY95-8 cruise in November, 1998.					
命名历史 Name History	由我国命名为犉羊海山,于 2014 年提交 SCUFN 审议通过。 This seamount was named Chunyang Seamount by China and approved by SCUFN in 2014.					
特征点坐标 Coordinates	18°19.70′N,178°15.90′W				长(km)×宽(km) Length (km) × Width (km)	67×38
最大水深(m) Max Depth(m)	4 000	最小水深(m) Min Depth(m)	1 600	高差(m) Total Relief(m)		2 400
地形特征 Feature Description	该海山长宽分别为 67 km 和 38 km。山顶平台水深约 1 800 m,最浅处水深 1 600 m,山麓水深 4 000 m,高差 2 400 m,顶部总体较平缓,斜坡地形陡峭(图 2-99)。 The length and width of this seamount are 67 km and 38 km respectively. The top platform depth is about 1 800 m. The minimum depth is 1 600 m. The piedmont depth is 4 000 m, which makes the total relief being 2 400 m. The top platform is flat while the slopes are steep (Fig.2-99).					
命名释义 Reason for Choice of Name	"犉羊"出自《诗经·小雅·无羊》。全诗描绘了一幅牛羊繁盛、生动活泼的放牧狩猎生活画卷。身着蓑衣、头戴斗笠的牧人放牧的牛羊遍布在山丘和池边,牧人轻轻挥鞭,成群的牛羊跃上坡顶。牧人梦见蝗虫都化为鱼,梦见了画着鹰隼的旗帜,太卜占梦之后,认为这个梦预示着来年五谷丰登、人畜兴旺,表现了古人对美好生活的追求与向往。此区域包含"维鱼"等多座海山,分别以"无羊"这首诗中的"蓑笠、犉羊、年丰、维鱼、阿池、麾肱、牧来"等词加以命名,"犉羊"意指大牛和群羊。 "Chunyang" comes from a poem named *Wuyang* in *Shijing · Xiaoya*. *Shijing* is a collection of ancient Chinese poems from 11th century B.C. to 6th century B.C. This poem describes the lively hunting life with prosperous cattle and sheep spreading over the hills and ponds. The shepherd wearing straw rain cape and bamboo hat gently waves his whip and the cattle and sheep jump onto the top of the hill. The shepherd dreams locusts turning into fish and flags with eagles. A seer tells him that the dream forebodes a great harvest the next year. This region has many seamounts such as the Weiyu Seamount. Six words in the poem *Wuyang* are used to name, the other six seamounts or ridges located in this region, namely "Suoli", "Chunyang", "Nianfeng", "Echi", "Huigong" and "Mulai". The dense seamounts in this region are like a flock of cattle and sheep. The term "Chunyang" represents the flock of cattle and sheep.					

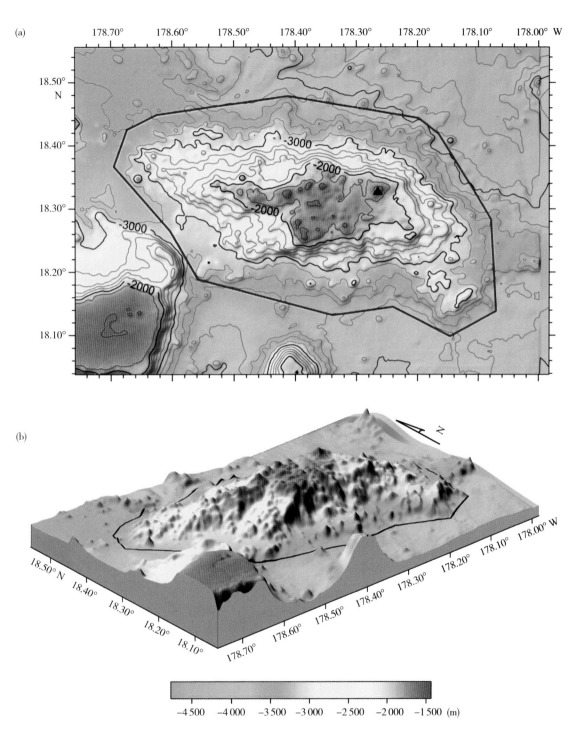

图 2-99　犉羊海山

(a) 地形图（等深线间隔 200 m）；(b) 三维图

Fig.2-99　Chunyang Seamount

(a) Bathymetric map (the contour interval is 200 m); (b) 3-D topographic map

2.6.29　年丰平顶海山
Nianfeng Guyot

中文名称 Chinese Name	年丰平顶海山 Nianfeng Pingdinghaishan	英文名称 English Name	Nianfeng Guyot	所在大洋 Ocean or Sea	中太平洋 Central Pacific Ocean
发现情况 Discovery Facts	此平顶海山于 1998 年 11 月由中国科考船"大洋一号"在执行 DY95-8 航次时调查发现。 This guyot was discovered by the Chinese R/V *Dayang Yihao* during the DY95-8 cruise in November, 1998.				
命名历史 Name History	由我国命名为年丰平顶海山，于 2014 年提交 SCUFN 审议通过。 This guyot was named Nianfeng Guyot by China and approved by SCUFN in 2014.				
特征点坐标 Coordinates	18°01.40′N，178°24.00′W			长(km)×宽(km) Length (km)× Width (km)	18 × 17
最大水深（m） Max Depth（m）	3 800	最小水深（m） Min Depth（m）	1 800	高差（m） Total Relief（m）	2 000
地形特征 Feature Description	该平顶海山俯视平面形态呈近圆形，长宽分别为 18 km 和 17 km。山顶平台水深约 2 000 m，最浅处水深 1 800 m，山麓水深 3 800 m，高差 2 000 m，山顶平坦，边坡陡峭（图 2–100）。 This guyot has a nearly round overlook plane shape. The length is 18 km and the width is 17 km. The top platform depth is about 2 000 m and the minimum depth is 1 800 m, while the piedmont depth is 3 800 m, which makes the total relief being 2 000 m. It has a flat top platform and steep slopes (Fig.2–100).				
命名释义 Reason for Choice of Name	"年丰"出自《诗经·小雅·无羊》。这首诗描绘了一幅牛羊繁盛、生动活泼的放牧狩猎生活画卷。身着蓑衣、头戴斗笠的牧人放牧的牛羊遍布在山丘和池边，牧人轻轻挥鞭，成群的牛羊跃上坡顶。牧人梦见蝗虫都化为鱼，梦见了画着鹰隼的旗帜，太卜占梦之后，认为这个梦预示着来年五谷丰登、人畜兴旺，表现了古人对美好生活的追求与向往。此区域包含"维鱼"等多座海山，分别以"无羊"这首诗中的"蓑笠、犉羊、年丰、维鱼、阿池、麀肱、牧来"等词加以命名，"年丰"意指风调雨顺、五谷丰登的好年景。 "Nianfeng" comes from a poem named *Wuyang* in *Shijing · Xiaoya*. *Shijing* is a collection of ancient Chinese poems from 11th century B.C. to 6th century B.C. This poem describes the lively hunting life with prosperous cattle and sheep spreading over the hills and ponds. The shepherd wearing straw rain cape and bamboo hat gently waves his whip and the cattle and sheep jump onto the top of the hill. The shepherd dreams locusts turning into fish and flags with eagles. A seer tells him that the dream forebodes a great harvest the next year. This region has many seamounts such as the Weiyu Seamount. Six words in the poem *Wuyang* are used to name, the other six seamounts or ridges located in this region, namely "Suoli", "Chunyang", "Nianfeng", "Echi", "Huigong" and "Mulai". The dense seamounts in this region are like a flock of cattle and sheep. "Nianfeng" represents a harvest year with good climate.				

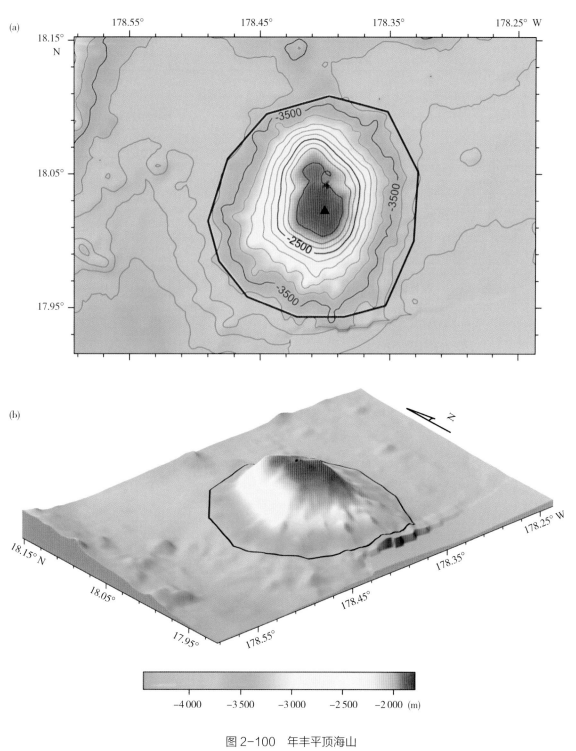

图 2-100 年丰平顶海山

(a) 地形图（等深线间隔 200 m）；(b) 三维图

Fig.2-100 Nianfeng Guyot

(a) Bathymetric map (the contour interval is 200 m); (b) 3-D topographic map

2.6.30 维鱼平顶海山
Weiyu Guyot

中文名称 Chinese Name	维鱼平顶海山 Weiyu Pingdinghaishan	英文名称 English Name	Weiyu Guyot	所在大洋 Ocean or Sea	中太平洋 Central Pacific Ocean
发现情况 Discovery Facts	此平顶海山于 1998 年 11 月由中国科考船 "大洋一号" 在执行 DY95-8 航次时调查发现。 This guyot was discovered by the Chinese R/V *Dayang Yihao* during the DY95-8 cruise in November, 1998.				
命名历史 Name History	由我国命名为维鱼平顶海山，于 2013 年提交 SCUFN 审议通过。 This guyot was named Weiyu Guyot by China and approved by SCUFN in 2013.				
特征点坐标 Coordinates	18°06.50′N，178°42.50′W			长(km)×宽(km) Length (km) × Width (km)	59×27
最大水深（m） Max Depth（m）	4750	最小水深（m） Min Depth（m）	1510	高差（m） Total Relief（m）	3240
地形特征 Feature Description	平顶海山俯视平面形态似鱼，头南尾北，长宽分别为 59 km 和 27 km。山顶平台水深约 1700 m，最浅处水深 1510 m，山麓水深 4750 m，高差 3240 m，顶部平坦，边坡陡峭（图 2–101）。 This guyot has a fish-like overlook plane shape with the head in south and the tail in north. The length is 59 km and the width is 27 km. The top platform depth is about 1700 m and the minimum depth is 1510 m, while the piedmont depth is 4750 m, which makes the total relief being 3240 m (Fig.2–101) the guyot has a flat top platform and steep slopes.				
命名释义 Reason for Choice of Name	"维鱼" 出自《诗经·小雅·无羊》。全诗描绘了一幅牛羊繁盛、生动活泼的放牧狩猎生活画卷。身着蓑衣、头戴斗笠的牧人放牧的牛羊遍布在山丘和池边，牧人轻轻挥鞭，成群的牛羊跃上坡顶。牧人梦见蝗虫都化为鱼，梦见了画着鹰隼的旗帜，太卜占梦之后，认为这个梦预示着来年五谷丰登、人畜兴旺，表现了古人对美好生活的追求与向往。此区域包含 "维鱼" 等多座海山，分别以 "无羊" 这首诗中的 "蓑笠、牸羊、年丰、维鱼、阿池、麾肱、牧来" 等词加以命名，"维鱼" 意指牧人梦见蝗虫都化为鱼。 "Weiyu" comes from a poem named *Wuyang* in *Shijing · Xiaoya*. *Shijing* is a collection of ancient Chinese poems from 11th century B.C. to 6th century B.C. This poem describes the lively hunting life with prosperous cattle and sheep spreading over the hills and ponds. The shepherd wearing straw rain cape and bamboo hat gently waves his whip and the cattle and sheep jump onto the top of the hill. The shepherd dreams locusts turning into fish and flags with eagles. He asked the fortune-teller about his dream and was told his dream auguring for a good harvest and the born of new babies. It representing ancient Chinese people's understanding of dreams and their expectation for a better life. This region has many seamounts, and Wuyang was used to name, the other six seamounts or ridges located in this region, namely "Suoli", "Chunyang", "Nianfeng", "Echi", "Huigong" and "Mulai". The dense seamounts in this region are like a flock of cattle and sheep. "Weiyu" means the shepherd dreamed that the locusts turned into fish.				

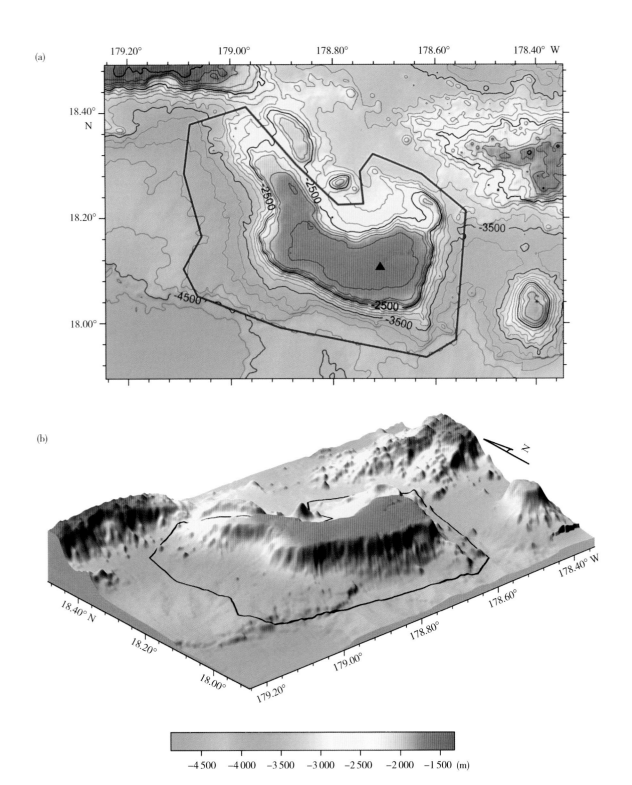

图 2-101　维鱼平顶海山

(a) 地形图 (等深线间隔 200 m)；(b) 三维图

Fig.2-101　Weiyu Guyot

(a) Bathymetric map (the contour interval is 200 m); (b) 3-D topographic map

2.6.31 麾肱海山
Huigong Seamount

中文名称 Chinese Name	麾肱海山 Huigong Haishan	英文名称 English Name	Huigong Seamount	所在大洋 Ocean or Sea	中太平洋 Central Pacific Ocean
发现情况 Discovery Facts					
命名历史 Name History					
特征点坐标 Coordinates	18°03.52′N，179°22.64′W			长(km)×宽(km) Length (km)× Width (km)	14×15
最大水深（m） Max Depth（m）	4 750	最小水深（m） Min Depth（m）	3 240	高差（m） Total Relief(m)	1 510

发现情况 / Discovery Facts: 此海山于 1998 年 11 月由中国科考船"大洋一号"在执行 DY95-8 航次时调查发现。

This seamount was discovered by the Chinese R/V *Dayang Yihao* during the DY95-8 cruise in November, 1998.

命名历史 / Name History: 由我国命名为麾肱海山，于 2015 年提交 SCUFN 审议通过。此海山为无羊海山群的一部分，无羊海山群在我国大洋航次调查报告中曾暂以代号命名。

This seamount was named Huigong Seamount by China and approved by SCUFN in 2015. This seamount is a part of Wuyang Seamounts. A code was used to name Wuyang Seamounts temporarily in Chinese cruise reports.

地形特征 / Feature Description: 此海山大致呈圆锥状，顶部发育 3 个小型山峰。长宽分别为 14 km 和 15 km。最浅处水深 3 240 m，山麓水深 4 750 m，高差 1 510 m（图 2–102）。

This seamount has a nearly conical shape and develops 3 small summits on the top. The length and width are 14 km and 15 km respectively. The minimum water depth is 3 240 m. The piedmont depth is 4 750 m, which makes the total relief being 1 510 m (Fig.2–102).

命名释义 / Reason for Choice of Name: "麾肱"出自《诗经·小雅·无羊》。全诗描绘了一幅牛羊繁盛、生动活泼的放牧狩猎生活画卷。身着蓑衣、头戴斗笠的牧人放牧的牛羊遍布在山丘和池边，牧人轻轻挥鞭，成群的牛羊跃上坡顶。牧人梦见蝗虫都化为鱼，梦见了画着鹰隼的旗帜，太卜占梦之后，认为这个梦预示着来年五谷丰登、人畜兴旺，表现了古人对美好生活的追求与向往。此区域包含"维鱼"等多座海山，分别以"无羊"这首诗中的"蓑笠、犉羊、年丰、维鱼、阿池、麾肱、牧来"等词加以命名，"麾肱"意指牧人挥臂甩鞭，驱赶牛羊。

"Huigong" comes from a poem named *Wuyang* in *Shijing · Xiaoya*. *Shijing* is a collection of ancient Chinese poems from 11th century B.C. to 6th century B.C. This poem describes the lively hunting life with prosperous cattle and sheep spreading over the hills and ponds. The shepherd wearing straw rain cape and bamboo hat gently waves his whip and the cattle and sheep jump onto the top of the hill. The shepherd dreams locusts turning into fish and flags with eagles. A seer tells him that the dream forebodes a great harvest the next year. This region has many seamounts such as the Weiyu Seamount. Six words in the poem *Wuyang* are used to name, the other six seamounts or ridges located in this region, namely "Suoli", "Chunyang", "Nianfeng", "Echi", "Huigong" and "Mulai". The dense seamounts in this region are like a flock of cattle and sheep. "Huigong" represents the shepherd waving his whip to drive flocks and herds.

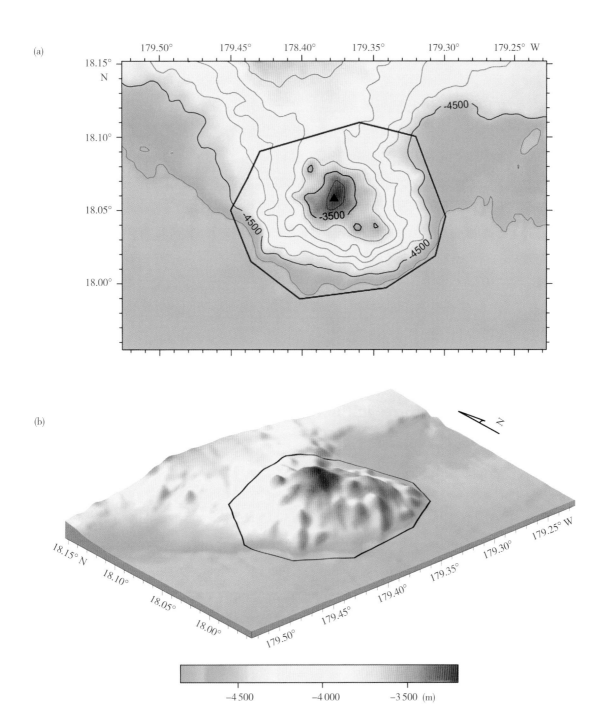

图 2-102　麈肱海山

(a) 地形图（等深线间隔 200 m）；(b) 三维图

Fig.2-102　Huigong Seamount

(a) Bathymetric map (the contour interval is 200 m); (b) 3-D topographic map

第 7 节　莱恩海岭区地理实体

　　莱恩海岭（又称莱恩海山链）位于中太平洋海山区西南，中太平洋海盆以东，由数十个海山构成。海岭大致上呈 NW—SE 向延伸，长度超过 1 500 km。此海岭的构成以小型海山为主，山顶呈尖峰状或发育小型平台。山顶水深 1 500 ～ 2 200 m，山麓水深 4 500 ～ 5 200 m。在形态上，一般由 3 个以上的海山相对集中构成海山群。

　　莱恩海岭由火山活动形成，火山活动早期主要受到北西向断裂活动控制，后期可能经受近东西断裂构造的干扰，经历多期海山活动影响。海山以尖峰状地貌为主，或山顶平台规模较小，推测是海山未曾出露海平面，或出露时间较短。此海岭少数海山仍出露海面，即为莱恩群岛。

　　在此海域我国共命名地理实体 18 个。其中海山群 3 个，包括彤弓海山群、谷陵海山群和柔木海山群；海山 15 个，包括北彤弓海山、彤弓海山、东彤弓海山、南彤弓海山、西谷海山、西陵海山、东陵海山、东谷海山、西水杉海山、水杉海山、南水杉海山、西柔木海山、柔木海山、北柔木海山和银杉海山（图 2-103）。

Section 7　Undersea Features in the Line Islands Chain

The Line Islands Chain is located in the southwest of the Central Pacific Seamount Area, east to the Central Pacific Basin and consists of dozens of seamounts. The ridge roughly runs NW to SE with the length over 1 500 km. This ridge is mainly made up of small seamounts, whose summit is spike-like or develops small platforms. The top platform depth is 1 500–2 200 m, while the piedmont depth is 4 500–5 200 m. Generally, more than 3 relatively gathering seamounts constitute seamount groups.

The formation of Line Islands Chain is caused by volcanic activities. The early volcanic activities were controlled mainly by NW to SE running fault activities, then may be interfered by nearly EW extending fault structure and have experienced several times of seamount activities. The geomorphology of seamounts is mainly spike-like, or the top platform scale is small. It is presumed that the seamounts had never reached out of the sea surface, or the exposure lasted a short time. A few seamounts of this ridge are still out of the sea surface, namely Line Islands.

In total, 18 undersea features have been named by China in the region of the Line Islands Chain. Among them, there are 3 seamounts, including Tonggong Seamounts, Guling Seamounts and Roumu Seamounts; 15 seamounts, including Beitonggong Seamount, Tonggong Seamount, Dongtonggong Seamount, Nantonggong Seamount, Xigu Seamount, Xiling Seamount, Dongling Seamount, Donggu Seamount, Xishuishan Seamount, Shuishan Seamount, Nanshuishan Seamount, Xiroumu Seamount, Roumu Seamount, Beiroumu Seamount and Yinshan Seamount (Fig. 2–103).

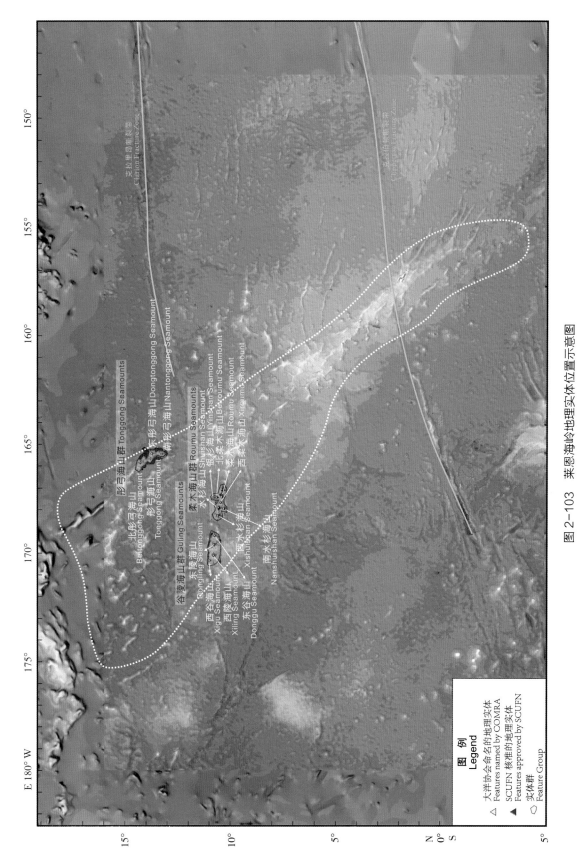

图 2-103　莱恩海岭地理实体位置示意图

Fig.2-103　Locations of the undersea features in the Line Islands Chain

2.7.1 彤弓海山群
Tonggong Seamounts

中文名称 Chinese Name	彤弓海山群 Tonggong Haishanqun	英文名称 English Name	Tonggong Seamounts	所在大洋 Ocean or Sea	中太平洋 Central Pacific Ocean
发现情况 Discovery Facts	此海山群于 2002 年 6—7 月由中国科考船 "海洋四号" 在执行 DY105-13 航次时调查发现。 The seamounts were discovered by the Chinese R/V *Haiyang Sihao* during the DY105-13 cruise from June to July, 2002.				
命名历史 Name History	由我国命名为彤弓海山群，于 2011 年提交 SCUFN 审议通过。该海山群由 6 个海山组成，其中 4 个海山规模较大，在我国大洋航次调查报告中曾暂以代号命名。 The name Tonggong was named by China and approved by SCUFN in 2011. The seamounts consist of 6 seamounts, the four of them are relatively large. A code name was used temporarily in Chinese cruise reports.				
特征点坐标 Coordinates	13°46.60′N，165°40.02′W（彤弓海山 / Tonggong Seamount） 14°13.80′N，165°51.60′W（北彤弓海山 / Beitonggong Seamount） 13°42.60′N，165°27.00′W（东彤弓海山 / Dongtonggong Seamount） 13°16.20′N，165°28.02′W（南彤弓海山 / Nantonggong Seamount）			长(km)×宽(km) Length (km)× Width (km)	180×63
最大水深（m） Max Depth（m）	5 410	最小水深（m） Min Depth（m）	1 290	高差（m） Total Relief（m）	4 120
地形特征 Feature Description	此海山群俯视平面形态狭长，近南北走向，长宽分别 180 km 和 63 km，其中最高峰位于中部的彤弓海山，峰顶水深为 1 290 m（图 2–104）。 The seamounts have an elongated overlook plane shape, running nearly N to S. The length is 180 km and the width is 63 km. The highest summit is Tonggong Seamount, which is located in the middle and the top depth is 1 290 m (Fig.2–104).				
命名释义 Reason for Choice of Name	"彤弓" 出自《诗经·小雅·彤弓》，彤弓指漆成红色的弓。此海山群平面形态似弓，故命名为彤弓海山群。海山群中的 4 个海山以群组化方式命名，分别为彤弓海山、北彤弓海山、东彤弓海山和南彤弓海山。 "Tonggong" comes from a poem named *Tonggong* in *Shijing · Xiaoya*. *Shijing* is a collection of ancient Chinese poems from 11th century B.C. to 6th century B.C. Tonggong refers to the bow with red paint. The poem reflects the ritual system from Zhou Dynasty, the Emperor usually rewarded meritorious subordinates with bows. Tonggong used to name the seamounts since the form of the seamounts look like a bow, and the four seamounts are named using group method as Tonggong Seamount, Beitonggong Seamount, Dongtonggong Seamount and Nantonggong Seamount respectively.				

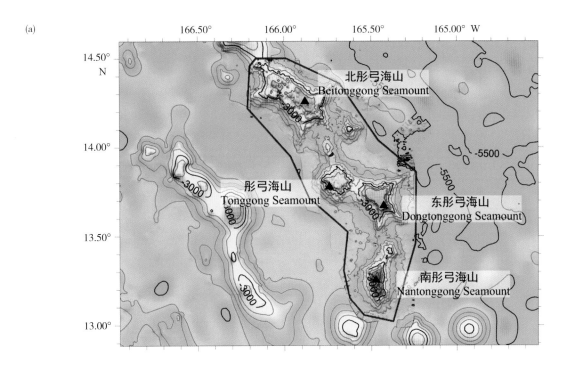

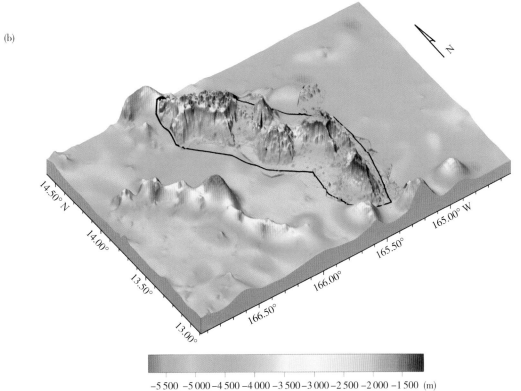

图 2-104　彤弓海山群

(a) 地形图（等深线间隔 500 m）；(b) 三维图

Fig.2-104　Tonggong Seamounts

(a) Bathymetric map (the contour interval is 500 m); (b) 3-D topographic map

2.7.2 北彤弓海山
Beitonggong Seamount

中文名称 Chinese Name	北彤弓海山 Beitonggong Haishan	英文名称 English Name	Beitonggong Seamount	所在大洋 Ocean or Sea	中太平洋 Central Pacific Ocean
发现情况 Discovery Facts	此海山于 2002 年 6—7 月由中国科考船"海洋四号"在执行 DY105-13 航次时调查发现。 This seamount was discovered by the Chinese R/V *Haiyang Sihao* during the DY105-13 cruise from June to July, 2002.				
命名历史 Name History	该海山是彤弓海山群的一部分，彤弓海山群在我国大洋航次调查报告中曾暂以代号命名。 This seamount is a part of Tonggong Seamounts. A code was used to name Tonggong Seamounts temporarily in Chinese cruise reports.				
特征点坐标 Coordinates	14°13.80′N，165°51.60′W			长(km)×宽(km) Length (km)×Width (km)	60×30
最大水深（m） Max Depth（m）	5 450	最小水深（m） Min Depth（m）	1 469	高差（m） Total Relief（m）	3 981
地形特征 Feature Description	北彤弓海山位于彤弓海山群北部。海山 SE—NW 走向，长宽分别 60 km 和 30 km，峰顶水深 1 469 m，山麓水深 5 450 m，高差 3 981 m。海山顶部地形相对平坦，山坡陡峭（图 2-105）。 Beitonggong Seamount is located in the north of Tonggong Seamounts and runs NW to SE. The length is 60 km and the width is 30 km. The top depth is 1 469 m and the piedmont depth is 5 450 m, which makes the total relief being 3 981 m. It has a relatively flat top and steep slopes (Fig.2-105).				
命名释义 Reason for Choice of Name	该海山位于彤弓海山群以北，故以此命名。 Bei means north in Chinese. This seamount is named Beitonggong since it is located in the north of Tonggong Seamounts.				

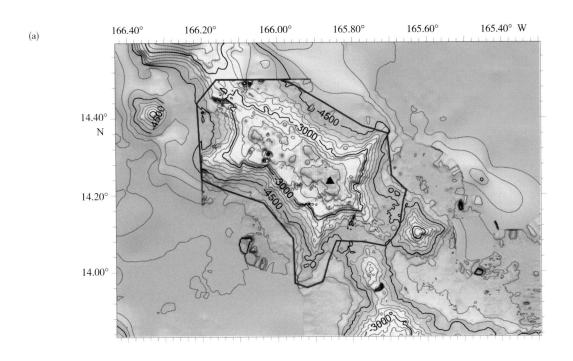

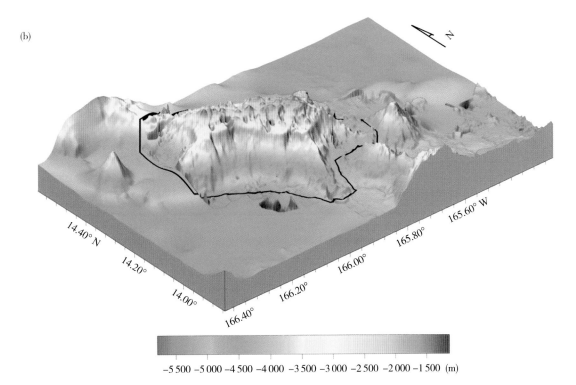

图 2-105　北彤弓海山

(a) 地形图（等深线间隔 300 m）；(b) 三维图

Fig.2-105　Beitonggong Seamount

(a) Bathymetric map (the contour interval is 300 m); (b) 3-D topographic map

2.7.3 彤弓海山
Tonggong Seamount

中文名称 Chinese Name	彤弓海山 Tonggong Haishan	英文名称 English Name	Tonggong Seamount	所在大洋 Ocean or Sea	中太平洋 Central Pacific Ocean
发现情况 Discovery Facts	此海山于 2002 年 6—7 月由中国科考船"海洋四号"在执行 DY105-13 航次时调查发现。 This seamount was discovered by the Chinese R/V *Haiyang Sihao* during the DY105-13 cruise from June to July, 2002.				
命名历史 Name History	该海山是彤弓海山群的一部分，彤弓海山群在我国大洋航次调查报告中曾暂以代号命名。 This seamount is a part of Tonggong Seamounts. A code was used to name Tonggong Seamounts temporarily in Chinese cruise reports.				
特征点坐标 Coordinates	13°46.60′N，165°40.02′W			长(km)×宽(km) Length (km)× Width (km)	33×33
最大水深（m） Max Depth（m）	5 280	最小水深（m） Min Depth（m）	1 980	高差（m） Total Relief（m）	3 300
地形特征 Feature Description	彤弓海山位于彤弓海山群中部，呈圆锥状，基座直径 33 km，最高峰峰顶水深 1 980 m，山麓水深 5 280 m，高差 3 300 m。海山顶部相对平坦，山坡陡峭（图 2–106）。 Tonggong Seamount is located in the middle of Tonggong Seamounts. It has a conical shape with a base diameter of 33 km. The top depth of the highest summit is 1 980 m and the piedmont depth is 5 280 m, which makes the total relief being 3 300 m. It has a relatively flat top and steep slopes (Fig.2–106).				
命名释义 Reason for Choice of Name	"彤弓"出自《诗经·小雅·彤弓》"彤弓弨兮，受言藏之"，彤弓指漆成红色的弓。周朝时天子以红色的弓赐予有功之臣，是一种重要的礼仪。此句意为天子赏赐长弓给有功之臣，臣子谨慎接受收藏，满心欢喜。 "Tonggong" comes from a poem named *Tonggong* in *Shijing · Xiaoya*. *Shijing* is a collection of ancient Chinese poems from 11th century B.C. to 6th century B.C. "The red gift-bows unbent were bestowed and stored." Tonggong refers to the bow with red paint. The poem reflects the ritual system from Zhou Dynasty, the emperor usually rewarded meritorious subordinates with bows. Tonggong used to name the seamounts since the shape of the seamounts look like a bow.				

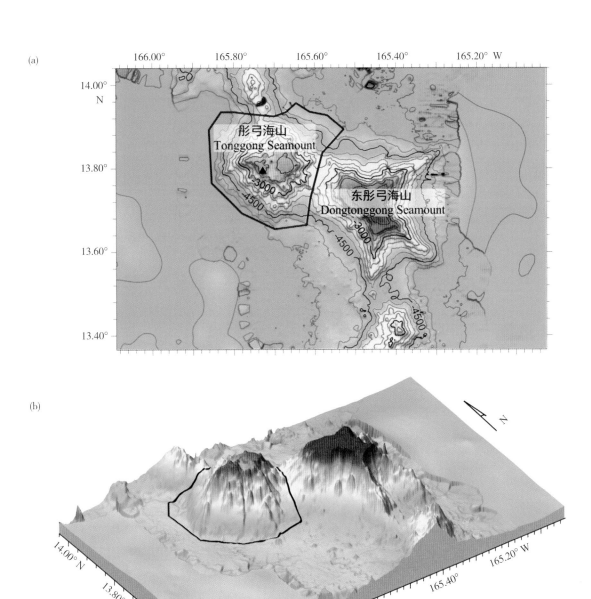

图 2-106　彤弓海山

(a) 地形图（等深线间隔 300 m）；(b) 三维图

Fig.2-106　Tonggong Seamount

(a) Bathymetric map (the contour interval is 300 m); (b) 3-D topographic map

2.7.4 东彤弓海山
Dongtonggong Seamount

中文名称 Chinese Name	东彤弓海山 Dongtonggong Haishan	英文名称 English Name	Dongtonggong Seamount	所在大洋 Ocean or Sea	中太平洋 Central Pacific Ocean
发现情况 Discovery Facts	此海山于2002年6—7月由中国科考船"海洋四号"在执行DY105-13航次时调查发现。 This seamount was discovered by the Chinese R/V *Haiyang Sihao* during the DY105-13 cruise from June to July, 2002.				
命名历史 Name History	该海山是彤弓海山群的一部分，彤弓海山群在我国大洋航次调查报告中曾暂以代号命名。 This seamount is a part of Tonggong Seamounts. A code was used to name Tonggong Seamounts temporarily in Chinese cruise reports.				
特征点坐标 Coordinates	13°42.60′N，165°27.00′W			长(km)×宽(km) Length (km)× Width (km)	50×50
最大水深（m） Max Depth（m）	5 080	最小水深（m） Min Depth（m）	1 563	高差（m） Total Relief(m)	3 517
地形特征 Feature Description	东彤弓海山位于彤弓海山群东部，海山发育北东、北西、南东和南西4条山脊，基座直径50 km。海山最高峰位于海山的中部，峰顶水深1 563 m，山麓水深5 080 m，高差3 517 m。海山顶部地形相对平坦，山坡陡峭（图2–107）。 Dongtonggong Seamount is located in the east of Tonggong Seamounts and develops four ridges, running to NE, NW, SE and SW, respectively. The base diameter is 50 km. The highest summit is in the middle of this seamount. The top depth is 1 563 m and the piedmont depth is 5 080 m, which makes the total relief being 3 517 m. It has a relatively flat top and steep slopes (Fig.2–107).				
命名释义 Reason for Choice of Name	该海山位于彤弓海山群东部，故以此命名。 Dong means east in Chinese. This seamount is named Dongtonggong since it is located in the east of Tonggong Seamounts.				

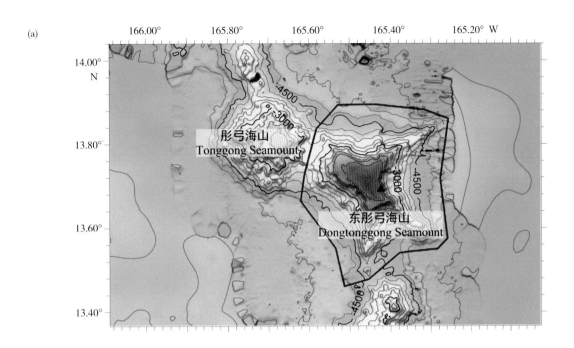

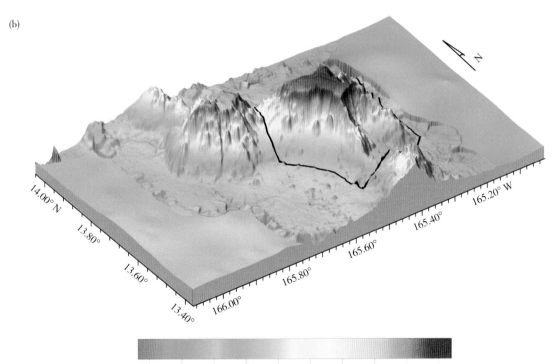

图 2-107　东彤弓海山

(a) 地形图 （等深线间隔 300 m）；(b) 三维图

Fig.2-107　Dongtonggong Seamount

(a) Bathymetric map (the contour interval is 300 m); (b) 3-D topographic map

2.7.5 南彤弓海山
Nantonggong Seamount

中文名称 Chinese Name	南彤弓海山 Nantonggong Haishan	英文名称 English Name	Nantonggong Seamount	所在大洋 Ocean or Sea	中太平洋 Central Pacific Ocean
发现情况 Discovery Facts	此海山于2002年6—7月由中国科考船"海洋四号"在执行DY105-13航次时调查发现。 This seamount was discovered by the Chinese R/V *Haiyang Sihao* during the DY105-13 cruise from June to July, 2002.				
命名历史 Name History	该海山是彤弓海山群的一部分，彤弓海山群在我国大洋航次调查报告中曾暂以代号命名。 This seamount is a part of Tonggong Seamounts. A code was used to name Tonggong Seamounts temporarily in Chinese cruise reports.				
特征点坐标 Coordinates	13°16.20′N，165°28.02′W		长(km)×宽(km) Length (km)× Width (km)		50×30
最大水深（m） Max Depth（m）	5 400	最小水深（m） Min Depth（m）	1 290	高差（m） Total Relief（m）	4 110
地形特征 Feature Description	南彤弓海山位于彤弓海山群南部，东彤弓海山以南。海山SSW—NNE走向，长宽分别为50 km和30 km。海山最高峰位于海山中部，峰顶水深1 290 m，山麓水深5 400 m，高差4 110 m（图2–108）。 Nantonggong Seamount is located in the south of Tonggong Seamounts, south to Dongtonggong Seamount, running NNE to SSW. The length is 50 km and the width is 30 km. The highest summit is in the middle of this seamount. The top depth is 1 290 m and the piedmont depth is 5 400 m, which makes the total relief being 4 110 m (Fig.2–108).				
命名释义 Reason for Choice of Name	该海山位于彤弓海山群南部，故以此命名。 Nan means south in Chinese. This seamount is named Nantonggong since it is located in the south of Tonggong Seamounts.				

(a)

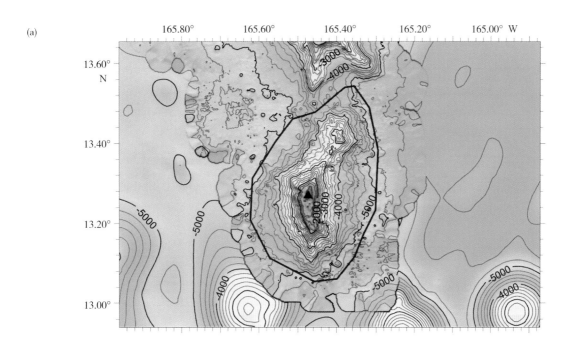

(b)

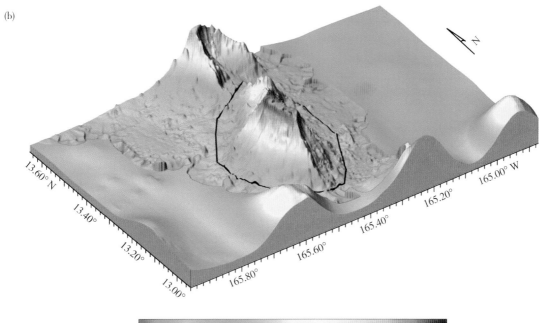

图 2-108　南彤弓海山

(a) 地形图（等深线间隔 200 m）；(b) 三维图

Fig.2-108　Nantonggong Seamount

(a) Bathymetric map (the contour interval is 200 m); (b) 3-D topographic map

2.7.6 谷陵海山群
Guling Seamounts

中文名称 Chinese Name	谷陵海山群 Guling Haishanqun		英文名称 English Name	Guling Seamounts	所在大洋 Ocean or Sea	中太平洋 Central Pacific Ocean
发现情况 Discovery Facts	此海山群于 2002 年 6—7 月由中国科考船 "海洋四号" 在执行 DY105-13 航次时调查发现，位于中太平洋莱恩海岭。 The seamounts were discovered by the Chinese R/V *Haiyang Sihao* during the DY105-13 cruise from June to July, 2002. It is located in Line Islands Chain of Central Pacific Ocean.					
命名历史 Name History	由我国命名为谷陵海山群，于 2013 年提交 SCUFN 审议通过。 The seamounts were named Guling Seamounts by China and approved by SCUFN in 2013.					
特征点坐标 Coordinates	10°57.10′N，170°22.20′W（西谷海山 / Xigu Seamount） 10°47.90′N，170°07.00′W（西陵海山 / Xiling Seamount） 10°44.80′N，169°36.50′W（东陵海山 / Dongling Seamount） 10°44.60′N，169°12.70′W（东谷海山 / Donggu Seamount）				长（km）× 宽（km） Length (km) × Width (km)	154 × 70
最大水深（m） Max Depth（m）	4 976	最小水深（m） Min Depth（m）	1 627		高差（m） Total Relief（m）	3 349
地形特征 Feature Description	该海山群主要由四座东西向排列的海山组成，长宽分别 154 km 和 70 km，其中最高峰位于东陵海山，峰顶水深 1 627 m（图 2–109）。 Guling Seamounts mainly consist of four seamounts arranged in E to W direction. The length is 154 km and the width is 70 km. The highest summit is in Dongling seamount and the depth of the top is 1 627 m (Fig.2–109).					
命名释义 Reason for Choice of Name	"谷陵" 出自《诗经·小雅·十月之交》"高岸为谷，深谷为陵"，指强烈地震使得高山变成深谷，深谷变成大山。该篇所述的日食、月食和强烈地震在古代视为不祥之兆。此海山群包含 4 个海山，均以其相对方位命名，分别为西谷、西陵、东谷、东陵海山。 "Guling" comes from a poem named *Shiyuezhijiao* in *Shijing · Xiaoya*. *Shijing* is a collection of ancient Chinese poems from 11th century B.C. "The hills subside into vales; The vales rise into hills." It means the violent earthquake. "Guling" is combined by two words "Gu" and "Ling" in Chinese. "Gu" means valleys and "Ling" means hills. In the poem, eclipse and earthquakes are disasters omens. Guling Seamounts consist of 4 seamounts which are all named according to their directions, namely Xigu, Xiling, Donggu and Dongling.					

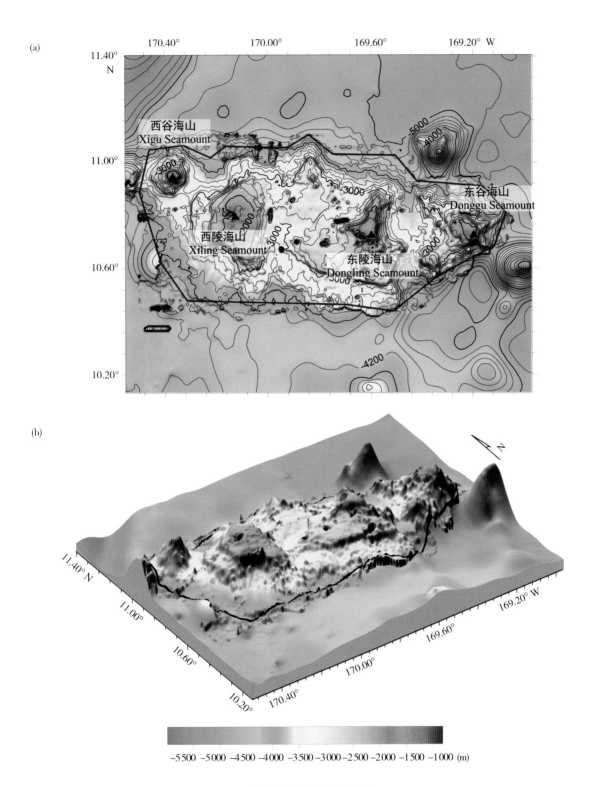

图 2-109　谷陵海山群

(a) 地形图（等深线间隔 200 m）；(b) 三维图

Fig.2-109　Guling Seamounts

(a) Bathymetric map (the contour interval is 200 m); (b) 3-D topographic map

2.7.7 西谷海山
Xigu Seamount

中文名称 Chinese Name	西谷海山 Xigu Haishan	英文名称 English Name	Xigu Seamount	所在大洋 Ocean or Sea	中太平洋 Central Pacific Ocean
发现情况 Discovery Facts	此海山于 2002 年 6—7 月由中国科考船"海洋四号"在执行 DY105-13 航次时调查发现。 This seamount was discovered by the Chinese R/V *Haiyang Sihao* during the DY105-13 cruise from June to July, 2002.				
命名历史 Name History	该海山是谷陵海山群的一部分,谷陵海山群在我国大洋航次调查报告中曾暂以代号命名。 This seamount is a part of Guling Seamounts. A code was used to name Guling Seamounts temporarily in Chinese cruise reports.				
特征点坐标 Coordinates	10°57.10′N,170°22.20′W			长(km)×宽(km) Length (km)× Width (km)	25×25
最大水深(m) Max Depth(m)	4 430	最小水深(m) Min Depth(m)	1 791	高差(m) Total Relief(m)	2 639
地形特征 Feature Description	西谷海山位于谷陵海山群的西部,基座直径 25 km。海山峰顶水深为 1 791 m,山麓水深 4 430 m,高差 2 639 m,与西陵海山以鞍部分隔(图 2–110)。 Xigu Seamount is located in the west of Guling Seamounts and the base diameter is 25 km. The top depth is 1 791 m and the piedmont depth is 4 430 m, which makes the total relief being 2 639 m. It is separated from Xiling Seamount by a saddle (Fig.2–110).				
命名释义 Reason for Choice of Name	该海山位于谷陵海山群的西部,故以此命名。 "Xi" means west in Chinese. It is named Xigu Seamount since it is located in the west of Guling Seamounts.				

(a)

(b)

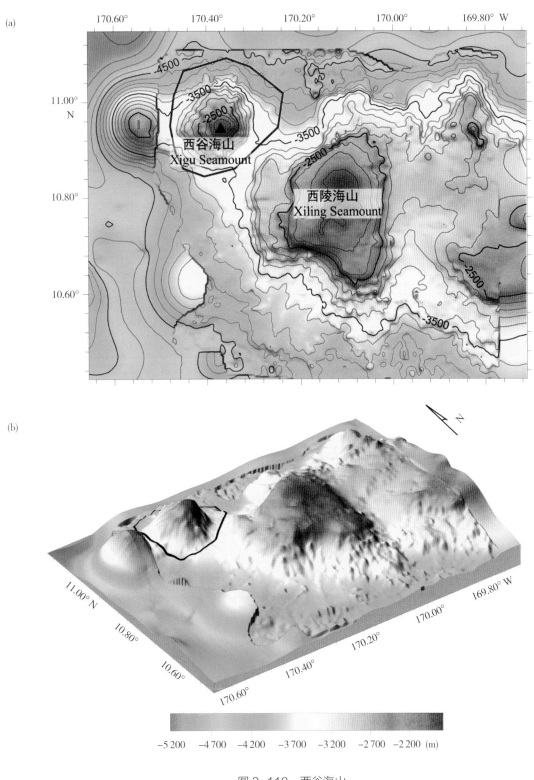

图 2-110　西谷海山

(a) 地形图（等深线间隔 200 m）；(b) 三维图

Fig.2-110　Xigu Seamount

(a) Bathymetric map (the contour interval is 200 m); (b) 3-D topographic map

2.7.8　西陵海山
Xiling Seamount

中文名称 Chinese Name	西陵海山 Xiling Haishan	英文名称 English Name	Xiling Seamount	所在大洋 Ocean or Sea	中太平洋 Central Pacific Ocean
发现情况 Discovery Facts	此海山于 2002 年 6—7 月由中国科考船"海洋四号"在执行 DY105-13 航次时调查发现。 This seamount was discovered by the Chinese R/V *Haiyang Sihao* during the DY105-13 cruise from June to July, 2002.				
命名历史 Name History	该海山是谷陵海山群的一部分，谷陵海山群在我国大洋航次调查报告中曾暂以代号命名。 This seamount is a part of Guling Seamounts. A code was used to name Guling Seamounts temporarily in Chinese cruise reports.				
特征点坐标 Coordinates	10°47.90′N，170°07.00′W			长(km)×宽(km) Length (km)× Width (km)	60×60
最大水深（m） Max Depth（m）	4 500	最小水深（m） Min Depth（m）	1 530	高差（m） Total Relief（m）	2 970
地形特征 Feature Description	西陵海山位于谷陵海山群的中西部，海山俯视平面形态呈圆形，基座直径 60 km。海山峰顶水深 1 530 m，山麓水深 4 500 m，高差 2 970 m，与西谷海山和东陵海山以平缓的鞍部分隔（图 2–111）。 Xiling Seamount is located in the midwest of Guling Seamounts. It has a nearly round overlook plane shape with the base diameter of 60 km. The top depth of the seamount is 1 530 m and the piedmont depth is 4 500 m, which makes the total relief being 2 970 m. It is separated from Xigu Seamount and Dongling Seamount by a saddle (Fig.2–111).				
命名释义 Reason for Choice of Name	该海山位于谷陵海山群西部，故以此命名。 "Xi" means west in Chinese. It is named Xiling Seamount since it is located in the west of Guling Seamounts.				

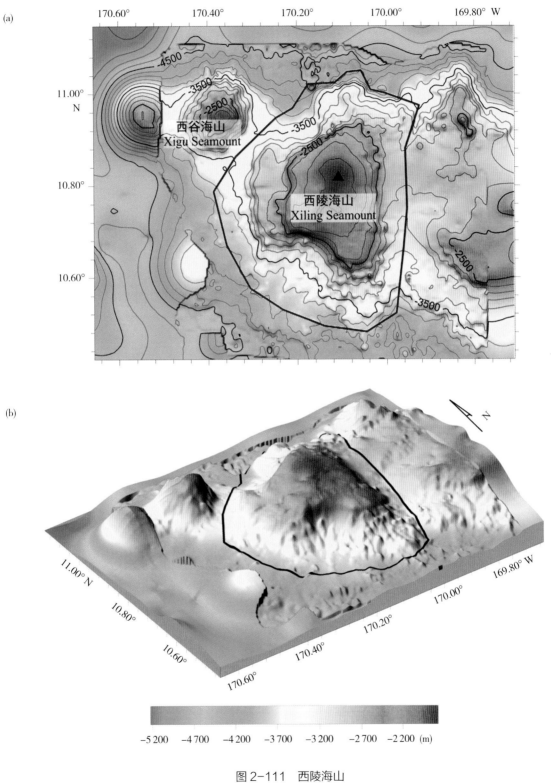

图 2-111　西陵海山

(a) 地形图（等深线间隔 200 m）；(b) 三维图

Fig.2-111　Xiling Seamount

(a) Bathymetric map (the contour interval is 200 m); (b) 3-D topographic map

2.7.9 东陵海山
Dongling Seamount

中文名称 Chinese Name	东陵海山 Dongling Haishan	英文名称 English Name	Dongling Seamount	所在大洋 Ocean or Sea	中太平洋 Central Pacific Ocean
发现情况 Discovery Facts	此海山于 2002 年 6—7 月由中国科考船"海洋四号"在执行 DY105-13 航次时调查发现。 This seamount was discovered by the Chinese R/V *Haiyang Sihao* during the DY105-13 cruise from June to July, 2002.				
命名历史 Name History	该海山是谷陵海山群的一部分，谷陵海山群在我国大洋航次调查报告中曾暂以代号命名。 This seamount is a part of Guling Seamounts. A code was used to name Guling Seamounts temporarily in Chinese cruise reports.				
特征点坐标 Coordinates	10°44.80′N，169°36.50′W			长(km)×宽(km) Length (km) × Width (km)	50 × 50
最大水深（m） Max Depth（m）	4 820	最小水深（m） Min Depth（m）	1 601	高差（m） Total Relief（m）	3 219
地形特征 Feature Description	东陵海山位于谷陵海山群的东部，形态不规则，基座直径 50 km。海山峰顶水深为 1 601 m，山麓水深 4 820 m，高差 3 219 m。此海山与西陵海山、东谷海山之间以鞍部分隔（图 2–112）。 Dongling Seamount is located in the east of Guling Seamounts. It has an irregular shape with the base diameter of 50 km. The top depth of the seamount is 1 601 m and the piedmont depth is 4 820 m, which makes the total relief being 3 219 m. It is separated from Xiling Seamount and Donggu Seamount by saddles (Fig.2–112).				
命名释义 Reason for Choice of Name	该海山位于谷陵海山群东部，故以此命名。 "Dong" means east in Chinese. It is named Dongling Seamount since it is located in the east of Guling Seamounts.				

(a)

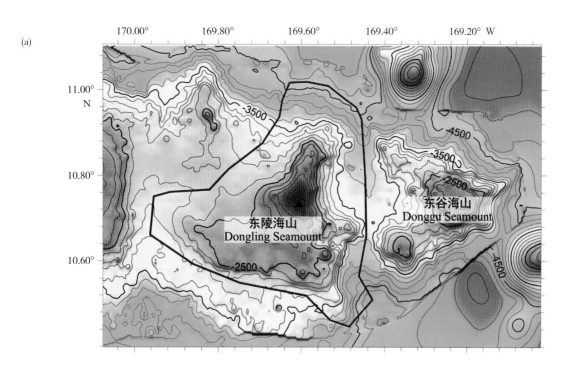

(b)

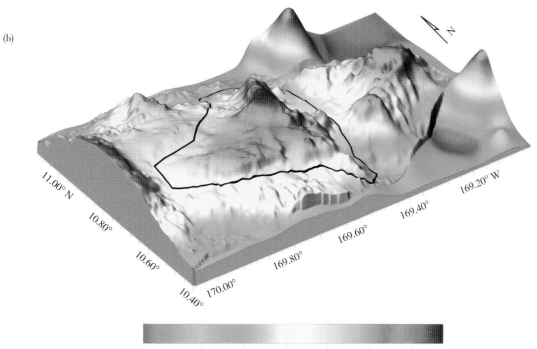

图 2-112　东陵海山

(a) 地形图（等深线间隔 200 m）；(b) 三维图

Fig.2-112　Dongling Seamount

(a) Bathymetric map (the contour interval is 200 m); (b) 3-D topographic map

2.7.10　东谷海山
Donggu Seamount

中文名称 Chinese Name	东谷海山 Donggu Haishan	英文名称 English Name	Donggu Seamount	所在大洋 Ocean or Sea	中太平洋 Central Pacific Ocean
发现情况 Discovery Facts	此海山于2002年6—7月由中国科考船"海洋四号"在执行DY105-13航次时调查发现。 This seamount was discovered by the Chinese R/V *Haiyang Sihao* during the DY105-13 cruise from June to July, 2002.				
命名历史 Name History	该海山是谷陵海山群的一部分，谷陵海山群在我国大洋航次调查报告中曾暂以代号命名。 This seamount is a part of Guling Seamounts. A code was used to name Guling Seamounts temporarily in Chinese cruise reports.				
特征点坐标 Coordinates	10°44.60′N，169°12.70′W			长(km)×宽(km) Length (km)× Width (km)	40×40
最大水深（m） Max Depth（m）	4 950	最小水深（m） Min Depth（m）	1 746	高差（m） Total Relief（m）	3 204
地形特征 Feature Description	东谷海山位于谷陵海山群的东部，基座直径40 km。海山峰顶水深为1 746 m，山麓水深4 950 m，高差3 204 m。此海山与东陵海山以鞍部分隔（图2–113）。 Donggu Seamount is located in the east of Guling Seamounts. The base diameter of the seamount is 40 km. The top depth of the seamount is 1 746 m and the piedmont depth is 4 950 m, which makes the total relief being 3 204 m. It is separated from Dongling Seamount by a saddle (Fig.2–113).				
命名释义 Reason for Choice of Name	该海山位于谷陵海山群东部，故以此命名。 "Dong" means east in Chinese. It is named Dongling Seamount since it is located in the east of Guling Seamounts.				

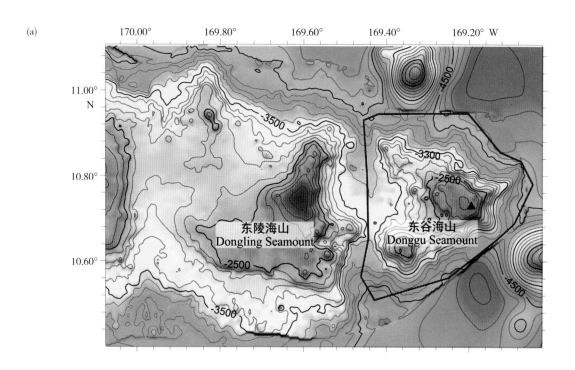

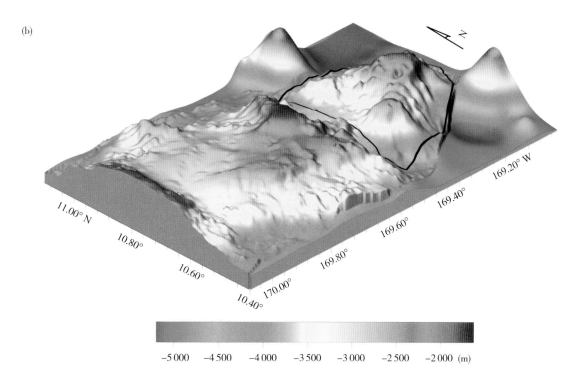

图 2-113　东谷海山

(a) 地形图（等深线间隔 200 m）；(b) 三维图

Fig.2-113　Donggu Seamount

(a) Bathymetric map (the contour interval is 200 m); (b) 3-D topographic map

2.7.11　柔木海山群
Roumu Seamounts

中文名称 Chinese Name	柔木海山群 Roumu Haishanqun	英文名称 English Name	Roumu Seamounts	所在大洋 Ocean or Sea	中太平洋 Central Pacific Ocean
发现情况 Discovery Facts	此海山群于 2002 年 6—7 月由中国科考船 "海洋四号" 在执行 DY105-13 航次时调查发现，位于中太平洋莱恩海岭。 The seamounts were discovered by the Chinese R/V *Haiyang Sihao* during the DY105-13 cruise from June to July, 2002. It is located in the Line Islands Chain in Central Pacific Ocean.				
命名历史 Name History	由我国命名为柔木海山群，于 2013 年提交 SCUFN 审议通过。 The seamounts were named Roumu Seamounts by China and approved by SCUFN in 2013.				
特征点坐标 Coordinates	10°17.60′N，167°59.40′W（南水杉海山 / Nanshuishan Seamount） 10°25.90′N，167°42.50′W（西柔木海山 /Xiroumu Seamount） 10°36.30′N，167°27.90′W（北柔木海山 /Beiroumu Seamount） 10°33.80′N，168°18.80′W（西水杉海山 /Xishuishan Seamount） 10°46.50′N，167°29.60′W（银杉海山 / Yinshan Seamount） 10°33.70′N，168°00.20′W（水杉海山 / Shuishan Seamount） 10°22.80′N，167°25.90′W（柔木海山 / Roumu Seamount）	长(km)×宽(km) Length (km)× Width (km)		166×80	
最大水深（m） Max Depth（m）	5 200	最小水深（m） Min Depth（m）	1 363	高差（m） Total Relief（m）	3 837
地形特征 Feature Description	此海山群由 7 个近东西向排列的山峰组成，长宽分别 166 km 和 80 km，其中最高峰位于中东部的柔木海山，峰顶水深 1 363 m（图 2–114）。 Roumu seamounts mainly consist of 7 seamounts arranged nearly in E to W direction. The length is 166 km and the width is 80 km. The highest summit is in Roumu Seamount in the middle east and its top depth is 1 363 m (Fig.2–114).				
命名释义 Reason for Choice of Name	"柔木" 出自《诗经·小雅·巧言》"荏染柔木，君子树之"，意为娇柔美丽的树木，由君子自己栽种，柔木为美丽的树木，稀有。此篇诗歌为作者揭露朝廷上某些人谗言误国的卑鄙行径，表达忧国忧民之情，也表达了洁身自好之意。该海山群命名为柔木海山群，含 7 个海山，选择柔木和我国稀有树种名称并按照其方位予以命名。 Roumu comes from a poem named *Qiaoyan* in *Shijing · Xiaoya*. *Shijing* is a collection of ancient Chinese poems from 11th century B.C. to 6th century B.C. "Pliant but long-lived trees, the wise king plants along the ways." "Roumu" means pliant but long-lived trees. The author use this poem to reveal that the country is risked by the slanderous talk of traitors and to show his intention to keep his integrity. This seamounts is named Roumu Seamounts including 7 seamounts, and they are named as Roumu and other Chinese rare trees according to their locations.				

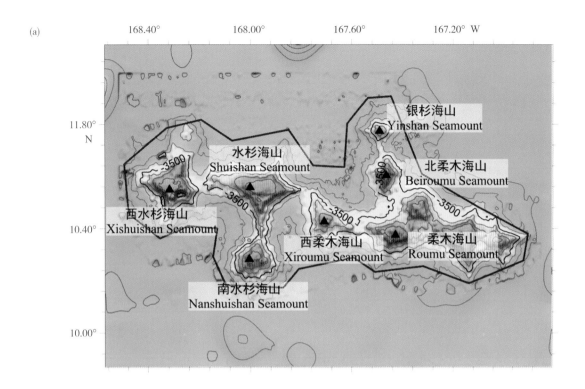

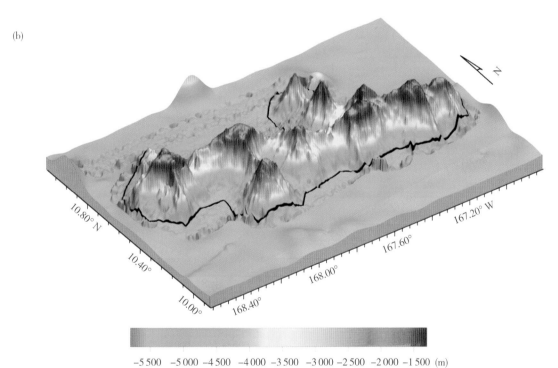

图 2-114　柔木海山群

(a) 地形图（等深线间隔 500 m）；(b) 三维图

Fig.2-114　Roumu Seamounts

(a) Bathymetric map (the contour interval is 500 m); (b) 3-D topographic map

2.7.12 西水杉海山
Xishuishan Seamount

中文名称 Chinese Name	西水杉海山 Xishuishan Haishan	英文名称 English Name	Xishuishan Seamount	所在大洋 Ocean or Sea	中太平洋 Central Pacific Ocean
发现情况 Discovery Facts	此海山于2002年6—7月由中国科考船"海洋四号"在执行DY105-13航次时调查发现。 This seamount was discovered by the Chinese R/V *Haiyang Sihao* during the DY105-13 cruise from June to July, 2002.				
命名历史 Name History	该海山是柔木海山群的一部分，柔木海山群在我国大洋航次调查报告中曾暂以代号命名。 This seamount is a part of Roumu Seamounts. A code was used to name Roumu Seamounts temporarily in Chinese cruise reports.				
特征点坐标 Coordinates	10°33.80′N，168°18.80′W			长(km)×宽(km) Length (km)× Width (km)	50×30
最大水深（m） Max Depth（m）	5 660	最小水深（m） Min Depth（m）	1 782	高差（m） Total Relief（m）	3 878
地形特征 Feature Description	西水杉海山位于柔木海山群西部，水杉海山以西。海山发育4条山脊。海山峰顶水深1 782 m，山麓水深5 660 m，高差3 878 m，山坡陡峭（图2–115）。 Xishuishan Seamount is located in the west of Roumu Seamounts, west to Shuishan Seamount. Four ridges are developed in this seamount. The top depth of the seamount is 1 782 m and piedmont depth is 5 660 m, which makes the total relief being 3 878 m. The seamount has steep slopes (Fig.2–115).				
命名释义 Reason for Choice of Name	该海山位于水杉海山以西，故以此命名。 Xi means west in Chinese. It is named Xishuishan Seamount since it is located in the west to Shuishan Seamount.				

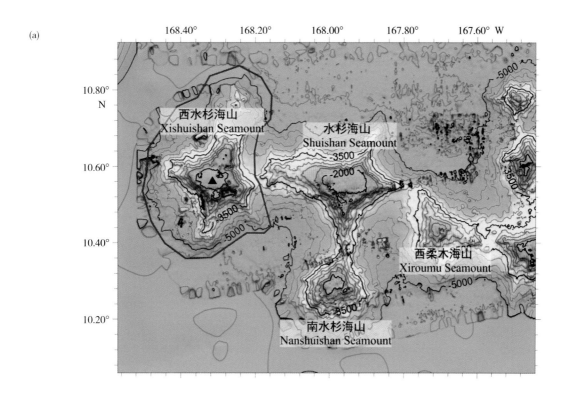

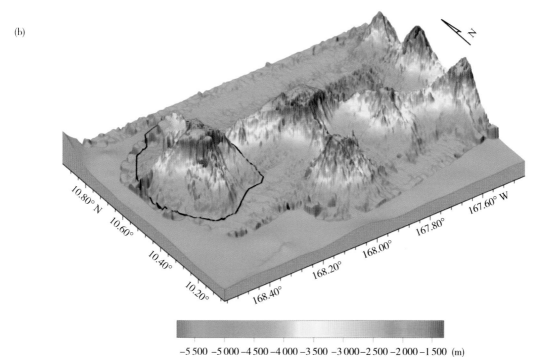

图 2-115　西水杉海山

(a) 地形图（等深线间隔 300 m）；(b) 三维图

Fig.2-115　Xishuishan Seamount

(a) Bathymetric map (the contour interval is 300 m); (b) 3-D topographic map

2.7.13 水杉海山
Shuishan Seamount

中文名称 Chinese Name	水杉海山 Shuishan Haishan	英文名称 English Name	Shuishan Seamount	所在大洋 Ocean or Sea	中太平洋 Central Pacific Ocean
发现情况 Discovery Facts	此海山于2002年6—7月由中国科考船"海洋四号"在执行DY105-13航次时调查发现。 This seamount was discovered by the Chinese R/V *Haiyang Sihao* during the DY105-13 cruise from June to July, 2002.				
命名历史 Name History	该海山是柔木海山群的一部分,柔木海山群在我国大洋航次调查报告中曾暂以代号命名。 This seamount is a part of Roumu Seamounts. A code was used to name Roumu Seamounts temporarily in Chinese cruise reports.				
特征点坐标 Coordinates	10°33.70′N,168°00.20′W			长(km)× 宽(km) Length (km) × Width (km)	40×40
最大水深(m) Max Depth (m)	5 360	最小水深(m) Min Depth (m)	1 528	高差(m) Total Relief (m)	3 832
地形特征 Feature Description	水杉海山位于柔木海山群西部,俯视平面形态呈放射状,发育东、南、西、北4条山脊,基座直径40 km。海山峰顶水深1 528 m,山麓水深5 360 m,高差3 832 m,山坡陡峭(图2–116)。 Shuishan Seamount is located in the west of Roumu Seamounts. It has an overlook plane shape like radiation pattern. It develops four ridges in east, south, west and north direction with the base diameter of 40 km. The top depth of the seamount is 1 528 m and the piedmont depth is 5 360 m, which makes the total relief being 3 832 m. The seamount has steep slopes (Fig.2–116).				
命名释义 Reason for Choice of Name	水杉为裸子植物,杉科,国家一级保护植物,白垩纪开始出现,第四纪冰期以后大量灭绝,对研究古植物、古气候和古地理以及裸子植物系统发育等有重要价值。此海山命名为水杉,体现了团组化命名方式。 "Shuishan" is the Chinese name of metasequoia which is gymnosperm, Taxodiaceae. It emerged in the Cretaceous and extincted in the later Quaternary glacial period. It is valuable for the study of ancient plants, ancient climate and ancient geography and phylogeny of gymnosperms. This seamount is named Shuishan Seamount embodying group naming method.				

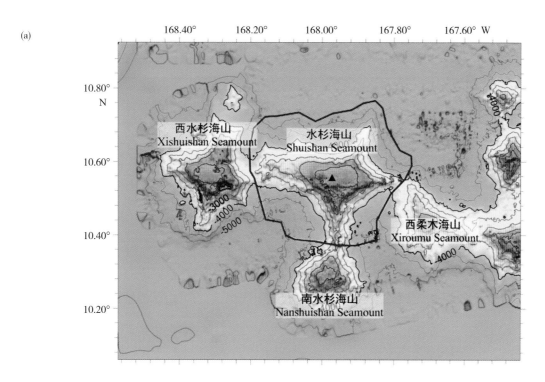

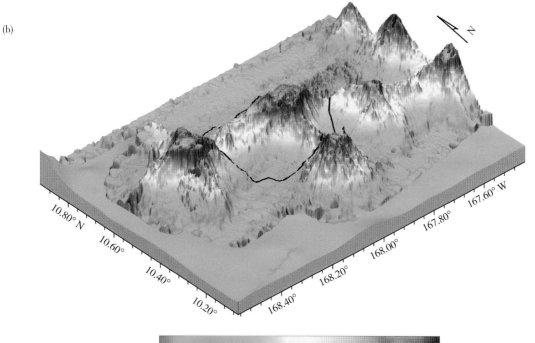

图 2-116　水杉海山

(a) 地形图（等深线间隔 500 m）；(b) 三维图

Fig.2-116　Shuishan Seamount

(a) Bathymetric map (the contour interval is 500 m); (b) 3-D topographic map

2.7.14 南水杉海山
Nanshuishan Seamount

中文名称 Chinese Name	南水杉海山 Nanshuishan Haishan	英文名称 English Name	Nanshuishan Seamount	所在大洋 Ocean or Sea	中太平洋 Central Pacific Ocean
发现情况 Discovery Facts	此海山于 2002 年 6—7 月由中国科考船"海洋四号"在执行 DY105-13 航次时调查发现。 This seamount was discovered by the Chinese R/V *Haiyang Sihao* during the DY105-13 cruise from June to July, 2002.				
命名历史 Name History	该海山是柔木海山群的一部分,柔木海山群在我国大洋航次调查报告中曾暂以代号命名。 This seamount is a part of Roumu Seamounts. A code was used to name Roumu Seamounts temporarily in Chinese cruise reports.				
特征点坐标 Coordinates	10°17.60′N,167°59.40′W			长 (km) × 宽 (km) Length (km) × Width (km)	30 × 30
最大水深（m） Max Depth（m）	5 250	最小水深（m） Min Depth（m）	1 717	高差（m） Total Relief（m）	3 533
地形特征 Feature Description	此海山呈圆锥状,基座直径 30 km,峰顶水深 1 717 m,山麓水深 5 250 m,高差 3 533 m。海山山坡陡峭,与水杉海山以鞍部分隔,鞍部水深 3 390 m(图 2–117)。 Nanshuishan Seamount has a conical overlook plane shape with the base diameter of 30 km. The top depth is 1 717 m and the piedmont depth is 5 250 m, which makes the total relief being 3 533 m. The seamount has steep slopes. It is separated from Shuishan Seamount by a saddle, where the depth is 3 390 m (Fig.2–117).				
命名释义 Reason for Choice of Name	该海山位于水杉海山以南,故以此命名。 Nan means south in Chinese. It is named Nanshuishan Seamount since it is located in the south to Shuishan Seamount.				

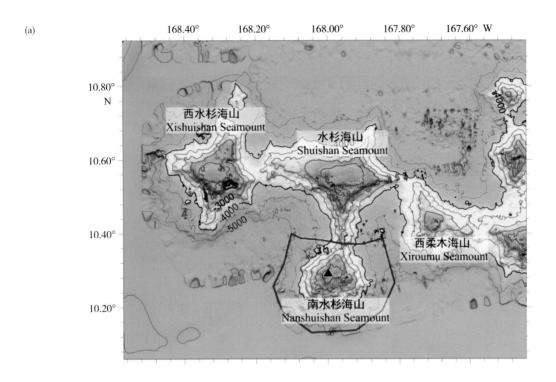

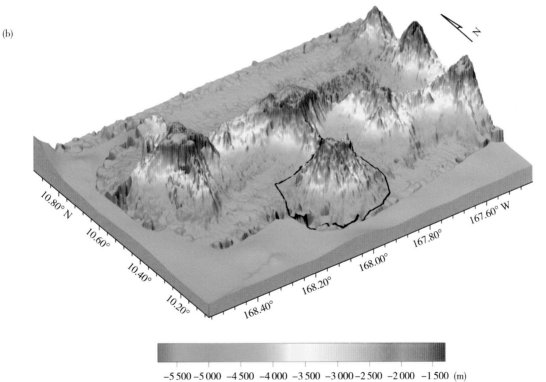

图 2-117 南水杉海山

(a) 地形图（等深线间隔 500 m）；(b) 三维图

Fig.2-117 Nanshuishan Seamount

(a) Bathymetric map (the contour interval is 500 m); (b) 3-D topographic map

2.7.15 西柔木海山
Xiroumu Seamount

中文名称 Chinese Name	西柔木海山 Xiroumu Haishan	英文名称 English Name	Xiroumu Seamount	所在大洋 Ocean or Sea	中太平洋 Central Pacific Ocean
发现情况 Discovery Facts	此海山于 2002 年 6—7 月由中国科考船"海洋四号"在执行 DY105-13 航次时调查发现。 This seamount was discovered by the Chinese R/V *Haiyang Sihao* during the DY105-13 cruise from June to July, 2002.				
命名历史 Name History	此海山是柔木海山群的一部分，柔木海山群在我国大洋航次调查报告中曾暂以代号命名。 This seamount is a part of Roumu Seamounts. A code was used to name Roumu Seamounts temporarily in Chinese cruise reports.				
特征点坐标 Coordinates	10°25.90′N，167°42.50′W			长(km)×宽(km) Length (km)× Width (km)	30×30
最大水深（m） Max Depth（m）	5 300	最小水深（m） Min Depth（m）	2 058	高差（m） Total Relief（m）	3 242
地形特征 Feature Description	此海山发育 3 条山脊，基座直径 30 km，峰顶水深 2 058 m，山麓水深为 5 300 m，高差 3 242 m。海山山坡陡峭，与水杉海山和柔木海山以鞍部分隔（图 2–118）。 Three ridges develope on the seamount. The base diameter is 30 km. The top depth is 2 058 m and the piedmont depth is 5 300 m, which makes the total relief being 3 242 m. The seamount has steep slopes and it is separated from Shuishan Seamount and Roumu Seamount by saddles (Fig.2–118).				
命名释义 Reason for Choice of Name	该海山位于柔木海山以西，故以此命名。 Xi means west in Chinese. It is named Xiroumu Seamount since it is located in the west to Roumu Seamount.				

(a)

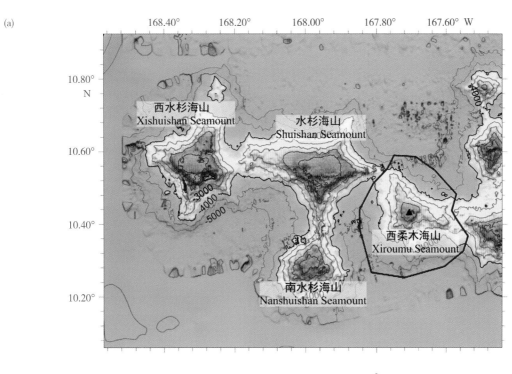

(b)

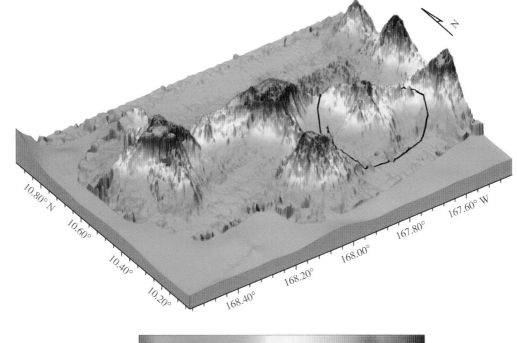

−5 500　−5 000　−4 500　−4 000　−3 500　−3 000　−2 500　−2 000　−1 500　(m)

图 2-118　西柔木海山

(a) 地形图（等深线间隔 500 m）；(b) 三维图

Fig.2-118　Xiroumu Seamount

(a) Bathymetric map (the contour interval is 500 m); (b) 3-D topographic map

2.7.16 柔木海山
Roumu Seamount

中文名称 Chinese Name	柔木海山 Roumu Haishan	英文名称 English Name	Roumu Seamount	所在大洋 Ocean or Sea	中太平洋 Central Pacific Ocean
发现情况 Discovery Facts	此海山于 2002 年 6—7 月由中国科考船"海洋四号"在执行 DY105-13 航次时调查发现。 This seamount was discovered by the Chinese R/V *Haiyang Sihao* during the DY105-13 cruise from June to July, 2002.				
命名历史 Name History	该海山是柔木海山群的一部分，柔木海山群在我国大洋航次调查报告中曾暂以代号命名。 This seamount is a part of Roumu Seamounts. A code was used to name Roumu Seamounts temporarily in Chinese cruise reports.				
特征点坐标 Coordinates	10°28.40′N，167°20.30′W 10°21.30′N，167°01.60′W 10°23.20′N，167°10.10′W 10°22.80′N，167°25.90′W（峰顶 /top）			长(km)× 宽(km) Length (km)× Width (km)	75×35
最大水深（m） Max Depth（m）	5 200	最小水深（m） Min Depth（m）	1 363	高差（m） Total Relief（m）	3 837
地形特征 Feature Description	海山发育 4 个东西向排列的山峰，长宽分别为 75 km 和 35 km。海山最高峰为西部山峰，峰顶水深 1 363 m，山麓水深 5 200 m，高差 3 837 m，海山边坡陡峭，与北柔木海山和西柔木海山以鞍部分隔（图 2–119）。 This seamount develops 4 summits arranged in E to W direction. The length and width of the seamount are 75 km and 35 km respectively. The highest summit in this seamount is the western one, its top depth is 1 363 m, and the piedmont depth is 5 200 m, which makes the total relief being 3 837 m. It has steep slopes and it is separated from Beiroumu Seamount and Xiroumu Seamount by a saddle (Fig.2–119).				
命名释义 Reason for Choice of Name	"柔木"出自《诗经·小雅·巧言》"荏染柔木，君子树之"，意为娇柔美丽的树木，由君子自己栽种，柔木为美丽的树木，稀有。此篇诗歌为作者揭露朝廷上某些人谗言误国的卑鄙行径，表达忧国忧民之情，也表达了洁身自好之意。此海山群含 7 个海山，选择柔木或我国稀有树种名称予以命名，该海山为柔木海山群的最高峰，规模也最大，故以此命名。 "Roumu" comes from a poem named *Qiaoyan* in *Shijing · Xiaoya*. *Shijing* is a collection of ancient Chinese poems from 11th century B.C. to 6th century B.C. "Pliant but long-lived trees, the wise king plants along the ways." "Roumu" means pliant but long-lived trees. The author use this poem to reveal that the country is risked by the slandous talk of traitors and to show his intention to keep his integrity. This seamounts is named Roumu Seamounts including 7 seamounts, and they are named using Roumu or other Chinese rare trees. This seamount is the highest and largest one among Roumu Seamounts, thus it is named Roumu Seamount.				

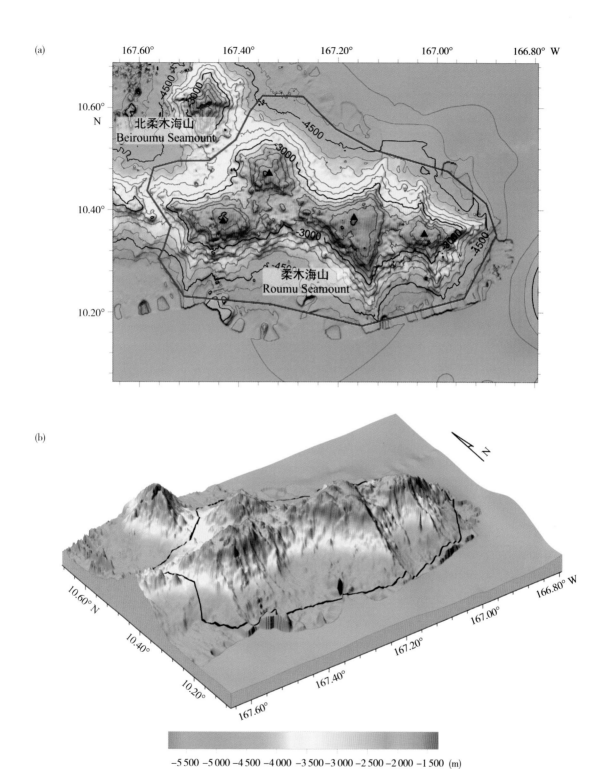

图 2-119　柔木海山

(a) 地形图（等深线间隔 300 m）；(b) 三维图

Fig.2-119　Roumu Seamount

(a) Bathymetric map (the contour interval is 300 m); (b) 3-D topographic map

2.7.17　北柔木海山
Beiroumu Seamount

中文名称 Chinese Name	北柔木海山 Beiroumu Haishan		英文名称 English Name	Beiroumu Seamount	所在大洋 Ocean or Sea	中太平洋 Central Pacific Ocean
发现情况 Discovery Facts	此海山于 2002 年 6—7 月由中国科考船"海洋四号"在执行 DY105-13 航次时调查发现。 This seamount was discovered by the Chinese R/V *Haiyang Sihao* during the DY105-13 cruise from June to July, 2002.					
命名历史 Name History	该海山是柔木海山群的一部分，柔木海山群在我国大洋航次调查报告中曾暂以代号命名。 This seamount is a part of Roumu Seamounts. A code was used to name Roumu Seamounts temporarily in Chinese cruise reports.					
特征点坐标 Coordinates	10°36.30′N，167°27.90′W			长(km)×宽(km) Length (km) × Width (km)		27 × 27
最大水深（m） Max Depth（m）	5 000	最小水深（m） Min Depth（m）		1 628	高差（m） Total Relief（m）	3 372
地形特征 Feature Description	此海山呈圆锥状，基座直径 27 km，峰顶水深 1 628 m，山麓水深 5 000 m，高差 3 372 m。海山山坡陡峭，与银杉海山和柔木海山以鞍部分隔（图 2–120）。 Beiroumu Seamount has a conical shape with the base diameter of 27 km. The top depth is 1 628 m and piedmont depth is 5 000 m, which makes the total relief being 3 372 m. The seamount has steep slopes. It is separated from Yinshan Seamount and Roumu Seamount by a saddle (Fig.2–120).					
命名释义 Reason for Choice of Name	该海山位于柔木海山以北，故以此命名。 Bei means north in Chinese. It is named Beiroumu Seamount since it is located in the north to Roumu Seamount.					

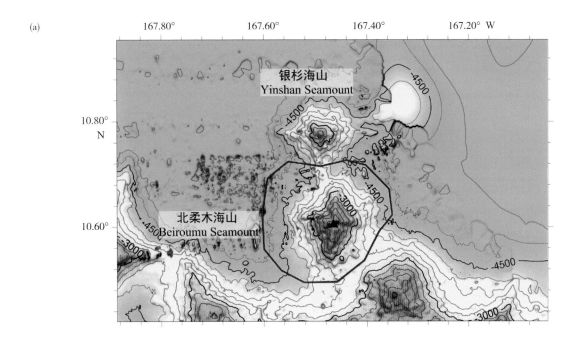

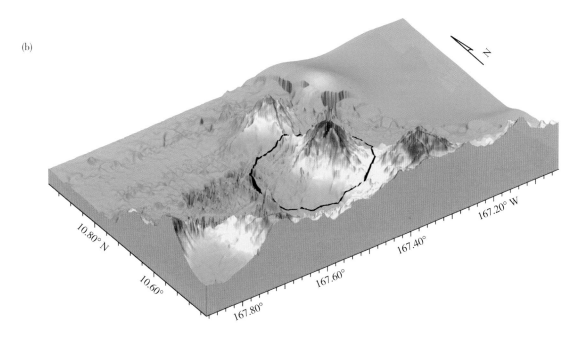

图 2-120　北柔木海山

(a) 地形图（等深线间隔 300 m）；(b) 三维图

Fig.2-120　Beiroumu Seamount

(a) Bathymetric map (the contour interval is 300 m); (b) 3-D topographic map

2.7.18　银杉海山
Yinshan Seamount

中文名称 Chinese Name	银杉海山 Yinshan Haishan	英文名称 English Name	Yinshan Seamount	所在大洋 Ocean or Sea	中太平洋 Central Pacific Ocean
发现情况 Discovery Facts	此海山于2002年6—7月由中国科考船"海洋四号"在执行DY105-13航次时调查发现。 This seamount was discovered by the Chinese R/V *Haiyang Sihao* during the DY105-13 cruise from June to July, 2002.				
命名历史 Name History	该海山是柔木海山群的一部分，柔木海山群在我国大洋航次调查报告中曾暂以代号命名。 This seamount is a part of Roumu Seamounts. A code was used to name Roumu Seamounts temporarily in Chinese cruise reports.				
特征点坐标 Coordinates	10°46.50′N，167°29.60′W			长(km)×宽(km) Length (km)× Width (km)	20×20
最大水深（m） Max Depth（m）	5 300	最小水深（m） Min Depth（m）	2 298	高差（m） Total Relief（m）	3 002
地形特征 Feature Description	海山呈圆锥状，基座直径20 km，峰顶水深2 298 m，山麓水深5 300 m，高差3 002 m。海山山坡陡峭，与北柔木海山以鞍部分隔（图2–121）。 This seamount has a conical shape with the base diameter of 20 km. The top depth is 2 298 m and the piedmont depth is 5 300 m, which makes the total relief being 3 002 m. This seamount has steep slopes and it is separated from Beiroumu Seamount by a saddle (Fig.2–121).				
命名释义 Reason for Choice of Name	银杉为裸子植物，松科，国家一级保护植物，是中国特有的第三纪残遗珍稀植物，对研究松科植物的系统发育、古植物区系、古地理及第四纪冰期气候等有重要价值。此海山命名为银杉，体现了团组化命名方式。 Yinshan is the Chinese name of *Cathaya argyrophylla*. It is a species of gymnospermae, Pinaceae. It is one of Chinese rare plant species and is valuable for study on Pinaceae phylogeny, ancient flora, paleogeography and Quaternary glacial climate. The seamount is named Yinshan as one of the seamount of Roumu Seamounts embodying group naming method.				

(a)

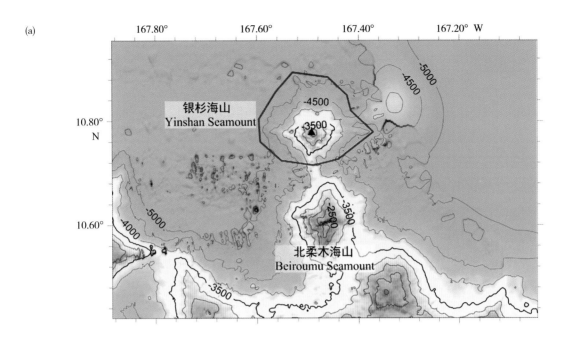

(b)

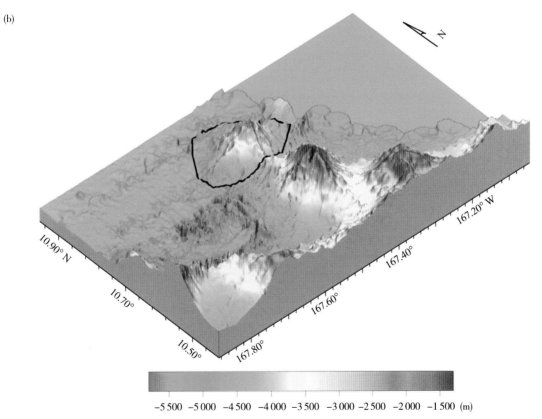

图 2-121　银杉海山

(a) 地形图（等深线间隔 500 m）；(b) 三维图

Fig.2-121　Yinshan Seamount

(a) Bathymetric map (the contour interval is 500 m); (b) 3-D topographic map

第 8 节　克拉里昂 – 克拉伯顿区地理实体

东北太平洋海盆发育一系列 NEE 向断裂带，横切整个海盆。这些断裂带是切割洋脊的转换断层，在地形地貌上主要表现为平行排列的深切槽谷或耸立的海山。我国的多金属结核调查区位于克拉里昂断裂带和克拉伯顿断裂带之间的深海平原，分为两个区块，暂命名为结核调查区东区和结核调查区西区。

结核东区范围为 7°15′—10°N，141°20′—148°W，整体上为一个大规模的断裂带，地貌上表现为 NNW 向相间排列的线状海脊和洼地。海脊顶部水深 5 075 ～ 5 150 m，洼地水深 5 250 ～ 5 300 m，相邻的洼地间距 15 ～ 20 km。结核东区的海山和海丘分散于断裂带上，顶部水深 4 150 ～ 4 900 m，山麓水深 5 200 ～ 5 300 m，高差 400 ～ 1 000 m。海山和海丘规模较小，基底直径 10 ～ 30 km。部分海山海丘相对集中构成海山群。

结核西区范围为 8°30′—11°N，151°—155°W，其地貌特征是广阔的深海平原上发育数条大型的海岭和多个海山群。深海平原为东太平洋海盆深海平原的一部分，地形平缓，水深 5 000 ～ 5 300 m。海岭由数十个海山和海丘紧密相连而构成，均呈近东西走向，最大的海岭在东区内长 325 km。海山和海丘一般呈椭圆形，顶部水深 4 500 ～ 4 600 m，高差 400 ～ 1 100 m。海山和海丘较为松散组合，构成海山群。

在此海域我国共命名地理实体 43 个。其中海山群 4 个，包括茑萝海丘群、郑庭芳海山群、陈伦炯海山群和刺螺平顶海山群；海山 12 个，包括清高海山、甘雨海山、斯寝海山、魏源海山、长庚海山、芳伯海山、如霆海山、朱应海山、楚茨海山、亿庾海山、怀允海山和启明海山；海丘和圆海丘 19 个，包括巩珍圆海丘群、巩珍圆海丘、达奚通圆海丘、维熊圆海丘、维蛇海丘、维罴圆海丘、维虺海丘、斯翼海丘、景福海丘、苏洵圆海丘、苏轼海丘、苏辙圆海丘、王勃圆海丘、杨炯海丘、卢照邻圆海丘、骆宾王圆海丘、天祐圆海丘、嘉卉圆海丘和圆鼓圆海丘；海岭和海山链 4 个，包括鉴真海岭、郑和海岭、张炳熹海岭和楚茨海山链；海底洼地 4 个，包括天祐南海底洼地、朱应西海底洼地、朱应北海底洼地和启明南海底洼地（图 2-122 和图 2-123）。

Section 8　Undersea Features in the Clarion-Clipperton Area

The Northeast Pacific Basin develops a series of NEE to SWW running fracture zones cutting the entire basin. These fracture zones are the transform faults cutting the ocean ridges, featured by deep troughs and high seamounts arranged parallel on topography and geomorphology. Chinese polymetallic nodule survey area is located in the abyssal plain, between the Clarion Fracture Zone and Clipperton Fracture Zone. It can be divided into two blocks, which are temporarily named the Eastern Nodule Area and Western Nodule Area.

The Eastern Nodule Area, located between 7°15′–10°N, 141°20′–148°W, is a large-scale fracture zone as a whole. The geomorphology is featured by NNW trend alternately arranged linear ridges and depressions. The depth of the ridges is 5 075–5 150 m, while the depth of depressions being 5 250–5 300 m, the distance between adjacent depressions being 15–20 km. The seamounts and hills of Eastern Nodule Area disperse on the fracture zone. Their top depth is 4 150–4 900 m, and the piedmont depth being 5 200–5 300 m, and the total relief being 400–1 000 m. The scale of seamounts and hills is small and the base diameter is 10–30 km. Some relatively gathering seamounts and hills constitute seamounts.

The Western Nodule Area located between 8°30′–11°N, 151°–155°W. The geomorphology is featured by developments of several large scale ridges and a number of seamounts on the vast abyssal plain. The abyssal plain is a part of the East Pacific Basin abyssal plain with smooth topography with water depth of 5 000–5 300 m. Dozens of seamounts and hills connect with each other tightly and constitute the ridges, which all extend towards nearly EW. The largest ridge in this area has a length of 325 km. The seamounts and hills are usually elliptic, and the top water depth is 4 500–4 600 m, the total relief being 400–1 100 m. Seamounts and hills which combine loosely constitute seamounts.

In total, 43 undersea features have been named by China in the region of the Clarion-Clipperton Area. Among them, there are 4 seamounts, including Niaoluo Hills, Zhengtingfang Seamounts, Chenlunjiong Seamounts and Ciluo Guyots; 12 seamounts, including Qinggao Seamount, Ganyu Seamount, Siqin Seamount, Weiyuan Seamount, Changgeng Seamount, Fangbo Seamount, Ruting Seamount, Zhuying Seamount, Chuci Seamount, Yiyu Seamount, Huaiyun Seamount and Qiming Seamount; 19 hills or knolls, including Gongzhen Knolls, Gongzhen Knoll, Daxitong Knoll, Weixiong Knoll, Weishe Hill, Weipi Knoll, Weihui Hill, Siyi Hill, Jingfu Hill, Suxun Knoll, Sushi Hill, Suzhe Knoll, Wangbo Knoll, Yangjiong Hill, Luzhaolin Knoll, Luobinwang Knoll, Tianhu Knoll, Jiahui Knoll and Yuangu Knoll; 4 ridges or seamount chains, including Jianzhen Ridge, Zhenghe Ridge, Zhangbingxi Ridge and Chuci Seamount Chain; 4 depressions, including Tianhunan Depression, Zhuyingxi Depression, Zhuyingbei Depression and Qimingnan Depression (Fig. 2–122, Fig. 2–123).

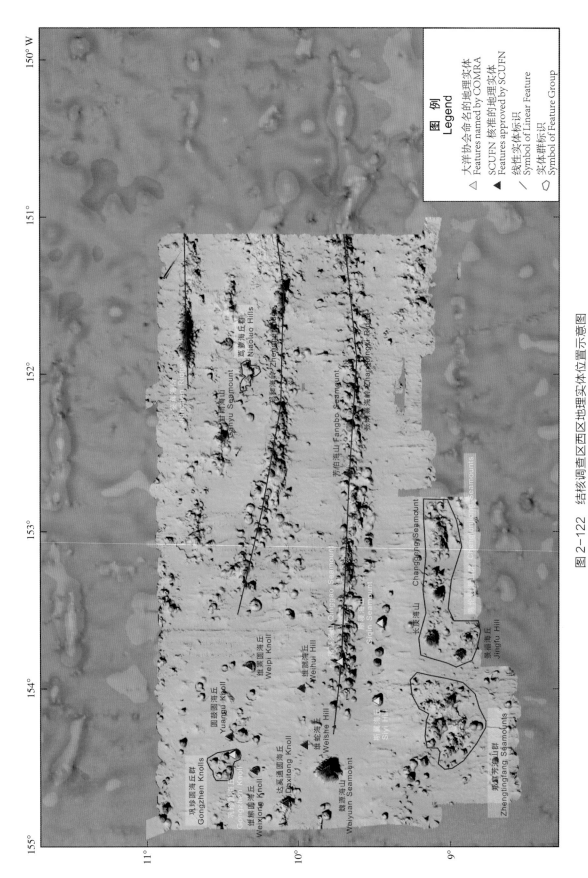

图 2-122　结核调查区西区地理实体位置示意图

Fig.2-122　Locations of the undersea features in the Western Nodule Area

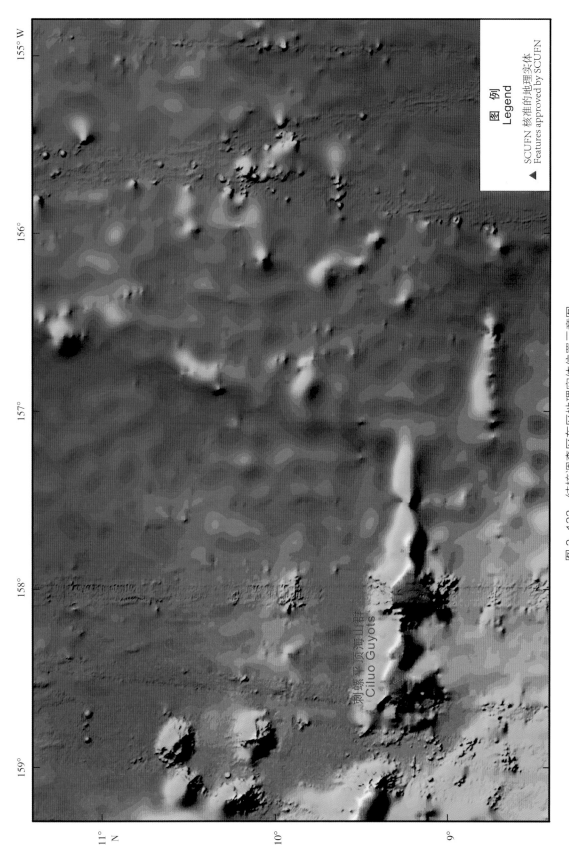

图 2-123　结核调查区东区地理实体位置示意图

Fig.2-123　Locations of the undersea features in the Eastern Nodule Area

2.8.1 鉴真海岭
Jianzhen Ridge

中文名称 Chinese Name	鉴真海岭 Jianzhen Hailing	英文名称 English Name	Jianzhen Ridge	所在大洋 Ocean or Sea	东太平洋 East Pacific Ocean
发现情况 Discovery Facts	此海岭于 1995 年 10 月由中国科考船"大洋一号"在执行 DY85-5 航次时发现。 This ridge was discovered by the Chinese R/V *Dayang Yihao* during the DY85-5 cruise in October, 1995.				
命名历史 Name History	在我国大洋航次调查报告和 GEBCO 地名辞典中未命名。 This ridge has not been named in the Chinese cruise reports or GEBCO gazetteer.				
特征点坐标 Coordinates	10°43.67′N，152°05.64′W 10°44.41′N，151°45.00′W 10°45.16′N，151°06.18′W			长 (km)×宽 (km) Length (km)× Width (km)	105×23
最大水深（m） Max Depth（m）	5 380	最小水深（m） Min Depth（m）	3 443	高差（m） Total Relief（m）	1 937
地形特征 Feature Description	该海岭形态狭长，东西走向，往东延伸情况不明，现有资料显示长宽分别为 105 km 和 23 km，最小水深为 3 443 m（图 2–124）。 This ridge has an elongated shape and runs E to W. Its extension towards east is remained unclear. According to current data, the length is 105 km and the width is 23 km with the minimum depth of 3 443 m (Fig.2–124).				
命名释义 Reason for Choice of Name	鉴真（公元 688—763 年），俗姓淳于，江苏扬州人，初盛唐时期著名高僧。鉴真先后六次东渡日本，历尽艰难险阻，在最后一次成功到达，在日本首都奈良建立唐招提寺，讲授佛经，传播中华文化，在中国航海史上影响深远。此海岭命名为"鉴真"，以纪念鉴真在我国航海史上的重要贡献。 Jian Zhen (A.D. 688–763), whose last name is Chunyu, was born in Yangzhou, Jiangsu Province, China. He was a famous monk in the early period of the Tang Dynasty. Jian Zhen had sailed to Japan six times, suffering from hardships and dangers and succeeded at last. He established Toshodai temple in Nara, the Japanese capital for Buddhist teaching, dissemination of Chinese culture. What he had done has left the eternity edifying anecdote in the history of Sino-Japanese navigation exchanges. This ridge is named after "Jianzhen", to commemorate his important contributions to the Chinese history of marine navigation.				

(a)

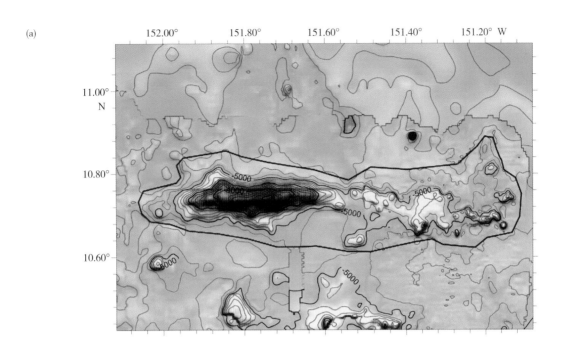

(b)

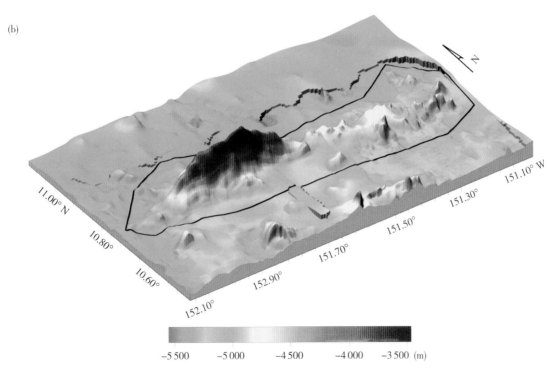

图 2-124　鉴真海岭

(a) 地形图（等深线间隔 100 m）；(b) 三维图

Fig.2-124　Jianzhen Ridge

(a) Bathymetric map (the contour interval is 100 m); (b) 3-D topographic map

2.8.2 甘雨海山
Ganyu Seamount

中文名称 Chinese Name	甘雨海山 Ganyu Haishan		英文名称 English Name	Ganyu Seamount	所在大洋 Ocean or Sea	东太平洋 East Pacific Ocean
发现情况 Discovery Facts	此海山于 1995 年 8 月由中国科考船"大洋一号"在执行 DY85-5 航次时发现。 This seamount was discovered by the Chinese R/V *Dayang Yihao* during the DY85-5 cruise in August, 1995.					
命名历史 Name History	由我国命名为甘雨海山，于 2013 年提交 SCUFN 审议通过。 Ganyu Seamount was named by China and approved by SCUFN in 2013.					
特征点坐标 Coordinates	10°30.20′N，152°24.20′W				长(km)×宽(km) Length (km) × Width (km)	22 × 14
最大水深（m） Max Depth（m）	5 200	最小水深（m） Min Depth（m）		3 950	高差（m） Total Relief（m）	1 250
地形特征 Feature Description	该海山东西走向，长宽分别为 22 km 和 14 km，峰顶水深 3 950 m，山麓水深 5 200 m，高差 1 250 m，山坡地形陡峭（图 2–125）。 Ganyu Seamount runs E to W. The length is 22 km and the width is 14 km. The top water depth is 3 950 m and the piedmont depth is 5 200 m, which makes the total relief being 1 250 m. It has steep slopes (Fig.2–125).					
命名释义 Reason for Choice of Name	"甘雨"出自《诗经·小雅·甫田》"以祈甘雨，以介我稷黍，以穀我士女"，甘雨为对农业耕作非常适宜的雨。此诗歌是周朝农民庆祝丰收时所唱，表达欢喜之情。中国古代是农耕社会，对天气非常依赖，此海山起名甘雨，以示中国古人对丰收的美好期望。 "Ganyu" comes from a poem named *Futian* in *Shijing · Xiaoya*. *Shijing* is a collection of ancient Chinese poems from 11th century B.C. to 6th century B.C. "They piously pray for showers sweet, for copious millet, sorghum and wheat, so that all have enough to eat." "Ganyu" means the rain suitable for agriculture. Ancient China is an agricultural society and very dependent on the weather. The seamount is named "Ganyu" to show ancient Chinese's expectations of good harvest.					

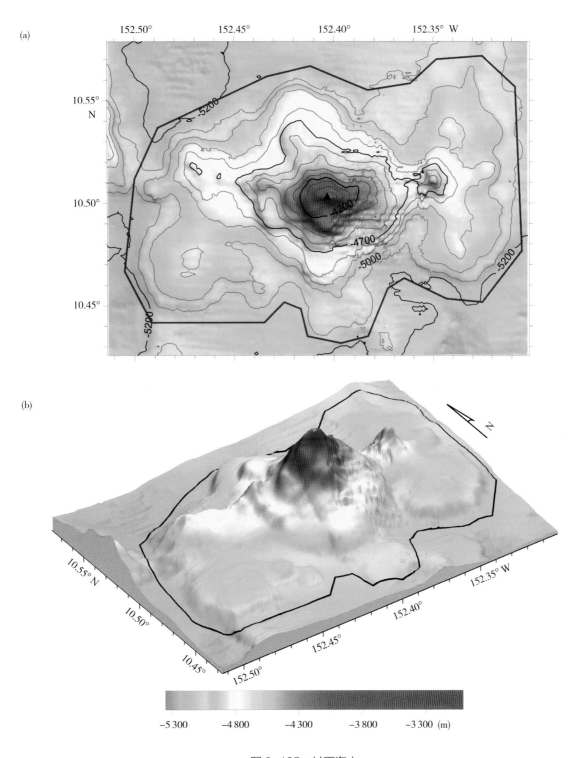

图 2-125　甘雨海山

(a) 地形图（等深线间隔 100 m）；(b) 三维图

Fig.2-125　Ganyu Seamount

(a) Bathymetric map (the contour interval is 100 m); (b) 3-D topographic map

2.8.3 莺萝海丘群
Niaoluo Hills

中文名称 Chinese Name	莺萝海丘群 Niaoluo Haiqiuqun	英文名称 English Name	Niaoluo Hills	所在大洋 Ocean or Sea	东太平洋 East Pacific Ocean
发现情况 Discovery Facts	此海丘群于 1995 年 10 月由中国科考船"大洋一号"在执行 DY85-5 航次时发现。 The hills were discovered by the Chinese R/V *Dayang Yihao* during the DY85-5 cruise in September, 1995.				
命名历史 Name History	由我国命名为莺萝海丘群，于 2015 年提交 SCUFN 审议通过。 Niaoluo Hills were named by China and approved by SCUFN in 2015.				
特征点坐标 Coordinates	10°19.03′N，152°01.08′W			长(km) × 宽(km) Length (km) × Width (km)	17 × 8
最大水深（m） Max Depth（m）	5 270	最小水深（m） Min Depth（m）	4 400	高差（m） Total Relief（m）	870
地形特征 Feature Description	该海丘群俯视平面形态不规则，最大长度和宽度分别为 17 km 和 8 km。海丘峰顶水深 4 400 m，山麓水深 5 270 m，高差 870 m（图 2–126）。 Niaoluo Hills has an irregular overlook plane shape. The maximum length and maximum width are 17 km and 8 km respectively. The top water depth of the hills is 4 400 m and the piedmont depth is 5 270 m, which makes the total relief being 870 m (Fig.2–126).				
命名释义 Reason for Choice of Name	"莺萝"出自《诗经·小雅·頍弁》"莺与女萝，施于松柏"，莺、女萝均为攀援性植物，意为莺与女萝攀附于松柏大树之上，暗喻周朝贵族之间互相攀附，生活奢华的场景。 "Niaoluo" comes from a poem named *Kuibian* in *Shijing · Xiaoya*. *Shijing* is a collection of ancient Chinese poems from 11th century B.C. to 6th century B.C. "As mistletoe and dodder twine, we stick to you, our cypress-pine." Niao and Luo are two kinds of commensal plants. The poem innuendoes that the aristocrats cling to each other and living in luxury.				

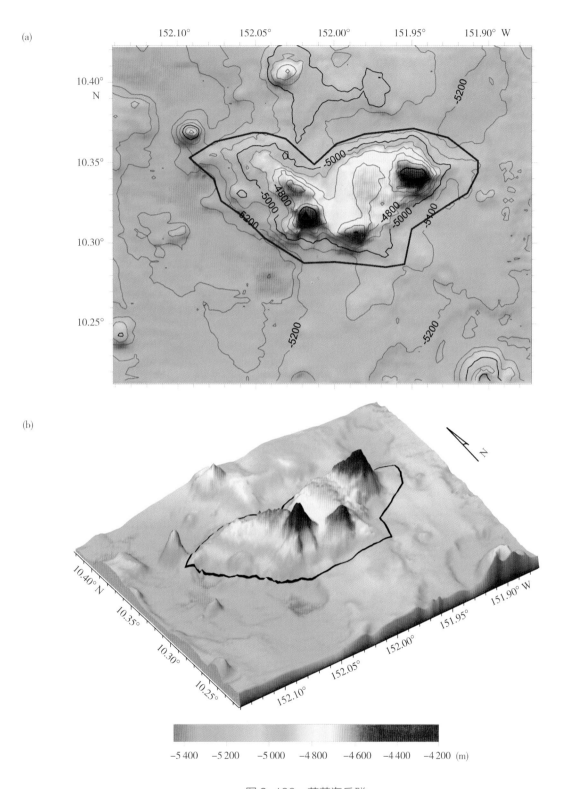

图 2-126　茑萝海丘群

(a) 地形图（等深线间隔 100 m）；(b) 三维图

Fig.2-126　Niaoluo Hills

(a) Bathymetric map (the contour interval is 100 m); (b) 3-D topographic map

2.8.4　郑和海岭
Zhenghe Ridge

中文名称 Chinese Name	郑和海岭 Zhenghe Hailing	英文名称 English Name	Zhenghe Ridge	所在大洋 Ocean or Sea	东太平洋 East Pacific Ocean
发现情况 Discovery Facts	此海岭于 1995 年 10 月由中国科考船"大洋一号"在执行 DY85-5 航次时发现。 This ridge was discovered by the Chinese R/V *Dayang Yihao* during the DY85-5 cruise in October, 1995.				
命名历史 Name History	由我国命名为郑和海岭，于 2016 年提交 SCUFN 审议通过。 Zhenghe Ridge was named by China and approved by SCUFN in 2016.				
特征点坐标 Coordinates	10°13.00′N，153°11.00′W 10°14.38′N，152°52.32′W 10°06.93′N，152°12.12′W 10°06.07′N，151°26.88′W 10°07.00′N，151°12.00′W			长 (km) × 宽 (km) Length (km) × Width (km)	220 × 25
最大水深（m） Max Depth（m）	5 450	最小水深（m） Min Depth（m）	3 612	高差（m） Total Relief（m）	1 838
地形特征 Feature Description	该海岭形态狭长，东西走向，往东延伸情况不明，现有资料显示长宽分别为 220 km 和 25 km，最高峰位于海岭东部，峰顶水深 3 612 m（图 2–127）。 This ridge has an elongated shape, and runs E to W. Its extension towards east is remained unclear. According to current data, the length and width are 220 km and 25 km respectively. The highest summit is in the east of the ridge, the top depth is 3 612 m (Fig.2–127).				
命名释义 Reason for Choice of Name	郑和（公元 1371—1433 年）是我国历史上伟大的航海家，曾 7 次下西洋，史料记载其到达的国家和地区达 36 个。此海岭命名为"郑和"，以纪念郑和在和平外交、传播中华文明、开拓航线和造船技术等方面的重要贡献。 Zheng He (A.D. 1371–1433) is the great navigator in the history of China. He had seven voyages to the western seas and arrived in 36 nations or areas, which made outstanding contributions to the exchange of Chinese and Western culture. This ridge is named after "Zhenghe", to commemorate his important contributions to Chinese navigation history.				

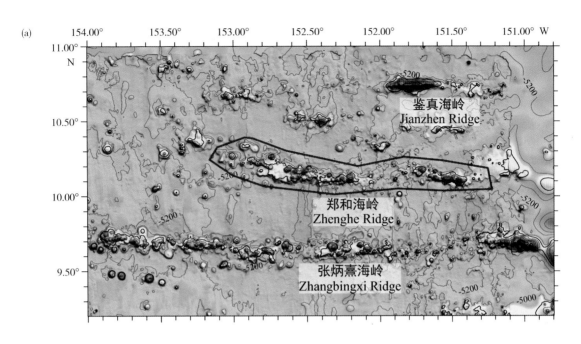

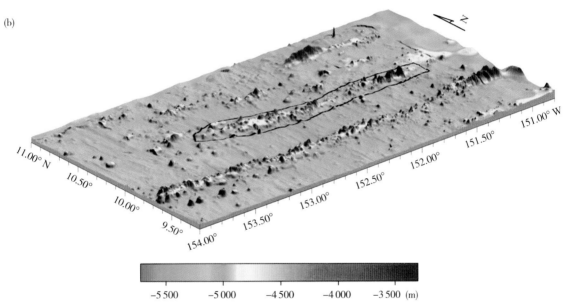

图 2-127　郑和海岭

(a) 地形图（等深线间隔 100 m）；(b) 三维图

Fig.2-127　Zhenghe Ridge

(a) Bathymetric map (the contour interval is 100 m); (b) 3-D topographic map

2.8.5 巩珍圆海丘群
Gongzhen Knolls

中文名称 Chinese Name	巩珍圆海丘群 Gongzhen Yuanhaiqiuqun	英文名称 English Name	Gongzhen Knolls	所在大洋 Ocean or Sea	东太平洋 East Pacific Ocean
发现情况 Discovery Facts					

此海丘群于 1995 年 9 月由中国科考船"大洋一号"在执行 DY85-5 航次时发现。
The knolls were discovered by the Chinese R/V *Dayang Yihao* during the DY85-5 cruise in September, 1995.

命名历史 Name History
由我国命名为巩珍圆海丘群，于 2015 年提交 SCUFN 审议通过。
Gongzhen Knolls were named by China and approved by SCUFN in 2015.

特征点坐标 Coordinates
10°33.07′N，154°27.60′W
10°31.12′N，154°32.88′W
10°29.66′N，154°28.20′W
10°29.34′N，154°32.88′W
10°28.91′N，154°25.38′W
10°27.61′N，154°30.84′W
10°25.34′N，154°25.26′W（巩珍圆海丘 / Gongzhen Knoll）

长(km)×宽(km) Length (km) × Width (km)：22 × 20

| 最大水深（m）Max Depth（m） | 5 300 | 最小水深（m）Min Depth（m） | 4 437 | 高差（m）Total Relief（m） | 863 |

地形特征 Feature Description
该圆海丘群由 7 座海丘（或圆海丘）组成。海丘群长宽分别为 22 km 和 20 km，最高峰位于巩珍圆海丘，峰顶水深 4 437 m（图 2-128）。
The knolls consist of 7 hills or knolls. The length and width of this knolls are 22 km and 20 km respectively. The highest summit is in Gongzhen Knoll, whose top water depth is 4 437 m (Fig.2-128).

命名释义 Reason for Choice of Name
巩珍（生卒年月不详，明代人），号养素生，公元 1431—1433 年随郑和下西洋，著《西洋番国志》，记录所经各国之风土人情。此海丘群命名为"巩珍"，以纪念巩珍在我国航海史上的重要贡献。
Gong Zhen (birth and death date unknown, living in Ming Dynasty), also named Yang Susheng, took part in Zheng He's sailing to the western seas during A.D. 1431–1433. He wrote a famous book, which described the custom of all the countries they went through. The knolls were named "Gongzhen" to memorize Gong Zhen's great contributions to the Chinese history of marine navigation.

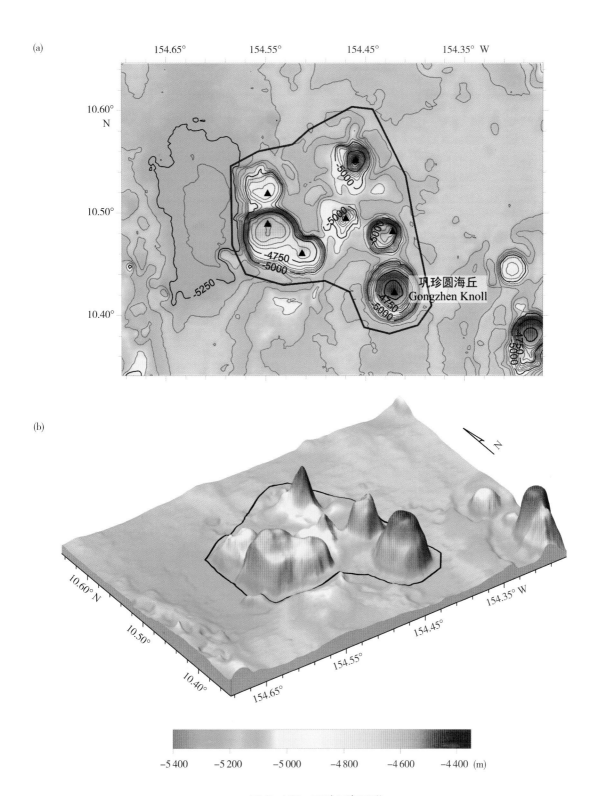

图 2-128　巩珍圆海丘群

(a) 地形图（等深线间隔 50 m）；(b) 三维图

Fig.2-128　Gongzhen Knolls

(a) Bathymetric map (the contour interval is 50 m); (b) 3-D topographic map

2.8.6 巩珍圆海丘
Gongzhen Knoll

中文名称 Chinese Name	巩珍圆海丘 Gongzhen Yuanhaiqiu		英文名称 English Name	Gongzhen Knoll	所在大洋 Ocean or Sea	东太平洋 East Pacific Ocean
发现情况 Discovery Facts	此圆海丘于 1995 年 9 月由中国科考船"大洋一号"在执行 DY85-5 航次时发现。 This knoll was discovered by the Chinese R/V *Dayang Yihao* during the DY85-5 cruise in September, 1995.					
命名历史 Name History	在我国大洋航次调查报告和 GEBCO 地名辞典中未命名。 This knoll has not been named in the Chinese cruise reports or GEBCO gazetteer.					
特征点坐标 Coordinates	10°25.34′N，154°25.26′W				长(km)×宽(km) Length (km) × Width (km)	8 × 7
最大水深（m） Max Depth（m）	5 220	最小水深（m） Min Depth（m）		4 437	高差（m） Total Relief（m）	783
地形特征 Feature Description	巩珍圆海丘位于巩珍海丘群南部，俯视平面形态呈圆形，基座直径 8 km，峰顶水深 4 437 m，山麓水深 5 220 m，高差 783 m（图 2–129）。 Gongzhen Knoll is located in the south of Gongzhen Knolls. It has a nearly round overlook plane shape with the base diameter of 8 km. The top depth is 4 437 m and the piedmont depth is 5 220 m, which makes the total relief being 783 m (Fig.2–129).					
命名释义 Reason for Choice of Name	巩珍（生卒年月不详，明代人）号养素生，公元 1431—1433 年随郑和下西洋，著《西洋番国志》，记录所经各国之风土人情。此圆海丘命名为"巩珍"，以纪念巩珍在我国航海史上的重要贡献。 Gong Zhen (birth and death date unknown, living in Ming Dynasty), also named Yang Susheng, took part in Zheng He's sailing to the western seas during A.D. 1431–1433. He wrote a famous book, which described the custom of all the countries they went through. This knoll is named "Gongzhen" to memorize Gong Zhen's great contributions to the Chinese history of marine navigation.					

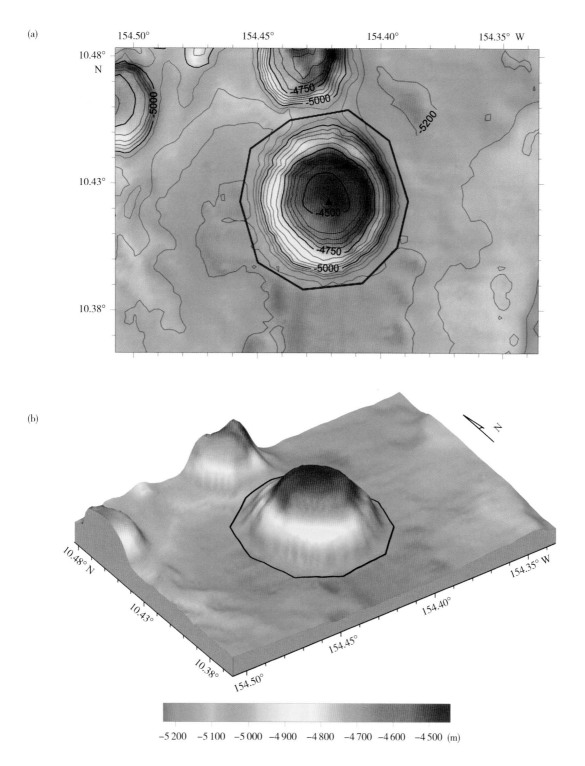

图 2-129　巩珍圆海丘

(a) 地形图（等深线间隔 50 m）；(b) 三维图

Fig.2-129　Gongzhen Knoll

(a) Bathymetric map (the contour interval is 50 m); (b) 3-D topographic map

2.8.7 维熊圆海丘
Weixiong Knoll

中文名称 Chinese Name	维熊圆海丘 Weixiong Yuanhaiqiu	英文名称 English Name	Weixiong Knoll	所在大洋 Ocean or Sea	东太平洋 East Pacific Ocean
发现情况 Discovery Facts					
命名历史 Name History					
特征点坐标 Coordinates	10°16.65′N，154°31.20′W		长 (km) × 宽 (km) Length (km) × Width (km)		7 × 5
最大水深（m） Max Depth（m）	5 200	最小水深（m） Min Depth（m）	4 538	高差（m） Total Relief（m）	662

发现情况 / Discovery Facts:

此圆海丘于 1995 年 9 月由中国科考船"大洋一号"在执行 DY85-5 航次时发现。

This knoll was discovered by the Chinese R/V *Dayang Yihao* during the DY85-5 cruise in September, 1995.

命名历史 / Name History:

由我国命名为维熊圆海丘，于 2017 年提交 SCUFN 审议通过。

Weixiong Knoll was named by China and approved by SCUFN in 2017.

地形特征 / Feature Description:

该圆海丘位于维蛇海丘西北 38 km，维罴圆海丘以西 68 km。圆海丘俯视平面形态近椭圆形，SN 走向，长宽分别 7 km 和 5 km，圆海丘顶部水深 4 538 m，山麓水深 5 200 m，高差 662 m（图 2–130）。

Weixiong Knoll is located in 38 km northwest to Weishe Hill, and 68 km west to Weipi Knoll. The knoll has a nearly elliptic overlook plane shape, running N to S. The length and width are 7 km and 5 km respectively. The top depth of the knoll is 4 538 m and the piedmont depth is 5 200 m, which makes the total relief being 662 m (Fig.2–130).

命名释义 / Reason for Choice of Name:

"维熊"出自《诗经·小雅·斯干》"吉梦维何？维熊维罴，维虺维蛇"。熊、罴皆为猛兽，虺、蛇皆为毒蛇。在古代梦到熊、蛇均为吉兆，梦见熊、罴可生男孩，梦见虺、蛇可生女孩。此诗歌为周朝贵族庆贺宫殿落成所唱，表现轻松快乐的生活场景。这 4 个海山海丘分别命名为维熊、维罴、维虺、维蛇，构成群组化系列地名。

"Weixiong" comes from a poem named *Sigan* in *Shijing · Xiaoya*. *Shijing* is a collection of ancient Chinese poems from 11th century B.C. to 6th century B.C. "What did you see in your good dream? I saw bears big and small; I saw reptiles, snake and all." This poem described a good dream which has the bear and snake, "Xiong" and "Pi" mean bears while "Hui" and "She" mean snakes in Chinese language, which is believed to be a good omen for bearing a baby. These 4 hills or knolls are named Weixiong, Weipi, Weihui, and Weishe respectively, constituting a group of feature names.

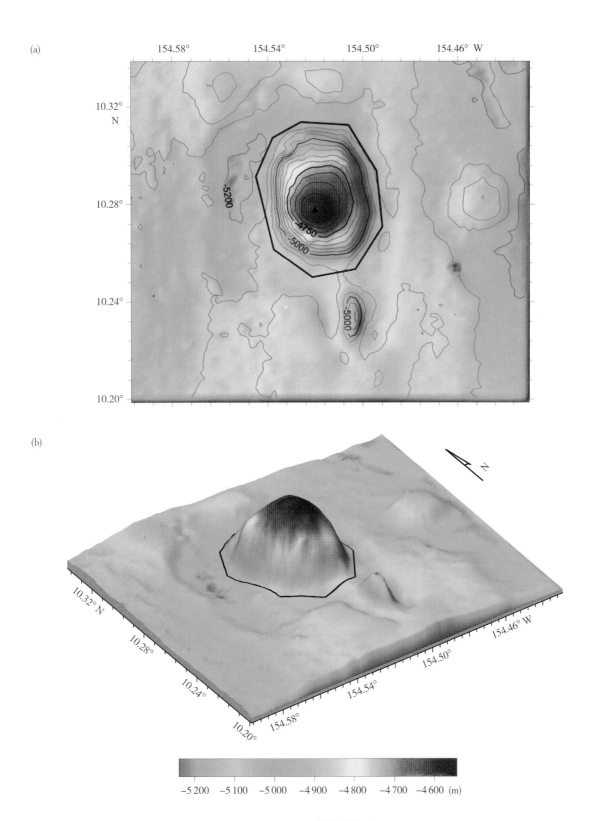

图 2-130　维熊圆海丘

(a) 地形图（等深线间隔 50 m）；(b) 三维图

Fig.2-130　Weixiong Knoll

(a) Bathymetric map (the contour interval is 50 m); (b) 3-D topographic map

2.8.8 维罴圆海丘
Weipi Knoll

中文名称 Chinese Name	维罴圆海丘 Weipi Yuanhaiqiu	英文名称 English Name	Weipi Knoll	所在大洋 Ocean or Sea	东太平洋 East Pacific Ocean
发现情况 Discovery Facts	此圆海丘于 1995 年 9 月由中国科考船"大洋一号"在执行 DY85-5 航次时发现。 This knoll was discovered by the Chinese R/V *Dayang Yihao* during the DY85-5 cruise in September, 1995.				
命名历史 Name History	由我国命名为维罴圆海丘，于 2016 年提交 SCUFN 审议通过。 Weipi Knoll was named by China and approved by SCUFN in 2016.				
特征点坐标 Coordinates	10°18.92′N，153°51.72′W			长(km)×宽(km) Length (km)× Width (km)	13×8
最大水深（m） Max Depth（m）	5 259	最小水深（m） Min Depth（m）	4 685	高差（m） Total Relief（m）	574
地形特征 Feature Description	该圆海丘位于维熊圆海丘以东 68 km，维虺海丘以北 38 km。圆海丘俯视平面形态呈椭圆形，近 SN 走向，长宽分别为 13 km 和 8 km，顶部水深 4 685 m，山麓水深 5 259 m，高差 574 m（图 2–131）。 Weipi Knoll is located in 68 km east to Weixiong Knoll and 38 km north to Weihui Hill. This knoll has an elliptic overlook plane shape, running nearly N to S. The length and width are 13 km and 8 km respectively. The top depth of this knoll is 4 685 m and the piedmont depth is 5 259 m, which makes the total relief being 574 m (Fig.2–131).				
命名释义 Reason for Choice of Name	"维罴"出自《诗经·小雅·斯干》"吉梦维何？维熊维罴,维虺维蛇"。熊、罴皆为猛兽,虺、蛇皆为毒蛇。此句意为梦到熊、蛇均为吉兆,梦见熊、罴可生男孩,梦见虺、蛇可生女孩。此诗歌为周朝贵族庆贺宫殿落成所唱,表现轻松快乐的生活场景。这 4 个海山海丘分别命名为维熊、维罴、维虺、维蛇,构成群组化系列地名。 "Weipi" comes from a poem named *Sigan* in *Shijing · Xiaoya*. *Shijing* is a collection of ancient Chinese poems from 11th century B.C. to 6th century B.C. "What did you see in your good dream? I saw bears big and small; I saw reptiles, snake and all." This poem described a good dream which has the bear and snake, "Xiong" and "Pi" mean bears while "Hui" and "She" mean snakes in Chinese language, which is believed to be a good omen for bearing a baby. These 4 hills or knolls are named Weixiong, Weipi, Weihui, and Weishe respectively, constituting a group of feature names.				

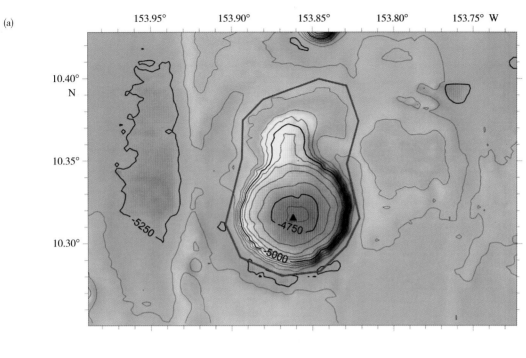

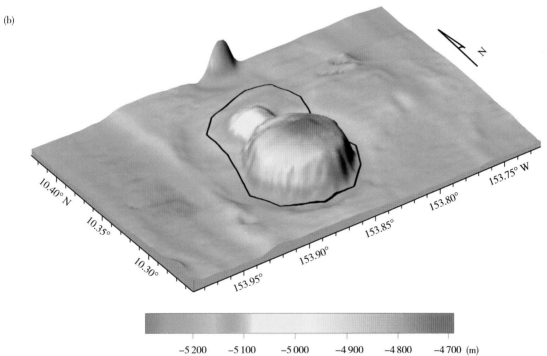

图 2-131　维罳圆海丘

(a) 地形图（等深线间隔 50 m）；(b) 三维图

Fig.2-131　Weipi Knoll

(a) Bathymetric map (the contour interval is 50 m); (b) 3-D topographic map

2.8.9 维虺海丘
Weihui Hill

中文名称 Chinese Name	维虺海丘 Weihui Haiqiu		英文名称 English Name	Weihui Hill	所在大洋 Ocean or Sea	东太平洋 East Pacific Ocean
发现情况 Discovery Facts	此海丘于 1995 年 9 月由中国科考船"大洋一号"在执行 DY85-5 航次时发现。 This hill was discovered by the Chinese R/V *Dayang Yihao* during the DY85-5 cruise in September, 1995.					
命名历史 Name History	由我国命名为维虺海丘，于 2017 年提交 SCUFN 审议通过。 Weihui Hill was named by China and approved by SCUFN in 2017.					
特征点坐标 Coordinates	9°58.62′N，154°00.00′W			长(km)×宽(km) Length (km) × Width (km)		6.5 × 5.7
最大水深（m） Max Depth（m）	5 231		最小水深（m） Min Depth（m）	4 746	高差（m） Total Relief（m）	485
地形特征 Feature Description	该海丘位于维罴圆海丘以南 38 km，维蛇海丘以东 38 km。海丘俯视平面形态近圆形，长宽分别 6.5 km 和 5.7 km，顶部水深 4 746 m，山麓水深 5 231 m，高差 485 m（图 2–132）。 Weihui Hill is located in 38 km south to Weipi Knoll，38 km east to Weishe Hill. This hill has a nearly round overlook plane shape. The length and width are 6.5 km and 5.7 km respectively. The top depth of this hill is 4 746 m and the piedmont depth is 5 231 m, which makes the total relief being 485 m (Fig.2–132).					
命名释义 Reason for Choice of Name	"维虺"出自《诗经·小雅·斯干》"吉梦维何？维熊维罴，维虺维蛇"。熊、罴皆为猛兽，虺、蛇皆为毒蛇。此句意为梦到熊、蛇均为吉兆，梦见熊、罴可生男孩，梦见虺、蛇可生女孩。此诗歌为周朝贵族庆贺宫殿落成所唱，表现轻松快乐的生活场景。这 4 个海山海丘分别命名为维熊、维罴、维虺、维蛇，构成群组化系列地名。 "Weihui" comes from a poem named *Sigan* in *Shijing · Xiaoya*. *Shijing* is a collection of ancient Chinese poems from 11th century B.C. to 6th century B.C. "What did you see in your good dream? I saw bears big and small; I saw reptiles, snake and all." This poem described a good dream which has the bear and snake, "Xiong" and "Pi" mean bears while "Hui" and "She" mean snakes in Chinese language, which is believed to be a good omen for bearing a baby. These 4 hills or knolls are named Weixiong, Weipi, Weihui, and Weishe respectively, constituting a group of feature names.					

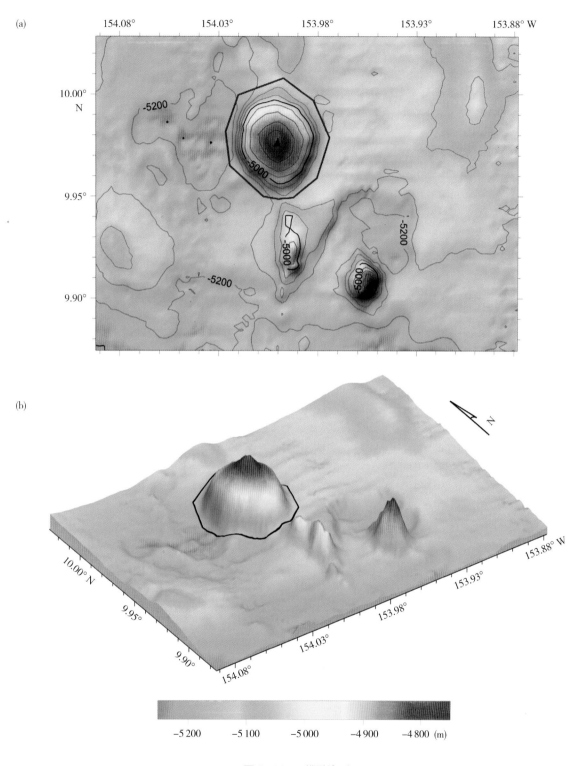

图 2-132　维妣海丘

(a) 地形图（等深线间隔 50 m）；(b) 三维图

Fig.2-132　Weihui Hill

(a) Bathymetric map (the contour interval is 50 m); (b) 3-D topographic map

2.8.10 维蛇海丘
Weishe Hill

中文名称 Chinese Name	维蛇海丘 Weishe Haiqiu		英文名称 English Name	Weishe Hill	所在大洋 Ocean or Sea	东太平洋 East Pacific Ocean
发现情况 Discovery Facts	此海丘于 1995 年 9 月由中国科考船"大洋一号"在执行 DY85-5 航次时发现。 This hill was discovered by the Chinese R/V *Dayang Yihao* during the DY85-5 cruise in September, 1995.					
命名历史 Name History	由我国命名为维蛇海丘，于 2017 年提交 SCUFN 审议通过。 Weishe Hill was named by China and approved by SCUFN in 2017.					
特征点坐标 Coordinates	9°56.89′N，154°21.54′W			长(km)×宽(km) Length (km)× Width (km)		12×8
最大水深（m） Max Depth（m）	5 240	最小水深（m） Min Depth（m）	4 903	高差（m） Total Relief（m）		337
地形特征 Feature Description	该海丘位于魏源海山北东 20 km。海丘俯视平面呈椭圆形，近 SN 走向，长宽分别为 12 km 和 8 km，顶部水深 4 903 m，山麓水深 5 240 m，高差 337 m（图 2–133）。 Weishe Hill is located in 20 km northeast to Weiyuan Seamount. The overlook plane shape of this hill is elliptic and it runs nearly N to S. The length and width is 12 km and 8 km respectively. The top depth and piedmont depth is 4 903 m and 5 240 m which makes the total relief being 337 m (Fig.2–133).					
命名释义 Reason for Choice of Name	"维蛇"出自《诗经·小雅·斯干》"吉梦维何？维熊维罴，维虺维蛇"。熊、罴皆为猛兽，虺、蛇皆为毒蛇。此句意为梦到熊、蛇均为吉兆，梦见熊、罴可生男孩，梦见虺、蛇可生女孩。此诗歌为周朝贵族庆贺宫殿落成所唱，表现轻松快乐的生活场景。这 4 个海山海丘分别命名为维熊、维罴、维虺、维蛇，构成群组化系列地名。 "Weishe" comes from a poem named *Sigan* in *Shijing · Xiaoya*. *Shijing* is a collection of ancient Chinese poems from 11th century B.C. to 6th century B.C. "What did you see in your good dream? I saw bears big and small; I saw reptiles, snake and all." This poem described a good dream which has the bear and snake, "Xiong" and "Pi" mean bears while "Hui" and "She" mean snakes in Chinese language, which is believed to be a good omen for bearing a baby. These 4 hills or knolls are named Weixiong, Weipi, Weihui, and Weishe respectively, constituting a group of feature names.					

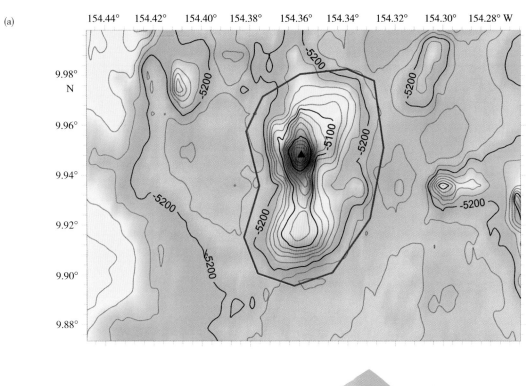

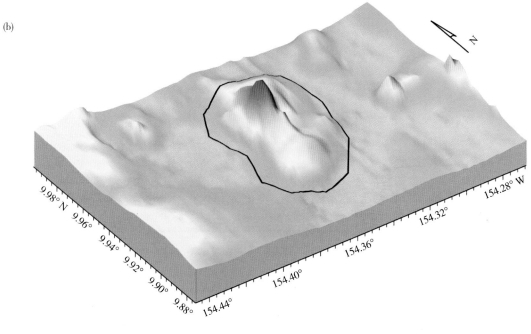

图 2-1313　维蛇海丘

(a) 地形图 (等深线间隔 20 m)；(b) 三维图

Fig.2-133　Weishe Hill

(a) Bathymetric map (the contour interval is 20 m); (b) 3-D topographic map

2.8.11 达奚通圆海丘
Daxitong Knoll

中文名称 Chinese Name	达奚通圆海丘 Daxitong Yuanhaiqiu	英文名称 English Name	Daxitong Knoll	所在大洋 Ocean or Sea	东太平洋 East Pacific Ocean
发现情况 Discovery Facts	此圆海丘于 1995 年 9 月由中国科考船 "大洋一号" 在执行 DY85-5 航次时发现。 This knoll was discovered by the Chinese R/V *Dayang Yihao* during the DY85-5 cruise in September, 1995.				
命名历史 Name History	由我国命名为达奚通圆海丘，于 2015 年提交 SCUFN 审议通过。 Daxitong Knoll was named by China and approved by SCUFN in 2015.				
特征点坐标 Coordinates	10°04.45′N，154°42.78′W			长(km)×宽(km) Length (km) × Width (km)	9 × 8
最大水深（m） Max Depth（m）	5 080	最小水深（m） Min Depth（m）	4 300	高差（m） Total Relief（m）	780
地形特征 Feature Description	该圆海丘俯视平面形态呈圆形，顶部地形相对平坦，周缘地形陡峭，峰顶水深 4 300 m，山麓水深 5 080 m，高差 780 m（图 2–134）。 This knoll has a nearly round overlook plane shape. The top of the knoll is flat while the sides are steep. The top depth is about 4 300 m and the piedmont depth is 5 080 m, which makes the total relief being 780 m (Fig.2–134).				
命名释义 Reason for Choice of Name	达奚通（生卒年月不详）是唐代武则天时人，曾航海西行经 36 国到达阿拉伯半岛南部，归国后曾著《海南诸蕃行记》。此圆海丘命名为 "达奚通"，以纪念达奚通在我国航海史上的重要贡献。 Da Xitong (birth and death date unknown), a Chinese navigator during the Tang Dynasty, in recognition of his significant contributions to the Chinese history of marine navigation. He sailed westbound, visiting 36 countries and went as far as the southern part of Arabian Peninsula.				

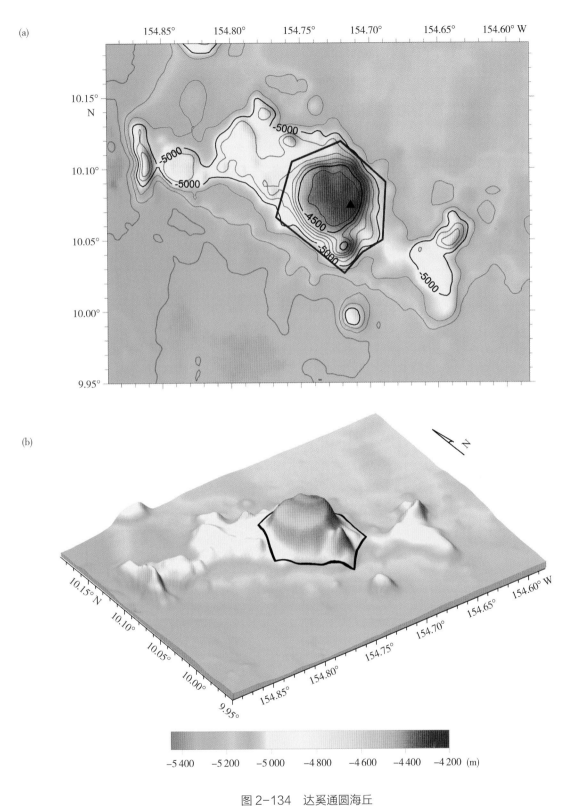

图 2-134　达奚通圆海丘

(a) 地形图（等深线间隔 100 m）；(b) 三维图

Fig.2-134　Daxitong Knoll

(a) Bathymetric map (the contour interval is 100 m); (b) 3-D topographic map

2.8.12 魏源海山
Weiyuan Seamount

中文名称 Chinese Name	魏源海山 Weiyuan Haishan	英文名称 English Name	Weiyuan Seamount	所在大洋 Ocean or Sea	东太平洋 East Pacific Ocean
发现情况 Discovery Facts	此海山于1995年9月由中国科考船"大洋一号"在执行 DY85-5 航次时发现。 This seamount was discovered by the Chinese R/V *Dayang Yihao* during the DY85-5 cruise in September, 1995.				
命名历史 Name History	由我国命名为魏源海山，于2012年提交 SCUFN 审议通过。 Weiyuan Seamount was named by China and approved by SCUFN in 2012.				
特征点坐标 Coordinates	9°48.40′N，154°31.80′W			长(km)×宽(km) Length (km)× Width (km)	20×17
最大水深（m） Max Depth（m）	5 200	最小水深（m） Min Depth（m）	2 950	高差（m） Total Relief（m）	2 250
地形特征 Feature Description	该海山近圆锥状，峰顶水深2 950 m，山麓水深5 200 m，高差2 250 m（图2–135）。 Weiyuan Seamount has a nearly conical shape. The top depth of this seamount is 2 950 m, and the piedmont depth is 5 200 m, which makes the total relief being 2 250 m (Fig.2–135).				
命名释义 Reason for Choice of Name	魏源（公元1794—1857年），清代著名思想家、政治家和文学家。著有《海国图志》，介绍各国地理、历史、科技，附世界地图、各大洲地图、分国地图，是我国最早的世界地理学著作之一。此海山命名为"魏源"，以纪念魏源在我国文化史上的重要贡献。 Wei Yuan (A.D. 1794–1857), a Chinese thinker, politician and litterateur during the Qing Dynasty in China. He wrote a book which is a geographical and historical monumental work in the late Qing Dynasty. This seamount is named Weiyuan to commemorate the great contributions in Chinese culture.				

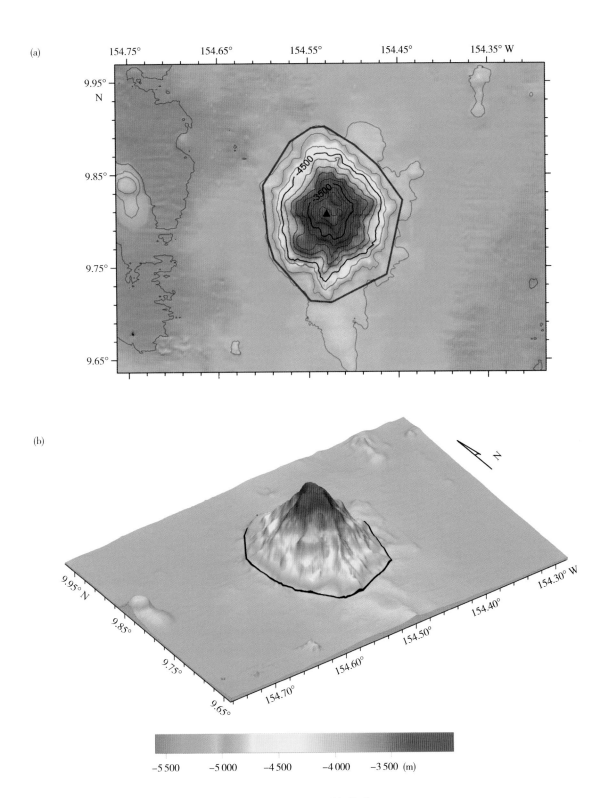

图 2-135　魏源海山

(a) 地形图（等深线间隔 200 m）；(b) 三维图

Fig.2-135　Weiyuan Seamount

(a) Bathymetric map (the contour interval is 200 m); (b) 3-D topographic map

2.8.13 张炳熹海岭
Zhangbingxi Ridge

中文名称 Chinese Name	张炳熹海岭 Zhangbinxi Hailing	英文名称 English Name	Zhangbingxi Ridge	所在大洋 Ocean or Sea	东太平洋 East Pacific Ocean
发现情况 Discovery Facts	此海岭于 1995 年 9 月由中国科考船"大洋一号"在执行 DY85-5 航次时发现。 This ridge was discovered by the Chinese R/V *Dayang Yihao* during the DY85-5 cruise in September, 1995.				
命名历史 Name History	由我国命名为张炳熹海岭，于 2015 年提交 SCUFN 审议通过。 Zhangbingxi Ridge was named by China and approved by SCUFN in 2015.				
特征点坐标 Coordinates	9°46.35′N，154°17.64′W 9°43.01′N，153°49.68′W 9°39.93′N，152°38.22′W 9°36.96′N，151°53.88′W 9°42.26′N，151°06.60′W			长(km)×宽(km) Length (km) × Width (km)	350×30
最大水深（m） Max Depth（m）	5 500	最小水深（m） Min Depth（m）	4 050	高差（m） Total Relief（m）	1 450
地形特征 Feature Description	该海岭形态狭长，东西走向，往东延伸情况不明，现有资料显示长宽分别为 350 km 和 30 km，最高峰位于西部的清高海山，峰顶水深 4 050 m（图 2–136）。 This ridge has an elongated shape, running E to W. Its extension towards east is remained unclear. According to current data, the length and width are 350 km and 30 km respectively. The highest summit of this ridge is Qinggao Seamount in the west, and the top depth is 4 050 m (Fig.2–136).				
命名释义 Reason for Choice of Name	张炳熹（公元 1919—2000 年），中国著名地质学家，中国科学院院士，国际地质科学联合会（IUGS）副主席、联合国国际海底管理局筹备委员会审查先驱投资申请专家小组成员。张炳熹为中国的地质教育、地质科学研究和地质矿产资源的普查勘探、制定国家科技发展规划和对外交流等都作出了重要的贡献。 Zhang Bingxi (1919–2000), a Chinese geologist, member of the Chinese Academy of Sciences. He was vice chairman of the International Union of Geological Science. He had a significant contribution to the geological education, geological research and exploration, making the strategic planning for national sciences and technology, and international collaborations.				

(a)

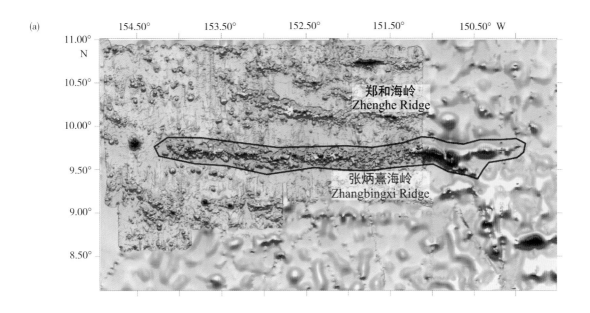

(b)

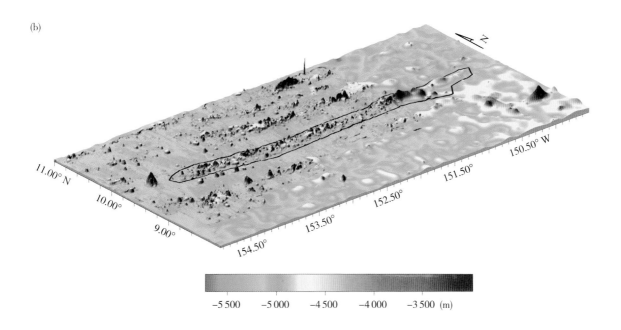

图 2-136 张炳熹海岭

(a) 地形图（等深线间隔 100 m）；(b) 三维图

Fig.2-136 Zhangbingxi Ridge

(a) Bathymetric map (the contour interval is 100 m); (b) 3-D topographic map

2.8.14　清高海山
Qinggao Seamount

中文名称 Chinese Name	清高海山 Qinggao Haishan		英文名称 English Name	Qinggao Seamount	所在大洋 Ocean or Sea	东太平洋 East Pacific Ocean
发现情况 Discovery Facts	此海山于 1995 年 9 月由中国科考船 "大洋一号" 在执行 DY85-5 航次时发现。 This seamount was discovered by the Chinese R/V *Dayang Yihao* during the DY85-5 cruise in September, 1995.					
命名历史 Name History	2013 年由中国大洋协会命名为清高海山，"蛟龙"号载人潜水器在该海山南侧下潜。 Qinggao Seamount was named by China Ocean Mineral Resources R & D Association in 2013. The manned submersible *Jiaolong* used to dive at the south of this seamount.					
特征点坐标 Coordinates	9°43.00′N，153°49.70′W				长(km)×宽(km) Length (km) × Width (km)	24×23
最大水深（m） Max Depth（m）	5 250	最小水深（m） Min Depth（m）		4 050	高差（m） Total Relief（m）	1 200
地形特征 Feature Description	该海山位于张炳熹海岭西端，长宽分别为 24 km 和 23 km，峰顶水深 4 050 m，山麓水深 5 250 m，高差 1 200 m（图 2–137）。 Qinggao Seamount is located in the west of Zhangbingxi Ridge. The length and width are 24 km and 23 km respectively. The top depth is 4 050 m and the piedmont depth is 5 250 m, which makes the total relief being 1 200 m (Fig.2–137).					
命名释义 Reason for Choice of Name	谢清高（公元 1765—1821 年），广东人，自 18 岁开始随着商船游历世界各国，后因双目失明方停止航海生涯。他根据记忆口述，请人笔录，著成《海录》一书，详细记录所到各国的政治制度、风土人情、物产、语言等方面的情况，其中包括当时的欧洲强国英国等国的风俗、政治制度。此书成为当时的中国人睁开眼睛看世界的最重要著作。此海山命名为"清高"，以纪念谢清高在我国航海史上的重要贡献。 Xie Qinggao (A.D. 1765–1821) was born in Guangdong Province of China. He travelled around the world from the age of 18 with business ship, but finished it because of his blindness. According to his memory, he wrote a famous book with someone's help, which recorded the policy, custom, products and language of every country where he had ever reached, including the European developed countries at that time such as Britain. It soon became an important work that helped the Chinese people recognized the whole world. This seamount was named Qinggao, in memory of his significant contributions to the Chinese history of marine navigation.					

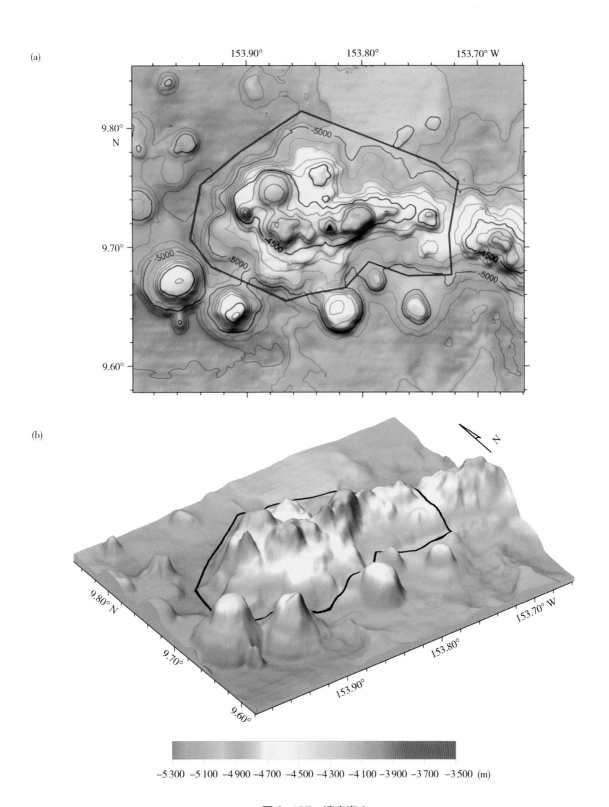

图 2-137　清高海山

(a) 地形图（等深线间隔 100 m）；(b) 三维图

Fig.2-137　Qinggao Seamount

(a) Bathymetric map (the contour interval is 100 m); (b) 3-D topographic map

2.8.15 芳伯海山
Fangbo Seamount

中文名称 Chinese Name	芳伯海山 Fangbo Haishan	英文名称 English Name	Fangbo Seamount	所在大洋 Ocean or Sea	东太平洋 East Pacific Ocean
发现情况 Discovery Facts	此海山于 1995 年 8 月由中国科考船"大洋一号"在执行 DY85-5 航次时发现。 This seamount was discovered by the Chinese R/V *Dayang Yihao* during the DY85-5 cruise in August, 1995.				
命名历史 Name History	由我国命名为芳伯海山,于 2014 年提交 SCUFN 审议通过。 Fangbo Seamount was named by China and approved by SCUFN in 2014.				
特征点坐标 Coordinates	9°40.10′N,152°38.50′W			长(km)× 宽(km) Length (km) × Width (km)	10 × 11
最大水深(m) Max Depth(m)	5 100	最小水深(m) Min Depth(m)	4 014	高差(m) Total Relief(m)	1 086
地形特征 Feature Description	该海山位于张炳熹海岭中部。海山呈圆锥形,基座直径 10 ~ 11 km,峰顶水深 4 014 m,山麓水深 5 100 m,高差 1 086 m,山坡地形陡峭(图 2–138)。 Fangbo Seamount is located in the middle of Zhangbingxi Ridge. It has a conical shape with the base diameter of 10–11 km. The top depth is 4 014 m and the piedmont depth is 5 100 m, which makes the total relief being 1 086 m. The seamount has steep slopes (Fig.2–138).				
命名释义 Reason for Choice of Name	罗芳伯(公元 1738—1795 年),与他的一百余名亲友漂洋过海,来到盛产金矿和钻石的婆罗洲(即加里曼丹岛)。他增进了早期中国与外国的交流,事迹被记载于谢清高所著的《海录》中。此海山命名为"芳伯",以纪念罗芳伯在我国航海史上的重要贡献。 Luo Fangbo (A.D. 1738–1795) used to navigate to Borneo (Kalimantan), which is abound in gold ores and diamonds with more than 100 relatives and friends. He promoted the communication between China and foreign countries. His stories are recorded in a famous book written by Xie Qinggao. This seamount is named after Fangbo, in memory of his significant contributions to the Chinese history of marine navigation.				

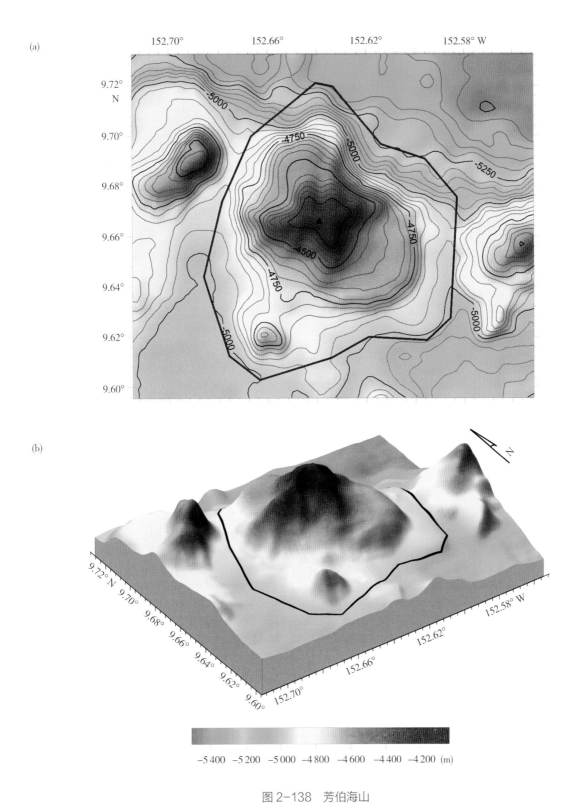

图 2-138　芳伯海山

(a) 地形图（等深线间隔 50 m）；(b) 三维图

Fig.2-138　Fangbo Seamount

(a) Bathymetric map (the contour interval is 50 m); (b) 3-D topographic map

2.8.16　斯翼海丘
Siyi Hill

中文名称 Chinese Name	斯翼海丘 Siyi Haiqiu		英文名称 English Name	Siyi Hill	所在大洋 Ocean or Sea	东太平洋 East Pacific Ocean
发现情况 Discovery Facts	此海丘于 1995 年 9 月由中国科考船"大洋一号"在执行 DY85-5 航次时发现。 This hill was discovered by the Chinese R/V *Dayang Yihao* during the DY85-5 cruise in September, 1995.					
命名历史 Name History	在我国大洋航次调查报告和 GEBCO 地名辞典中未命名。 This hill has not been named in the Chinese cruise reports or GEBCO gazetteer.					
特征点坐标 Coordinates	9°28.43′N，154°03.78′W				长(km)×宽(km) Length (km)× Width (km)	12×12
最大水深（m） Max Depth（m）	5 288	最小水深（m） Min Depth（m）	4 301		高差（m） Total Relief（m）	987
地形特征 Feature Description	该海丘位于斯寝海山以西 50 km。海丘俯视平面形态呈圆形，基座直径 12 km，顶部水深 4 301 m，山麓水深 5 288 m，高差 987 m（图 2–139）。 Siyi Hill is located in 50 km west to Siqin Seamount. This hill has a nearly round overlook plane shape with the base diameter of 12 km. The top depth is 4 301 m and the piedmont depth is 5 288 m, which makes the total relief being 987 m (Fig.2–139).					
命名释义 Reason for Choice of Name	"斯翼"出自《诗经·小雅·斯干》"如跂斯翼，如矢斯棘"。本篇为周朝贵族祝贺宫殿落成的歌辞，表现轻松快乐的生活场景。意为宫殿布局方正工整，设计美轮美奂，如跂斯翼意为宫殿雄伟大方，如人肃立。 "Siyi" comes from a poem named *Sigan* in *Shijing · Xiaoya*. *Shijing* is a collection of ancient Chinese poems from 11th century B.C. to 6th century B.C. "As steady as a man standing, as straight as an arrow flying." This poem was written for the celebration of a new palace. It described the scene of a relaxed and happy life. "As steady as a man standing" means the palace is brilliant, neat and well designed.					

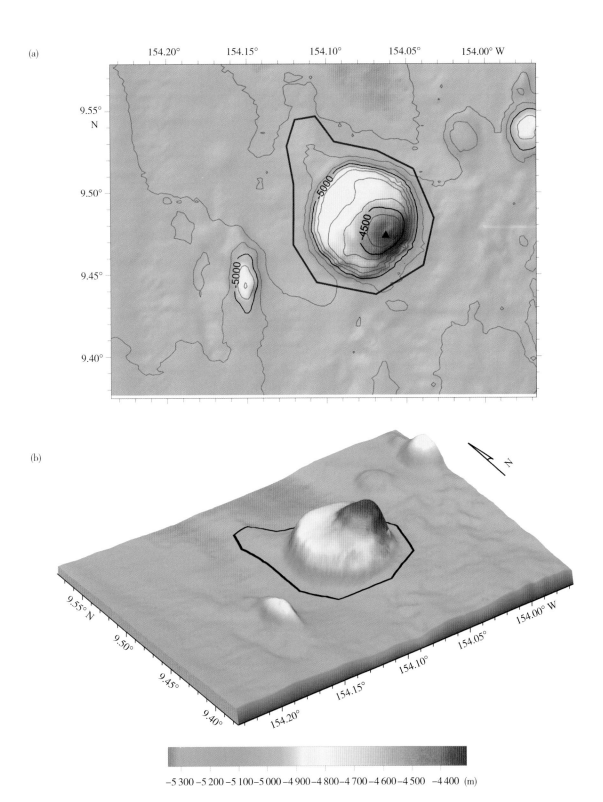

图 2-139　斯翼海丘

(a) 地形图（等深线间隔 100 m）；(b) 三维图

Fig.2-139　Siyi Hill

(a) Bathymetric map (the contour interval is 100 m); (b) 3-D topographic map

2.8.17 斯寝海山
Siqin Seamount

中文名称 Chinese Name	斯寝海山 Siqin Haishan	英文名称 English Name	Siqin Seamount	所在大洋 Ocean or Sea	东太平洋 East Pacific Ocean
发现情况 Discovery Facts	此海山于 1995 年 9 月由中国科考船 "大洋一号" 在执行 DY85-5 航次时发现。 This seamount was discovered by the Chinese R/V *Dayang Yihao* during the DY85-5 cruise in September, 1995.				
命名历史 Name History	在我国大洋航次调查报告和 GEBCO 地名辞典中未命名。 This seamount has not been named in the Chinese cruise reports or GEBCO gazetteer.				
特征点坐标 Coordinates	9°27.46′N，153°34.44′W			长(km)×宽(km) Length (km) × Width (km)	15×9
最大水深（m） Max Depth（m）	5 230	最小水深（m） Min Depth（m）	4 132	高差（m） Total Relief（m）	1 098
地形特征 Feature Description	该海山位于斯翼海丘以东 50 km。海山东西走向，长宽分别 15 km 和 9 km，峰顶水深 4 132 m，山麓水深为 5 230 m，高差 1 098 m（图 2–140）。 Siqin Seamount is located in 50 km east to Siyi Hill, running E to W. The length and width are 15 km and 9 km respectively. The top depth is 4 132 m and the piedmont depth is 5 230 m, which makes the total relief being 1 098 m (Fig.2–140).				
命名释义 Reason for Choice of Name	"斯寝" 出自《诗经·小雅·斯干》"下莞上簟，乃安斯寝"，指睡眠。本篇为周朝贵族祝贺宫殿落成的歌辞，表现轻松快乐的生活场景。此句描述卧室的床上陈设，下面铺垫蒲席，上面铺簟，非常舒适。 "Siqin" comes from a poem named *Sigan* in *Shijing · Xiaoya*. *Shijing* is a collection of ancient Chinese poems from 11th century B.C. to 6th century B.C. "Bamboo mat above, rush mat below, the king sleeps soundly till cock-crow." This poem was written for the celebration of a new palace. It described the scene of a relaxed and happy life. "Siqin" means the bedroom is comfortable and suitable for a good rest.				

(a)

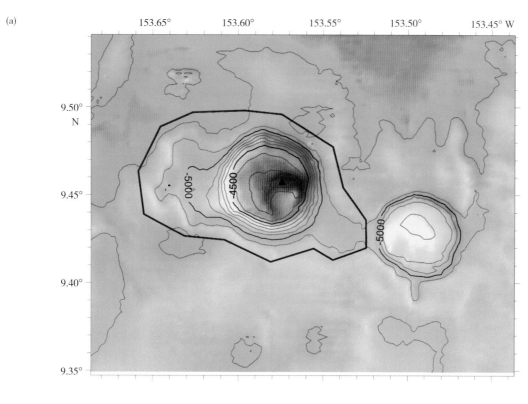

(b)

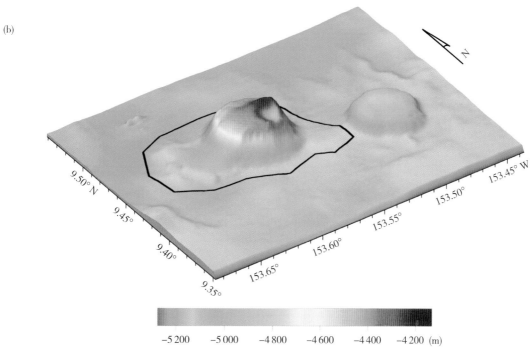

图 2-140 斯寝海山

(a) 地形图（等深线间隔 100 m）；(b) 三维图

Fig.2-140 Siqin Seamount

(a) Bathymetric map (the contour interval is 100 m); (b) 3-D topographic map

2.8.18 郑庭芳海山群
Zhengtingfang Seamounts

中文名称 Chinese Name	郑庭芳海山群 Zhengtingfang Haishanqun	英文名称 English Name	Zhengtingfang Seamounts	所在大洋 Ocean or Sea	东太平洋 East Pacific Ocean
发现情况 Discovery Facts	此海山群于 1995 年 9 月由中国科考船"大洋一号"在执行 DY85-5 航次时发现。 The seamounts were discovered by the Chinese R/V *Dayang Yihao* during the DY85-5 cruise in September, 1995.				
命名历史 Name History	由我国命名为郑庭芳海山群，于 2015 年提交 SCUFN 审议通过。 Zhengtingfang Seamounts were named by China and approved by SCUFN in 2015.				
特征点坐标 Coordinates	9°06.46′N，154°13.68′W 9°00.52′N，154°04.14′W(峰顶 /top)			长(km)×宽(km) Length (km)× Width (km)	32×11
最大水深（m） Max Depth（m）	5 200	最小水深（m） Min Depth（m）	3 920	高差（m） Total Relief（m）	1 280
地形特征 Feature Description	该海山群包含多个海山海丘，俯视平面形态不规则，长宽分别 32 km 和 11 km，最高峰位于海山群中部，峰顶水深 3 920 m（图 2–141）。 Zhengtingfang Seamounts include several seamounts or hills, which has an irregular overlook plane shape. The length and width are 32 km and 11 km respectively. The highest summit is in the middle of the seamounts, and its top depth is 3 920 m (Fig.2–141).				
命名释义 Reason for Choice of Name	郑庭芳（公元 1857—1920 年），海南人，自幼随父开始远洋航行与海洋捕捞，成人后长期担任船长，闯荡西南沙与东南亚各国，远至非洲、欧洲等地。他从长期航海生涯中掌握了高超的航海技术，包括海洋气候的预报等技术。他是开发南海西南沙的先驱者之一。此海山群命名为"郑庭芳"，以纪念郑庭芳在我国航海史上的重要贡献。 Zheng Tingfang (A.D. 1857–1920) was born in Hainan Province of China. He began his voyage with his father in his childhood time, and had become a captain when he grew up. He had ever been to Xisha Islands and Southeast Asia, even to Europe and Africa. From the long term of sailing, he mastered outstanding navigation techniques including forecasting marine climate during his voyage, also, he is one of the pioneers in development in the South China Sea and Xisha Islands. The seamounts were named after Zhengtingfang, in memory of his significant contributions to the Chinese history of marine navigation.				

(a)

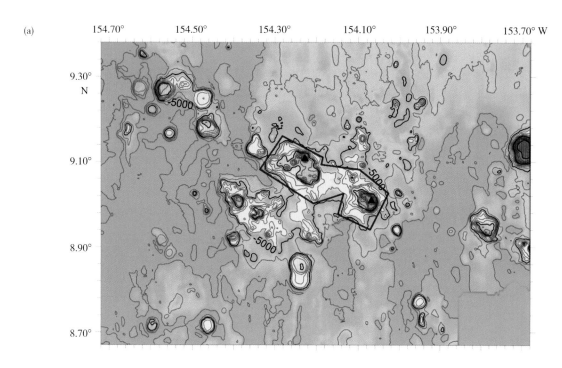

(b)

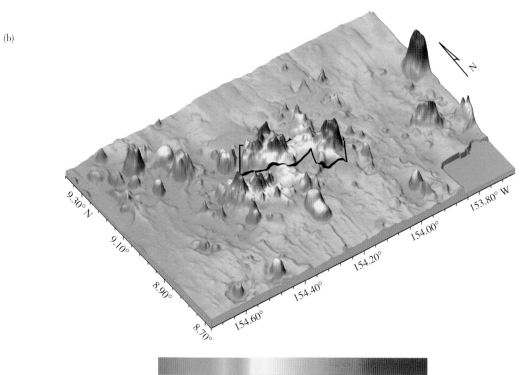

−5 600　−5 400　−5 200　−5 000　−4 800　−4 600　−4 400　−4 200　−4 000　−3 800　−3 600　(m)

图 2-141　郑庭芳海山群

(a) 地形图（等深线间隔 100 m）；(b) 三维图

Fig.2-141　Zhengtingfang Seamounts

(a) Bathymetric map (the contour interval is 100 m); (b) 3-D topographic map

2.8.19 陈伦炯海山群
Chenlunjiong Seamounts

中文名称 Chinese Name	陈伦炯海山群 Chenlunjiong Haishanqun	英文名称 English Name	Chenlunjiong Seamounts	所在大洋 Ocean or Sea	东太平洋 East Pacific Ocean
发现情况 Discovery Facts	此海山群于 1995 年 9 月由中国科考船"大洋一号"在执行 DY85-5 航次时发现。 The seamounts were discovered by the Chinese R/V *Dayang Yihao* during the DY85-5 cruise in September, 1995.				
命名历史 Name History	在我国大洋航次调查报告和 GEBCO 地名辞典中未命名。 The seamounts have not been named in the Chinese cruise reports or GEBCO gazetteer.				
特征点坐标 Coordinates	9°08.50′N，153°41.60′W（长庚海山 / Changgeng Seamount） 9°05.59′N，153°24.06′W（峰顶 /top） 9°03.86′N，153°03.90′W 8°56.10′N，153°47.30′W（景福海丘 / Jingfu Hill） 8°53.93′N，153°40.62′W			长 (km) × 宽 (km) Length (km) × Width (km)	117 × 26
最大水深（m） Max Depth（m）	5 421	最小水深（m） Min Depth（m）	3 684	高差（m） Total Relief（m）	1 737
地形特征 Feature Description	该海山群包含多个海山海丘，其中有景福海丘和长庚海山。海山群东西走向，往东延伸情况不明，现有资料显示长宽分别为 117 km 和 26 km，峰顶水深 3 684 m（图 2–142）。 Chenlunjiong Seamounts consist of 7 seamounts or hills, including Jingfu Hill and Changgeng Seamount. The seamounts run E to W and the extension towards east is remained unclear. According to current data, the length and width are 117 km and 26 km respectively, while the top depth is 3 684 m (Fig.2–142).				
命名释义 Reason for Choice of Name	陈伦炯（公元约 1685—1748），从小随父出海，曾至日本。陈伦炯著《海国闻见录》一书，叙述中外海洋地理风貌，范围包括亚洲、欧洲与非洲，又概述了当时我国海岸的地貌、水文、航海资料，具有重要的航海指南价值。此海山群命名为"陈伦炯"，以纪念陈伦炯在我国航海史上的重要贡献。 Chen Lunjiong (A.D. 1685–1784) began his voyage with his father from his childhood time and ever had been to Japan. He wrote a famous book,which stated the marine geographical features, including Asia, Europe and Africa. It also summarized the topographic features, hydrology of China, which has a great value in guidance of voyage. The seamounts were named after Chenlunjiong, in memory of his significant contributions to the Chinese history of marine navigation.				

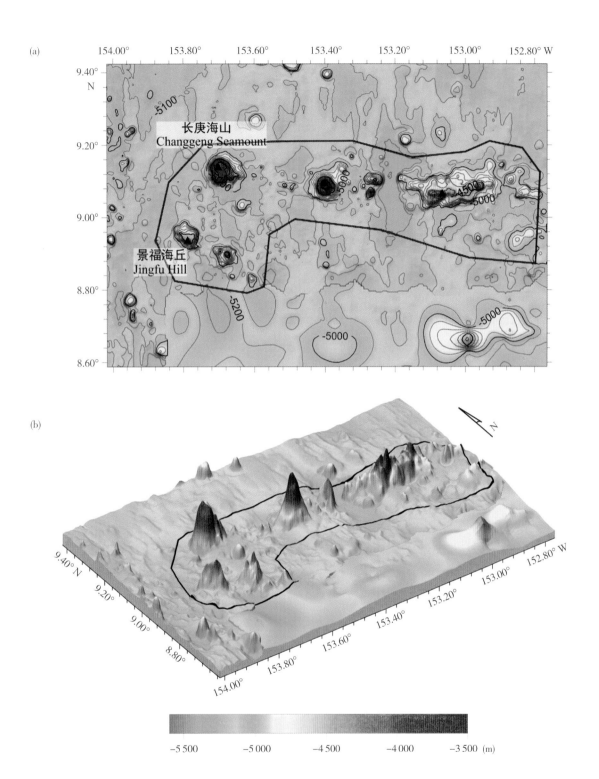

图 2-142　陈伦炯海山群

(a) 地形图（等深线间隔 100 m）；(b) 三维图

Fig.2-142　Chenlunjiong Seamounts

(a) Bathymetric map (the contour interval is 100 m); (b) 3-D topographic map

2.8.20　长庚海山
Changgeng Seamount

中文名称 Chinese Name	长庚海山 Changgeng Haishan	英文名称 English Name	Changgeng Seamount	所在大洋 Ocean or Sea	东太平洋 East Pacific Ocean
发现情况 Discovery Facts	此海山于 1995 年 8 月由中国科考船 "大洋一号" 在执行 DY85-5 航次时发现。 This Seamount was discovered by the Chinese R/V *Dayang Yihao* during the DY85-5 cruise in August, 1995.				
命名历史 Name History	由我国命名为长庚海山，于 2013 年提交 SCUFN 审议通过。 Changgeng Seamount was named by China and approved by SCUFN in 2013.				
特征点坐标 Coordinates	9°08.50′N，153°41.60′W			长 (km)×宽 (km) Length (km)× Width (km)	14×14
最大水深（m） Max Depth（m）	5 200	最小水深（m） Min Depth（m）	3 826	高差（m） Total Relief（m）	1 374
地形特征 Feature Description	该海山俯视平面形态呈圆形，基座直径 14 km，峰顶水深 3 826 m，山麓水深 5 200 m，高差 1 374 m。海山西坡地形较陡，东坡地形较缓（图 2–143）。 Changgeng Seamount has a nearly round overlook plane shape with the base diameter of 14 km. The top depth is 3 826 m and the piedmont depth is 5 200 m, which makes the total relief being 1 374 m. The western slope of the seamount is steep while the eastern slope is relatively flat (Fig.2–143).				
命名释义 Reason for Choice of Name	长庚出自《诗经·小雅·大东》，"东有启明，西有长庚"。此诗歌描述中国东部人民遭受周朝的残酷统治、生活困苦的情景，表达作者忧愤抗争的激情。金星天亮前出现在东方，称为 "启明星"；黄昏时出现在西方，称为 "长庚星"。启明海山和长庚海山分别位于中国多金属结核合同区的东区和西区，故东区的海山命名为启明，以喻日出东方；西区的海山命名为长庚，以喻日落西方。 "Changgeng" comes from a poem named *Dadong* in *Shijing · Xiaoya. Shijing* is a collection of ancient Chinese poems from 11th century B.C. to 6th century B.C. "The east presents the morning star; The west presents the evening star." Venus is named "Qimingxing" (the morning star) when it appears in the east before dawn. However it is named "Changgengxing" (the evening star) when it appears in the west after dusk. Qiming Seamount is located in the eastern part of Chinese polymetallic nodules contract area while Changgeng Seamount is located in the western part. Hence the seamount in the eastern part is named "Qiming" to symbolize the east sunrise while the seamount in the western part is named "Changgeng" to symbolize the west sunset.				

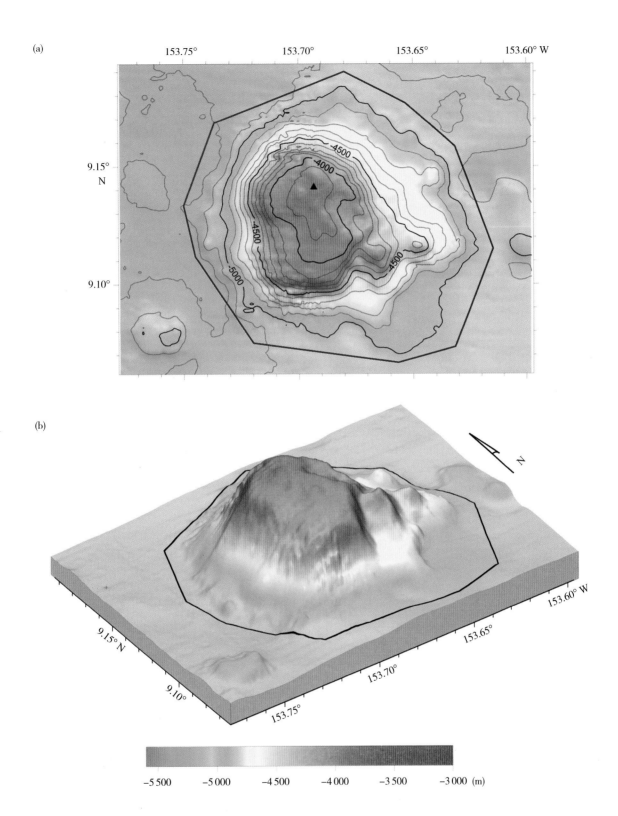

图 2-143　长庚海山

(a) 地形图（等深线间隔 100 m）；(b) 三维图

Fig.2-143　Changgeng Seamount

(a) Bathymetric map (the contour interval is 100 m); (b) 3-D topographic map

2.8.21 景福海丘
Jingfu Hill

中文名称 Chinese Name	景福海丘 Jingfu Haiqiu	英文名称 English Name	Jingfu Hill	所在大洋 Ocean or Sea	东太平洋 East Pacific Ocean
发现情况 Discovery Facts	此海丘于 1995 年 8 月由中国科考船"大洋一号"在执行 DY85-5 航次时发现。 This hill was discovered by the Chinese R/V *Dayang Yihao* during the DY85-5 cruise in August, 1995.				
命名历史 Name History	由我国命名为景福海丘,于 2014 年提交 SCUFN 审议通过。 Jingfu Hill was named by China and approved by SCUFN in 2014.				
特征点坐标 Coordinates	8°56.10′N,153°47.30′W			长(km)× 宽(km) Length (km)× Width (km)	12 × 8
最大水深(m) Max Depth(m)	5 100	最小水深(m) Min Depth(m)	4 233	高差(m) Total Relief(m)	867
地形特征 Feature Description	该海丘 NW—SE 走向,长宽分别 12 km 和 8 km,峰顶水深 4 233 m,山麓水深 5 100 m,高差 867 m。海丘北坡地形较缓,南坡地形较陡(图 2–144)。 Jingfu Hill runs NW to SE. The length and width are 12 km and 8 km respectively. The top depth is 4 233 m and the piedmont depth is 5 100 m, which makes the total relief being 867 m. The northern slope of the hill is gentle while the southern slope is steep (Fig.2–144).				
命名释义 Reason for Choice of Name	"景福"出自《诗经·小雅·大田》"以享以祀,以介景福",大福之意。此诗歌描述周王巡视秋收的场景,此句为古人祭拜上天和祖先,希望得到祝福和保佑。以"景福"命名海丘,寄寓着人们对美好生活的向往。 "Jingfu" comes from a poem named *Datian* in *Shijing · Xiaoya*. *Shijing* is a collection of ancient Chinese poems from 11th century B.C. to 6th century B.C. "The sacrificial rites we observe, to pray for blessings we deserve." This verse describes how ancient people worshipped the heaven and their ancestors and prayed for their blessings. "Jingfu" represents the wish for a better life of ancient Chinese people.				

(a)

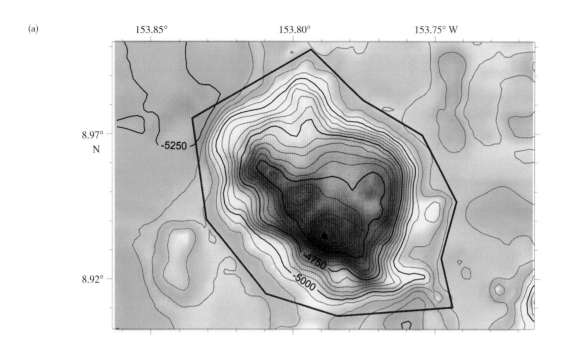

(b)

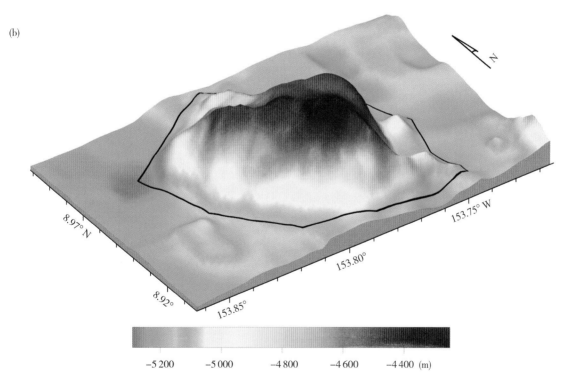

图 2-144　景福海丘

地形图（等深线间隔 50 m）；(b) 三维图

Fig.2-144　Jingfu Hill

(a) Bathymetric map (the contour interval is 50 m); (b) 3-D topographic map

2.8.22 如霆海山
Ruting Seamount

中文名称 Chinese Name	如霆海山 Ruting Haishan	英文名称 English Name	Ruting Seamount	所在大洋 Ocean or Sea	东太平洋 East Pacific Ocean
发现情况 Discovery Facts	此海山于 1995 年 10 月由中国科考船"大洋一号"在执行 DY85-5 航次时发现。 This seamount was discovered by the Chinese R/V *Dayang Yihao* during the DY85-5 cruise in October, 1995.				
命名历史 Name History	在我国大洋航次调查报告和 GEBCO 地名辞典中未命名。 This seamount has not been named in the Chinese cruise reports or GEBCO gazetteer.				
特征点坐标 Coordinates	8°31.03′N，148°21.18′W			长(km)×宽(km) Length (km)× Width (km)	28×10
最大水深（m） Max Depth（m）	5 290	最小水深（m） Min Depth（m）	4 173	高差（m） Total Relief（m）	1 117
地形特征 Feature Description	该海山 NW—SE 走向，长宽分别为 28 km 和 10 km，海山顶部水深 4 173 m，山麓水深为 5 290 m，高差 1 117 m，山坡地形陡峭（图 2–145）。 Ruting Seamount runs NW to SE. The length and width are 28 km and 10 km respectively. The top depth of the seamount is 4 173 m and the piedmont depth is 5 290 m, which makes the total relief being 1 117 m. Its slopes are steep (Fig.2–145).				
命名释义 Reason for Choice of Name	"如霆"出自《诗经·小雅·采芑》"戎车啴啴，啴啴焞焞，如霆如雷"。此诗歌描绘周朝将军为威慑荆蛮而进行军事演习的画面。此句描绘战车行进时的轰响声，表现军事演习的声势浩大、场面壮观的场景。 "Ruting" comes from a poem named *Caiqi* in *Shijing · Xiaoya*. *Shijing* is a collection of ancient Chinese poems from 11th century B.C. to 6th century B.C. "The war-chariots rumble on; They rumble and roll on; As if thunder is roaring anon." This poem described the military maneuver of Zhou Dynasty to deter the enemies. The sentence described the loud sound from the chariot and the great scene of military manoeuvre.				

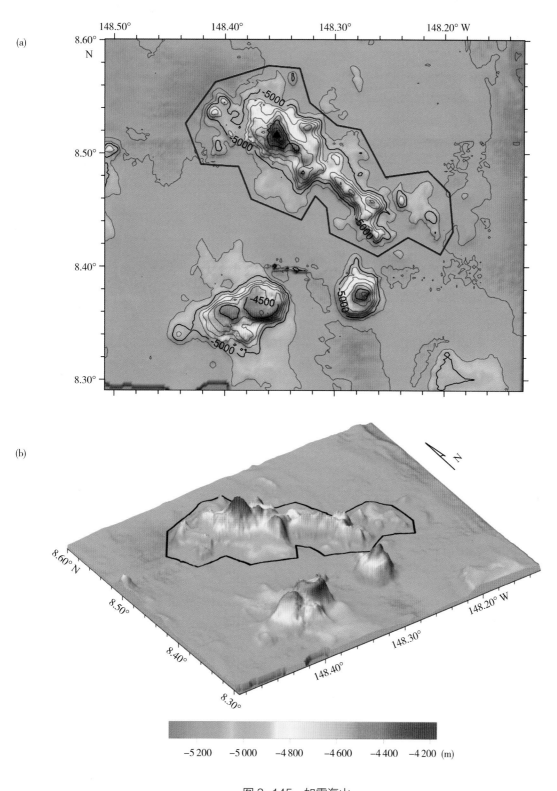

图 2-145　如霆海山

(a) 地形图（等深线间隔 100 m）；(b) 三维图

Fig.2-145　Ruting Seamount

(a) Bathymetric map (the contour interval is 100 m); (b) 3-D topographic map

2.8.23 苏洵圆海丘
Suxun Knoll

中文名称 Chinese Name	苏洵圆海丘 Suxun Yuanhaiqiu		英文名称 English Name	Suxun Knoll	所在大洋 Ocean or Sea	东太平洋 East Pacific Ocean
发现情况 Discovery Facts	此圆海丘于 1995 年 10 月由中国科考船"大洋一号"在执行 DY85-5 航次时发现。 This knoll was discovered by the Chinese R/V *Dayang Yihao* during the DY85-5 cruise in October, 1995.					
命名历史 Name History	由我国命名为苏洵圆海丘，于 2015 年提交 SCUFN 审议通过。 Suxun Knoll was named by China and approved by SCUFN in 2015.					
特征点坐标 Coordinates	8°10.35′N，146°43.20′W			长(km)×宽(km) Length (km) × Width (km)		11 × 8
最大水深（m） Max Depth（m）	5 408	最小水深（m） Min Depth（m）	4 676	高差（m） Total Relief（m）		732
地形特征 Feature Description	该海丘位于苏轼海丘西侧，苏辙圆海丘以西 20 km。海丘俯视平面形态呈椭圆形，SN 走向，长宽分别为 11 km 和 8 km，海丘峰顶水深 4 676 m，山麓水深 5 408 m，高差 732 m（图 2–146）。 This knoll is located in the west to Sushi Hill and 20 km west to Suzhe Knoll. It has an elliptic overlook plane shape, running N to S. The length and width are 11 km and 8 km respectively. The top depth of this knoll is 4 676 m and the piedmont depth is 5 480 m, which makes the total relief being 732 m (Fig.2–146).					
命名释义 Reason for Choice of Name	苏洵（公元 1009—1066 年），字明允，北宋文学家，其代表作有《六国论》等，与其子苏轼、苏辙并以文学著称于世，世称"三苏"，均被列入"唐宋八大家"。此圆海丘命名为"苏洵"，以纪念苏洵在我国文学史上的重要贡献。 Su Xun (A.D. 1009–1066), whose another name taken at the age of twenty was Mingyun, a writer and poet in Song Dynasty. His two sons, Su Zhe, Su Shi and he were well known because of their literary achievements and called "San Su" together. They were ranked into a group called "Eight Great Prose Masters of the Tang and Song Dynasty". This knoll is named after Suxun, in memory of his great contributions to the Chinese history of literature.					

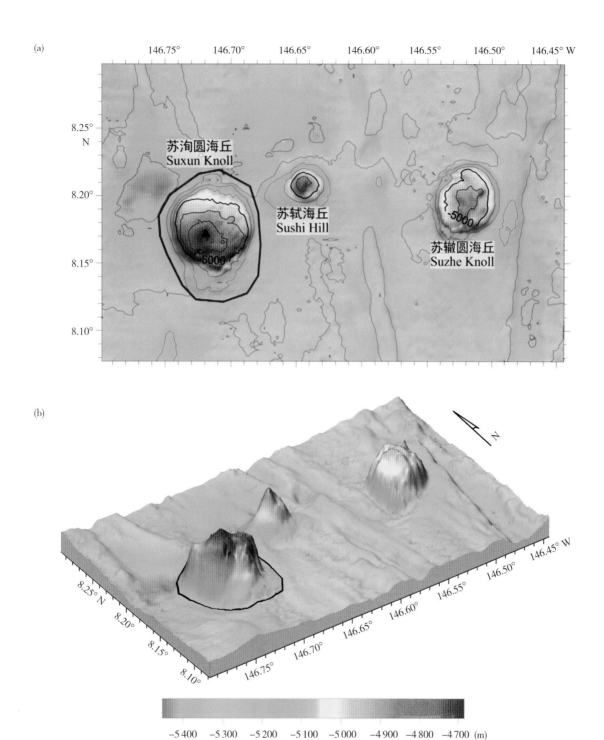

图 2-146　苏洵圆海丘

(a) 地形图（等深线间隔 100 m）；(b) 三维图

Fig.2-146　Suxun Knoll

(a) Bathymetric map (the contour interval is 100 m); (b) 3-D topographic map

2.8.24 苏轼海丘
Sushi Hill

中文名称 Chinese Name	苏轼海丘 Sushi Haiqiu	英文名称 English Name	Sushi Hill	所在大洋 Ocean or Sea	东太平洋 East Pacific Ocean
发现情况 Discovery Facts	此海丘于 1995 年 10 月由中国科考船"大洋一号"在执行 DY85-5 航次时发现。 This hill was discovered by the Chinese R/V *Dayang Yihao* during the DY85-5 cruise in October, 1995.				
命名历史 Name History	由我国命名为苏轼海丘，于 2015 年提交 SCUFN 审议通过。 Sushi Hill was named by China and approved by SCUFN in 2015.				
特征点坐标 Coordinates	8°12.62′N，146°38.70′W			长 (km) × 宽 (km) Length (km) × Width (km)	8 × 7
最大水深（m） Max Depth（m）	5 404	最小水深（m） Min Depth（m）	4 815	高差（m） Total Relief（m）	589
地形特征 Feature Description	该海丘位于苏洵圆海丘以东 2 km，苏辙圆海丘以西 13 km。海丘俯视平面形态近椭圆形，东西走向，长宽分别为 8 km 和 7 km，峰顶水深 4 815 m，山麓水深 5 404 m，高差 589 m（图 2–147）。 Sushi Hill is located in 2 km east to Suxun Knoll and 13 km west to Suzhe Knoll. It has a nearly elliptic overlook plane shape, running E to W. The length and width are 8 km and 7 km respectively. The top depth of the hill is 4 815 m while the piedmont depth is 5 404 m, which makes the total relief being 589 m (Fig.2–147).				
命名释义 Reason for Choice of Name	苏轼（公元 1037—1101 年），号东坡居士，宋代重要的文学家，宋代文学最高成就的代表。其代表作品有《赤壁赋》《石钟山记》等，与其父苏洵、其弟苏辙并以文学著称于世，世称"三苏"，均被列入"唐宋八大家"。此海丘命名为"苏轼"，以纪念苏轼在我国文学史上的重要贡献。 Su Shi (A.D. 1037–1101), was a Chinese writer (litterateur) during the Song Dynasty. Su Shi, together with his father Su Xun and his brother Su Zhe, were well known because of their literary achievements and called "San Su" together. They were ranked into a group called "Eight Great Prose Masters of the Tang and Song Dynasty". This hill is named after Sushi, in memory of his great contributions to the Chinese history of literature.				

(a)

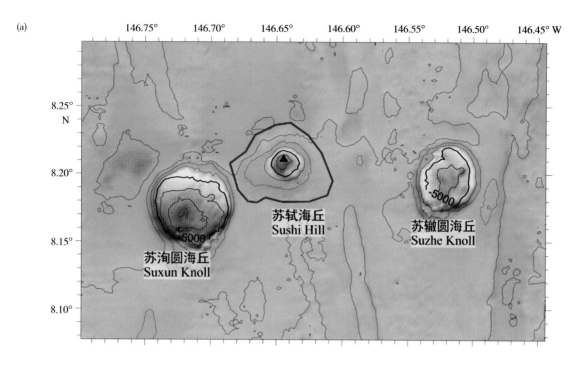

(b)

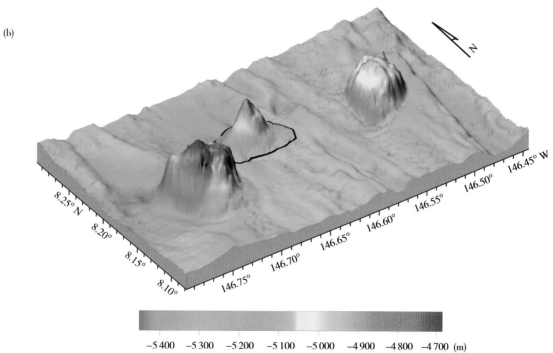

图 2-147　苏轼海丘

(a) 地形图（等深线间隔 100 m）；(b) 三维图

Fig.2-147　Sushi Hill

(a) Bathymetric map (the contour interval is 100 m); (b) 3-D topographic map

2.8.25 苏辙圆海丘
Suzhe Knoll

中文名称 Chinese Name	苏辙圆海丘 Suzhe Yuanhaiqiu	英文名称 English Name	Suzhe Knoll	所在大洋 Ocean or Sea	东太平洋 East Pacific Ocean
发现情况 Discovery Facts	此圆海丘于 1995 年 10 月由中国科考船"大洋一号"在执行 DY85-5 航次时发现。 This knoll was discovered by the Chinese R/V *Dayang Yihao* during the DY85-5 cruise in October, 1995.				
命名历史 Name History	由我国命名为苏辙圆海丘，于 2015 年提交 SCUFN 审议通过。 Suzhe Knoll was named by China and approved by SCUFN in 2015.				
特征点坐标 Coordinates	8°12.18′N，146°30.54′W			长(km)× 宽(km) Length (km) × Width (km)	8 × 7
最大水深（m） Max Depth（m）	5 358	最小水深（m） Min Depth（m）	4 848	高差（m） Total Relief（m）	510
地形特征 Feature Description	该圆海丘位于苏轼海丘以东 13 km。海丘呈圆锥形，基座直径 7 ~ 8 km，顶部水深为 4 848 m，山麓水深 5 358 m，高差 510 m（图 2–148）。 This knoll is located in 13 km east to Sushi hill. It has a conical shape with the base diameter of 7–8 km. The top depth of the knoll is 4 848 m and the piedmont depth is 5 358 m, which makes the total relief being 510 m (Fig.2–148).				
命名释义 Reason for Choice of Name	苏辙（公元 1039—1112 年），字子由，北宋文学家、诗人、唐宋八大家之一，其代表作有《春秋传》、《栾城集》等，与其父苏洵、其兄苏轼并以文学著称于世，世称"三苏"，均被列入"唐宋八大家"。此圆海丘命名为"苏辙"，以纪念苏辙在我国文学史上的重要贡献。 Su Zhe (A.D. 1039–1112), was a Chinese writer (litterateur and poet) during the Song Dynasty. Su Zhe, together with his father Su Xun and his brother Su Shi, were well known because of their literary achievements and called "San Su" together. They were ranked into a group called "Eight Great Prose Masters of the Tang and Song Dynasty". This knoll is named after Suzhe, in memory of his great contributions to the Chinese history of literature.				

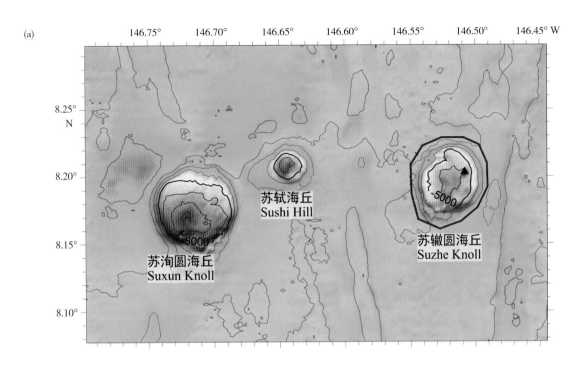

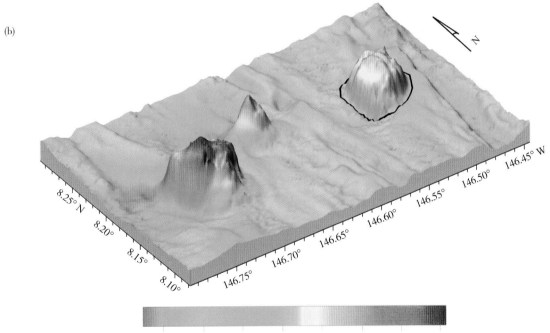

图 2-148　苏辙圆海丘

(a) 地形图（等深线间隔 100 m）；(b) 三维图

Fig.2-148　Suzhe Knoll

(a) Bathymetric map (the contour interval is 100 m); (b) 3-D topographic map

2.8.26 王勃圆海丘
Wangbo Knoll

中文名称 Chinese Name	王勃圆海丘 Wangbo Yuanhaiqiu		英文名称 English Name	Wangbo Knoll	所在大洋 Ocean or Sea	东太平洋 East Pacific Ocean
发现情况 Discovery Facts	此圆海丘于 1995 年 10 月由中国科考船"大洋一号"在执行 DY85-5 航次时发现。 This knoll was discovered by the Chinese R/V *Dayang Yihao* during the DY85-5 cruise in October, 1995.					
命名历史 Name History	由我国命名为王勃圆海丘，于 2016 年提交 SCUFN 审议通过。 Wangbo Knoll was named by China and approved by SCUFN in 2016.					
特征点坐标 Coordinates	7°49.67′N，146°06.42′W			长(km)×宽(km) Length (km) × Width (km)		9 × 8
最大水深（m） Max Depth（m）	5 393	最小水深（m） Min Depth（m）	4 709	高差（m） Total Relief（m）		684
地形特征 Feature Description	该圆海丘位于杨炯海丘西侧。圆海丘近 NE—SW 走向，长宽分别为 9 km 和 8 km，顶部水深 4 709 m，山麓水深 5 393 m，高差 684 m。圆海丘顶部地形平缓，周缘地形较陡（图 2–149）。 Wangbo Knoll is located in the west to Yangjiong Hill. This knoll runs nearly NE to SW. The length and width are 9 km and 8 km respectively. The top depth of this knoll is 4 709 m and the piedmont depth is 5 393 m, which makes the total relief being 684 m. The top of this knoll is flat while the sides are steep (Fig.2–149).					
命名释义 Reason for Choice of Name	王勃（公元 650—676 年），字子安，"初唐四杰"之首，擅长诗歌五律、五绝和骈文，代表作品有《送杜少府之任蜀州》、《滕王阁序》等。此圆海丘命名为"王勃"，以纪念王勃在我国文学史上的重要贡献。 Wang Bo (A.D. 650–676), whose another name taken at the age of twenty was Zian, the head of "Chu Tang Si Jie" (Four Paragons of the Early Tang Dynasty), was good at several kinds of poems，which all are the names of type of poems and articles. This knoll is named after Wangbo, in memory of his great contributions to the Chinese history of literature.					

(a)

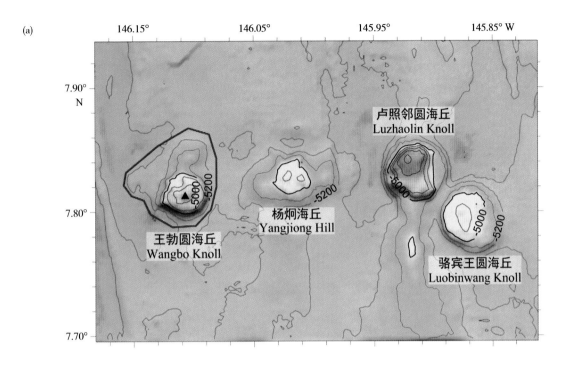

(b)

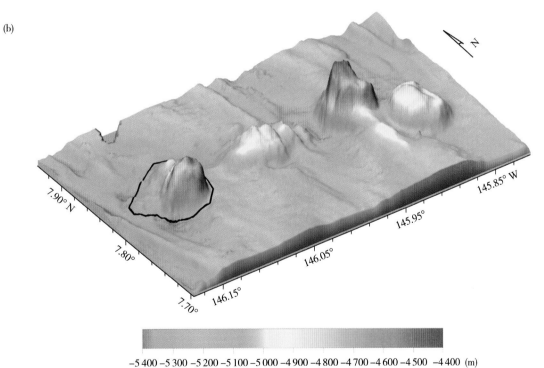

−5 400　−5 300　−5 200　−5 100　−5 000　−4 900　−4 800　−4 700　−4 600　−4 500　−4 400　(m)

图 2-149　王勃圆海丘

(a) 地形图（等深线间隔 100 m）；(b) 三维图

Fig.2-149　Wangbo Knoll

(a) Bathymetric map (the contour interval is 100 m); (b) 3-D topographic map

2.8.27 杨炯海丘
Yangjiong Hill

中文名称 Chinese Name	杨炯海丘 Yangjiong Haiqiu	英文名称 English Name	Yangjiong Hill	所在大洋 Ocean or Sea	东太平洋 East Pacific Ocean
发现情况 Discovery Facts	此海丘于 1995 年 10 月由中国科考船"大洋一号"在执行 DY85-5 航次时发现。 This hill was discovered by the Chinese R/V *Dayang Yihao* during the DY85-5 cruise in October, 1995.				
命名历史 Name History	由我国命名为杨炯海丘，于 2016 年提交 SCUFN 审议通过。 Yangjiong Hill was named by China and approved by SCUFN in 2016.				
特征点坐标 Coordinates	7°49.72′N，146°01.45′W			长(km)×宽(km) Length (km)× Width (km)	12×6
最大水深（m） Max Depth（m）	5 353	最小水深（m） Min Depth（m）	4 881	高差（m） Total Relief（m）	472
地形特征 Feature Description	该海丘位于王勃圆海丘以东 10 km，卢照邻圆海丘以西 7 km。海丘东西走向，长宽分别为 12 km 和 6 km，峰顶水深为 4 881 m，山麓水深 5 353 m，高差 472 m，海丘周缘地形较缓（图 2–150）。 Yangjiong Hill is located in 10 km east to Wangbo Knoll and 7 km west to Luzhaolin Knoll. This hill runs E to W and the length and width are 12 km and 6 km respectively. The top depth of the hill is 4 881 m and the piedmont depth is 5 353 m, which makes the total relief being 472 m. The sides of the knoll are relatively flat (Fig.2–150).				
命名释义 Reason for Choice of Name	杨炯（公元 650—692 年），"初唐四杰"之一，以边塞征战诗著名，所作如《从军行》《出塞》《战城南》等，表现了为国立功的战斗精神，气势轩昂，风格豪放。此海丘命名为"杨炯"，以纪念杨炯在我国文学史上的重要贡献。 Yang Jiong (A.D. 650–692), was one of the Four Paragons of the Early Tang Dynasty. He was well known for the "battle-poem", which described the brave and honor from the soldiers. This hill is named "Yangjiong", in memory of his great contributions to the Chinese history of literature.				

(a)

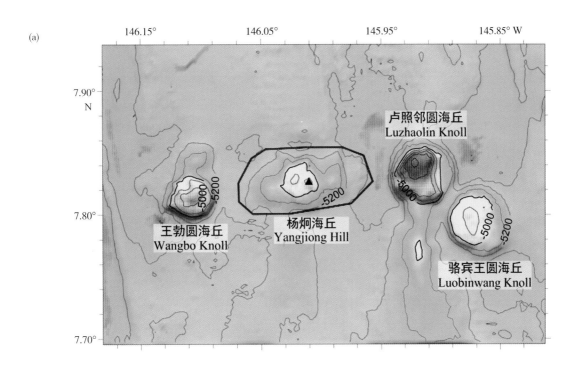

(b)

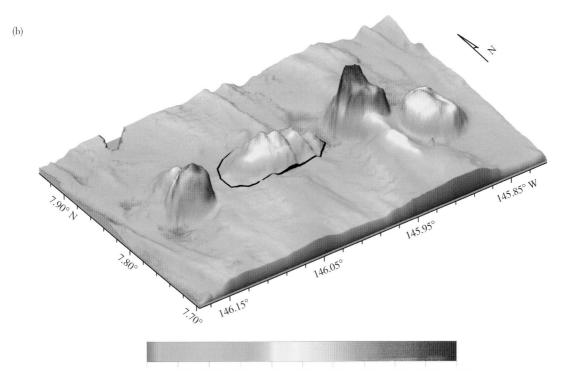

−5 400 −5 300 −5 200 −5 100 −5 000 −4 900 −4 800 −4 700 −4 600 −4 500 −4 400 (m)

图 2-150　杨炯海丘

(a) 地形图（等深线间隔 100 m）；(b) 三维图

Fig.2-150　Yangjiong Hill

(a) Bathymetric map (the contour interval is 100 m); (b) 3-D topographic map

2.8.28 卢照邻圆海丘
Luzhaolin Knoll

中文名称 Chinese Name	卢照邻圆海丘 Luzhaolin Yuanhaiqiu	英文名称 English Name	Luzhaolin Knoll	所在大洋 Ocean or Sea	东太平洋 East Pacific Ocean
发现情况 Discovery Facts	此圆海丘于 1995 年 10 月由中国科考船 "大洋一号" 在执行 DY85-5 航次时发现。 This knoll was discovered by the Chinese R/V *Dayang Yihao* during the DY85-5 cruise in October, 1995.				
命名历史 Name History	由我国命名为卢照邻圆海丘, 于 2016 年提交 SCUFN 审议通过。 Luzhaolin Knoll was named by China and approved by SCUFN in 2016.				
特征点坐标 Coordinates	7°50.53′N, 145°55.32′W			长(km) × 宽(km) Length (km) × Width (km)	8 × 7
最大水深（m） Max Depth（m）	5379	最小水深（m） Min Depth（m）	4578	高差（m） Total Relief（m）	801
地形特征 Feature Description	该圆海丘位于骆宾王圆海丘西北侧, 杨炯海丘以东 7 km。圆海丘俯视平面形态近圆形, 基座直径 7 ~ 8 km, 顶部水深 4 578 m, 山麓水深 5 379 m, 高差 801 m, 圆海丘北坡地形较陡, 南坡地形较缓 (图 2–151)。 Luzhaolin Knoll is located in the northwest to Luobinwang Knoll, 7 km east to Yangjiong Hill. The knoll has a nearly round overlook plane shape with the base diameter of 7–8 km. The top depth of the knoll is 4 578 m and the piedmont depth is 5 379 m, which makes the total relief being 801 m. The northern slope of the knoll is steep while the southern slope is flat (Fig.2–151).				
命名释义 Reason for Choice of Name	卢照邻(公元约 635—689 年), 字升之, "初唐四杰" 之一, 擅长诗歌骈文。其作品名句 "得成比目何辞死, 愿作鸳鸯不羡仙" 等被后人誉为经典。此圆海丘命名为 "卢照邻", 以纪念卢照邻在我国文学史上的重要贡献。 Lu Zhaolin (about A.D. 635–689), whose another name taken at the age of twenty was Shengzhi, was one of the Four Paragons of the Early Tang Dynasty. He is good at poem and parallel prose. The knoll is named after Luzhaolin, in memory of his great contributions in the Chinese history of literature.				

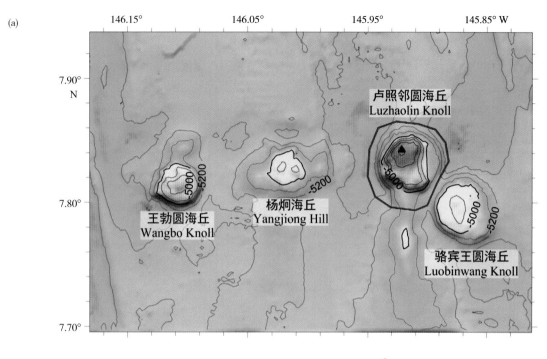

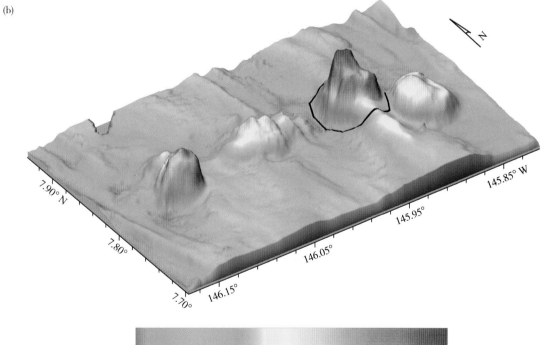

图 2-151　卢照邻圆海丘

(a) 地形图（等深线间隔 100 m）；(b) 三维图

Fig.2-151　Luzhaolin Knoll

(a) Bathymetric map (the contour interval is 100 m); (b) 3-D topographic map

2.8.29 骆宾王圆海丘
Luobinwang Knoll

中文名称 Chinese Name	骆宾王圆海丘 Luobinwang Yuanhaiqiu	英文名称 English Name	Luobinwang Knoll	所在大洋 Ocean or Sea	东太平洋 East Pacific Ocean
发现情况 Discovery Facts	此圆海丘于 1995 年 10 月由中国科考船"大洋一号"在执行 DY85-5 航次时发现。 This knoll was discovered by the Chinese R/V *Dayang Yihao* during the DY85-5 cruise in October, 1995.				
命名历史 Name History	由我国命名为骆宾王圆海丘,于 2016 年提交 SCUFN 审议通过。 Luobinwang Knoll was named by China and approved by SCUFN in 2016.				
特征点坐标 Coordinates	7°47.99′N,145°52.56′W			长(km)×宽(km) Length (km)× Width (km)	8×7
最大水深(m) Max Depth(m)	5 359	最小水深(m) Min Depth(m)	4 837	高差(m) Total Relief(m)	522
地形特征 Feature Description	该圆海丘位于卢照邻圆海丘东侧。圆海丘俯视平面形态近圆形,基座直径 7 ~ 8 km,顶部水深 4 837 m,山麓水深 5 359 m,高差 522 m,圆海丘顶部地形平缓,周缘地形较陡(图 2–152)。 Luobinwang Knoll is located in the east to Luzhaolin Knoll. The knoll has a nearly round overlook plane shape with the diameter of 7–8 km. The top depth of the knoll is 4 837 m and the piedmont depth is 5 359 m, which makes the total relief being 522 m. The top of the knoll is flat and the sides are steep (Fig.2–152).				
命名释义 Reason for Choice of Name	骆宾王(公元 619—687 年),字观光,初唐四杰之一,其诗辞采华赡,格律谨严,代表作有《帝京篇》《咏鹅》等。此圆海丘命名为"骆宾王",以纪念他在我国文学史上的重要贡献。 Luo Binwang (about A.D. 619–687), whose another name taken at the age of twenty was Guanguang, was one of the Four Paragons of the Early Tang Dynasty. His poems are full of gorgeous words and follow strict metrics. The knoll is named after Luobinwang, in memory of his great contributions in the Chinese history of literature.				

(a)

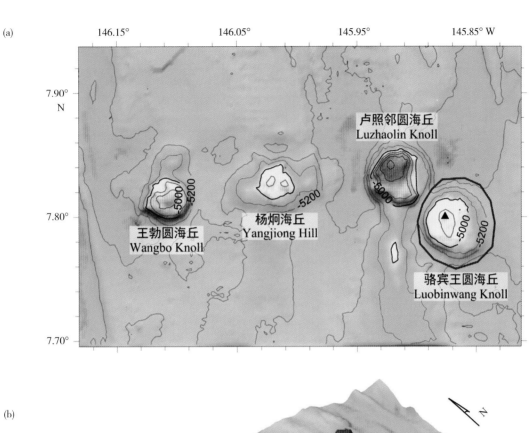

(b)

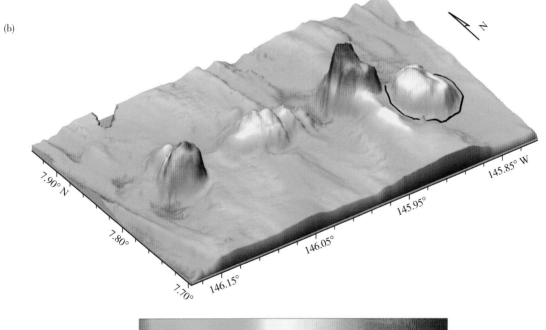

−5 400 −5 300 −5 200 −5 100 −5 000 −4 900 −4 800 −4 700 −4 600 −4 500 −4 400 (m)

图 2-152　骆宾王圆海丘

(a) 地形图（等深线间隔 100 m）；(b) 三维图

Fig.2-152　Luobinwang Knoll

(a) Bathymetric map (the contour interval is 100 m); (b) 3-D topographic map

2.8.30 天祐圆海丘
Tianhu Knoll

中文名称 Chinese Name	天祐圆海丘 Tianhu Yuanhaiqiu	英文名称 English Name	Tianhu Knoll	所在大洋 Ocean or Sea	东太平洋 East Pacific Ocean
发现情况 Discovery Facts	此海丘于 1995 年 8 月由中国科考船"大洋一号"在执行 DY85-5 航次时发现。 This knoll was discovered by the Chinese R/V *Dayang Yihao* during the DY85-5 cruise in August, 1995.				
命名历史 Name History	由我国命名为天祐圆海丘,于 2014 年提交 SCUFN 审议通过。 Tianhu Knoll was named by China and approved by SCUFN in 2014.				
特征点坐标 Coordinates	8°28.30′N,145°44.70′W		长(km)× 宽(km) Length (km)× Width (km)		7 × 7
最大水深(m) Max Depth(m)	5 400	最小水深(m) Min Depth(m)	4 480	高差(m) Total Relief(m)	920
地形特征 Feature Description	该圆海丘位于骆宾王圆海丘以北 80 km。圆海丘整体呈圆锥形,基座直径 7 km,顶部发育开口朝北的破火山口。圆海丘峰顶水深 4 480 m,山麓水深 5 400 m,高差为 920 m(图 2–153)。 Tianhu Knoll is located in 80 km north to Luobinwang Knoll. This knoll has a conical shape as a whole with the base diameter of 7 km. The top develops a caldera open towards north. The top depth of this knoll is 4 480 m and the piedmont depth is 5 400 m, which makes the total relief being 920 m (Fig.2–153).				
命名释义 Reason for Choice of Name	"天祐"出自《诗经·小雅·信南山》"曾孙寿考,受天之祐"。此诗歌描绘周朝贵族在秋收后拜祭天地的场景。这两句指子孙后代生生不息,皆受上天保佑之福。"天祐"意为受上天保佑和祝福,以此命名圆海丘,表示中国古人对子孙后代的美好祝福。 "Tianhu" comes from a poem named *Xinnanshan* in *Shijing · Xiaoya*. *Shijing* is a collection of ancient Chinese poems from 11th century B.C. to 6th century B.C. "We the descendants are of Yu, blessed by Heaven long and true." It means that the endless generations will be blessed by God. "Tianhu" means being blessed by God, The knoll is named "Tianhu", representing Chinese people's best wishes for the posterity.				

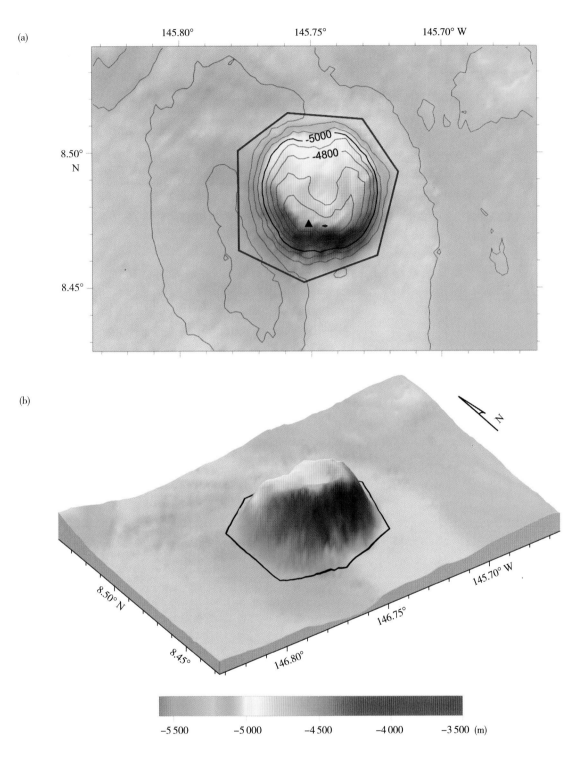

图 2-153　天祜圆海丘

(a) 地形图 （等深线间隔 100 m）； (b) 三维图

Fig.2-153　Tianhu Knoll

(a) Bathymetric map (the contour interval is 100 m); (b) 3-D topographic map

2.8.31 天祐南海底洼地
Tianhunan Depression

中文名称 Chinese Name	天祐南海底洼地 Tianhunan Haidiwadi	英文名称 English Name	Tianhunan Depression	所在大洋 Ocean or Sea	东太平洋 East Pacific Ocean
发现情况 Discovery Facts	此海底洼地于 1995 年 10 月由中国科考船"大洋一号"在执行 DY85-5 航次时发现。 This depression was discovered by the Chinese R/V *Dayang Yihao* during the DY85-5 cruise in October, 1995.				
命名历史 Name History	在我国大洋航次调查报告和 GEBCO 地名辞典中未命名。 This knoll has not been named in Chinese cruise reports or GEBCO gazetteer.				
特征点坐标 Coordinates	8°24.66′N，145°45.42′W 8°18.60′N，145°43.92′W 8°09.09′N，145°42.60′W 8°00.06′N，145°41.82′W（最深点 /deepest point） 7°52.85′N，145°39.96′W			长(km)× 宽(km) Length (km)× Width (km)	69×3
最大水深（m） Max Depth（m）	5 426	最小水深（m） Min Depth（m）	5 030	高差（m） Total Relief（m）	396
地形特征 Feature Description	该海底洼地位于天祐圆海丘以南，NNW 走向，长宽分别为 69 km 和 3 km。海底洼地底部起伏不平，总体南北两侧浅，中部深。海底洼地水深最深 5 426 m，水深最浅 5 030 m，高差 396 m（图 2–154）。 Tianhunan Depression is located in the south to Tianhu Knoll, running NNW to SSE. The length is 69 km and the width is 3 km. The bottom of the depression is fluctuant. Generally, both southern and northern side are shallow while the middle is deep. The maximum depth of the depression is 5 426 m and the minimum depth is 5 030 m, which makes the total relief being 396 m (Fig.2–154).				
命名释义 Reason for Choice of Name	该海底洼地位于天祐圆海丘以南，故以此命名。 Nan means south in Chinese. This depression is named Tianhunan because it is at the south to Tianhu Knoll.				

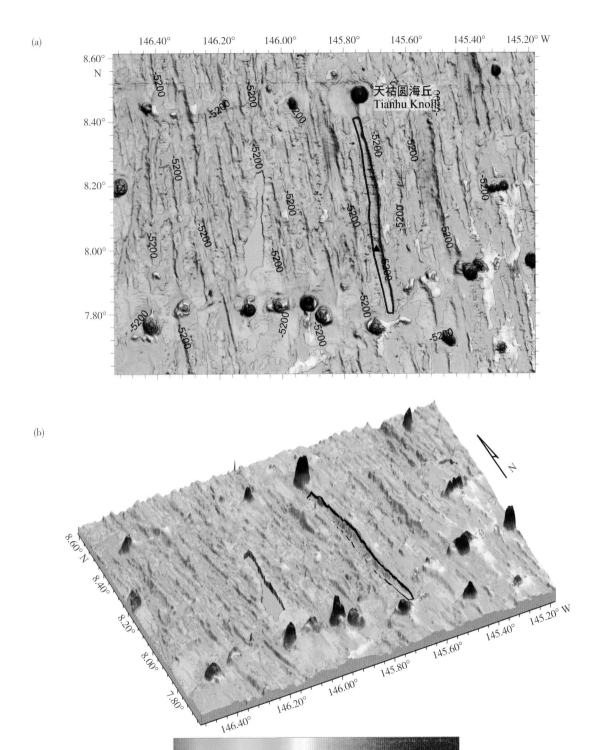

图 2-154　天祐南海底洼地

(a) 地形图（等深线间隔 100 m）；(b) 三维图

Fig.2-154　Tianhunan Depression

(a) Bathymetric map (the contour interval is 100 m); (b) 3-D topographic map

2.8.32 朱应海山
Zhuying Seamount

中文名称 Chinese Name	朱应海山 Zhuying Haishan	英文名称 English Name	Zhuying Seamount	所在大洋 Ocean or Sea	东太平洋 East Pacific Ocean
发现情况 Discovery Facts	此海山于 1995 年 8 月由中国科考船"大洋一号"在执行 DY85-5 航次时发现。 This seamount was discovered by the Chinese R/V *Dayang Yihao* during the DY85-5 cruise in August, 1995.				
命名历史 Name History	由我国命名为朱应海山,于 2013 年提交 SCUFN 审议通过。 Zhuying Seamount was named by China and approved by SCUFN in 2013.				
特征点坐标 Coordinates	8°41.00′N,144°12.60′W		长(km)× 宽(km) Length (km) × Width (km)		16 × 14
最大水深(m) Max Depth(m)	5 350	最小水深(m) Min Depth(m)	3 882	高差(m) Total Relief(m)	1 468
地形特征 Feature Description	该海山 SE—NW 走向,长宽分别为 16 km 和 14 km,海山峰顶水深 3 882 m,山麓水深 5 350 m,高差 1 468 m,海山山坡地形陡峭(图 2–155)。 Zhunying Seamount runs NW to SE. The length is 16 km and the width is 14 km. The top depth of the seamount is 3 882 m and the piedmont depth is 5 350 m, which makes the total relief being 1 468 m. The slopes of the seamount are steep (Fig.2–155).				
命名释义 Reason for Choice of Name	朱应(公元 220—280 年),三国时代吴国人,曾组织船队出访南海周边国家,是中国古代有历史记载以来最早航海到达东南亚的旅行家。朱应著有《扶南异物志》与《吴时外国传》,记载了当时航海和出访国家的一些珍贵史料。现以朱应命名此海山,以纪念他在航海历史上的贡献。 Zhu Ying (A.D. 220–280) was a Chinese traveller from the State of Wu during the Three Kingdoms period in Chinese history. He led a fleet to visit countries around the South China Sea and was the first traveller sailing to Southeast Asia throughout Chinese history. Zhu Ying wrote two famous books recording his sailing and his visits to those countries. The seamount is named after Zhuying, to commemorate his contributions to the Chinese history of marine navigation.				

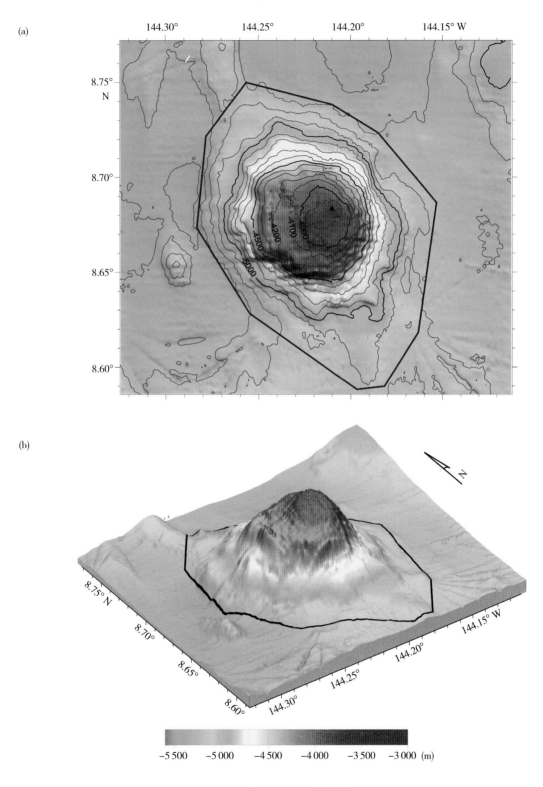

图 2-155　朱应海山

(a) 地形图（等深线间隔 100 m）；(b) 三维图

Fig.2-155　Zhuying Seamount

(a) Bathymetric map (the contour interval is 100 m); (b) 3-D topographic map

2.8.33 朱应西海底洼地
Zhuyingxi Depression

中文名称 Chinese Name	朱应西海底洼地 Zhuyingxi Haidiwadi	英文名称 English Name	Zhuyingxi Depression	所在大洋 Ocean or Sea	东太平洋 East Pacific Ocean
发现情况 Discovery Facts	此海底洼地于 1995 年 11 月由中国科考船"大洋一号"在执行 DY85-5 航次时发现。 This depression was discovered by the Chinese R/V *Dayang Yihao* during the DY85-5 cruise in November, 1995.				
命名历史 Name History	在我国大洋航次调查报告和 GEBCO 地名辞典中未命名。 This depression has not been named in Chinese cruise reports or GEBCO gazetteer.				
特征点坐标 Coordinates	8°58.16′N，144°19.26′W 8°53.72′N，144°22.20′W 8°48.87′N，144°23.16′W（最深点 / deepest point）			长(km) × 宽(km) Length (km) × Width (km)	43 × 8
最大水深（m） Max Depth（m）	5 557	最小水深（m） Min Depth（m）	5 080	高差（m） Total Relief（m）	477
地形特征 Feature Description	该海底洼地位于朱应海山以西。海底洼地南北走向，长宽分别为 43 km 和 8 km。海底洼地底部南浅北深，中部发育小型隆起。海底洼地水深最深 5 557 m，水深最浅 5 080 m，高差 477 m（图 2–156）。 Zhuyingxi Depression is located in the west to Zhuying Seamount, running N to S. The length is 43 km and the width is 8 km. The bottom of depression is shallow in the south and deep in the north. There are some small ridges developed in the middle. The maximum depth of depression is 5 557 m and the minimum depth is 5 080 m, which makes the total relief being 477 m (Fig.2–156).				
命名释义 Reason for Choice of Name	该海底洼地位于朱应海山以西，故以此命名。 Xi means west in Chinese. The depression is named Zhuyingxi because it is at the west to Zhuying Seamount.				

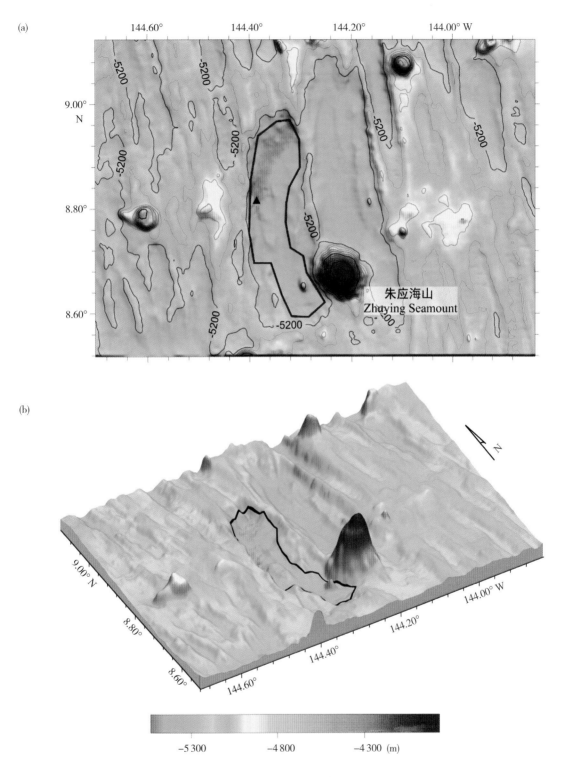

图 2-156 朱应西海底洼地

(a) 地形图（等深线间隔 100 m）；(b) 三维图

Fig.2-156 Zhuyingxi Depression

(a) Bathymetric map (the contour interval is 100 m); (b) 3-D topographic map

2.8.34 朱应北海底洼地
Zhuyingbei Depression

中文名称 Chinese Name	朱应北海底洼地 Zhuyingbei Haidiwadi	英文名称 English Name	Zhuyingbei Depression	所在大洋 Ocean or Sea	东太平洋 East Pacific Ocean
发现情况 Discovery Facts	此海底洼地于 1995 年 11 月由中国科考船"大洋一号"在执行 DY85-5 航次时发现。 This depression was discovered by the Chinese R/V *Dayang Yihao* during the DY85-5 cruise in November, 1995.				
命名历史 Name History	在我国大洋航次调查报告和 GEBCO 地名辞典中未命名。 This depression has not been named in Chinese cruise reports or GEBCO gazetteer.				
特征点坐标 Coordinates	8°47.09′N，144°11.46′W（最深点 / deepest point） 8°55.97′N，144°14.82′W 9°02.27′N，144°13.62′W			长(km)×宽(km) Length (km) × Width (km)	34×11
最大水深（m） Max Depth（m）	5 455	最小水深（m） Min Depth（m）	5 100	高差（m） Total Relief（m）	355
地形特征 Feature Description	该海底洼地位于朱应海山北侧。海底洼地南北走向，长宽分别为 34 km 和 11 km。海底洼地底部平缓，水深最深 5 455 m，水深最浅 5 100 m，高差 355 m（图 2-157）。 Zhuyingbei Depression is located in the north to Zhuying Seamount, running N to S. The length is 34 km and the width is 11 km. The bottom of depression is flat. The maximum depth is 5 455 m and the minimum depth is 5 100 m, which makes the total relief being 355 m (Fig.2-157).				
命名释义 Reason for Choice of Name	该海底洼地位于朱应海山以北，故以此命名。 "Bei" means north in Chinese. The depression is named Zhuyingbei because it is at the north to Zhuying Seamount.				

(a)

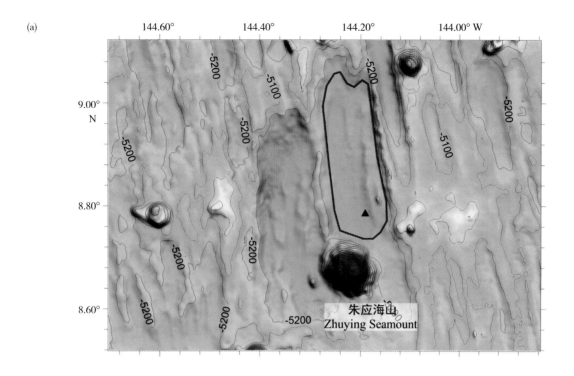

(b)

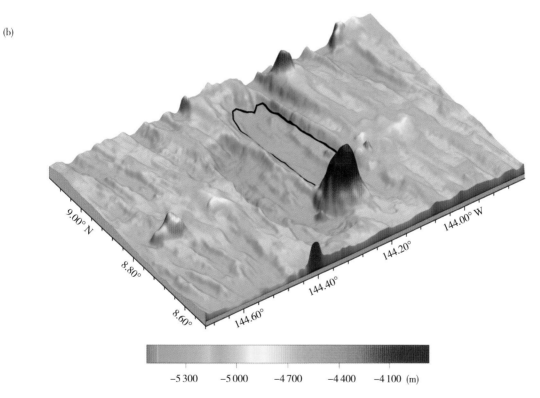

图 2-157 朱应北海底洼地

(a) 地形图（等深线间隔 100 m）；(b) 三维图

Fig.2-157 Zhuyingbei Depression

(a) Bathymetric map (the contour interval is 100 m); (b) 3-D topographic map

2.8.35 嘉卉圆海丘
Jiahui Knoll

中文名称 Chinese Name	嘉卉圆海丘 Jiahui Yuanhaiqiu	英文名称 English Name	Jiahui Knoll	所在大洋 Ocean or Sea	东太平洋 East Pacific Ocean
发现情况 Discovery Facts	此圆海丘于 1997 年 8 月由中国科考船"大洋一号"在执行 DY85-5 航次时发现。 This knoll was discovered by the Chinese R/V *Dayang Yihao* during the DY85-5 cruise in August, 1997.				
命名历史 Name History	由我国命名为嘉卉圆海丘，于 2014 年提交 SCUFN 审议通过。 Jiahui Knoll was named by China and approved by SCUFN in 2014.				
特征点坐标 Coordinates	8°29.80′N，144°24.30′W			长 (km) × 宽 (km) Length (km) × Width (km)	7 × 6
最大水深（m） Max Depth（m）	5 200	最小水深（m） Min Depth（m）	4 265	高差（m） Total Relief（m）	935
地形特征 Feature Description	该圆海丘俯视平面形态呈圆形，顶部地形相对平坦，周缘地形陡峭，峰顶水深 4 265 m，山麓水深 5 200 m，高差 935 m（图 2–158）。 This knoll has a round overlook plane shape. The top is relatively flat while the sides are steep. The top depth is about 4 265 m and the piedmont depth is 5 200 m, which makes the total relief being 935 m (Fig.2–158).				
命名释义 Reason for Choice of Name	"嘉卉"出自《诗经·小雅·四月》"山有嘉卉，侯栗侯梅。废为残贼，莫知其尤！"嘉卉指美丽的鲜花，意为山坡上百花盛开，栗树和梅树茂密生长，却被摧残，表达了人们对破坏自然美景的愤慨之情。 "Jiahui" comes from a poem named *Siyue* in *Shijing · Xiaoya*. *Shijing* is a collection of ancient Chinese poems from 11th century B.C. to 6th century B.C. "On mountains, trees grow grand and fair; Chestnuts here, and plum trees there. The trees are levelled to the ground; Despoilers can nowhere be found." "Jiahui" means beautiful flowers. The poem means that flowers are blooming all over the hill, the chestnut and plum trees are destroyed showing his irritation for those who destroy the nature.				

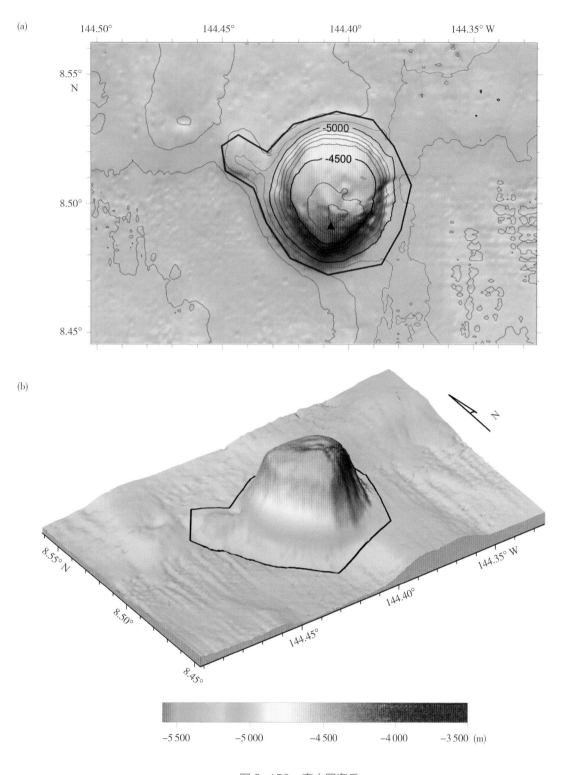

图 2-158　嘉卉圆海丘

(a) 地形图（等深线间隔 100 m）；(b) 三维图

Fig.2-158　Jiahui Knoll

(a) Bathymetric map (the contour interval is 100 m); (b) 3-D topographic map

2.8.36 楚茨海山
Chuci Seamount

中文名称 Chinese Name	楚茨海山 Chuci Haishan	英文名称 English Name	Chuci Seamount	所在大洋 Ocean or Sea	东太平洋 East Pacific Ocean
发现情况 Discovery Facts	此海山于 1995 年 11 月由中国科考船"大洋一号"在执行 DY85-5 航次时发现。 The seamount was discovered by the Chinese R/V *Dayang Yihao* during the DY85-5 cruise in November, 1995.				
命名历史 Name History	由我国命名为楚茨海山，于 2014 年提交 SCUFN 审议通过。 Chuci Seamount was named by China and approved by SCUFN in 2014.				
特征点坐标 Coordinates	7°49.20′N，144°31.70′W			长(km)×宽(km) Length (km) × Width (km)	15×10
最大水深（m） Max Depth（m）	5 250	最小水深（m） Min Depth（m）	3 783	高差（m） Total Relief（m）	1 467
地形特征 Feature Description	该海山位于楚茨海山链西部，东西走向，长宽分别为 15 km 和 10 km。海山顶部地形较缓，周缘地形较陡，峰顶水深 3 783 m，山麓水深 5 250 m，高差 1 467 m（图 2–159）。 Chuci Seamount is located in the western part of Chuci Seamount Chain, running E to W. The length is 15 km and the width is 10 km. The top of seamount is flat while the sides are steep. The top depth is 3 783 m and the piedmont depth is 5 250 m, which makes the total relief being 1 467 m (Fig.2–159).				
命名释义 Reason for Choice of Name	"楚茨"出自《诗经·小雅·楚茨》"楚楚者茨，言抽其棘"，指茂密的荆棘。此句是指田野上的蒺藜十分茂密，需要清除它们来播种粮食。此篇诗歌是一首祭祖拜神的乐歌，描述了祭拜的全过程。 "Chuci" comes from a poem named *Chuci* in *Shijing · Xiaoya*. *Shijing* is a collection of ancient Chinese poems from 11th century B.C. to 6th century B.C. "Where in clusters grew the thorn, by the roots they've all been torn." It means that the tribulus on the ground are so thick that we need to clear them away to plant crops. The seamount is named "Chuci" in order to symbolize an industrious and pioneering spirit of ancient Chinese. This poem is a eulogy to sing praise of the sacrificial rites held by the noblemen in the Zhou Dynasty.				

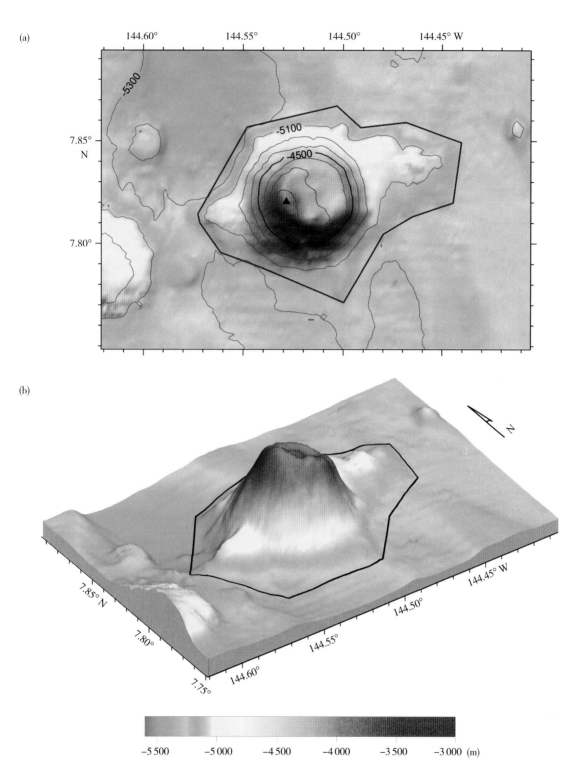

图 2-159　楚茨海山

(a) 地形图（等深线间隔 200 m）；(b) 三维图

Fig.2-159　Chuci Seamount

(a) Bathymetric map (the contour interval is 200 m); (b) 3-D topographic map

2.8.37 楚茨海山链
Chuci Seamount Chain

中文名称 Chinese Name	楚茨海山链 Chuci Haishanlian	英文名称 English Name	Chuci Seamount Chain	所在大洋 Ocean or Sea	东太平洋 East Pacific Ocean
发现情况 Discovery Facts	此海山链于 1995 年 11 月由中国科考船"大洋一号"在执行 DY85-5 航次时发现。 This seamount chain was discovered by the Chinese R/V *Dayang Yihao* during the DY85-5 cruise in November, 1995.				
命名历史 Name History	在我国大洋航次调查报告和 GEBCO 地名辞典中未命名。 This seamount chain has not been named in Chinese cruise reports or GEBCO gazetteer.				
特征点坐标 Coordinates	7°47.51′N，144°41.16′W 7°49.20′N，144°31.70′W（楚茨海山 / Chuci Seamount） 7°50.86′N，144°20.16′W 7°57.12′N，144°16.20′W 7°58.52′N，144°07.74′W 7°59.22′N，144°00.66′W 7°57.33′N，143°52.20′W			长（km）× 宽（km） Length (km) × Width (km)	120 × 15
最大水深（m） Max Depth（m）	5 520	最小水深（m） Min Depth（m）	3 783	高差（m） Total Relief（m）	1 737
地形特征 Feature Description	该海山链位于嘉卉圆海丘以南 65 km，海山链由 7 座近北东东向排列的海山（或海丘）组成，长宽分别为 120 km 和 15 km，最高峰位于海山链西部的楚茨海山，峰顶水深 3 783 m（图 2–160）。 Chuci Seamount Chain is located in 65 km south to Jiahui Knoll. This seamount chain consists of seven seamounts (or hills) nearly arranged in NNE–SSW direction. The length of this seamount chain is 120 km and the width is 15 km. The highest summit is at Chuci Seamount in the west of the seamount chain. The top depth is 3 783 km (Fig.2–160).				
命名释义 Reason for Choice of Name	以海山链中最高的楚茨海山的名字命名。 This seamount chain is named after the highest seamount in the chain, namely Chuci Seamount.				

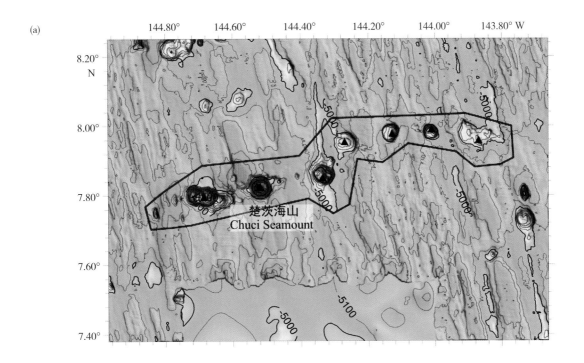

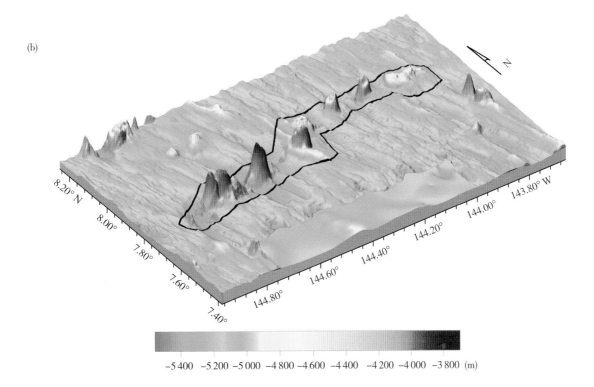

图 2-160　楚茨海山链

(a) 地形图（等深线间隔 100 m）；(b) 三维图

Fig.2-160　Chuci Seamount Chain

(a) Bathymetric map (the contour interval is 100 m); (b) 3-D topographic map

2.8.38 亿庾海山
Yiyu Seamount

中文名称 Chinese Name	亿庾海山 Yiyu Haishan	英文名称 English Name	Yiyu Seamount	所在大洋 Ocean or Sea	东太平洋 East Pacific Ocean
发现情况 Discovery Facts	此海山于 1995 年 11 月由中国科考船"大洋一号"在执行 DY85-5 航次时发现。 This seamount was discovered by the Chinese R/V *Dayang Yihao* during the DY85-5 cruise in November, 1995.				
命名历史 Name History	在我国大洋航次调查报告和 GEBCO 地名辞典中未命名。 This seamount has not been named in Chinese cruise reports or GEBCO gazetteer.				
特征点坐标 Coordinates	8°25.19′N，143°08.70′W			长 (km) × 宽 (km) Length (km) × Width (km)	9 × 8
最大水深（m） Max Depth（m）	5 318	最小水深（m） Min Depth（m）	4 205	高差（m） Total Relief（m）	1 113
地形特征 Feature Description	该海山俯视平面形态呈圆形，基座直径 8 ~ 9 km，海山顶部地形相对平缓，周缘地形较陡，峰顶水深 4 205 m，山麓水深 5 318 m，高差 1 113 m（图 2–161）。 Yiyu Seamount has a round overlook plane shape with the base diameter of 8–9 km. The top of the seamount is flat while the sides are steep. The top depth is 4 205 m and the piedmont depth is 5 318 m, which makes the total relief being 1 113 m (Fig.2–161).				
命名释义 Reason for Choice of Name	"亿庾"出自《诗经·小雅·楚茨》"我仓既盈，我庾维亿。以为酒食，以享以祀，以妥以侑，以介景福"。"亿庾"指粮食喜获丰收，粮仓里堆满了粮食。此篇是一首祭祖拜神的乐歌，描述了祭拜的场景。 "Yiyu" comes from a poem named *Chuci* in *Shijing · Xiaoya*. *Shijing* is a collection of ancient Chinese poems from 11th century B.C. to 6th century B.C. "Here stand our barns in the fields; There stand our stacks in the fields. As meat and drink they serve; As sacrificial food they serve. The sacrificial rites we observe, to pray for blessings we deserve." "Yiyu" means that the barn is full of food, which represent a harvest. This poem is a eulogy to sing praise of the sacrificial rites held by the noblemen in the Zhou Dynasty.				

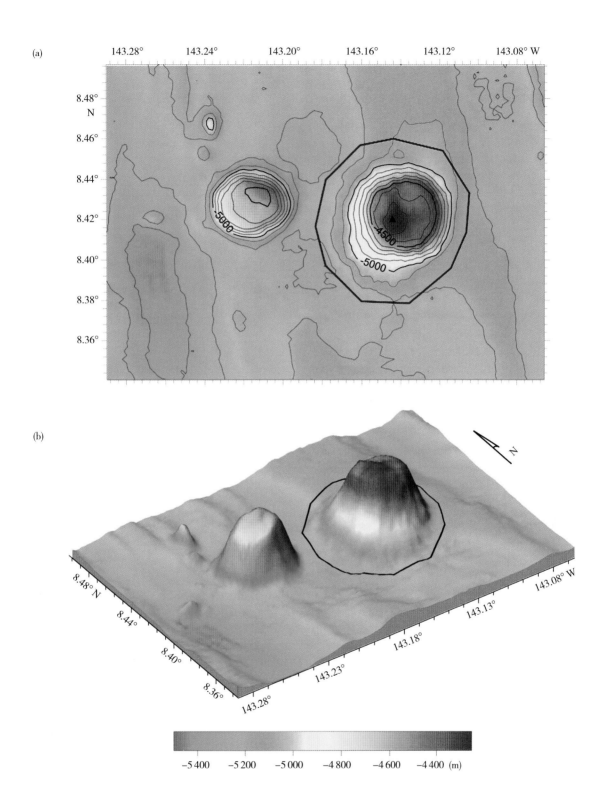

图 2-161　亿庾海山

(a) 地形图（等深线间隔 100 m）；(b) 三维图

Fig.2-161　Yiyu Seamount

(a) Bathymetric map (the contour interval is 100 m); (b) 3-D topographic map

2.8.39 怀允海山
Huaiyun Seamount

中文名称 Chinese Name	怀允海山 Huaiyun Haishan	英文名称 English Name	Huaiyun Seamount	所在大洋 Ocean or Sea	东太平洋 East Pacific Ocean
发现情况 Discovery Facts	此海山于 1995 年 11 月由中国科考船"大洋一号"在执行 DY85-5 航次时发现。 This seamount was discovered by the Chinese R/V *Dayang Yihao* during the DY85-5 cruise in November, 1995.				
命名历史 Name History	在我国大洋航次调查报告和 GEBCO 地名辞典中未命名。 This seamount has not been named in Chinese cruise reports or GEBCO gazetteer.				
特征点坐标 Coordinates	9°22.32′N，142°26.76′W		长(km)×宽(km) Length (km)× Width (km)	8 × 7	
最大水深（m） Max Depth（m）	5 262	最小水深（m） Min Depth（m）	4 135	高差（m） Total Relief（m）	1 127
地形特征 Feature Description	该海山位于朱应海山北东 200 km。海山峰顶水深 4 135 m，山麓水深 5 262 m，高差 1 127 m（图 2–162）。 Huaiyun Seamount is located at 200 km northeast to Zhuying Seamount. The top depth of the seamount is 4 135 m and the piedmont depth is 5 262 m, which makes the total relief being 1 127 m (Fig.2–162).				
命名释义 Reason for Choice of Name	"怀允"出自《诗经·小雅·鼓钟》"鼓钟将将，淮水汤汤，忧心且伤。淑人君子，怀允不忘。" 此句意为耳闻钟鼓铿锵，面临滔滔淮河，悲从中来，怀念君子的美德。"怀允"意指要记住君子的善行和美德。这是一首描写聆听音乐，怀念善人君子的诗歌。 "Huaiyun" comes from a poem named *Guzhong* in *Shijing · Xiaoya*. *Shijing* is a collection of ancient Chinese poems from 11th century B.C. to 6th century B.C. "When I hear the ringing bells beside the river Huai that swells, in my heart deep sorrow dwells. O for my lord, the former king, whose honours in my ears e'er ring" This poem records the poet's recollection of the virtuous gentleman when he hears the music from the court. "Huaiyun" means to keep the gentleman's good deed and virtue in mind.				

(a)

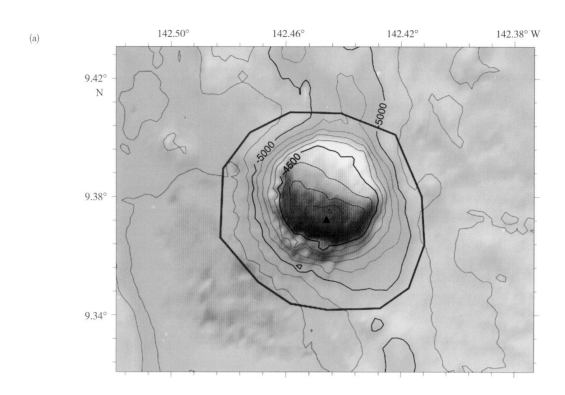

(b)

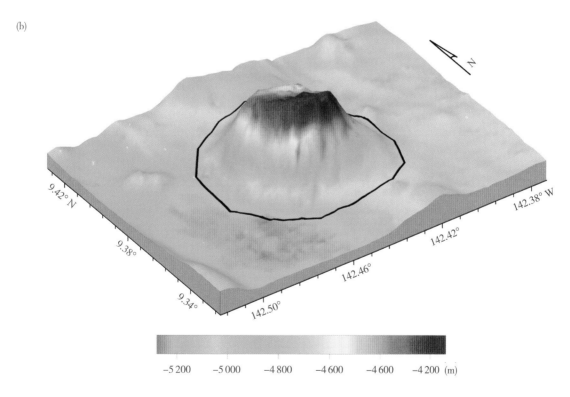

图 2-162　怀允海山

(a) 地形图（等深线间隔 100 m）；(b) 三维图

Fig.2-162　Huaiyun Seamount

(a) Bathymetric map (the contour interval is 100 m); (b) 3-D topographic map

2.8.40　启明海山
Qiming Seamount

中文名称 Chinese Name	启明海山 Qiming Haishan	英文名称 English Name	Qiming Seamount	所在大洋 Ocean or Sea	东太平洋 East Pacific Ocean
发现情况 Discovery Facts	此海山于 1995 年 8 月由中国科考船"大洋一号"在执行 DY85-5 航次时发现。 This seamount was discovered by the Chinese R/V *Dayang Yihao* during the DY85-5 cruise in August, 1995.				
命名历史 Name History	由我国命名为启明海山，于 2013 年提交 SCUFN 审议通过。 Qiming Seamount was named by China and approved by SCUFN in 2013.				
特征点坐标 Coordinates	8°20.30′N，142°16.50′W			长 (km) × 宽 (km) Length (km) × Width (km)	7.3 × 7.3
最大水深（m） Max Depth（m）	5 260	最小水深（m） Min Depth（m）	4 255	高差（m） Total Relief（m）	1 005
地形特征 Feature Description	该海山大致呈圆锥形，基座直径 7.3 km，峰顶水深 4 255 m，山麓水深 5 260 m，高差 1 005 m。海山西坡地形较陡，东坡地形较缓（图 2–163）。 Qiming Seamount has a nearly conical shape with the base diameter of 7.3 km. The top depth is 4 255 m and the piedmont depth is 5 260 m, which makes the total relief being 1 005 m. The western slope of the seamount is steep while the east is flat (Fig.2–163).				
命名释义 Reason for Choice of Name	启明出自《诗经·小雅·大东》，"东有启明，西有长庚"。此诗歌描述中国东部人民遭受周朝的残酷统治、生活困苦的情景，表达作者忧愤抗争的激情。金星天亮前出现在东方，称为"启明星"；黄昏时出现在西方，称为"长庚星"。启明海山和长庚海山分别位于中国多金属结核合同区的东区和西区，故东区的海山命名为启明，以喻日出东方；西区的海山命名为长庚，以喻日落西方。 "Qiming" comes from a poem named *Dadong* in *Shijing · Xiaoya*. *Shijing* is a collection of ancient Chinese poems from 11th century B.C. to 6th century B.C. "The east presents the morning star; The west presents the evening star." Venus is named "Qimingxing" (the morning star) when it appears in the east before dawn. However it is named "Changgengxing" (the evening star) when it appears in the west after dusk. Qiming Seamount is located in the eastern part of Chinese polymetallic nodules contract area while Changgeng Seamount is located in the western part. Hence the seamount in the eastern part is named "Qiming" to symbolize the east sunrise while the seamount in the western part is named "Changgeng" to symbolize the west sunset.				

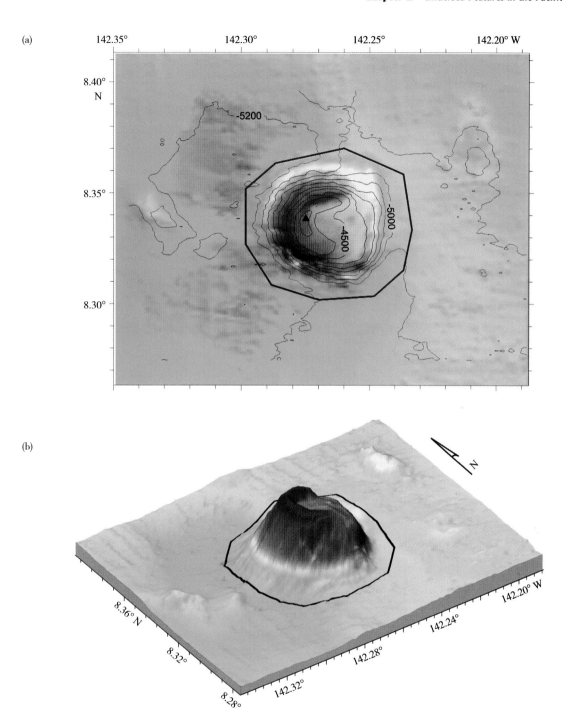

图 2-163　启明海山

(a) 地形图（等深线间隔 100 m）；(b) 三维图

Fig.2-163　Qiming Seamount

(a) Bathymetric map (the contour interval is 100 m); (b) 3-D topographic map

2.8.41 启明南海底洼地
Qimingnan Depression

中文名称 Chinese Name	启明南海底洼地 Qimingnan Haidiwadi	英文名称 English Name	Qimingnan Depression	所在大洋 Ocean or Sea	东太平洋 East Pacific Ocean
发现情况 Discovery Facts	此海底洼地于 1997 年 8 月由中国科考船"大洋一号"在执行 DY85-5 航次时发现。 This depression was discovered by the Chinese R/V *Dayang Yihao* during the DY85-5 cruise in August, 1997.				
命名历史 Name History	在我国大洋航次调查报告和 GEBCO 地名辞典中未命名。 This depression has not been named in Chinese cruise reports or GEBCO gazetteer.				
特征点坐标 Coordinates	7°51.88′N，142°12.18′W 8°04.37′N，142°16.26′W（最深点 /deepest point） 8°19.73′N，142°18.30′W		长(km)×宽(km) Length (km)× Width (km)		53×6
最大水深（m） Max Depth（m）	5 370	最小水深（m） Min Depth（m）	5 015	高差（m） Total Relief（m）	355
地形特征 Feature Description	该海底洼地位于启明海山以南，NNW—SSE 走向，长宽分别为 53 km 和 6 km，海底洼地南北两侧深，中部浅，水深最深 5 370 m，水深最浅 5 015 m，高差 355 m（图 2–164）。 Qimingnan Depression is located in the south to Qiming Seamounts, running NNW to SSE. The length is 53 km and the width is 6 km. The southern and northern sides of this depression is deep while the middle is shallow. The maximum depth is 5 370 m and the minimum depth is 5 015 m, which makes the total relief being 355 m (Fig.2–164).				
命名释义 Reason for Choice of Name	该海底洼地位于启明海以南，故以此命名。 "Nan" means south in Chinese. The depression is named Qimingnan because it is at the south to Qiming Seamount.				

(a)

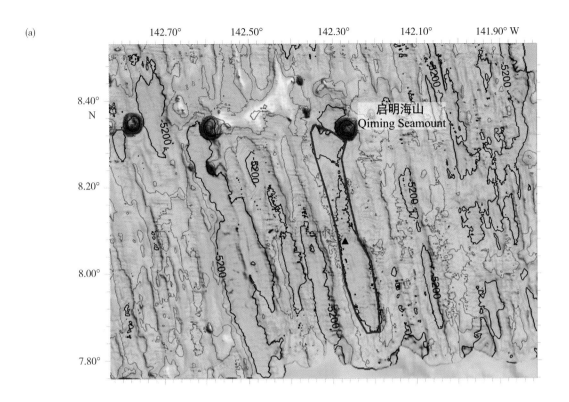

(b)

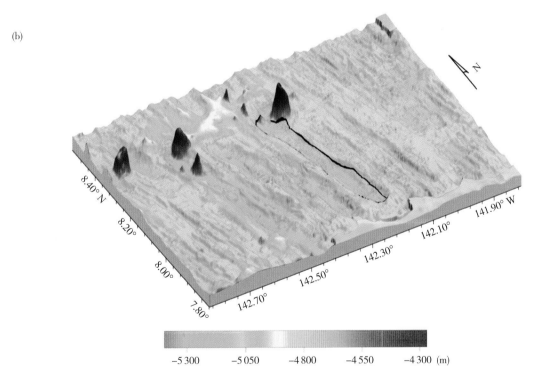

图 2-164　启明南海底洼地

(a) 地形图（等深线间隔 100 m）；(b) 三维图

Fig.2-164　Qimingnan Depression

(a) Bathymetric map (the contour interval is 100 m); (b) 3-D topographic map

2.8.42　圆鼓圆海丘
Yuangu Knoll

中文名称 Chinese Name	圆鼓圆海丘 Yuangu Yuanqiu	英文名称 English Name	Yuangu Knoll	所在大洋 Ocean or Sea	东太平洋 East Pacific Ocean
发现情况 Discovery Facts	此圆丘于 2017 年由中国科考船"向阳红 03 号"在执行测量任务时调查发现。 This knoll was discovered by the Chinese R/V *Xiangyanghong* 03 during the survey in 2017.				
命名历史 Name History	由我国命名为圆鼓圆海丘，于 2018 年提交 SCUFN 审议通过。 This feature was named Yuangu Knoll by China and the name was approved by SCUFN in 2018.				
特征点坐标 Coordinates	10°26.7′N，154°18.0′W			长 (km) × 宽 (km) Length (km) × Width (km)	4.0 × 4.0
最大水深（m） Max Depth（m）	5 150	最小水深（m） Min Depth（m）	4 844	高差（m） Total Relief（m）	306
地形特征 Feature Description	圆鼓圆海丘位于太平洋中央海盆，在 Gongzhen knolls 东方向 14 km，大体形状呈圆形呈圆形（图 2–165）。 Yuangu Knoll is located in the Central Pacific Basin and 14 km east of the Gongzhen Knolls. Its overall shape is round (Fig.2–165).				
命名释义 Reason for Choice of Name	该海丘形似圆鼓，故名。 This hill has a shape like a round drum, so it is named with the Chinese name of drum, "Yuangu".				

(a)

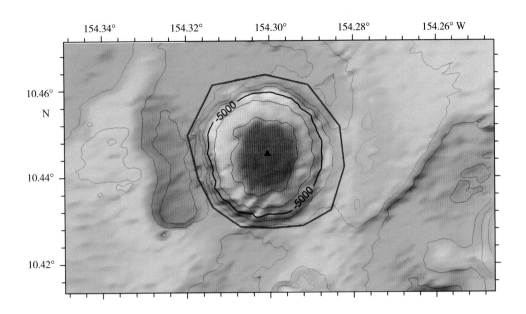

(b)

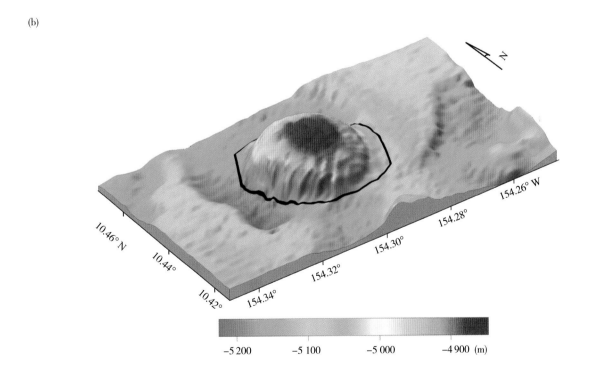

图 2-165　圆鼓圆海丘

(a) 地形图（等深线间隔 50 m）；(b) 三维图

Fig.2-165　Yuangu Knoll

(a) Bathymetric map (the contour interval is 50 m); (b) 3-D topographic map

2.8.43 刺螺平顶海山群
Ciluo Guyots

中文名称 Chinese Name	刺螺平顶海山群 Ciluo Pingdinghaishanqun	英文名称 English Name	Ciluo Guyots	所在大洋 Ocean or Sea	东太平洋 East Pacific Ocean
发现情况 Discovery Facts	此平顶海山群于 2017 年由中国科考船"向阳红 03 号"在执行测量任务时调查发现。 This guyots was discovered by the Chinese R/V *Xiangyanghong* 03 during the survey in 2017.				
命名历史 Name History	由我国命名为刺螺平顶海山群，于 2018 年提交 SCUFN 审议通过。 This feature was named Ciluo Guyots by China and the name was approved by SCUFN in 2018.				
特征点坐标 Coordinates	9°17.0′N，158°31.9′W 9°12.4′N，157°59.2′W			长(km)× 宽(km) Length (km) × Width (km)	178 × 43
最大水深（m） Max Depth（m）	4 800	最小水深（m） Min Depth（m）	1 290	高差（m） Total Relief（m）	3 510
地形特征 Feature Description	刺螺平顶海山群位于太平洋中央海盆，在 Ironwood seamout 东南方向 200 km，有 2 个山峰（图 2–166）。 Ciluo Guyots is located in the Central Pacific Basin, 200 km southeast to Ironwood seamount. It has two peaks (Fig.2–166).				
命名释义 Reason for Choice of Name	刺螺，也叫维纳斯骨螺，一种骨螺科动物，分布于东印度洋与西太平洋。此海山群形似刺螺，故名。 Ciluo, also named Venus comb murex, a kind of Muridae aniamls, are distributed in East Indian Ocean and West Pacific Ocean. This guyots is named after Ciluo because it looks like it.				

(a)

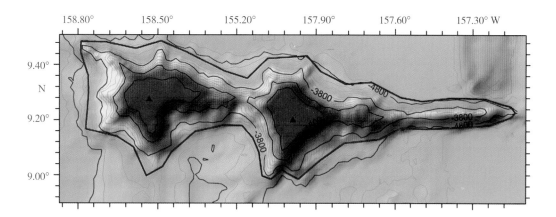

(b)

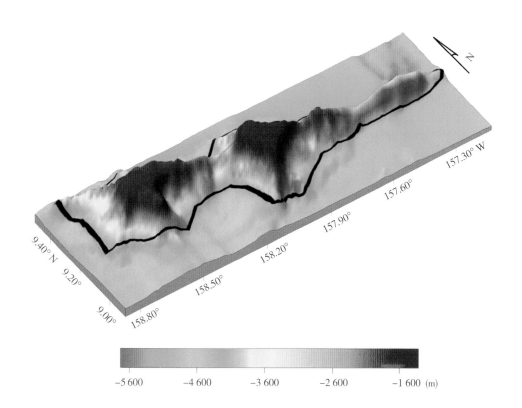

图 2-166 刺螺平顶海山群

(a) 地形图 (等深线间隔 200 m)；(b) 三维图

Fig.2-166 Ciluo Guyots

(a) Bathymetric map (the contour interval is 200 m); (b) 3-D topographic map

347 ■

第 9 节 东太平洋海隆地理实体和热液区

东太平洋海隆是一条位于东太平洋快速扩张的洋中脊（Haymon et al., 1991），全扩张速率大于 88 mm/a，南起 55°S，130°W 附近，与太平洋 – 南极洲洋脊相接，北止于加利福尼亚湾索尔顿海盆。东太平洋海隆轴部不发育大西洋型或印度洋型扩张中轴裂谷，海隆轴部沿走向发育狭长断裂地堑或扩张裂隙（Menard, 1966）。我国调查区域主要集中在 5°N—8°S 区的洋中脊扩张隆起区，东侧为太平洋板块，北东侧为科科斯板块，南东侧为纳兹卡板块，水深范围总体上在 2 500 ～ 3 500 m 之间。

此海域我国共命名地理实体 14 个，热液区 18 个。其中海山 2 个，包括维翰海山和颁首平顶海山；海丘 9 个，包括鸟巢海丘、西鸳鸯海丘、东鸳鸯海丘、硕人海丘、白云海丘、啸歌海丘、那居圆海丘、依蒲海丘和白华海丘；海脊 2 个，包括白茅海脊和太白海脊；海渊 1 个，即太白海渊。

18 个热液区分别是太白北 1 号热液区、太白北 2 号热液区、太白北 3 号热液区、太白北 4 号热液区、太白南 1 号热液区、太白南 2 号热液区、太白南 3 号热液区、鸟巢北 1 号热液区、鸟巢热液区、鸟巢南 1 号热液区、鸟巢南 2 号热液区、鸟巢南 3 号热液区、鸟巢南 4 号热液区、宝石 1 号热液区、角弓 1 号热液区、徽猷 1 号热液区、鱼藻 1 号热液区和鱼藻 2 号热液区（图 2-167）。

Section 9   Undersea Features and Hydrothermal Fields in the East Pacific Rise

The East Pacific Rise is a fast-spreading mid-ocean ridge in the East Pacific Ocean with the full spreading rate being over 88 mm/a. It starts from around 55°S, 130°W in the south, and connects with the Pacific-Antarctic Ridge, ending at the Salton Basin in the Gulf of California. The East Pacific Rise shaft does not develop Atlantic or Indian type spreading axial rifts, but develops narrow fracture grabens or spreading cracks along the strike direction. Chinese investigation focused on the area of 5°N–8°S, which is the area of ridge spreading and rising. The area has the Pacific Plate in the east, the Cocos Plate in the northeast, the Nazca Plate in the southeast, with the water depth range generally being 2 500–3 500 m.

In total, 14 undersea features and 18 hydrothermal fields have been named by China in the region of the East Pacific Rise. Among them, there are 2 seamounts, including Weihan Seamount and Fenshou Guyot; 9 hills or knoll, including Niaochao Hill, Xiyuanyang Hill, Dongyuanyang Hill, Shuoren Hill, Baiyun Hill, Xiaoge Hill, Nuoju Knoll, Yipu Hill and Baihua Hill; 2 ridges, including Baimao Ridge and Taibai Ridge; 1 deep, that is Taibai Deep.

The 18 hydrothermal fields respectively are Taibaibei-1 Hydrothermal Field, Taibaibei-2 Hydrothermal Field, Taibaibei-3 Hydrothermal Field, Taibaibei-4 Hydrothermal Field, Taibainan-1 Hydrothermal Field, Taibainan-2 Hydrothermal Field, Taibainan-3 Hydrothermal Field, Niaochaobei-1 Hydrothermal Field, Niaochao Hydrothermal Field, Niaochaonan-1 Hydrothermal Field, Niaochaonan-2 Hydrothermal Field, Niaochaonan-3 Hydrothermal Field, Niaochaonan-4 Hydrothermal Field, Baoshi-1 Hydrothermal Field, Jiaogong-1 Hydrothermal Field, Huiyou-1 Hydrothermal Field, Yuzao-1 Hydrothermal Field and Yuzao-2 Hydrothermal Field (Fig. 2–167).

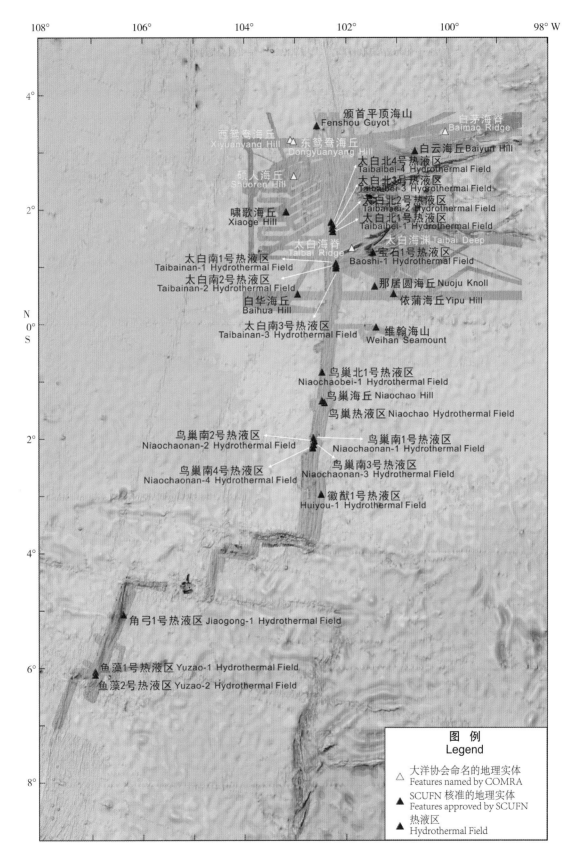

图 2-167　东太平洋海隆地理实体位置示意图

Fig.2-167　Locations of the undersea features in the East Pacific Rise

2.9.1　颁首平顶海山
Fenshou Guyot

中文名称 Chinese Name	颁首平顶海山 Fenshou Pingdinghaishan	英文名称 English Name	Fenshou Guyot	所在大洋 Ocean or Sea	东太平洋 East Pacific Ocean
发现情况 Discovery Facts		此平顶海山于 2009 年由中国科考船"大洋一号"在执行 DY115-21 航次时调查发现。 This guyot was discovered by the Chinese R/V *Dayang Yihao* during the DY115-21 cruise in 2009.			
命名历史 Name History	由我国命名为颁首平顶海山，于 2016 年提交 SCUFN 审议通过。 Fenshou Guyot was named by China and approved by SCUFN in 2016.				
特征点坐标 Coordinates	3°01.28′N，101°55.21′W			长 (km) × 宽 (km) Length (km) × Width (km)	8 × 8
最大水深（m） Max Depth（m）	3 400	最小水深（m） Min Depth（m）	2 250	高差（m） Total Relief（m）	1 150
地形特征 Feature Description	该平顶海山位于东太平洋海隆，平面形态呈圆形，基底直径约 8 km。平顶海山顶部水深 2 250 m，山麓水深 3 400 m，高差 1 150 m（图 2–168）。 This guyot is located in the East Pacific Rise. It has a round plane shape with the base diameter of 8 km. The top depth of the guyot is 2 250 m and the piedmont depth is 3 400 m, which makes the total relief being 1 150 m (Fig.2–168).				
命名释义 Reason for Choice of Name	"颁首"出自《诗经·小雅·鱼藻》"鱼在在藻，有颁其首。王在在镐，岂乐饮酒。"此诗描述了鱼在水藻中自由游动，大王在欢饮美酒的场景。 "Fenshou" comes from a poem named *Yuzao* in *Shijing · Xiaoya*. *Shijing* is a collection of ancient Chinese poems from 11th century B.C. to 6th century B.C. "Where is the fish? Among the algae is the fish, its head so large and fine. Where is the king? Here in Hao is the king, hale and healthy, drinking wine." "Fenshou" means a big head of fish. The poem shows the scene that the fish is swimming leisurely among the algae and the King is drinking good wine.				

(a)

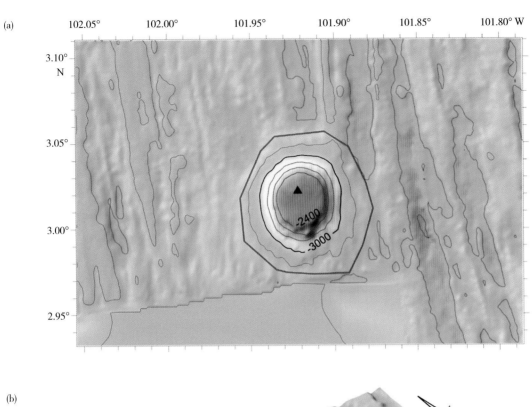

(b)

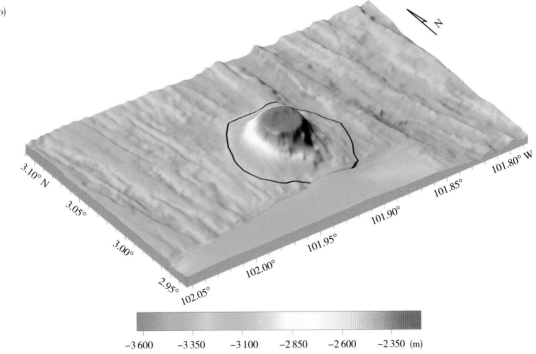

图 2-168　颁首平顶海山

(a) 地形图 （等深线间隔 200 m）；(b) 三维图

Fig.2-168　Fenshou Guyot

(a) Bathymetric map (the contour interval is 200 m); (b) 3-D topographic map

2.9.2 西鸳鸯海丘
Xiyuanyang Hill

中文名称 Chinese Name	西鸳鸯海丘 Xiyuanyang Haiqiu		英文名称 English Name	Xiyuanyang Hill	所在大洋 Ocean or Sea	东太平洋 East Pacific Ocean
发现情况 Discovery Facts	此海丘于 2009 年由中国科考船"大洋一号"在执行 DY115-21 航次时调查发现。 Xiyuanyang Hill was discovered by the Chinese R/V *Dayang Yihao* during the DY115-21 cruise in 2009.					
命名历史 Name History	在我国大洋航次调查报告和 GEBCO 地名辞典中未命名。 This hill has not been named in Chinese cruise reports or GEBCO gazetteer.					
特征点坐标 Coordinates	3°10.23′N，103°02.18′W			长(km)× 宽(km) Length (km)× Width (km)		3.0 × 2.2
最大水深（m） Max Depth（m）	3 400	最小水深（m） Min Depth（m）	2 970	高差（m） Total Relief(m)		430
地形特征 Feature Description	该海丘位于东太平洋海隆，平面形态呈椭圆形，长宽分别为 3.0 km 和 2.2 km，顶部水深 2 970 m，山麓水深 3 400 m，高差 430 m。与东鸳鸯海丘遥相呼应（图 2–169）。 Xiyuanyang Hill is located in the East Pacific Rise. It has a nearly elliptic plane shape. The length is 3.0 km and the width is 2.2 km. The top depth of the hill is 2 970 m and the piedmont depth is 3 400 m, which makes the total relief being 430 m. It is echoed to Dongyuanyang Hill at a distance (Fig.2–169).					
命名释义 Reason for Choice of Name	"鸳鸯"出自《诗经·小雅·鸳鸯》"鸳鸯于飞，毕之罗之"，意思是鸳鸯轻轻飞翔。"鸳鸯"也是中国文化中象征美好爱情的鸟类。此处两座海丘紧邻，成因相同，故按照其相对位置分别命名为西鸳鸯海丘、东鸳鸯海丘。 "Yuanyang" comes from a poem named *Yuanyang* in *Shijing · Xiaoya*. *Shijing* is a collection of ancient Chinese poems from 11th century B.C. to 6th century B.C. "The mandarin ducks fly in pairs; They are caught with nets and snares." "Yuanyang" means mandarin ducks. In traditional Chinese culture, mandarin ducks are believed to be lifelong couples, unlike other species of ducks. Hence they are regarded as a symbol of conjugal affection and fidelity, and are frequently featured in Chinese art. These two adjoining hills with same origin look like a couple, thus the one in the west is named Xiyuanyang Hill and the other one in the east is named Dongyuanyang hill.					

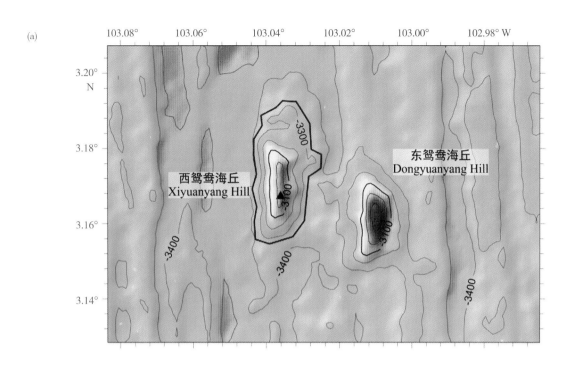

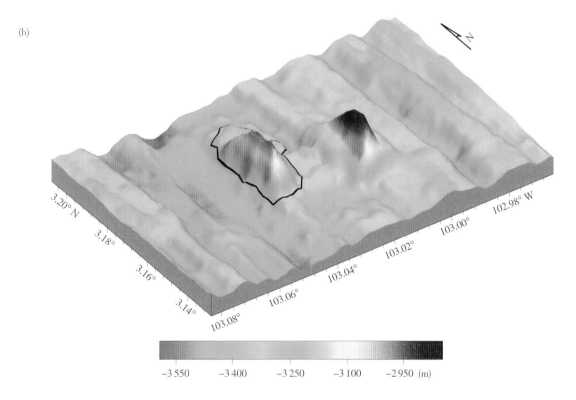

图 2-169　西鸳鸯海丘

(a) 地形图 （等深线间隔 100 m）；(b) 三维图

Fig.2-169　Xiyuanyang Hill

(a) Bathymetric map (the contour interval is 100 m); (b) 3-D topographic map

355

2.9.3 东鸳鸯海丘
Dongyuanyang Hill

中文名称 Chinese Name	东鸳鸯海丘 Dongyuanyang Haiqiu	英文名称 English Name	Dongyuanyang Hill	所在大洋 Ocean or Sea	东太平洋 East Pacific Ocean
发现情况 Discovery Facts	此海丘于 2009 年由中国科考船"大洋一号"在执行 DY115-21 航次时调查发现。 This hill was discovered by the Chinese R/V *Dayang Yihao* during the DY115-21 cruise in 2009.				
命名历史 Name History	在我国大洋航次调查报告和 GEBCO 地名辞典中未命名。 This hill has not been named in Chinese cruise reports or GEBCO gazetteer.				
特征点坐标 Coordinates	3°09.76′N，103°00.65′W		长(km)× 宽(km) Length (km) × Width (km)		3.8 × 2.5
最大水深（m） Max Depth（m）	3 350	最小水深（m） Min Depth（m）	2 850	高差（m） Total Relief（m）	500
地形特征 Feature Description	该海丘位于东太平洋海隆，长宽分别为 3.8 km 和 2.5 km，顶部水深 2 850 m，山麓水深 3 350 m，最大高差 500 m，与西鸳鸯海丘遥相呼应（图 2–170）。 This hill is located in the East Pacific Rise. The length is 3.8 km and the width is 2.5 km. The top depth of the hill is 2 850 m and the piedmont depth is 3 350 m, which makes the total relief being 500 m. It is echoed to Xiyuanyang Hill at a distance (Fig.2–170).				
命名释义 Reason for Choice of Name	"鸳鸯"出自《诗经·小雅·鸳鸯》"鸳鸯于飞，毕之罗之"，意思是鸳鸯轻轻飞翔。"鸳鸯"也是中国文化中象征美好爱情的鸟类。此处两座海丘紧邻，成因相同，故按照其相对位置分别命名为西鸳鸯海丘、东鸳鸯海丘。 "Yuanyang" comes from a poem named *Yuanyang* in *Shijing · Xiaoya*. *Shijing* is a collection of ancient Chinese poems from 11th century B.C. to 6th century B.C. "The mandarin ducks fly in pairs; They are caught with nets and snares." "Yuanyang" means mandarin ducks. In traditional Chinese culture, mandarin ducks are believed to be lifelong couples, unlike other species of ducks. Hence they are regarded as a symbol of conjugal affection and fidelity, and are frequently featured in Chinese art. These two adjoining hills with same origin look like a couple, thus the one in the west is named Xiyuanyang Hill and the other one in the east is named Dongyuanyuang hill.				

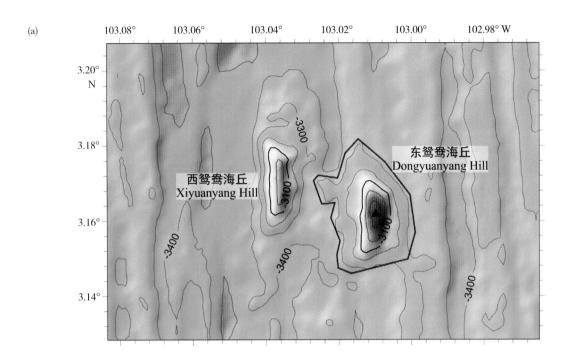

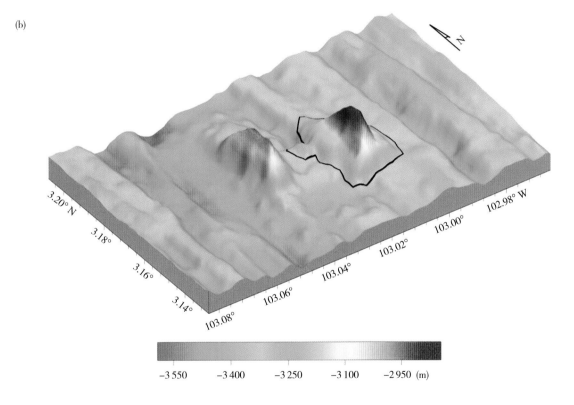

图 2-170　东鸳鸯海丘

(a) 地形图（等深线间隔 100 m）；(b) 三维图

Fig.2-170　Dongyuanyang Hill

(a) Bathymetric map (the contour interval is 100 m); (b) 3-D topographic map

2.9.4 白茅海脊
Baimao Ridge

中文名称 Chinese Name	白茅海脊 Baimao Haiji		英文名称 English Name	Baimao Ridge	所在大洋 Ocean or Sea	东太平洋 East Pacific Ocean
发现情况 Discovery Facts	此海脊于 2009 年由中国科考船"大洋一号"在执行 DY115-21 航次时调查发现。 Baimao Ridge was discovered by the Chinese R/V *Dayang Yihao* during the DY115-21 cruise in 2009.					
命名历史 Name History	在我国大洋航次调查报告和 GEBCO 地名辞典中未命名。 This ridge has not been named in Chinese cruise reports or GEBCO gazetteer.					
特征点坐标 Coordinates	3°21.56′N，100°04.92′W 3°21.71′N，100°00.98′W 3°21.33′N，99°56.84′W			长(km) × 宽(km) Length (km) × Width (km)		12.4 × 6.5
最大水深（m） Max Depth（m）	3 400	最小水深（m） Min Depth（m）	2 790	高差（m） Total Relief（m）		610
地形特征 Feature Description	该海脊位于东太平洋海隆，呈西宽东窄的长条状，长度 12.4 km，宽为 6.5 km。海脊顶部最小水深 2 790 m，海脊周边水深约 3 400 m，高差 610 m，海脊两侧地形较陡（图 2–171）。 Baimao Ridge is located in the East Pacific Rise. It has an elongated shape, the west is wide while the east is narrow. The length is 12.4 km and the width is 6.5 km. The minimum top depth of the ridge is 2 790 m and the surrounding depth is 3 400 m, which makes the total relief being 610 m. Both sides of the ridge are steep (Fig.2–171).					
命名释义 Reason for Choice of Name	"白茅"出自《诗经·小雅·白华》"白华菅兮，白茅束兮。之子之远，俾我独兮"，指白色的茅草。该篇以白茅把开白花的菅草捆成束映射夫妇之间相亲相爱。但现在丈夫远去，只留妻子独守空房，为本篇奠定了凄婉悲剧的基调。 Baimao comes from a poem named *Baihua* in *Shijing · Xiaoya*. *Shijing* is a collection of ancient Chinese poems from 11th century B.C. to 6th century B.C. "The reeds with flowers small and white with cogon vines are bound up tight. My lord has gone far, far away; Alone at home I daily stay." "Baimao" means white cogon. The poem uses bounding up the reed with cogon vines to imply the love between husband and wife. However, the husband went far away and left the wife home alone and the wife missed her husband very much.					

(a)

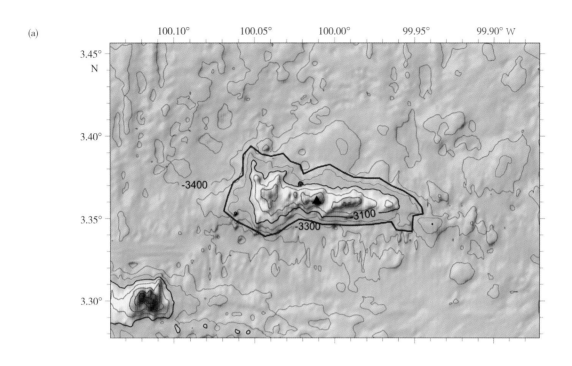

(b)

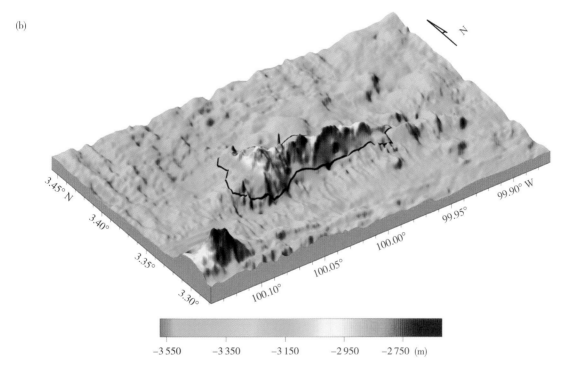

图 2-171　白茅海脊

(a) 地形图（等深线间隔 100 m）；(b) 三维图

Fig.2-171　Baimao Ridge

(a) Bathymetric map (the contour interval is 100 m); (b) 3-D topographic map

2.9.5 白云海丘
Baiyun Hill

中文名称 Chinese Name	白云海丘 Baiyun Haiqiu	英文名称 English Name	Baiyun Hill	所在大洋 Ocean or Sea	东太平洋 East Pacific Ocean
发现情况 Discovery Facts	此海丘于 2009 年由中国科考船"大洋一号"在执行 DY115-21 航次时调查发现。 Baiyun Hill was discovered by the Chinese R/V *Dayang Yihao* during the DY115-21 cruise in 2009.				
命名历史 Name History	由我国命名为白云海丘，于 2017 年提交 SCUFN 审议通过。 Baiyun Hill was named by China and approved by SCUFN in 2017.				
特征点坐标 Coordinates	2°59.21′N，100°37.59′W			长 (km)× 宽 (km) Length (km) × Width (km)	4.6 × 2.8
最大水深（m） Max Depth（m）	3 350	最小水深（m） Min Depth（m）	2 480	高差（m） Total Relief（m）	870
地形特征 Feature Description	该海丘位于东太平洋海隆，长宽分别为 4.6 km 和 2.8 km。海丘顶部水深 2 480 m，山麓水深 3 350 m，高差 870 m，地形坡度较陡（图 2–172）。 Baiyun Hill is located in the East Pacific Rise. The length is 4.6 km and the width is 2.8 km. The top depth of the hill is 2 480 m and the piedmont depth is 3 350 m, which makes the total relief being 870 m. The slopes are relatively steep (Fig.2–172).				
命名释义 Reason for Choice of Name	"白云"出自《诗经·小雅·白华》"英英白云，露彼菅茅。天步艰难，之子不犹。"白云指白色的云彩。以白云普降甘露滋润那些菅草和茅草，来反衬和怨恨丈夫违背常理，不能与妻子休戚与共。 "Baiyun" comes from a poem named *Baihua* in *Shijing · Xiaoya*. *Shijing* is a collection of ancient Chinese poems from 11th century B.C. to 6th century B.C. "When thick clouds spread across the sky, on reeds and cogons heavy dews lie. The way of fate is harsh and blind; My lord has thrown me out of mind." "Baiyun" means the white clouds. The sentence means white clouds lying dews on the reeds, which serves as a foil to the husband who could not take care of his wife well.				

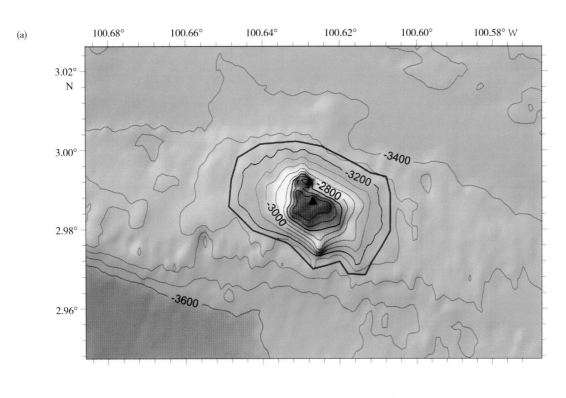

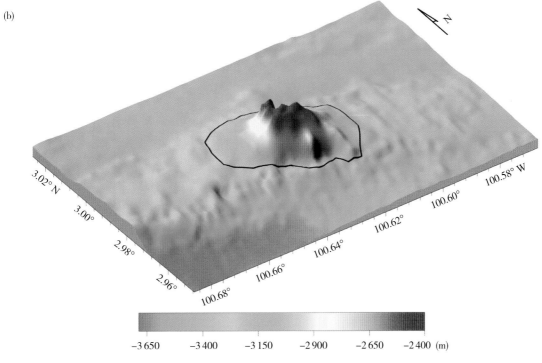

图 2-172　白云海丘

(a) 地形图（等深线间隔 100 m）；(b) 三维图

Fig.2-172　Baiyun Hill

(a) Bathymetric map (the contour interval is 100 m); (b) 3-D topographic map

2.9.6 硕人海丘
Shuoren Hill

中文名称 Chinese Name	硕人海丘 Shuoren Haiqiu	英文名称 English Name	Shuoren Hill	所在大洋 Ocean or Sea	东太平洋 East Pacific Ocean
发现情况 Discovery Facts					

表格内容如下：

发现情况 Discovery Facts	此海丘于 2009 年由中国科考船"大洋一号"在执行 DY115-21 航次时调查发现。 Shuoren Hill was discovered by the Chinese R/V *Dayang Yihao* during the DY115-21 cruise in 2009.				
命名历史 Name History	在我国大洋航次调查报告和 GEBCO 地名辞典中未命名。 This hill has not been named in Chinese cruise reports or GEBCO gazetteer.				
特征点坐标 Coordinates	2°33.50′N，102°59.54′W			长(km)×宽(km) Length (km)× Width (km)	6.2×4.5
最大水深（m） Max Depth（m）	3 290	最小水深（m） Min Depth（m）	2 780	高差（m） Total Relief（m）	510
地形特征 Feature Description	该海丘位于东太平洋海隆，长宽分别为 6.2 km 和 4.5 km。海丘顶部水深 2 780 m，山麓水深 3 290 m，高差 510 m（图 2–173）。 This hill is located in the East Pacific Rise. The length is 6.2 km and the width is 4.5 km. The top depth of the hill is 2 780 m and the piedmont depth is 3 290 m, which makes the total relief being 510 m (Fig.2–173).				
命名释义 Reason for Choice of Name	"硕人"出自《诗经·小雅·白华》"樵彼桑薪，昂烘于煁。维彼硕人，实劳我心。""硕人"指高大健美的男子。此诗表达了女子因丈夫不在身边而产生的哀怨和失落感。 "Shuoren" comes from a poem named *Baihua* in *Shijing · Xiaoya*. *Shijing* is a collection of ancient Chinese poems from 11th century B.C. to 6th century B.C. "The mulberry tree is felled and hewn to heat the tiny stove at noon. When I recall the man I follow, my heart is filled with woe and sorrow." "Shuoren" means tall, strong and handsome men. This poem expresses the woman's sadness and sense of loss because of absence of her husband.				

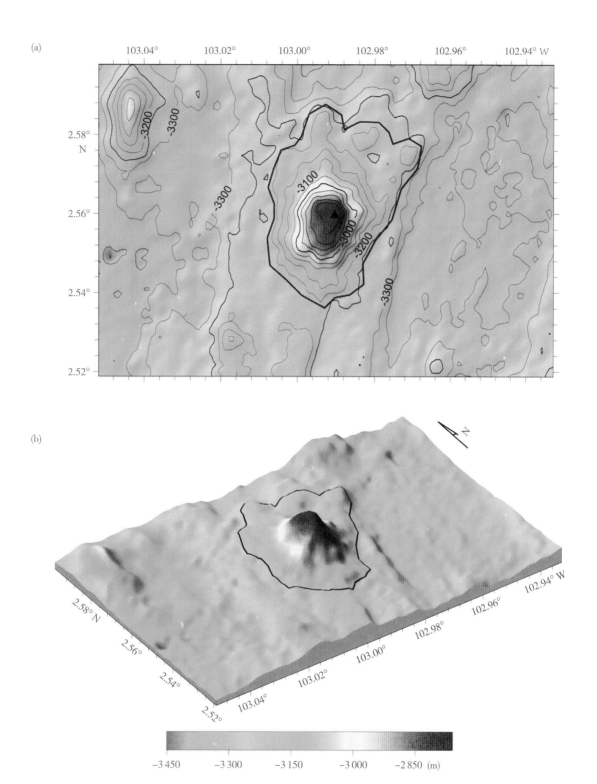

图 2-173 硕人海丘

(a) 地形图（等深线间隔 50 m）；(b) 三维图

Fig.2-173 ShuorenHill

(a) Bathymetric map (the contour interval is 50 m); (b) 3-D topographic map

2.9.7 啸歌海丘
Xiaoge Hill

中文名称 Chinese Name	啸歌海丘 Xiaoge Haiqiu	英文名称 English Name	Xiaoge Hill	所在大洋 Ocean or Sea	东太平洋 East Pacific Ocean
发现情况 Discovery Facts	此海丘于 2009 年由中国科考船"大洋一号"在执行 DY115-21 航次时调查发现。 Xiaoge Hill was discovered by the Chinese R/V *Dayang Yihao* during the DY115-21 cruise in 2009.				
命名历史 Name History	由我国命名为啸歌海丘，于 2017 年提交 SCUFN 审议通过。 Xiaoge Hill was named by China and approved by SCUFN in 2017.				
特征点坐标 Coordinates	1°55.27′N，103°08.95′W			长 (km) × 宽 (km) Length (km) × Width (km)	7.5 × 5
最大水深（m） Max Depth（m）	3 360	最小水深（m） Min Depth（m）	2 700	高差（m） Total Relief（m）	660
地形特征 Feature Description	该海丘位于东太平洋海隆，呈南北向延伸的长条状，长宽分别为 7.5 km 和 5 km。海丘顶部水深 2 700 m，山麓水深 3 360 m，高差 660 m（图 2–174）。 Xiaoge Hill is located in the East Pacific Rise. It has an elongated shape, running N to S. The length is 7.5 km and the width is 5 km. The top depth of the hill is 2 700 m and the piedmont depth is 3 360 m, which makes the total relief being 660 m (Fig.2–174).				
命名释义 Reason for Choice of Name	"啸歌"出自《诗经·小雅·白华》"滮池北流，浸彼稻田。啸歌伤怀，念彼硕人。""啸歌"出指长啸吟咏。此诗表达了女子因丈夫不在身边而产生的哀怨和失落感。 "Xiaoge" comes from a poem named *Baihua* in *Shijing · Xiaoya*. *Shijing* is a collection of ancient Chinese poems from 11th century B.C. to 6th century B.C. "The Biao Stream quietly northward flows, moistening fields in which rice grows. My sighs become a song of sorrow when I recall the man I follow." "Xiaoge" is a way to intone with rhythm and sorrow. This poem expresses the woman's sadness and sense of loss because of absence of her husband.				

(a)

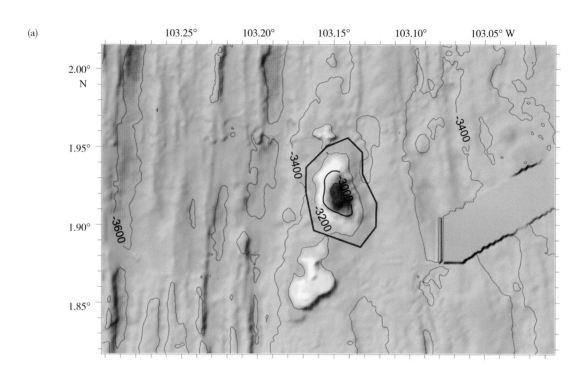

(b)

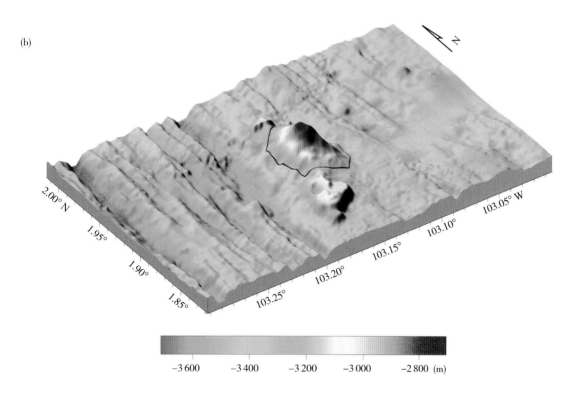

图 2-174　啸歌海丘

(a) 地形图（等深线间隔 200 m）；(b) 三维图

Fig.2-174　Xiaoge Hill

(a) Bathymetric map (the contour interval is 200 m); (b) 3-D topographic map

2.9.8 太白海脊
Taibai Ridge

中文名称 Chinese Name	太白海脊 Taibai Haiji	英文名称 English Name	Taibai Ridge	所在大洋 Ocean or Sea	东太平洋 East Pacific Ocean
发现情况 Discovery Facts	此海脊于 2009 年由中国科考船"大洋一号"在执行 DY115-21 航次时调查发现。 Taibai Ridge was discovered by the Chinese R/V *Dayang Yihao* during the DY115-21 cruise in 2009.				
命名历史 Name History	在我国大洋航次调查报告和 GEBCO 地名辞典中未命名。国外文献中用名"Dietz Ridge",被翻译为"迪茨海脊"。 This hill has not been named in Chinese cruise reports or GEBCO gazetteer. It was referred as Dietz Ridge in some foreign literature, which was translated as "Dici Haiji" in Chinese.				
特征点坐标 Coordinates	1°12.69′N,102°07.42′W 1°19.07′N,101°52.46′W 1°24.45′N,101°40.41′W			长(km)×宽(km) Length (km) × Width (km)	58 × 10
最大水深(m) Max Depth(m)	3 200	最小水深(m) Min Depth(m)	2 300	高差(m) Total Relief(m)	900
地形特征 Feature Description	该海脊位于东太平洋海隆,NE—SW 走向,长宽分别为 58 km 和 10 km。海脊顶部水深 2 300 m,山麓水深 3 200 m,高差 900 m。海脊东侧为太白海渊(图 2–175)。 Taibai Ridge is located in the East Pacific Rise, running NE to SW. The length is 58 km and the width is 10 km. The top depth is 2 300 m and the piedmont depth is 3 200 m, which makes the total relief being 900 m. The east to Taibai Ridge is Taibai Deep (Fig.2–175).				
命名释义 Reason for Choice of Name	李白(公元701—762年),字太白,号青莲居士,唐代著名的浪漫主义诗人,被后人誉为"诗仙"。李白存世诗文千余篇,有《李太白集》传世。该海脊取名太白海脊,以纪念他在文学史上的杰出贡献。 Li Bai (A.D. 701–762), whose another name taken at the age of twenty was Taibai, was also known as the Hermit of Green Lotus. He is the most romantic poet in Tang Dynasty and called "poetic genius" by the descendants. There are thousands of poems preserved nowadays. This ridge is named after Taibai, in memory of his brilliant contributions in the Chinese history of literature.				

(a)

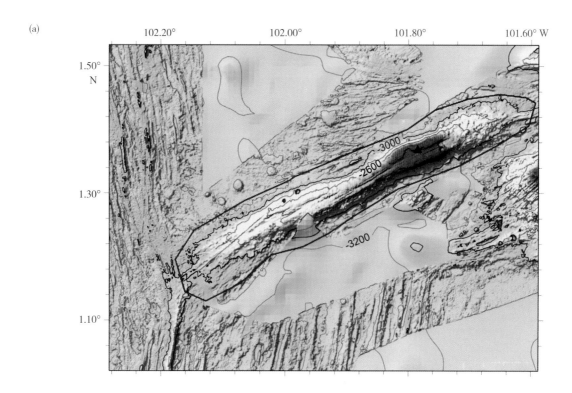

(b)

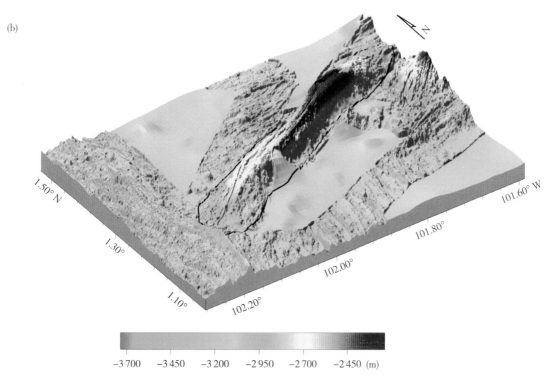

图 2-175　太白海脊

(a) 地形图（等深线间隔 200 m）；(b) 三维图

Fig.2-175　Taibai Ridge

(a) Bathymetric map (the contour interval is 200 m); (b) 3-D topographic map

2.9.9 太白热液区
Taibai Hydrothermal Fields

中文名称 Chinese Name	太白热液区 Taibai Reyequ	英文名称 English Name	Taibai Hydrothermal Fields	所在大洋 Ocean or Sea	东太平洋 East Pacific Ocean
发现情况 Discovery Facts	太白热液区于 2011 年由中国科考船"大洋一号"在执行 DY125-22 航次调查时，通过海底观察及水体热液异常探测发现。太白北 1-4 号热液区位于东太平洋海隆上的太白海脊北侧，围岩为玄武岩，水深约 3 400 m，太白南 1-3 号热液区位于东太平洋海隆上的太白海脊南侧。 Taibai Hydrothermal Fields were discovered by the Chinese R/V *Dayang Yihao* during the DY115-22 cruise in 2011. They were confirmed by undersea observation and hydrothermal anomaly detection. Taibaibei-1 to Taibaibei-4 Hydrothermal Field are located in the north to Taibai Ridge in East Pacific Rise, where the surrounding rocks are basalt and the depth is about 3 400 m. Taibainan-1 to Taibainan-3 Hydrothermal Field are located in the south to Taibai Ridge in East Pacific Rise.				
命名历史 Name History	太白北 4 号热液区在 DY125-22 航次现场报告中曾使用"国庆"名称，其余热液区在我国大洋航次调查报告中未命名。 Taibaibei-4 Hydrothermal Field was temporarily named "Guoqing" in the Chinese DY115-22 cruise field report. The other hydrothermal fields have not been named in the Chinese cruise reports.				
特征点坐标 Coordinates	1°36.90′N，102°15.60′W（太白北 1 号 / Taibaibei-1） 1°39.60′N，102°16.38′W（太白北 2 号 / Taibaibei-2） 1°40.80′N，102°16.20′W（太白北 3 号 / Taibaibei-3） 1°44.40′N，102°17.40′W（太白北 4 号 / Taibaibei-4） 1°03.60′N，102°10.20′W（太白南 1 号 / Taibainan-1） 1°02.40′N，102°10.80′W（太白南 2 号 / Taibainan-2） 1°00.00′N，102°11.40′W（太白南 3 号 / Taibainan-3）		长 (km) × 宽 (km) Length (km) × Width (km)	3 400	
命名释义 Reason for Choice of Name	采用热液区附近的"太白海脊"的名字命名，命名为"太白热液区"。为便于使用，以太白海脊为参照，按照地理位置由北向南依次命名为太白北 1 号、太白北 2 号、太白北 3 号、太白北 4 号、太白南 1 号、太白南 2 号和太白南 3 号热液区（图 2–176）。 The hydrothermal fields are named Taibai Hydrothermal Fields after the nearby Taibai Ridge. For convenience, considering Taibai Ridge as a reference, these hydrothermal fields were named Taibaibei-1, Taibaibei-2, Taibaibei-3, Taibaibei-4, Taibainan-1, Taibainan-2 and Taibainan-3 Hydrothermal Field from north to south according to their geographical locations (Fig.2–176).				

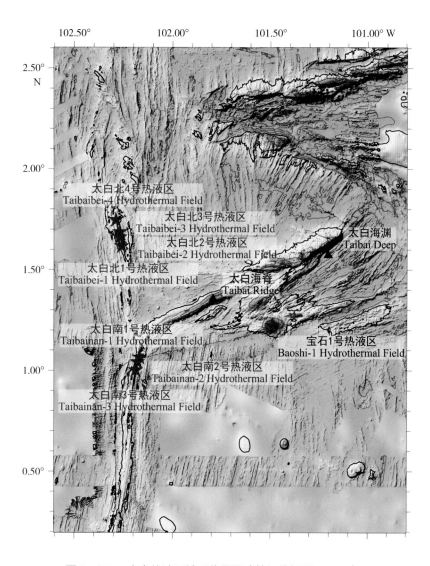

图 2-176　太白热液区地理位置图（等深线间隔 500 m）

Fig.2-176　Bathymetric map of the Taibai Hydrothermal Field (The contour interval is 500 m)

2.9.10 太白海渊
Taibai Deep

中文名称 Chinese Name	太白海渊 Taibai Haiyuan	英文名称 English Name	Taibai Deep	所在大洋 Ocean or Sea	东太平洋 East Pacific Ocean
发现情况 Discovery Facts	此海渊于 2009 年由中国科考船"大洋一号"在执行 DY115-21 航次时调查发现。 Taibai Deep was discovered by the Chinese R/V *Dayang Yihao* during the DY115-21 cruise in 2009.				
命名历史 Name History	在我国大洋航次调查报告和 GEBCO 地名辞典中未命名。国外文献中用名"Dietz Deep",被翻译为"迪茨海渊"。 This deep has not been named in Chinese cruise reports or GEBCO gazetteer. It was referred as Dietz Deep in some foreign literature, which was translated as "Dici Haiyuan" in Chinese.				
特征点坐标 Coordinates	1°33.60′N, 101°13.20′W			长(km)×宽(km) Length (km) × Width (km)	54×20
最大水深(m) Max Depth (m)	4 890	最小水深(m) Min Depth (m)	2 260	高差(m) Total Relief (m)	2 630
地形特征 Feature Description	该海渊位于东太平洋海隆,长宽分别为 54 km 和 20 km,边缘水深 2 260 m,最深处约 4 890 m,高差 2 630 m。海渊西侧为太白海脊(图 2–177)。 Taibai Deep is located in the East Pacific Rise. The length is 54 km and the width is 20 km. The marginal depth is 2 260 m and the maximum depth is 4 890 m, which makes the total relief being 2 630 m. The west to the deep is Taibai Ridge (Fig.2–177).				
命名释义 Reason for Choice of Name	李白(公元 701—762 年),字太白,号青莲居士,唐代著名的浪漫主义诗人,被后人誉为"诗仙"。李白存世诗文千余篇,有《李太白集》传世。该海渊取名太白海渊,以纪念他在文学史上的杰出贡献。 Li Bai (A.D. 701–762), whose courtesy name is Taibai, was also known as the Hermit of Green Lotus. He is the most romantic poet in Tang Dynasty and called "poetic genius" by the descendants. There are thousands of poems preserved nowadays. This deep is named after Taibai, in memory of his brilliant contributions in the Chinese history of literature.				

(a)

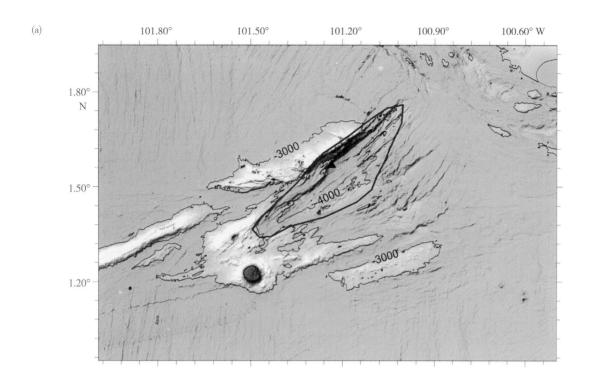

(b)

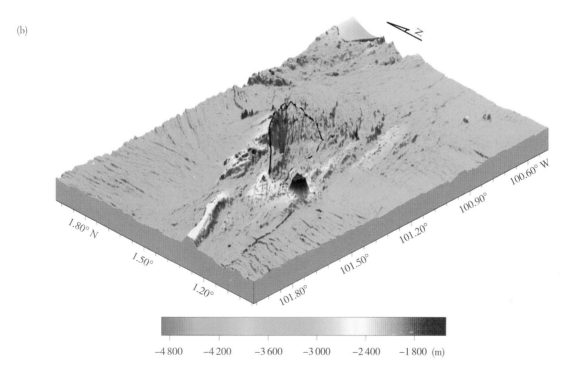

图 2-177 太白海渊

(a) 地形图（等深线间隔 1 000 m）；(b) 三维图

Fig.2-177 Taibai Deep

(a) Bathymetric map (the contour interval is 1 000 m); (b) 3-D topographic map

2.9.11　宝石 1 号热液区
Baoshi-1 Hydrothermal Field

中文名称 Chinese Name	宝石 1 号热液区 Baoshi-1Reyequ	英文名称 English Name	Baoshi-1 Hydrothermal Field	所在大洋 Ocean or Sea	东太平洋 East Pacific Ocean
发现情况 Discovery Facts	此热液区于 2009 年由中国科考船"大洋一号"在执行 DY115-21 航次调查时，通过海底观察、水体热液异常探测及取样证实。该热液区位于东太平洋海隆上的太白海渊西南侧，水深约 1 450 ~ 1 700 m。 This hydrothermal field was discovered by the Chinese R/V *Dayang Yihao* during the DY115-21 cruise in 2009. It was confirmed by undersea observation, hydrothermal anomaly detection and samples. It is located in the southwest of Taibai Deep in the East Pacific Rise with the water depth of about 1 450–1 700 m.				
命名历史 Name History	在大洋航次现场调查报告中曾用名为"宝石山"热液区，英文曾用名为"Precious Stone Hydrothermal Field"。 It was temporarily named "Baoshishan" in Chinese and "Precious Stone Hydrothermal Field" in English in the Chinese cruise reports.				
特征点坐标 Coordinates	1°13.20′S，101°29.40′W			水深 (m) Depth (m)	1 450 ~ 1 700 m
命名释义 Reason for Choice of Name	宝石山位于杭州西湖的北里湖北岸，山体在日光映照下如流霞缤纷，熠熠闪光，因此得名。以宝石命名该热液区，寓意海底热液活动强烈。为便于使用，更名为宝石 1 号热液区，英文名称更名为 Baoshi-1 Hydrothermal Field（图 2–178）。 Baoshi Mountain is located in the north shore of the West Lake in Hangzhou, Zhejiang Province of China. Like a gem, the mountain shines colorfully under sunlight and was named "Baoshi", which means gem in Chinese. This hydrothermal field was named "Baoshi" to denote its strong activeness. For convenience, it was renamed Baoshi-1 Hydrothermal Field (Fig.2–178).				

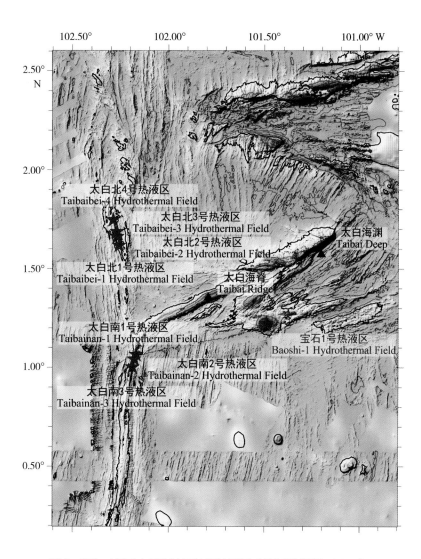

图 2-178　宝石 1 号热液区地理位置图（等深线间隔 500 m）

Fig.2-178　Bathymetric map of the Baoshi-1 Hydrothermal Field (The contour interval is 500 m)

2.9.12 那居圆海丘
Nuoju Knoll

中文名称 Chinese Name	那居圆海丘 Nuoju Yuanhaiqiu	英文名称 English Name	Nuoju Knoll	所在大洋 Ocean or Sea	东太平洋 East Pacific Ocean
发现情况 Discovery Facts					
命名历史 Name History					
特征点坐标 Coordinates			长(km)×宽(km) Length (km)× Width (km)		7.5×7.2
最大水深（m） Max Depth（m）	3 200	最小水深（m） Min Depth（m）	2 650	高差（m） Total Relief（m）	550

发现情况 / Discovery Facts: 此圆海丘于 2009 年由中国科考船"大洋一号"在执行 DY115-21 航次时调查发现。
Nuoju Knoll was discovered by the Chinese R/V *Dayang Yihao* during the DY115-21 cruise in 2009.

命名历史 / Name History: 由我国命名为那居圆海丘，于 2017 年提交 SCUFN 审议通过。
Nuoju Knoll was named by China and approved by SCUFN in 2017.

特征点坐标 / Coordinates: 0°38.36′N，101°25.43′W

地形特征 / Feature Description: 该圆海丘位于东太平洋海隆，规模较小，平面形态近圆形，长宽分别为 7.5 km 和 7.2 km。圆海丘顶部水深 2 650 m，山麓水深 3 200 m，高差 550 m（图 2–179）。
Nuoju Knoll is located in the East Pacific Rise with a relatively small scale. It has a nearly round plane shape. The length is 7.5 km and the width is 7.2 km. The top depth of Nuoju Knoll is 2 650 m and the piedmont depth is 3 200 m, which makes the total relief being 550 m (Fig.2–179).

命名释义 / Reason for Choice of Name: "那居"出自《诗经·小雅·鱼藻》"鱼在在藻，依于其蒲。王在在镐，有那其居"，描述了鱼紧贴相依水草、安详静立，大王居住在安乐的场景，借喻人们安居乐业的和谐氛围。
"Nuoju" comes from a poem named *Yuzao* in *Shijing · Xiaoya*. *Shijing* is a collection of ancient Chinese poems from 11th century B.C. to 6th century B.C. "Where is the fish? Among the algae is the fish, sheltered by the weeds. Where is the king? Here in Hao is the king; A peaceful life he leads." "Nuoju" means that the King leads a peaceful life.

(a)

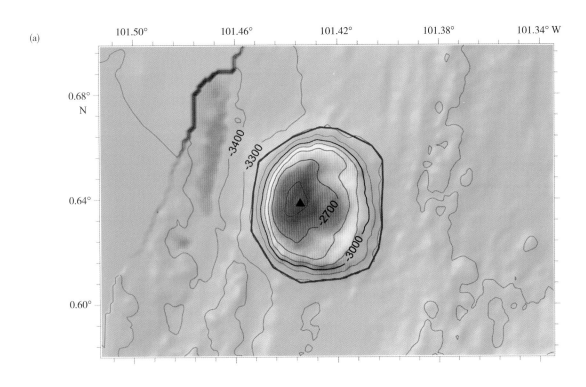

(b)

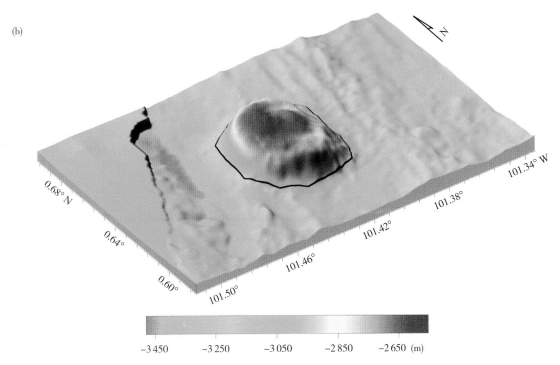

图 2-179　那居圆海丘

(a) 地形图（等深线间隔 100 m）；(b) 三维图

Fig.2-179　Nuoju Knoll

(a) Bathymetric map (the contour interval is 100 m); (b) 3-D topographic map

2.9.13　依蒲海丘
Yipu Hill

中文名称 Chinese Name	依蒲海丘 Yipu Haiqiu		英文名称 English Name	Yipu Hill	所在大洋 Ocean or Sea	东太平洋 East Pacific Ocean
发现情况 Discovery Facts	此海丘于 2009 年由中国科考船"大洋一号"在执行 DY115-21 航次时调查发现。 Yipu Hill was discovered by the Chinese R/V *Dayang Yihao* during the DY115-21 cruise in 2009.					
命名历史 Name History	由我国命名为依蒲海丘，于 2016 年提交 SCUFN 审议通过。 Yipu Hill was named by China and approved by SCUFN in 2016.					
特征点坐标 Coordinates	0°30.62′N，101°02.97′W				长(km)×宽(km) Length (km) × Width (km)	14×9
最大水深（m） Max Depth（m）	3 400	最小水深（m） Min Depth（m）	2 500		高差（m） Total Relief（m）	900
地形特征 Feature Description	该海丘位于东太平洋海隆，平面形态呈椭圆形，长宽分别为 14 km 和 9 km。海丘顶部水深 2 500 m，山麓水深 3 400 m，高差 900 m（图 2–180）。 Yipu Hill is located in the East Pacific Rise. It has an elliptic plane shape. The length and width are 14 km and 9 km respectively. The top depth of the hill is 2 500 m and the piedmont depth is 3 400 m, which makes the total relief being 900 m (Fig.2–180).					
命名释义 Reason for Choice of Name	"依蒲"出自《诗经·小雅·鱼藻》"鱼在在藻，依于其蒲。王在在镐，有那其居"，描述了鱼紧贴相依水草、安详静立，大王居住在安乐的场景，借喻人们安居乐业的和谐氛围。 "Yipu" comes from a poem named *Yuzao* in *Shijing · Xiaoya*. *Shijing* is a collection of ancient Chinese poems from 11th century B.C. to 6th century B.C. "Where is the fish? Among the algae is the fish, sheltered by the weeds. Where is the king? Here in Hao is the king; A peaceful life he leads." "Yipu" means that the fish clings to the weeds.					

(a)

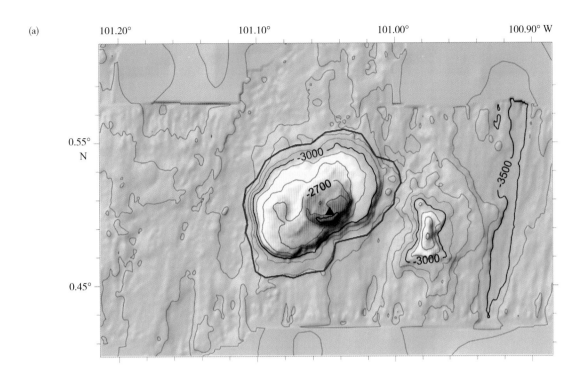

(b)

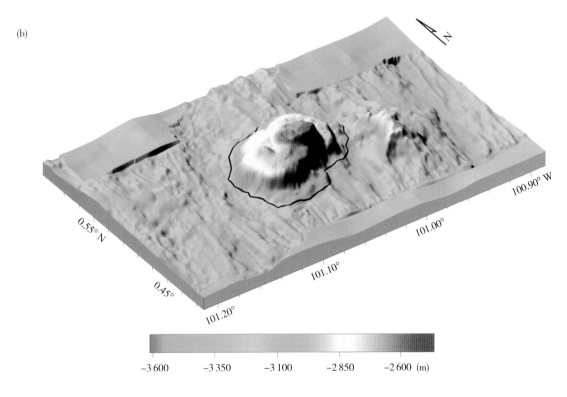

图 2-180　依蒲海丘

(a) 地形图（等深线间隔 100 m）；(b) 三维图

Fig.2-180　Yipu Hill

(a) Bathymetric map (the contour interval is 100 m); (b) 3-D topographic map

2.9.14 白华海丘
Baihua Hill

中文名称 Chinese Name	白华海丘 Baihua Haiqiu	英文名称 English Name	Baihua Hill	所在大洋 Ocean or Sea	东太平洋 East Pacific Ocean
发现情况 Discovery Facts	此海丘于 2009 年由中国科考船 "大洋一号" 在执行 DY115-21 航次时调查发现。 This hill was discovered by the Chinese R/V *Dayang Yihao* during the DY115-21 cruise in 2009.				
命名历史 Name History	由我国命名为白华海丘, 于 2016 年提交 SCUFN 审议通过。 Baihua Hill was named by China and approved by SCUFN in 2016.				
特征点坐标 Coordinates	0°29.63′N, 102°56.10′W		长(km)×宽(km) Length (km)× Width (km)		11.5×6.4
最大水深（m） Max Depth（m）	3 400	最小水深（m） Min Depth（m）	2 800	高差（m） Total Relief(m)	600
地形特征 Feature Description	该海丘位于东太平洋海隆, 长宽分别为 11.5 km 和 6.4 km, 顶部水深 2800m, 山麓水深 3 400 m, 高差 600 m（图 2–181）。 This hill is located in the East Pacific Rise. The length is 11.5 km and the width is 6.4 km. The top depth of the hill is 2 800 m and the piedmont depth is 3 400 m, which makes the total relief being 600 m (Fig.2–181).				
命名释义 Reason for Choice of Name	"白华" 出自《诗经·小雅·白华》 "白华菅兮, 白茅束兮。之子之远, 俾我独兮。" 白华指开白色花的菅草。以白茅把开白花的菅草捆成束映射夫妇之间相亲相爱。但现在丈夫远去, 只留妻子独守空房。为本篇奠定了全诗凄婉悲剧的基调。 "Baihua" comes from a poem named *Baihua* in *Shijing · Xiaoya*. *Shijing* is a collection of ancient Chinese poems from 11th century B.C. to 6th century B.C. "The reeds with flowers small and white with cogon vines are bound up tight. My lord has gone far, far away; Alone at home I daily stay." "Baihua" means reed with white flowers. The poem uses bounding up the reed with cogon vines to imply the love between husband and wife. However, the husband went far away and left the wife home alone and the wife missed her husband very much.				

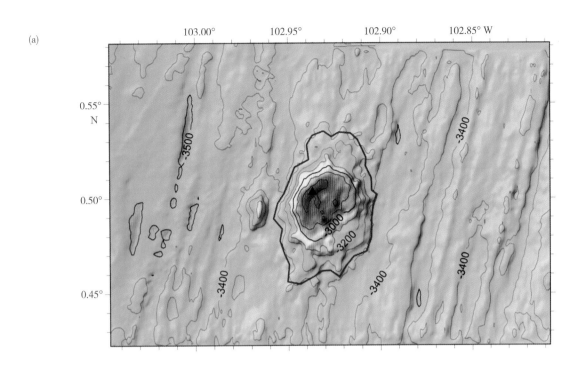

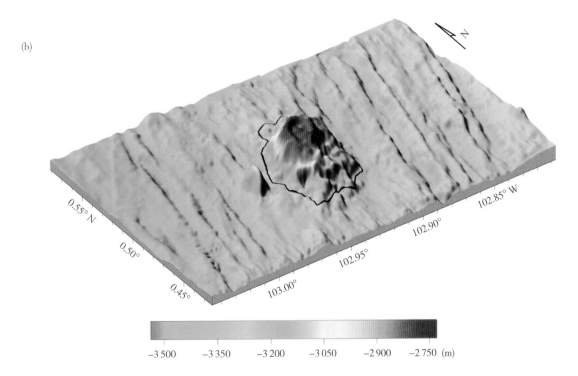

图 2-181　白华海丘

(a) 地形图（等深线间隔 100 m）；(b) 三维图

Fig.2-181　Baihua Hill

(a) Bathymetric map (the contour interval is 100 m); (b) 3-D topographic map

2.9.15 维翰海山
Weihan Seamount

中文名称 Chinese Name	维翰海山 Weihan Haishan	英文名称 English Name	Weihan Seamount	所在大洋 Ocean or Sea	东太平洋 East Pacific Ocean
发现情况 Discovery Facts	此海山于 2009 年 10 月由中国科考船"大洋一号"在执行 DY115-21 航次时调查发现。 This seamount was discovered by the Chinese R/V *Dayang Yihao* during the DY115-21 cruise in October, 2009.				
命名历史 Name History	由我国命名为维翰海山，于 2012 年提交 SCUFN 审议通过。 Weihan Seamount was named by China and approved by SCUFN in 2012.				
特征点坐标 Coordinates	0°05.20′S，101°24.20′W			长(km)×宽(km) Length (km) × Width (km)	8 × 8
最大水深（m） Max Depth（m）	3 200	最小水深（m） Min Depth（m）	2 200	高差（m） Total Relief（m）	1 000
地形特征 Feature Description	该海山位于东太平洋海隆，海山顶部东南高西北低，平面形态呈圆形，基座直径约 8 km。海山顶部水深 2 200 m，山麓水深 3 200 m，高差 1 000 m（图 2–182）。 Weihan Seamount is located in the East Pacific Rise. The southeastern side of the seamount top is high while the northwestern side is low. It has a round plane shape with the base diameter of 8 km. The top depth of this seamount is 2 200 m and the piedmont depth is 3 200 m, which makes the total relief being 1 000 m (Fig.2–182).				
命名释义 Reason for Choice of Name	"维翰"出自《诗经·大雅·板》"大邦维屏，大宗维翰，怀德维宁，宗子维城"，意为诸侯大国是国家的屏障，宗族是国家的栋梁，怀德是国家平安的保证，嫡系是国家的城垒，反映了当时的治国理念。翰，寓意栋梁。 "Weihan" comes from a poem named *Ban* in *Shijing · Daya. Shijing* is a collection of ancient Chinese poems from 11th century B.C. to 6 th century B.C. "Great states form screens protecting you; Close kin are forts supporting you. Good virtue brings repose to all; Close kinsmen are your strengthened wall." The poem reveals the former concept of governing a country. "Han" means the backbone of the nation.				

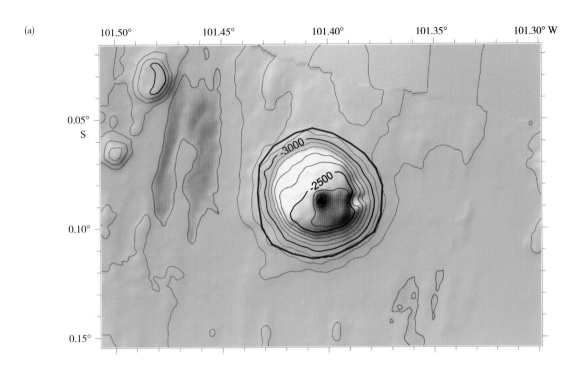

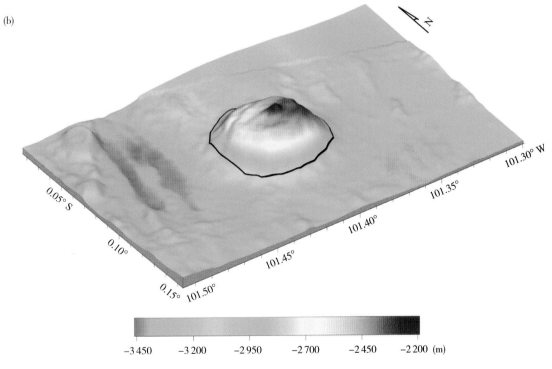

图 2-182　维瀚海山

(a) 地形图（等深线间隔 100 m）；(b) 三维图

Fig.2-182　Weihan Seamount

(a) Bathymetric map (the contour interval is 100 m); (b) 3-D topographic map

2.9.16　鸟巢海丘
Niaochao Hill

中文名称 Chinese Name	鸟巢海丘 Niaochao Haiqiu		英文名称 English Name	Niaochao Hill	所在大洋 Ocean or Sea	东太平洋 East Pacific Ocean
发现情况 Discovery Facts	此海丘于 2008 年 8 月由中国科考船"大洋一号"在执行 DY115-20 航次时调查发现。 This hill was discovered by the Chinese R/V *Dayang Yihao* during DY115-20 cruise in August, 2008.					
命名历史 Name History	由我国命名为鸟巢海丘，于 2011 年提交 SCUFN 审议通过。 Niaochao Hill was named by China and approved by SCUFN in 2011.					
特征点坐标 Coordinates	1°22.00′S，102°27.50′W				长(km)× 宽(km) Length (km) × Width (km)	3.5×3.5
最大水深（m） Max Depth（m）	2 875		最小水深（m） Min Depth（m）	2 625	高差（m） Total Relief（m）	250
地形特征 Feature Description	该海丘位于东太平洋海隆，外观呈环形，环形火山口边缘高于中心点 120 m，东侧略高于西侧。海丘基座直径约 3.5 km，顶部水深 2 625 m，山麓水深 2 875 m，高差 250 m（图 2–183）。 This hill is located in the East Pacific Rise and has an annular shape. The edges of annular crater is 120 m higher than the middle, and the eastern side is a little higher than the western side. The base diameter of the hill is 3.5 km. The top depth is 2 625 m and the piedmont depth is 2 875 m, which makes the total relief being 250 m (Fig.2–183).					
命名释义 Reason for Choice of Name	该处海底塌陷火山口形状酷似奥运体育馆"鸟巢"，又因该海丘被发现时正值北京举办第 28 届奥运会，故以此命名。 The hill was discovered when the 28th Olympic Games were being held in Beijing, China. The crater looks like the building of the Beijing National Stadium known as the "Bird's nest" for its architecture. "Bird's nest" is "Niaochao" in Chinese.					

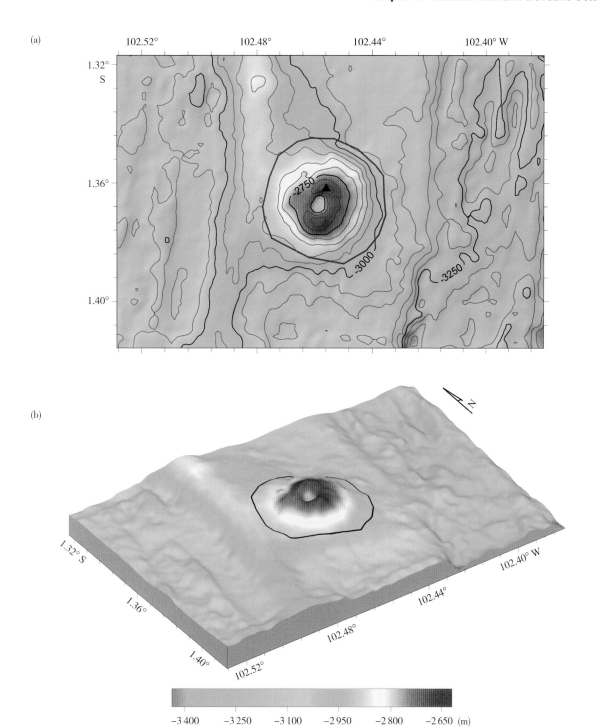

图 2-183　鸟巢海丘

(a) 地形图（等深线间隔 50 m）；(b) 三维图

Fig.2-183　Niaochao Hill

(a) Bathymetric map (the contour interval is 50 m); (b) 3-D topographic map

2.9.17　鸟巢热液区
Niaochao Hydrothermal Field

中文名称 Chinese Name	鸟巢热液区 Niaochao Reyequ	英文名称 English Name	Niaochao Hydrothermal Field	所在大洋 Ocean or Sea	东太平洋 East Pacific Ocean
发现情况 Discovery Facts	此热液区于 2008 年 8 月由中国科考船"大洋一号"在执行 DY115-20 航次调查时，通过海底观察、水体热液异常探测及取样证实。该热液区位于赤道附近的东太平洋海隆一离轴火山上，围岩为玄武岩，水深约 2 900 m。 This hydrothermal field was discovered by the Chinese R/V *Dayang Yihao* during the DY115-20 cruise in August, 2008. It was confirmed by undersea observation, hydrothermal anomaly detection and samples. It is located in an off-axis volcano of the East Pacific Rise near equator with the water depth of about 2 900 m. The surrounding rocks are basalt.				
命名历史 Name History	DY115-20 航次现场报告中曾使用"EPR 1°22′S"热液区名称。 This hydrothermal field was once named "EPR 1°22′S" in the Chinese DY115-20 cruise field report.				
特征点坐标 Coordinates	0°50.22′S，102°27.00′W（鸟巢北 1 号 / Niaochaobei-1） 1°22.08′S，102°27.00′W（鸟巢 / Niaochao） 2°01.20′S，102°37.20′W（鸟巢南 1 号 / Niaochaonan-1） 2°04.62′S，102°36.60′W（鸟巢南 2 号 / Niaochaonan-2） 2°09.06′S，102°38.40′W（鸟巢南 3 号 / Niaochaonan-3） 2°13.14′S，102°39.00′W（鸟巢南 4 号 / Niaochaonan-4）			水深（m） Depth(m)	2 900
命名释义 Reason for Choice of Name	采用热液区附近的"鸟巢海丘"的名字命名，命名为"鸟巢热液区"。为便于使用，以鸟巢海丘为参照，按照地理位置由北向南依次命名为鸟巢北 1 号、鸟巢、鸟巢南 1 号、鸟巢南 2 号、鸟巢南 3 号和鸟巢南 4 号热液区（图 2–184）。 This hydrothermal field was named Niaochao Hydrothermal Field after the nearby Niaochao Hill. For convenience, considering Niaochao Hill as a reference, these hydrothermal fields were named Niaochaobei-1, Niaochao, Niaochaonan-1, Niaochaonan-2, Niaochaonan-3 and Niaochaonan-4 Hydrothermal Field from north to south according to their geographical locations (Fig.2–184).				

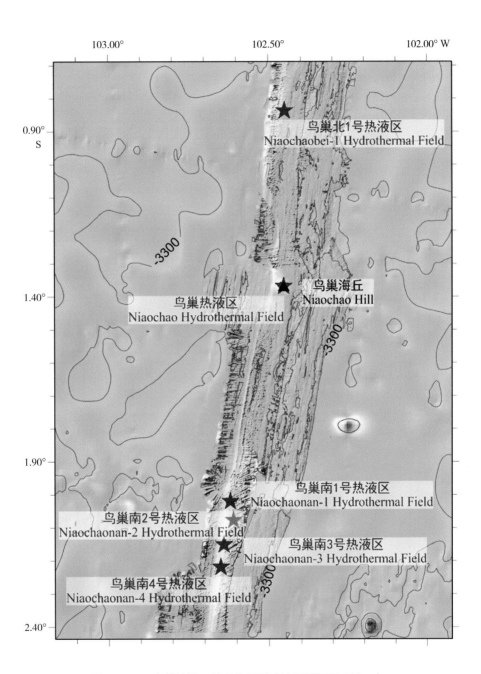

图 2-184　鸟巢热液区地理位置图（等深线间隔 500 m）

Fig.2-184　Bathymetric map of the Niaochao Hydrothermal Field

(The contour interval is 500 m)

2.9.18　徽猷 1 号热液区
Huiyou-1 Hydrothermal Field

中文名称 Chinese Name	徽猷 1 号热液区 Huiyou-1 Reyequ	英文名称 English Name	Huiyou-1 Hydrothermal Field	所在大洋 Ocean or Sea	东太平洋 East Pacific Ocean
发现情况 Discovery Facts	此热液区于 2011 年在太平洋由中国科考船"大洋一号"在执行 DY125-22 航次调查时，通过海底观察、水体热液异常探测及取样验证。该热液区位于东太平洋海隆，围岩为玄武岩，水深约 3 000 m。 This hydrothermal field was discovered by the Chinese R/V *Dayang Yihao* during the DY125-22 cruise in 2011. It was confirmed by undersea observation, hydrothermal anomaly detection and samples. It is located in the East Pacific Rise with the water depth of about 3 000 m and surrounding rocks are basalt.				
命名历史 Name History	在我国大洋航次调查报告中未命名。 This hydrothermal field has not been named in Chinese cruise reports.				
特征点坐标 Coordinates	3°06.26′S，102°33.18′W			水深 (m) Depth(m)	3 000
命名释义 Reason for Choice of Name	"徽猷"出自《诗经·小雅·角弓》"君子有徽猷，小人与属"，意为君子如果有美德，小民自然会来依附，改变恶习，相亲为善的。徽猷指美善之道 (图 2–185)。 "Huiyou" comes from a poem named *Jiaogong* in *Shijing · Xiaoya*. *Shijing* is a collection of ancient Chinese poems from 11th century B.C. to 6th century B.C. "If rulers noble virtues grow, inferior folk will surely follow." The poem means that the folks will surely follow the gentlemen if they have noble virtues. "Huiyou" means noble virtues (Fig.2–185).				

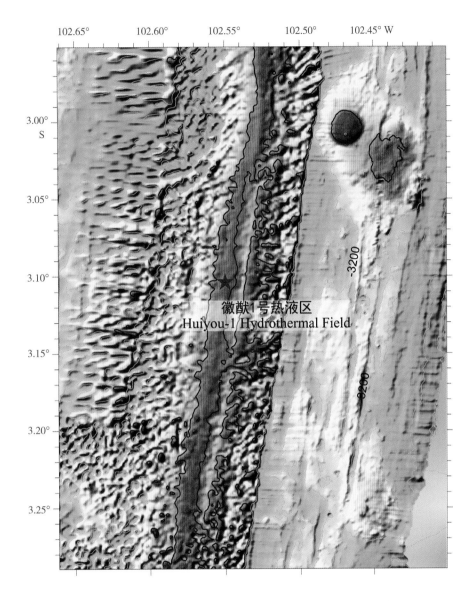

图 2-185 徽猷 1 号热液区地理位置图（等深线间隔 100 m）

Fig.2-185 Bathymetric map of the Huiyou-1 Hydrothermal Field
(The contour interval is 100 m)

2.9.19 角弓 1 号热液区
Jiaogong-1 Hydrothermal Field

中文名称 Chinese Name	角弓 1 号热液区 Jiaogong-1 Reyequ	英文名称 English Name	Jiaogong-1 Hydrothermal Field	所在大洋 Ocean or Sea	东太平洋 East Pacific Ocean
发现情况 Discovery Facts	此热液区于 2011 年在太平洋由中国科考船"大洋一号"在执行 DY125-22 航次调查时，通过海底观察、水体热液异常探测及取样证实。该热液区位于东太平洋海隆，围岩为玄武岩，水深约 2 700 m。 This hydrothermal field was discovered by the Chinese R/V *Dayang Yihao* during the DY115-22 cruise in 2011. It was confirmed by undersea observation, hydrothermal anomaly detection and samples. It is located in the East Pacific Rise with the water depth of about 2 700 m and surrounding rocks are basalt.				
命名历史 Name History	在我国大洋航次调查报告中未命名。 This hydrothermal field has not been named in Chinese cruise reports.				
特征点坐标 Coordinates	5°18.00′S，106°28.92′W			水深 (m) Depth(m)	2 700
命名释义 Reason for Choice of Name	"角弓"出自《诗经·小雅·角弓》"骍骍角弓，翩其反矣"，意为角弓调整很方便，弓弦放松向外反，喻兄弟之间不可疏远，要团结一致。角弓指用动物的角装饰的弓 (图 2–186)。 "Jiaogong" comes from a poem named *Jiaogong* in *Shijing · Xiaoya. Shijing* is a collection of ancient Chinese poems from 11th century B.C. to 6th century B.C. "Tight-bent bows adorned with horn will rebound when strings are loosened." The poem means that tight-bent bows will rebound easily if the strings are loosened implying that siblings should always unite together. "Jiaogong" means bows decorated with horns (Fig.2–186).				

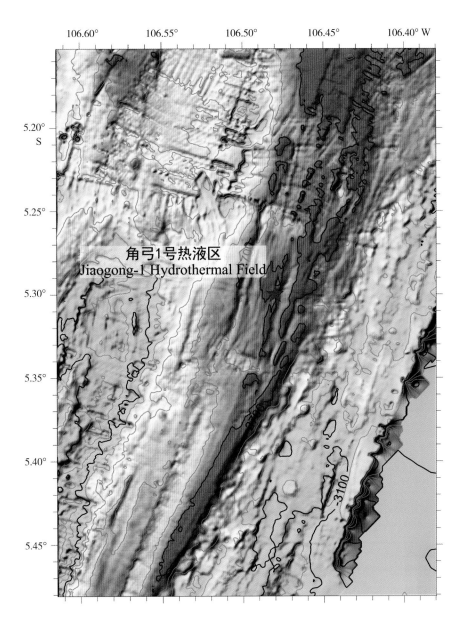

图 2-186　角弓 1 号热液区地理位置图（等深线间隔 100 m）

Fig.2-186　Bathymetric map of the Jiaogong-1 Hydrothermal Field

（The contour interval is 100 m）

2.9.20　鱼藻 1 号热液区、鱼藻 2 号热液区
Yuzao-1 Hydrothermal Field and Yuzao-2 Hydrothermal Field

中文名称 Chinese Name	鱼藻 1 号热液区 鱼藻 2 号热液区 Yuzao-1 Reyequ Yuzao-2 Reyequ	英文名称 English Name	Yuzao-1 Hydrothermal Field Yuzao-2 Hydrothermal Field	所在大洋 Ocean or Sea	东太平洋 East Pacific Ocean
发现情况 Discovery Facts					
命名历史 Name History					
特征点坐标 Coordinates				水深 (m) Depth(m)	2 700
命名释义 Reason for Choice of Name					

发现情况 Discovery Facts

　　鱼藻 1 号热液区和鱼藻 2 号热液区于 2011 年在太平洋由中国科考船"大洋一号"在执行 DY125-22 航次调查时，通过海底观察、水体热液异常探测及取样证实。两个热液区均位于东太平洋海隆，围岩为玄武岩，水深约 2 700 m。

Yuzao-1 and Yuzao-2 Hydrothermal Field were discovered by the Chinese R/V *Dayang Yihao* during the DY125-22 cruise in 2011. They were confirmed by undersea observation, hydrothermal anomaly detection and samples. Both fields are located in the East Pacific Rise with the water depth of about 2 700 m and surrounding rocks are basalt.

命名历史 Name History

在我国大洋航次调查报告中未命名。

These two hydrothermal fields have not been named in Chinese cruise reports.

特征点坐标 Coordinates

6°03.60′S，106°50.76′W（鱼藻 1 号 / Yuzao-1）
6°04.80′S，106°50.94′W（鱼藻 2 号 / Yuzao-2）

命名释义 Reason for Choice of Name

　　"鱼藻"出自《诗经·小雅·鱼藻》的篇名，指鱼在水草间游来游去。本篇以咏鱼得其所乐，喻百姓安居乐业之和谐气氛（图 2–187）。

"Yuzao" comes from a poem named *Yuzao* in *Shijing · Xiaoya*. *Shijing* is a collection of ancient Chinese poems from 11th century B.C. to 6th century B.C. This poem described that fish swam among the water weeds, implying the harmonious atmosphere of a prosperous and contented life (Fig.2–187).

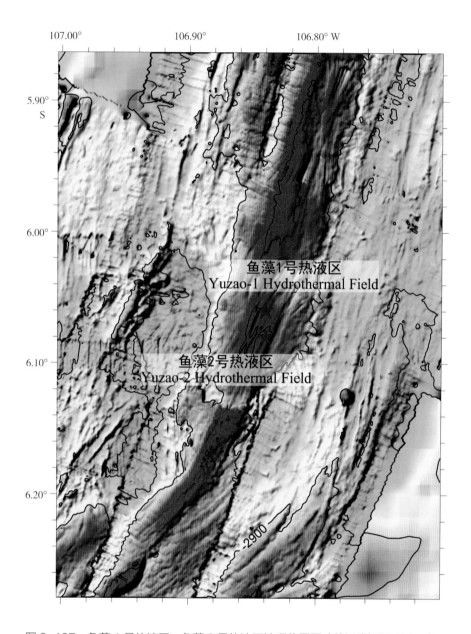

图 2-187　鱼藻 1 号热液区、鱼藻 2 号热液区地理位置图（等深线间隔 100 m）

Fig.2-187　Bathymetric map of the Yuzao-1 and Yuzao-2 Hydrothermal Field

(The contour interval is 100 m)

第3章
大西洋地理实体
Chapter 3
Undersea Features in the
Atlantic Ocean

第1节　命名概况

大西洋位于欧洲、非洲和南北美洲之间，是地球上的第二大洋。大西洋东西向狭窄，南北向略呈"S"形延伸，自北至南全长约 1.6 万千米，其赤道区域宽度最窄，最短距离仅约 2 400 km。大西洋海底地貌分为大陆架、大陆坡、大陆隆、大洋中脊和大洋盆地 5 大类基本地貌单元。

大西洋大陆架的宽度从几十千米到上千千米不等，以东北部波罗的海和北海以及西北欧大不列颠岛周围和挪威海沿岸海域的大陆架最宽广，最宽处达 1 000 km 以上；北美大陆东岸、南美大陆北岸加勒比海沿岸大陆架都较狭窄。大陆架外缘是大陆坡，面积约 768 万平方千米，其中沿欧非大陆架的大陆坡地形比较陡，宽度一般只有 20 ~ 30 km；沿美洲的大陆坡地形比较平缓，宽度超过 50 ~ 80 km。墨西哥海盆西缘和阿根廷东侧的大陆坡形态呈阶梯状，大致从水深 100 ~ 200 m 逐级降至水深 5 000 m 以上。大陆坡上发育大量海山、盆地、海底峡谷等地理实体。大陆坡与洋盆之间的局部海域发育大陆隆，如格陵兰－冰岛隆起、布莱克隆起和马尔维纳斯隆起等。

大陆坡与洋盆之间发育一系列岛弧、边缘海盆、海底高地及深海沟。岛弧带和深海沟有两条：一条是由大、小安的列斯群岛组成的双列岛弧带和其北侧的波多黎各海沟，另一条是南美南端与南极半岛之间由南乔治亚岛、南桑威奇群岛和南奥克尼群岛组成的向东延伸的岛弧带及岛弧东缘的南桑威奇海沟。

大西洋大洋中脊北起冰岛，纵贯大西洋，南至布韦岛，然后转向东北与西南印度洋中脊相连，全长约 1.6 万千米，宽度 1 500 ~ 2 000 km。大西洋中脊形似"S"，由一系列狭窄和被断裂分割的平行岭脊组成，脊顶水深 2 500 ~ 3 000 m。少数峰脊凸出海面成为岛屿，如冰岛、亚速尔群岛等。位于赤道附近的罗曼什断裂带，把大西洋中脊截成南北两段并错开 1 000 余千米，北段称北大西洋中脊，纵贯大西洋北部，长约 1.05 万千米。南段称南大西洋中脊，纵贯大西洋南部，长约 4 500 km。

大西洋盆地由大洋中脊分隔为东西两列海盆。东侧自北而南有西欧罗巴海盆、伊比利亚海盆、加那利海盆、佛得角海盆、几内亚海盆、安哥拉海盆、开普海盆；西侧有北亚美利加海盆、巴西海盆和阿根廷海盆。这些海盆平均深度 4 000 ~ 6 300 m，地形平缓，发育一些小型的海山和洼地（图 3-1）。

Section 1　Overview of the Naming

The Atlantic Ocean is located between Europe, Africa, South America and North America, which is the world's second largest ocean considering its scale. The Atlantic is narrow in the EW direction and runs N to S slightly with the shape of S. The whole length from north to south is about 16 000 km. The width of its equatorial zone is the narrowest, with the shortest distance being about 2 400 km. The Atlantic seabed geomorphology can be divided into 5 categories of fundamental geomorphic units, which are continental shelf, continental slope, continental rise, mid-ocean ridge and ocean basin.

The width of the Atlantic Continental Shelf ranges from tens to thousands of kilometers. The northeast shelf of the Baltic Sea and the Northern Sea and shelf around the Island of Great Britain in Northwest Europe and the Norwegian Sea coastal area are the widest. The largest width is over 1 000 km. However, the shelf of east coast of the North American Continent, north coast of the South American Continent and coast of the Caribbean Sea is respectively narrow. Outer the continental shelf is the continental slope with an area of about 7.68 million square kilometers. The continental slope topography is steep along the Euro-African Continental Shelf with the width generally being only 20–30 km, while the continental slope topography is gentle along the American Continental Shelf with the width being over 50–80 km. The continental slope of western margin of the Mexico Basin and eastern side of Argentina has a stair-step shape. The water depth roughly decreases from 100–200 m to over 5 000 m step by step. The continental slope develops many undersea features like seamounts, basins, canyons, etc. The local area between the continental slope and ocean basin develops the continental rise, such as Greenland-Iceland Rise, Black Rise and Malvinas Rise.

Between the continental slope and ocean basin, it develops a series of island arcs, marginal sea basins, undersea elevations and deep trenches. There are two island arc zones and deep trenches, one is the double island arc zone made up by the Greater, Lesser Antilles, and the Puerto Rico Trench north to it, the other is the east extending island arc zone made up by South Georgia Island, South Sandwich Islands and South Orkney Islands between the southern side of South America and the Antarctic Peninsula, and the South Sandwich Trench at the eastern margin of the arc zone.

The Atlantic Ridge starts from Iceland in the north, running across the Atlantic Ocean to the Bouvet Island in the south, and then turns towards northeast connecting the Southwest Indian Ridge. The total length is about 16 000 km and the width is 1 500–2 000 km. The Atlantic Ridge has a S-like shape and consists of a series of narrow and parallel ridges split by fractures, with the water depth of ridge summits being 2 500–3 000 m. A few ridge summits project out of the sea surface and become islands, such as the Iceland, the Azores, etc. The La Romanche Fracture Zone near the equator divides the Atlantic Ridge into the southern and northern segments with the distance of over 1 000 km. The northern segment is called the North Atlantic Ridge, which extends across the northern part of Atlantic Ocean with a length of about 10 500 km. The southern segment is called the South Atlantic Ridge, which extends across the southern part of Atlantic Ocean with a length of about 4 500 km.

The Atlantic Basin is divided into eastern and western basins by the mid-ocean ridge. In the east, there are West Europa Basin, Iberian Basin, Canary Basin, Verde Basin, Guinea Basin, Angola Basin, Cape Basin; In the west, there are North America Basin, Brazil Basin, Argentina Basin. These sea basins have an average depth of 4 000–6 300 m, with smooth topography and development of some small seamounts and depressions（Fig. 3–1）.

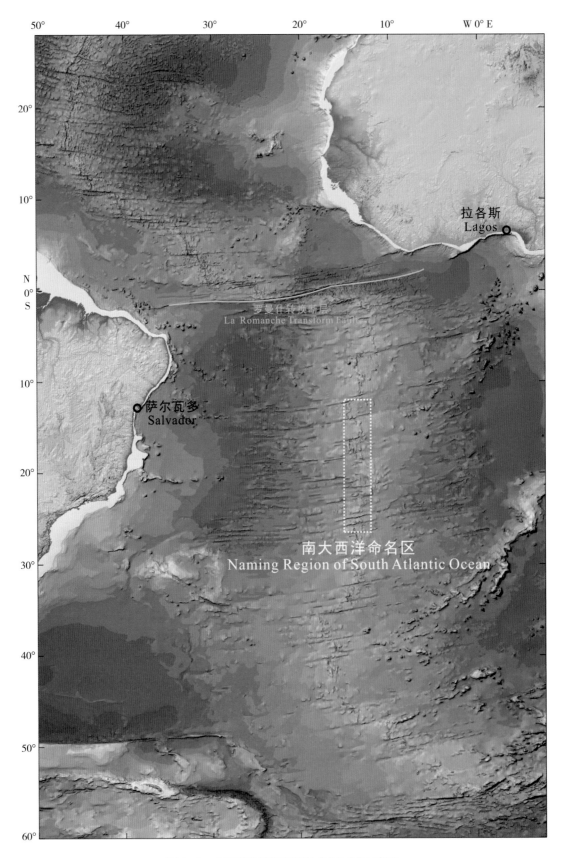

图 3-1　大西洋海底地理实体命名区域示意图

Fig.3-1　The Atlantic Ocean undersea features naming regions

第 2 节　南大西洋洋中脊地理实体和热液区

　　南大西洋洋中脊总长约 4 500 km，北起赤道附近的罗曼什转换断层，南到印度洋布维岛附近的三联点，南部延伸于南极地带的大西洋中，约占了大西洋洋中脊总长度的三分之一，为慢速扩张洋中脊（全扩张速率为 20 ～ 50 mm/a），一般认为其地质和地球物理特征与其他慢速扩张洋中脊相似（Dick et al., 2003; Devey et al., 2010）。洋中脊被很多断层所切割，较大的断层为转换断层，转换断层之间还有非转换断层，转换断层和非转换断层发育一系列数十千米长的分离扩张段（陈永顺，2003；李兵，2014）。在安哥拉－巴西段出现了密集的被断续的断层所破坏的区域，具有宽的，斜对方向和经线方向排列的特征，看起来像大的裂隙网。非转换断层长度通常为几十千米，它们从多角度来切割洋脊，而转换断层只从横向上切割洋脊为洋脊段。洋中脊发育的地理实体类型主要为裂谷、海脊、海山和海丘等。

　　南大西洋洋底深度是变化的，大体深度为 4 000 ～ 5 000 m。洋中脊轴部是 1 ～ 1.5 km 深、4 ～ 15 km 宽的大裂谷，裂谷两侧的脊状山脉相距 20 ～ 40 km。轴部裂谷谷壁是由使地壳向上运动而形成这些脊状山脉的大断层组成。轴部裂谷谷底是洋壳生成的原始地点，多数脊段都具有冲向中央谷底中心的轴向火山脊。这些轴向火山脊本身由较小的山脊、环形丘体及不同的特征地形构成，轴向火山脊宽达 2 ～ 4 km、高 100 ～ 600 m，个别的火山局部可以接近海水表面，有时比海水表面要高，其发现的火山规模比在以平缓熔岩流为特征的快速扩张脊处大得多。

　　在南大西洋洋中脊，共命名地理实体 13 个，热液区 7 个。其中海山 4 个，包括采蘩海山、宵征海山、凯风海山和方舟海山；海丘 4 个，包括安吉海丘、赤狐海丘、黑乌海丘和洵美海丘；海脊 4 个，包括驺虞海脊、如玉海脊、唐棣海脊和文王海脊；裂谷 1 个，即德音裂谷。热液区 7 个，包括驺虞 1 号热液区、驺虞 2 号热液区、采蘩 1 号热液区、采蘩 2 号热液区、太极 1 号热液区、德音 1 号热液区和洵美 1 号热液区（图 3-2）。

Section 2　Undersea Features and Hydrothermal Fields in the South Atlantic Ridge

The South Atlantic Ridge has a total length of 4 500 km. It starts from the La Romanche Transform Fault near the equator in the north, south to the triple point near the Bouvet Island in the Indian Ocean, stretching to the Atlantic Ocean in the Antarctic in the south. It accounts for 1/3 the total length of the Atlantic Ridge and it is a slow-spreading ridge (full-spreading rate being 20–50 mm/a). It is generally believed that the geological and geophysical characteristics are similar to other slow-spreading ridges (Dick et al., 2003; Devey et al., 2010). The ridge is cut by several faults and the larger one is the transform fault, between which there are non-transform faults. Transform and non-transform faults develop a series of separated extending segments with the length of tens of kilometers (Chen, 2003; Li, 2014). In Angola-Brazil segment, it appears dense region destroyed by intermittent faults, which has characteristics of broad, oblique and longitudinal arrangement and looks like a large network of fissures. The length of non-transform faults is usually a few tens of kilometers and they cut the ridge from multiple angles, while transform faults only cut the ridge to segments crosswise. The undersea feature types developed by the ridge are mainly gaps, ridges, seamounts, knolls, etc.

The depth of South Atlantic seabed is changing with the depth, roughly being 4 000–5 000 m. The axial portion of the ridge is the Great Rift Valley with the depth of 1–1.5 km and width of 4–15 km. The ridge-like mountains on both sides of the rift are away from 20–40 km. The axial rift walls consist of large faults which made the crust moving upward and formed these ridge-like mountains. The axial rift floor is the original generation place of ocean crust. Most of the ridge segments have axial volcanic ridges which crash towards the middle of the central rift floor. These axial volcanic ridges themselves are made up of smaller ridges, annular hills and different characteristic topography. The width of axial volcanic ridges is up to 2–4 km, while the height is 100–600 m. Parts of several volcanoes can be close to the sea surface, sometimes higher than it. The discovered volcano scale is much bigger than fast-spreading ridges featured by gentle stream of lava.

In total, 13 undersea features and 7 hydrothermal fields have been named by China in the region of the South Atlantic Ridge. Among them, there are 4 seamounts, including Caifan Seamount, Xiaozheng Seamount, Kaifeng Seamount and Fangzhou Seamount; 4 hills, including Anji Hill, Chihu Hill, Heiwu Hill and Xunmei Hill; 4 ridges, including Zouyu Ridge, Ruyu Ridge, Tangdi Ridge and Wenwang Ridge; 1 gap, that is Deyin Gap. There are 7 hydrothermal fields, including Zouyu-1 Hydrothermal Field, Zouyu-2 Hydrothermal Field, Caifan-1 Hydrothermal Field, Caifan-2 Hydrothermal Field, Taiji-1 Hydrothermal Field, Deyin-1 Hydrothermal Field and Xunmei-1 Hydrothermal Field（Fig. 3–2）.

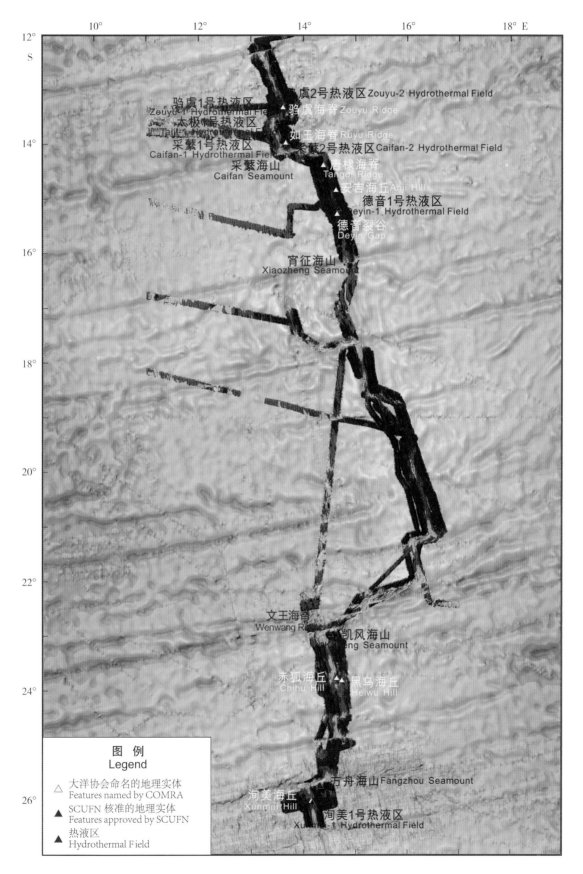

图 3-2　南大西洋洋中脊地理实体位置示意图

Fig.3-2　Locations of the undersea features in the South Atlantic Ridge

3.2.1 驺虞海脊
Zouyu Ridge

中文名称 Chinese Name	驺虞海脊 Zouyu Haiji		英文名称 English Name	Zouyu Ridge	所在大洋 Ocean or Sea	南大西洋 South Atlantic Ocean
发现情况 Discovery Facts	此海脊于 2011 年由中国科考船 "大洋一号" 在执行 DY125-22 航次调查时发现。 This ridge was discovered by the Chinese R/V *Dayang Yihao* during the DY125-22 cruise in 2011.					
命名历史 Name History	在我国大洋航次调查报告和 GEBCO 地名辞典中未命名。 This ridge has not been named in Chinese cruise reports or GEBCO gazetteer.					
特征点坐标 Coordinates	13°10.02′S，14°25.80′W 13°19.80′S，14°24.00′W 13°32.04′S，14°23.40′W			长 (km) × 宽 (km) Length (km) × Width (km)		42 × 9
最大水深（m） Max Depth（m）	3 580	最小水深（m） Min Depth（m）	2 310	高差（m） Total Relief（m）		1 270
地形特征 Feature Description	此海脊位于南大西洋洋中脊，走向与中央裂谷近似一致，长宽分别为 42 km 和 9 km。海脊顶部最浅处水深约 2 310 m，底部水深约 3 580 m，高差约 1 270 m（图 3-3）。 Zouyu Ridge is located in the South Atlantic Ridge, nearly parallel with the central rift valley. The length and width are 42 km and 9 km respectively. The minimum top depth of the ridge is 2 310 m and the piedmont depth is 3 580 m, which makes the total relief being 1 270 m (Fig.3–3).					
命名释义 Reason for Choice of Name	"驺虞" 出自《诗经·国风·驺虞》"彼茁者葭，壹发五豝，于嗟乎驺虞。" "驺虞" 意为射技高强的猎人。驺虞也是我国古代神话传说中的仁兽，生性仁慈。 "Zouyu" comes from a poem named *Zouyu* in *Shijing · Guofeng*. *Shijing* is a collection of ancient Chinese poems from 11th century B.C. to 6th century B.C. "Amid the green reed sprouts, he has killed five wild sows. How well the forester shoots!" "Zouyu" means hunters with excellent skills. It is also referred to a benevolence beast in Chinese ancient legends.					

(a)

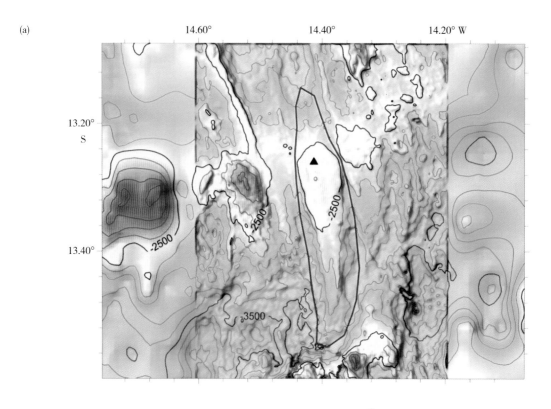

(b)

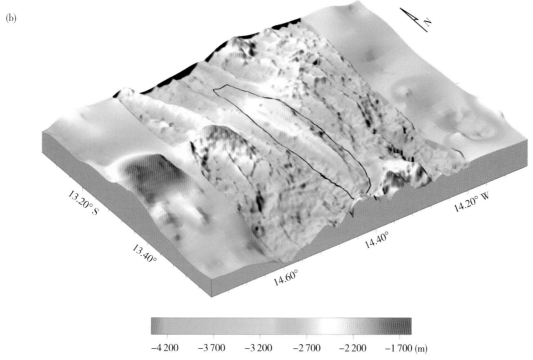

图 3-3 驺虞海脊

(a) 地形图（等深线间隔 200 m）；(b) 三维图

Fig.3-3 Zouyu Ridge

(a) Bathymetric map (the contour interval is 200 m); (b) 3-D topographic map

3.2.2 驺虞 1 号热液区、驺虞 2 号热液区
Zouyu-1 Hydrothermal Field and Zouyu-2 Hydrothermal Field

中文名称 Chinese Name	驺虞 1 号热液区 驺虞 2 号热液区 Zouyu-1 Reyequ Zouyu-2 Reyequ	英文名称 English Name	Zouyu-1 Hydrothermal Field Zouyu-2 Hydrothermal Field	所在大洋 Ocean or Sea	南大西洋 South Atlantic Ocean
发现情况 Discovery Facts	驺虞 1 号热液区于 2009 年由中国科考船"大洋一号"在执行 DY115-21 航次调查时发现，通过海底观察、水体热液异常探测和取样证实。 驺虞 2 号热液区于 2011 年由中国科考船"大洋一号"在执行 DY125-22 航次调查时发现，通过海底观察、水体热液异常探测和取样证实。 驺虞 1 号和驺虞 2 号热液区位于南大西洋洋中脊的新生洋脊上，围岩为玄武岩，水深约 2 200 ～ 2 300 m。 Zouyu-1 Hydrothermal Field was discovered by the Chinese R/V *Dayang Yihao* during the DY115-21 cruise in 2009. It was confirmed by undersea observation, hydrothermal anomaly detection and samples. Zouyu-2 Hydrothermal Field was discovered by the Chinese R/V *Dayang Yihao* during the DY125-22 cruise in 2011. It was confirmed by undersea observation, hydrothermal anomaly detection and samples. Zouyu-1 and Zouyu-2 Hydrothermal Fields are located in the newly born ridge in the South Atlantic Ridge with the water depth of 2 200–2 300 m. Their surrounding rocks are basalt.				
命名历史 Name History	驺虞 1 号热液区在 DY115-21 航次现场报告中曾用名为"贝利珠热液区"，英文曾用名为"Baily's Beads Hydrothermal Field"。 驺虞 2 号热液区在 DY125-22 航次现场报告中曾用名为"情人谷热液区"，英文曾用名"Valentine Valley Hydrothermal Field"。 Zouyu-1 Hydrothermal Field was once named "Beilizhu Reyequ" in Chinese and "Baily's Beads Hydrothermal Field" in English in the Chinese DY115-21 cruise field report. Zouyu-2 Hydrothermal Field was once named "Qingrengu Reyequ" in Chinese and "Valentine Valley Hydrothermal Field" in English in the Chinese DY125-22 cruise field report.				
特征点坐标 Coordinates	13°17.15′S，14°24.79′W（驺虞 1 号 / Zouyu-1） 13°15.26′S，14°24.77′W（驺虞 2 号 / Zouyu-2)			水深 (m) Depth(m)	2200 ～ 2300
命名释义 Reason for Choice of Name	采用热液区附近的"驺虞海脊"的名字命名，分别命名为"驺虞 1 号热液区"和"驺虞 2 号热液区"（图 3-4）。 These two hydrothermal fields are named Zouyu-1 and Zouyu-2 Hydrothermal Field after the nearby Zouyu Ridge (Fig.3–4).				

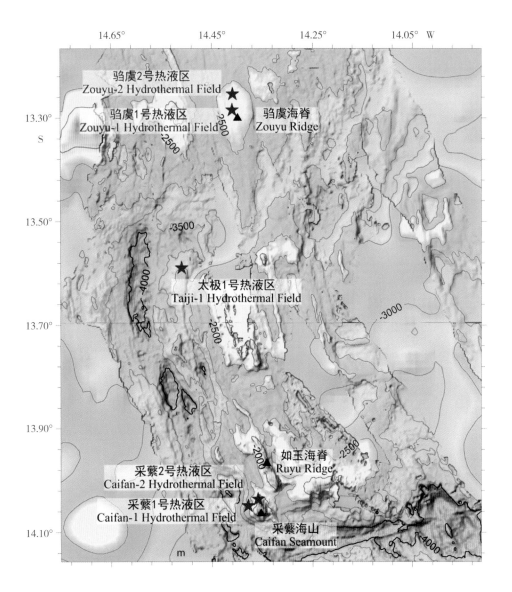

图 3-4　驺虞热液区地理位置图（等深线间隔 500 m）

Fig.3-4　Bathymetric map of the Zouyu Hydrothermal Field (The contour interval is 500 m)

3.2.3　太极 1 号热液区
Taiji-1 Hydrothermal Field

中文名称 Chinese Name	太极 1 号热液区 Taiji-1 Reyequ	英文名称 English Name	Taiji-1 Hydrothermal Field	所在大洋 Ocean or Sea	南大西洋 South Atlantic Ocean
发现情况 Discovery Facts	此热液区于 2011 年由中国科考船"大洋一号"在执行 DY125-22 航次调查时发现，通过海底观察、水体热液异常探测及取样证实。 该热液区位于南大西洋洋中脊中央裂谷与转换断层的交汇处，围岩为玄武岩，水深约 2 400 m。 This hydrothermal field was discovered by the Chinese R/V *Dayang Yihao* during the DY125-22 cruise in 2011. It was confirmed by undersea observation, hydrothermal anomaly detection and samples. It is located in the intersection of the central rift valley and the transform fault of South Atlantic Ridge with the water depth of about 2 400 m. Its surrounding rocks are basalt.				
命名历史 Name History	在 DY115-22 航次现场报告中曾用名为"太极"热液区。 This hydrothermal field was once named Taiji Hydrothermal Field in the Chinese DY115-22 ocean cruise field report.				
特征点坐标 Coordinates	13°35.42′S，14°31.01′W			水深 (m) Depth(m)	2 400
命名释义 Reason for Choice of Name	调查该热液区时，发现海底有黑白两条鱼在游动，形似太极图，故取名为"太极"（图 3–5）。 When investigating this hydrothermal field, two fishes, black and white respectively, were discovered swimming nearby. And this scene looked like the shape of China's Tai Chi, so it got the name of Taiji (Fig.3–5).				

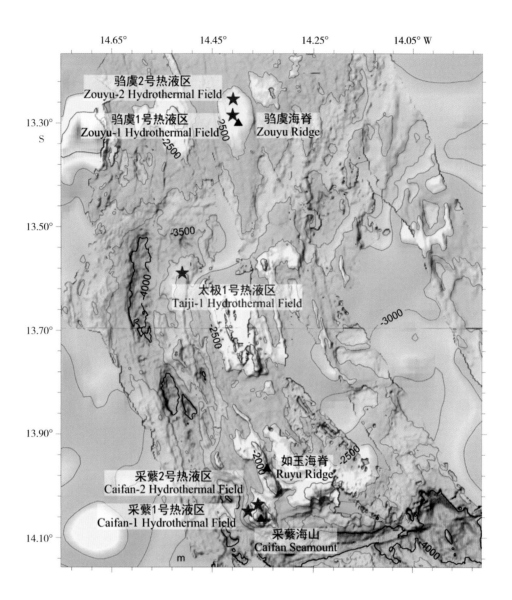

图 3-5　太极 1 号热液区地理位置图（等深线间隔 500 m）

Fig.3-5　Bathymetric map of the Taiji-1 Hydrothermal Field (The contour interval is 500 m)

3.2.4 如玉海脊
Ruyu Ridge

中文名称 Chinese Name	如玉海脊 Ruyu Haiji	英文名称 English Name	Ruyu Ridge	所在大洋 Ocean or Sea	南大西洋 South Atlantic Ocean
发现情况 Discovery Facts	此海脊于 2011 年由中国科考船"大洋一号"在执行 DY125-22 航次调查时发现。 This ridge was discovered by the Chinese R/V *Dayang Yihao* during the DY125-22 cruise in 2011.				
命名历史 Name History	在我国大洋航次调查报告和 GEBCO 地名辞典中未命名。 This ridge has not been named in Chinese cruise reports or GEBCO gazetteer.				
特征点坐标 Coordinates	13°56.13′S，14°19.40′W			长(km)×宽(km) Length (km)×Width (km)	14×5
最大水深（m） Max Depth（m）	2 550	最小水深（m） Min Depth（m）	1 570	高差（m） Total Relief（m）	980
地形特征 Feature Description	此海脊位于南大西洋洋中脊，走向与中央裂谷近似一致，长宽分别为 14 km 和 5 km。海脊顶部最浅处水深约 1 570 m，底部水深 2 550 m，高差约 980 m（图 3–6）。 Ruyu Ridge is located in the South Atlantic Ridge, nearly parallel with the central rift valley. The length is 14 km and the width is 5 km. The minimum top depth of the ridge is 1 570 m and the piedmont depth is 2 550 m, which makes the total relief being 980 m (Fig.3–6).				
命名释义 Reason for Choice of Name	"如玉"出自《诗经·国风·野有死麕》"白茅纯束，有女如玉"。如玉，像玉一样洁白。在中国文学中也多以"玉"来形容女性的肤润貌美，姿容出众。玉是中国传统文化的一个重要组成部分，以玉为中心载体的玉文化，深深影响了古代中国人的思想观念，成为中国文化不可缺少的一部分。 "Ruyu" comes from a poem named *Yeyousijun* in *Shijing · Guofeng*. *Shijing* is a collection of ancient Chinese poems from 11th century B.C. to 6th century B.C. "The lass is fair as jade; White cogon wreathes her head." "Ruyu" means pure and white like a jade. "Ruyu" is often used to describe women's smooth skin and outstanding beauty. Jade is an important part of Chinese traditional culture. The culture of jade deeply influenced the ideas of the ancient Chinese people and become an indispensable part of Chinese culture.				

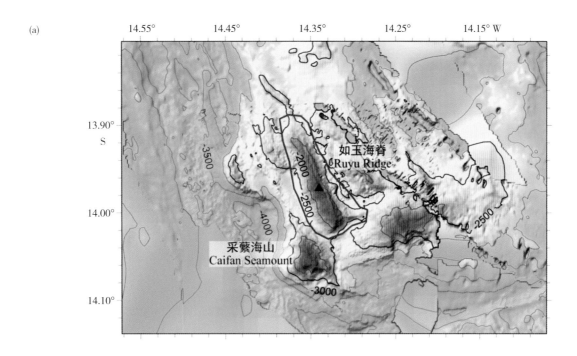

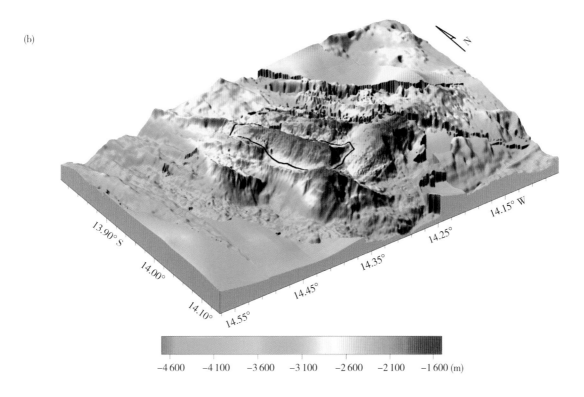

图 3-6　如玉海脊

(a) 地形图（等深线间隔 500 m）；(b) 三维图

Fig.3-6　Ruyu Ridge

(a) Bathymetric map (the contour interval is 500 m); (b) 3-D topographic map

3.2.5 采蘩海山
Caifan Seamount

中文名称 Chinese Name	采蘩海山 Caifan Haishan	英文名称 English Name	Caifan Seamount	所在大洋 Ocean or Sea	南大西洋 South Atlantic Ocean
发现情况 Discovery Facts	此海山于 2009 年 12 月由中国科考船"大洋一号"在执行 DY115-21 航次调查时发现。 This seamount was discovered by the Chinese R/V *Dayang Yihao* during the DY115-21 cruise in December, 2009.				
命名历史 Name History	由我国命名为采蘩海山,于 2012 年提交 SCUFN 审议通过。 Caifan Seamount was named by China and approved by SCUFN in 2012.				
特征点坐标 Coordinates	14°03.10'S, 14°21.10'W		长(km)×宽(km) Length (km)× Width (km)		18×10
最大水深(m) Max Depth(m)	3 800	最小水深(m) Min Depth(m)	1 600	高差(m) Total Relief(m)	2 200
地形特征 Feature Description	采蘩海山位于南大西洋洋中脊,近似穹隆状,长和宽分别为 18 km 和 10 km。海山顶部最浅处水深约 1 600 m,山麓水深约 3 800 m,高差约 2 200 m(图 3–7)。 Caifan Seamount is located in the South Atlantic Ridge and has a dome-like shape. The length is 18 km and the width is 10 km. The minimum top depth of the seamount is 1 600 m and the piedmont depth is 3 800 m, which makes the total relief being 2 200 m (Fig.3–7).				
命名释义 Reason for Choice of Name	"采蘩"出自《诗经·国风·采蘩》"于以采蘩,于涧之中"。蘩指白蒿,根茎可食,古代常用祭祀用品。此篇主要描述采办祭祀所需用来燎烧的蒿草以及寻找和采办各项祭祀用品,完成祭祀的辛劳。 "Caifan" comes from a poem named *Caifan* in *Shijing · Guofeng*. *Shijing* is a collection of ancient Chinese poems from 11th century B.C. to 6th century B.C. "Where to pick wormwood leaves? In the stream in the valleys." "Fan" means white artemisia, its root and stem are edible and it is commonly used in ancient sacrifice. The poem mainly describes collecting artemisia used for burning during the sacrifice ceremony and looking for other sacrifice objects, to show the hardship finishing the ceremony.				

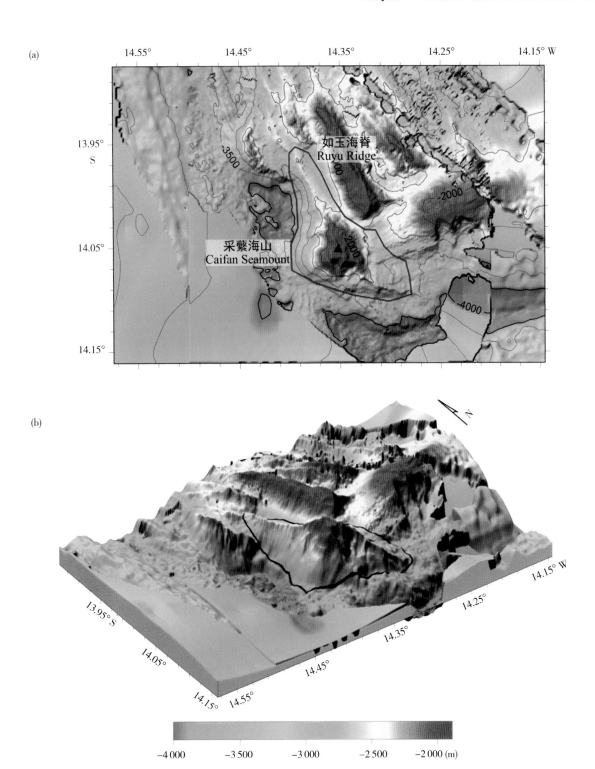

图 3-7　采藻海山

(a) 地形图（等深线间隔 500 m）；(b) 三维图

Fig.3-7　Caifan Seamount

(a) Bathymetric map (the contour interval is 500 m); (b) 3-D topographic map

3.2.6 采矾 1 号热液区、采矾 2 号热液区
Caifan-1 Hydrothermal Field and Caifan-2 Hydrothermal Field

中文名称 Chinese Name	采矾 1 号热液区 采矾 2 号热液区 Caifan-1 Reyequ Caifan-2 Reyequ	英文名称 English Name	Caifan-1 Hydrothermal Field Caifan-2 Hydrothermal Field	所在大洋 Ocean or Sea	南大西洋 South Atlantic Ocean
发现情况 Discovery Facts	采矾 1 号热液区于 2009 年在大西洋由中国科考船"大洋一号"在执行 DY115-21 航次调查时发现,通过海底观察、水体热液异常探测和取样证实; 采矾 2 号热液区于 2011 年在大西洋由中国科考船"大洋一号"在执行 DY125-22 航次时发现,通过海底观察及取样证实; 采矾 1 号和采矾 2 号热液区位于南大西洋洋中脊中央裂谷西侧新生洋脊的斜坡上,围岩为玄武岩。采矾 1 号热液区水深约 2 000 m,采矾 2 号热液区水深约 2 400 m。 Caifan-1 Hydrothermal Field was discovered by the Chinese R/V *Dayang Yihao* during the DY115-21 cruise in 2009. It was confirmed by undersea observation, hydrothermal anomaly detection and samples. Caifan-2 Hydrothermal Field was discovered by the Chinese R/V *Dayang Yihao* during the DY125-22 cruise in 2011. It was confirmed by undersea observation and samples. Caifan-1 and Caifan-2 Hydrothermal Fields are located in the newly born ridge slopes, west of the central rift valley in South Atlantic Ridge and the surrounding rocks are basalt. The depth of Caifan-1 Hydrothermal Field is about 2 000 m and the depth of Caifan-2 Hydrothermal Field is about 2 400 m.				
命名历史 Name History	采矾 1 号热液区在 DY115-21 航次现场报告中曾用名为"彩虹湾南"热液区,英文曾用名为"Rainbow Bay South"Hydrothermal Field。 采矾 2 号热液区在 DY125-22 航次现场报告中曾用名为"彩虹湾北"热液区,英文曾用名为"Rainbow Bay North"Hydrothermal Field。 Caifan-1 Hydrothermal Field was once named "Caihongwannan Reyequ" in Chinese and "Rainbow Bay South Hydrothermal Field" in English in the Chinese DY115-21 cruise field report. Caifan-2 Hydrothermal Field was once named "Caihongwanbei Reyequ" in Chinese and "Rainbow Bay North Hydrothermal Field" in English in the Chinese DY125-22 cruise field report.				
特征点坐标 Coordinates	14°02.87′S,14°22.79′W(采矾 1 号热液区 / Caifan-1 Hydrothermal Field) 14°02.18′S,14°21.83′W(采矾 2 号热液区 / Caifan-2 Hydrothermal Field)			水深 (m) Depth (m)	2 000 ~ 2 400
命名释义 Reason for Choice of Name	采用热液区附近的"采矾海山"的名字命名,分别命名为"采矾 1 号热液区"和"采矾 2 号热液区"(图 3–8)。 The two hydrothermal fields are named Caifan-1 Hydrothermal Field and Caifan-2 Hydrothermal Field respectively after the nearby Caifan Seamount (Fig.3–8).				

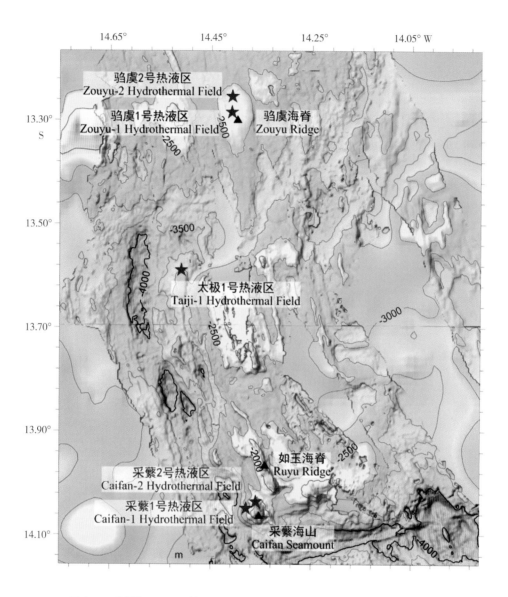

图 3-8　采蘩 1 号和采蘩 2 号热液区地理位置图（等深线间隔 500 m）

Fig.3-8　Bathymetric map of the Caifan-1 and Caifan-2 Hydrothermal Field (The contour interval is 500 m)

3.2.7　唐棣海脊
Tangdi Ridge

中文名称 Chinese Name	唐棣海脊 Tangdi Haiji		英文名称 English Name	Tangdi Ridge	所在大洋 Ocean or Sea	南大西洋 South Atlantic Ocean
发现情况 Discovery Facts	此海脊于 2011 年由中国科考船 "大洋一号" 在执行 DY125-22 航次调查时发现。 This ridge was discovered by the Chinese R/V *Dayang Yihao* during the DY125-22 cruise in 2011.					
命名历史 Name History	在我国大洋航次调查报告和 GEBCO 地名辞典中未命名。 This ridge has not been named in Chinese cruise reports or GEBCO gazetteers.					
特征点坐标 Coordinates	14°28.09'S，13°40.64'W			长(km)×宽(km) Length (km) × Width (km)		17×4
最大水深（m） Max Depth（m）	2 400		最小水深（m） Min Depth（m）	1 924	高差 (m) Total Relief (m)	476
地形特征 Feature Description	此海脊位于南大西洋洋中脊，近南北向展布，长和宽分别为 17 km 和 4 km。海脊顶部最浅处水深约 1 924 m，山麓水深约 2 400 m，高差约 476 m（图 3–9）。 Tangdi Ridge is located in the South Atlantic Ridge, running nearly N to S. The length is 17 km and the width is 4 km. The minimum top depth of the ridge is 1 924 m and the piedmont depth is 2 400 m, which makes the total relief being 476 m (Fig.3–9) .					
命名释义 Reason for Choice of Name	"唐棣" 出自《诗经·国风·何彼襛矣》"何彼襛矣，唐棣之华"。唐棣，树木名，其果实形似李，可食，其花鲜艳。此诗描绘了君王之女出嫁时车服的豪华奢侈和结婚场面的气派和排场，就像唐棣之花盛开的场景。 "Tangdi" comes from a poem named *Hebinongyi* in *Shijing · Guofeng*. *Shijing* is a collection of ancient Chinese poems from 11th century B.C. to 6th century B.C. "Why so pretentious is the carriage? Its showy curtains lend the image." "Tangdi" means sparrow-plume, a tree species with plume-like fruite and bright-colored flowers. This poem describes the luxuries of king's daughter on her wedding day, just like the bloom of sparrow-plume flowers.					

(a)

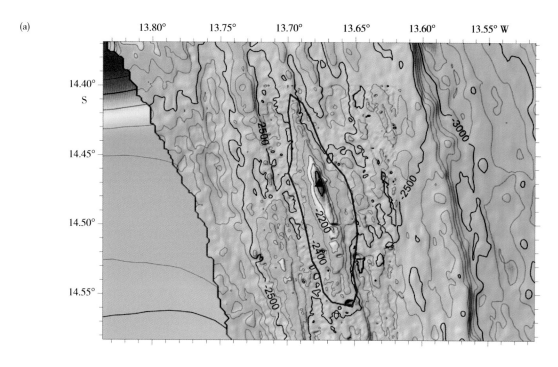

(b)

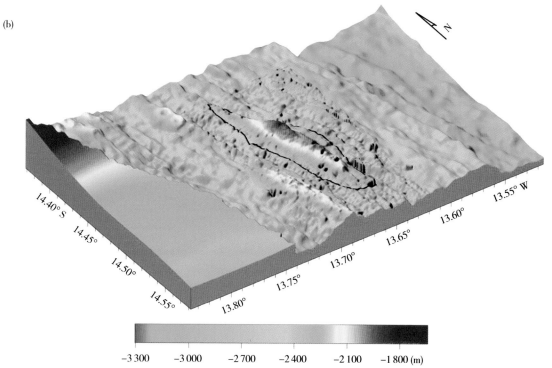

图 3-9　唐棣海脊

(a) 地形图（等深线间隔 100 m）；(b) 三维图

Fig.3-9　Tangdi Ridge

(a) Bathymetric map (the contour interval is 100 m); (b) 3-D topographic map

3.2.8 安吉海丘
Anji Hill

中文名称 Chinese Name	安吉海丘 Anji Haiqiu	英文名称 English Name	Anji Hill	所在大洋 Ocean or Sea	南大西洋 South Atlantic Ocean
发现情况 Discovery Facts	此海丘于 2011 年由中国科考船 "大洋一号" 在执行 DY125-22 航次调查时发现。 This hill was discovered by the Chinese R/V *Dayang Yihao* during the DY125-22 cruise in 2011.				
命名历史 Name History	在我国大洋航次调查报告和 GEBCO 地名辞典中未命名。 This hill has not been named in Chinese cruise reports or GEBCO gazetteer.				
特征点坐标 Coordinates	14°52.20′S，13°22.50′W			长(km)×宽(km) Length (km)× Width (km)	5×4
最大水深（m） Max Depth（m）	2600	最小水深（m） Min Depth（m）	1890	高差（m） Total Relief（m）	710
地形特征 Feature Description	安吉海丘位于南大西洋洋中脊中央裂谷，俯视平面形态呈圆形，长宽分别为 5 km 和 4 km。海丘顶部最浅处水深约 1890 m，山麓水深约 2600 m，高差约 710 m（图 3-10）。 Anji Hill is located in the central rift valley in the South Atlantic Ridge. It has a round overlook plane shape. The length is 5 km and the width is 4 km. The minimum top depth of the hill is 1890 m and the piedmont depth is 2600 m, which makes the total relief being 710 m（Fig.3-10）.				
命名释义 Reason for Choice of Name	"安吉" 出自《诗经·唐风·无衣》"岂曰无衣？七兮。不如子之衣，安且吉兮。" 形容衣服舒适美丽。本句意为 "谁说我没有衣服穿，七件。却都不及你亲手做的，又舒适又美观。" 睹物思人，表达了丈夫对妻子的思念之情。 "Anji" comes from a poem named *Wuyi* in *Shijing · Tangfeng*. *Shijing* is a collection of ancient Chinese poems from 11th century B.C. to 6th century B.C. "Without a robe of the seventh rank? Your robe is better made than mine, a robe so soft, a robe so fine!" The sentence means "Although I have seven robes, none of them is sewed by you. Your robe is so soft and fine." It Shows a husband longing for his wife.				

(a)

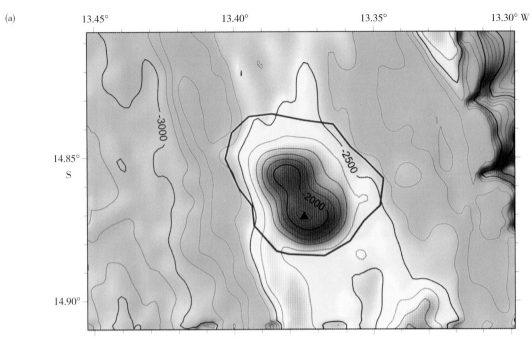

(b)

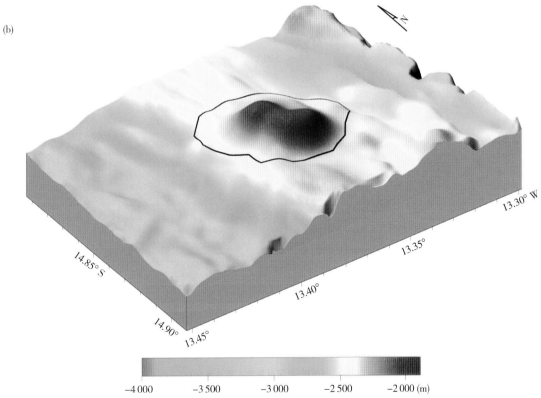

图 3-10　安吉海丘

(a) 地形图（等深线间隔 100 m）；(b) 三维图

Fig.3-10　Anji Hill

(a) Bathymetric map (the contour interval is 100 m); (b) 3-D topographic map

3.2.9 德音裂谷
Deyin Gap

中文名称 Chinese Name	德音裂谷 Deyin Liegu	英文名称 English Name	Deyin Gap	所在大洋 Ocean or Sea	南大西洋 South Atlantic Ocean
发现情况 Discovery Facts	此裂谷于 2011 年由中国科考船 "大洋一号" 在执行 DY125-22 航次调查时发现。 This gap was discovered by the Chinese R/V *Dayang Yihao* during the DY125-22 cruise in 2011.				
命名历史 Name History	在我国大洋航次调查报告和 GEBCO 地名辞典中未命名。 This gap has not been named in Chinese cruise reports or GEBCO gazetteer.				
特征点坐标 Coordinates	15°15.76′S，13°21.86′W			长 (km) × 宽 (km) Length (km) × Width (km)	16 × 3
最大水深（m） Max Depth（m）	3 206	最小水深（m） Min Depth（m）	3 000	高差（m） Total Relief（m）	206
地形特征 Feature Description	此裂谷位于南大西洋洋中脊，近南北向展布，裂谷长宽分别约为 16 km 和 3 km，最大水深为 3 206 m（图 3-11）。 Deyin Gap is located in the South Atlantic Ridge, running nearly N to S. The length of Deyin Gap is 16 km and the width is 3 km. The maximum depth of the gap is 3 206 m (Fig.3–11).				
命名释义 Reason for Choice of Name	"德音"出自《诗经·国风·谷风》"德音莫违，及尔同死"。"德音"指誓言，意为"不要背弃往日的誓言，与你生死相依两不忘。" "Deyin" comes from a poem named *Gufeng* in *Shijing · Guofeng*. *Shijing* is a collection of ancient Chinese poems from 11th century B.C. to 6th century B.C. "Your sweet words filled my ears: 'We'll stay a hundred years'. " *Gufeng* is a poem written by a deserted women, whose heartless husband loves his new wife. "Deyin" means pledge. The poem means while I do nothing contrary to my good name, I would keep loving you for all of my life and live with you till our death.				

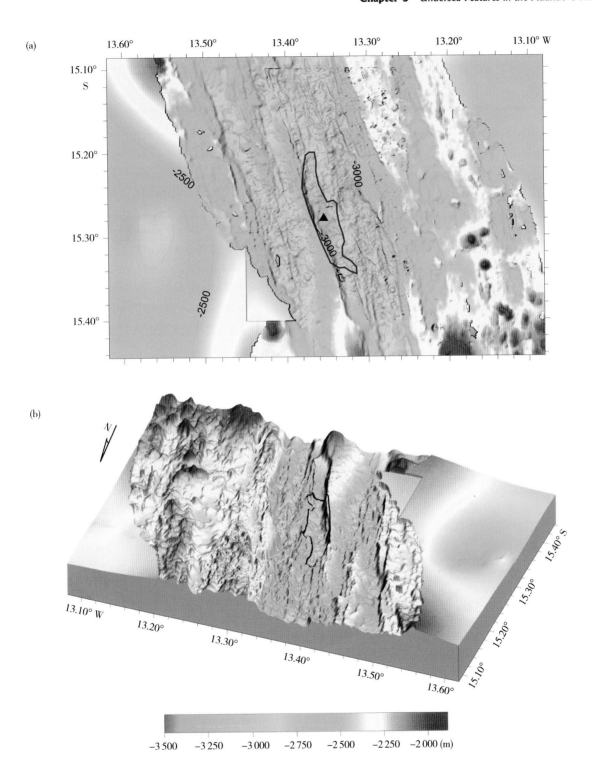

图 3-11　德音裂谷

(a) 地形图 （等深线间隔 200 m）；(b) 三维图

Fig.3-11　Deyin Gap

(a) Bathymetric map (the contour interval is 200 m); (b) 3-D topographic map

3.2.10 德音 1 号热液区
Deyin-1 Hydrothermal Field

中文名称 Chinese Name	德音 1 号热液区 Deyin-1 Reyequ	英文名称 English Name	Deyin-1 Hydrothermal Field	所在大洋 Ocean or Sea	南大西洋 South Atlantic Ocean
发现情况 Discovery Facts	此热液区于 2011 年由中国科考船"大洋一号"在执行 DY125-22 航次调查时，通过海底观察和取样证实。该热液区位于南大西洋中央裂谷中新火山脊与次级火山带错段交汇处，围岩为玄武岩，水深约 2 793 m。 This hydrothermal field was discovered by the Chinese R/V *Dayang Yihao* during the DY125-22 cruise in 2011. It was confirmed by undersea observation and samples. Deyin-1 Hydrothermal Field is located in the intersection of new volcanic ridges and secondary volcanic belts of the central rift valley of the South Atlantic Ridge, with the water depth of about 2 793 m. Its surrounding rocks are basalt.				
命名历史 Name History	在 DY115-22 航次现场报告中曾用名为"15°S 热液区"。 It was once named "15°S Hydrothermal Field" in the Chinese DY115-22 cruise field report.				
特征点坐标 Coordinates	15°09.94'S，13°21.38'W			水深 (m) Depth (m)	2 793
命名释义 Reason for Choice of Name	采用热液区附近的"德音裂谷"的名字命名，命名为"德音 1 号热液区"（图 3–12）。 It was named Deyin-1 Hydrothermal Field after the nearby Deyin Gap (Fig.3–12).				

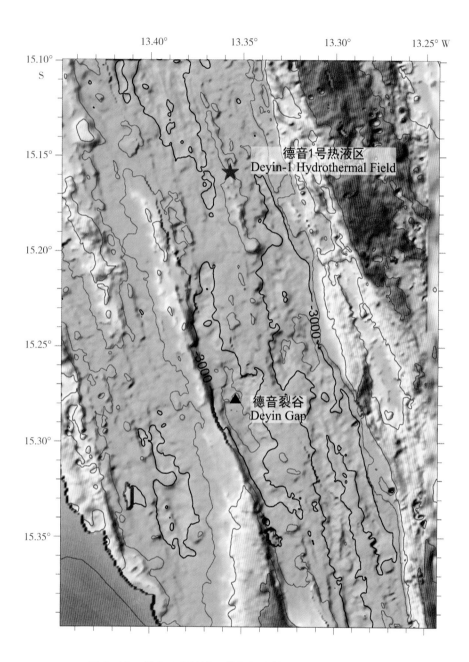

图 3-12　德音 1 号热液区位置图（等深线间隔 200 m）

Fig.3-12　Bathymetric map of the Deyin-1 Hydrothermal Field (The contour interval is 200 m)

3.2.11　宵征海山
Xiaozheng Seamount

中文名称 Chinese Name	宵征海山 Xiaozheng Haishan	英文名称 English Name	Xiaozheng Seamount	所在大洋 Ocean or Sea	南大西洋 South Atlantic Ocean
发现情况 Discovery Facts	此海山于 2011 年 4 月由中国科考船"大洋一号"在执行 DY125-22 航次调查时发现。 This seamount was discovered by the Chinese R/V *Dayang Yihao* during the DY125-22 cruise in April, 2011.				
命名历史 Name History	由我国命名为宵征海山，于 2012 年提交 SCUFN 审议通过。 Xiaozheng Seamount was named by China and approved by SCUFN in 2012.				
特征点坐标 Coordinates	16° 12.9' S，13° 06.5' W			长(km)×宽(km) Length (km)× Width (km)	15×12
最大水深（m） Max Depth（m）	3 600	最小水深（m） Min Depth（m）	2 150	高差（m） Total Relief（m）	1 450
地形特征 Feature Description	此海山位于南大西洋洋中脊，长和宽分别为 15 km 和 12 km，顶部最浅处水深约 2 150 m，山麓水深约 3 600 m，高差约 1 450 m（图 3-13）。 Xiaozheng Seamount is located in the South Atlantic Ridge. The length is 15 km and the width is 12 km. The minimum top depth of the seamount is 2 150 m and piedmont depth is 3 600 m, which makes the total relief being 1 450 m (Fig.3-13).				
命名释义 Reason for Choice of Name	"宵征"出自《诗经·国风·召南》"肃肃宵征，夙夜在公"。"宵征"意为夜间行走。此海山调查发现时正值深夜时分，取此命名，体现出大洋科考人员夜以继日，不辞辛苦和自强的精神。 "Xiaozheng" comes from a poem named *Zhaonan* in *Shijing · Guofeng*. *Shijing* is a collection of ancient Chinese poems from 11th century B.C. to 6th century B.C. "As my man serves the state, so hard he works and late." "Xiaozheng" means marching at night. This seamount is named "Xiaozheng" since it was discovered during a survey at night. It also reflects that the scientists worked hard day and night to achieve their goals.				

(a)

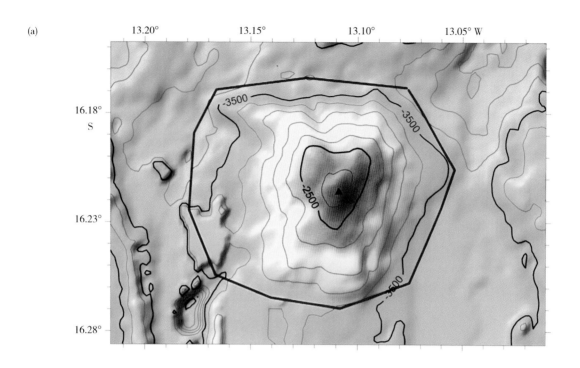

(b)

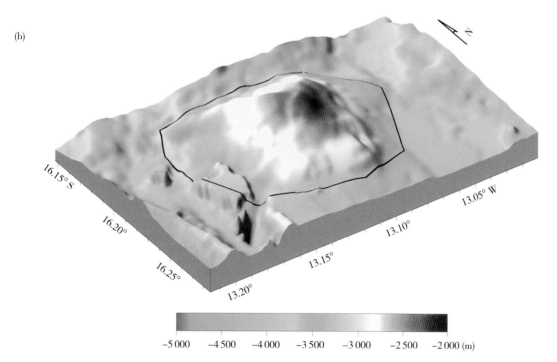

图 3-13　宵征海山

(a) 地形图 （等深线间隔 200 m）；(b) 三维图

Fig.3-13　Xiaozheng Seamount

(a) Bathymetric map (the contour interval is 200 m); (b) 3-D topographic map

3.2.12 凯风海山
Kaifeng Seamount

中文名称 Chinese Name	凯风海山 Kaifeng Haishan	英文名称 English Name	Kaifeng Seamount	所在大洋 Ocean or Sea	南大西洋 South Atlantic Ocean
发现情况 Discovery Facts	此海山于 2011 年 5 月由中国科考船"大洋一号"在执行 DY125-22 航次调查时发现。 This seamount was discovered by the Chinese R/V *Dayang Yihao* during the DY125-22 cruise in May, 2011.				
命名历史 Name History	由我国命名为凯风海山，于 2012 年提交 SCUFN 审议通过。 Kaifeng Seamount was named by China and approved by SCUFN in 2012.				
特征点坐标 Coordinates	22°56.6′S，13°25.9′W			长(km)×宽(km) Length (km)× Width (km)	27×21
最大水深（m） Max Depth（m）	4 200	最小水深（m） Min Depth（m）	1 700	高差（m） Total Relief（m）	2 500
地形特征 Feature Description	此海山位于南大西洋洋中脊，长和宽分别为 27 km 和 21 km，顶部最浅处水深约 1 700 m，山麓水深约 4 200 m，高差约 2 500 m（图 3–14）。 Kaifeng Seamount is located in the South Atlantic Ridge. The length is 27 km and the width is 21 km. The minimum top depth of the seamount is 1 700 m and the piedmont depth is 4 200 m, which makes the total relief being 2 500 m (Fig.3–14).				
命名释义 Reason for Choice of Name	"凯风"出自《诗经·国风·凯风》"凯风自南，吹彼棘心"。凯风表示温暖的风，在此海山调查发现时，海面平静，和风习习，体现出大洋科考人员此时的愉悦心情，故以此命名。 "Kaifeng" comes from a poem named *Kaifeng* in *Shijing · Guofeng*. *Shijing* is a collection of ancient Chinese poems from 11th century B.C. to 6th century B.C. "From the south comes the breeze, caressing tender jujube trees." "Kaifeng" means warm wind. This seamount was discovered while the sea breeze blowing gently, thus it is named "Kaifeng" to reflect the delighted mood of ocean researchers at that time.				

(a)

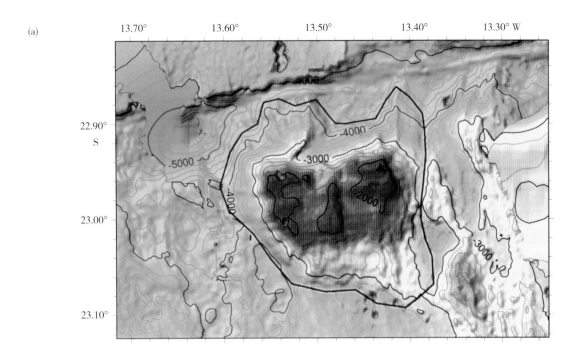

(b)

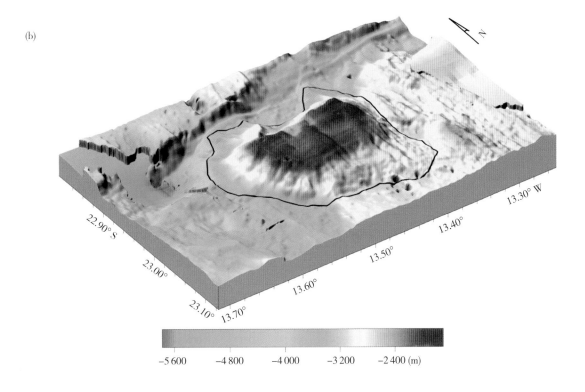

图 3-14　凯风海山

(a) 地形图（等深线间隔 200 m）；(b) 三维图

Fig.3-14　Kaifeng Seamount

(a) Bathymetric map (the contour interval is 200 m); (b) 3-D topographic map

3.2.13 赤狐海丘
Chihu Hill

中文名称 Chinese Name	赤狐海丘 Chihu Haiqiu	英文名称 English Name	Chihu Hill	所在大洋 Ocean or Sea	南大西洋 South Atlantic Ocean
发现情况 Discovery Facts	此海丘于 2011 年由中国科考船"大洋一号"在执行 DY125-22 航次调查时发现。 This hill was discovered by the Chinese R/V *Dayang Yihao* during the DY125-22 cruise in 2011.				
命名历史 Name History	在我国大洋航次调查报告和 GEBCO 地名辞典中未命名。 This hill has not been named in Chinese cruise reports or GEBCO gazetteer.				
特征点坐标 Coordinates	23°44.65'S，13°21.94′W			长(km)×宽(km) Length (km) × Width (km)	16.6 × 8.6
最大水深（m） Max Depth（m）	3 300	最小水深（m） Min Depth（m）	2 380	高差（m） Total Relief（m）	920
地形特征 Feature Description	此海丘位于南大西洋洋中脊，长和宽分别为 16.6 km 和 8.6 km。海丘顶部最浅处水深约 2 380 m，山麓水深约 3 300 m，高差约 920 m（图 3–15）。 Chihu Hill is located in the South Atlantic Ridge. The length is 16.6 km and the width is 8.6 km. The minimum top depth of the seamount is 2 380 m and the piedmont depth is 3 300 m, which makes the total relief being 920 m (Fig.3–15).				
命名释义 Reason for Choice of Name	"赤狐"出自《诗经·国风·北风》"莫赤匪狐，莫黑匪乌"。"赤狐"指红色狐狸，意为"没有不是红色的狐狸，也没有不是黑色的乌鸦"。 "Chihu" comes from a poem named *Beifeng* in *Shijing · Guofeng*. *Shijing* is a collection of ancient Chinese poems from 11th century B.C. to 6th century B.C. "All foxes are red and all crows are black." "Chihu" means red fox.				

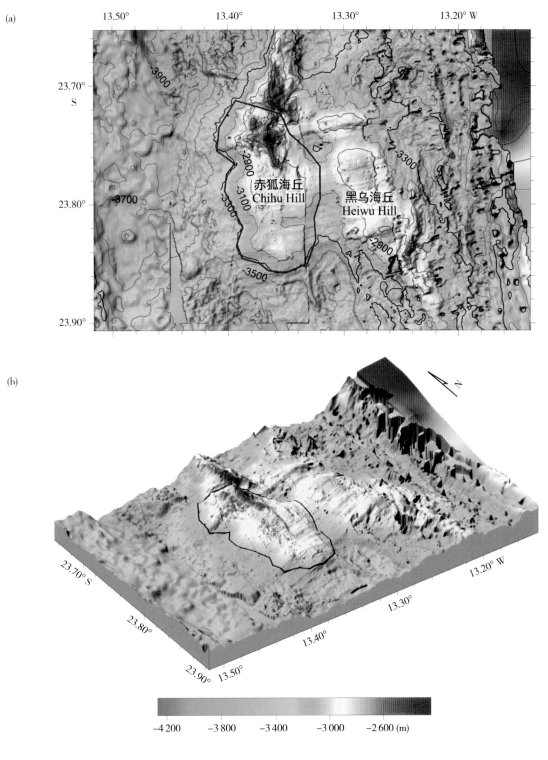

图 3-15　赤狐海丘

(a) 地形图（等深线间隔 200 m）；(b) 三维图

Fig.3-15　Chihu Hill

(a) Bathymetric map (the contour interval is 200 m); (b) 3-D topographic map

3.2.14 黑乌海丘
Heiwu Hill

中文名称 Chinese Name	黑乌海丘 Heiwu Haiqiu	英文名称 English Name	Heiwu Hill	所在大洋 Ocean or Sea	南大西洋 South Atlantic Ocean
发现情况 Discovery Facts	此海丘于 2011 年由中国科考船"大洋一号"在执行 DY125-22 航次调查时发现。 This hill was discovered by the Chinese R/V *Dayang Yihao* during the DY125-22 cruise in 2011.				
命名历史 Name History	在我国大洋航次调查报告和 GEBCO 地名辞典中未命名。 This hill has not been named in Chinese cruise reports or GEBCO gazetteer.				
特征点坐标 Coordinates	23°49.26'S，13°15.78'W			长(km)×宽(km) Length (km)× Width (km)	11×6
最大水深（m） Max Depth（m）	3 700	最小水深（m） Min Depth（m）	3 139	高差（m） Total Relief（m）	561
地形特征 Feature Description	此海丘位于南大西洋洋中脊，长和宽分别为 11 km 和 6 km。海丘顶部最浅处水深约 3 139 m，山麓水深约 3 700 m，高差约 561 m（图 3–16）。 Heiwu Hill is located in the South Atlantic Ridge. The length is 11 km and the width is 6 km. The minimum top depth of the hill is 3 139 m and the piedmont depth is 3 700 m, which makes the total relief being 561 m (Fig.3–16).				
命名释义 Reason for Choice of Name	"黑乌"出自《诗经·国风·北风》"莫赤匪狐，莫黑匪乌"。"黑乌"指乌鸦，意为"没有不是红色的狐狸，也没有不是黑色的乌鸦"。 "Heiwu" comes from a poem named *Beifeng* in *Shijing · Guofeng*. *Shijing* is a collection of ancient Chinese poems from 11th century B.C. to 6th century B.C. "All foxes are red and all crows are black." "Heiwu" means black crow.				

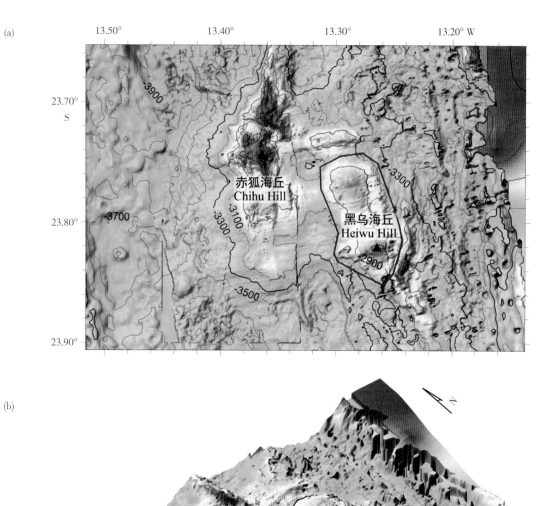

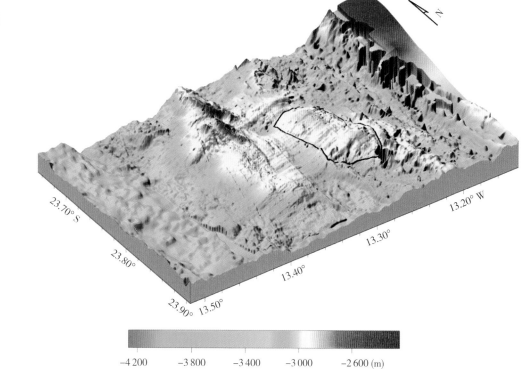

图 3-16　黑乌海丘

(a) 地形图（等深线间隔 200 m）；(b) 三维图

Fig.3-16　Heiwu Hill

(a) Bathymetric map (the contour interval is 200 m); (b) 3-D topographic map

3.2.15 方舟海山
Fangzhou Seamount

中文名称 Chinese Name	方舟海山 Fangzhou Haishan	英文名称 English Name	Fangzhou Seamount	所在大洋 Ocean or Sea	南大西洋 South Atlantic Ocean
发现情况 Discovery Facts	此海山于 2011 年 3 月由中国科考船"大洋一号"在执行 DY125-22 航次调查时发现。 The seamount was discovered by the Chinese R/V *Dayang Yihao* during the DY125-22 cruise in March, 2011.				
命名历史 Name History	该海山由我国命名为方舟海山，于 2014 年提交 SCUFN 审议通过。 Fangzhou Seamount was named by China and approved by SCUFN in 2014.				
特征点坐标 Coordinates	25°46.8′S，13°38.8′W			长 (km) × 宽 (km) Length (km) × Width (km)	24 × 10
最大水深（m） Max Depth（m）	4 400	最小水深（m） Min Depth（m）	1 600	高差（m） Total Relief（m）	2 800
地形特征 Feature Description	方舟海山位于南大西洋中脊右侧离轴部位，北部毗邻转换断层，长宽分别为 24 km 和 10 km，海山顶部最小水深 1 600 m，最大高差 2 800 m（图 3–17）。 Fangzhou Seamount is located in the right off-axis portion of the South Atlantic Ridge. The northern part adjoins the transform faults. The length is 24 km and the width is 10 km. The minimum top depth of the seamount is 1 600 m and the total relief is 2 800 m (Fig.3–17).				
命名释义 Reason for Choice of Name	"方舟"出自《诗经·国风·谷风》"就其深矣，方之舟之，就其浅矣，游之泳之"，指处理事情要像渡河一样，河水很深时，要乘坐竹筏舟船渡过，如果河水很浅，就游过去。象征处理事情时，要根据不同情况，采用不同的方式。 "Fangzhou" comes from a poem named *Gufeng* in *Shijing · Guofeng*. *Shijing* is a collection of ancient Chinese poems from 11th century B.C. to 6th century B.C. "When river water is deep and wide, I'll take a boat upon the tide. Where river water is low and slow, I'll swim or wade across the flow." This poem means that one should respond appropriately to the situation just like crossing rivers. We may cross the river by a raft or a boat where it is deep, while swimming across it where it is shallow. In other words, we may adopt different methods under different circumstances when we solve problems. The seamount is named "Fangzhou" in order to demonstrate flexible means to handling problems.				

(a)

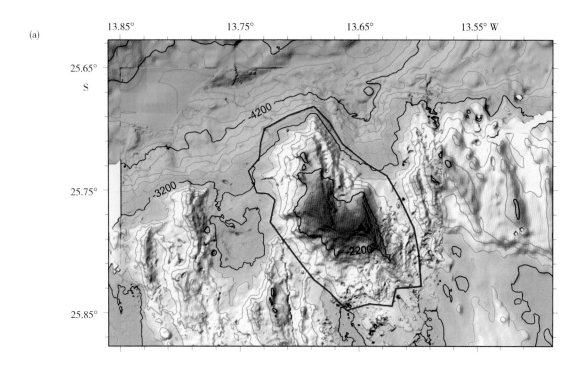

(b)

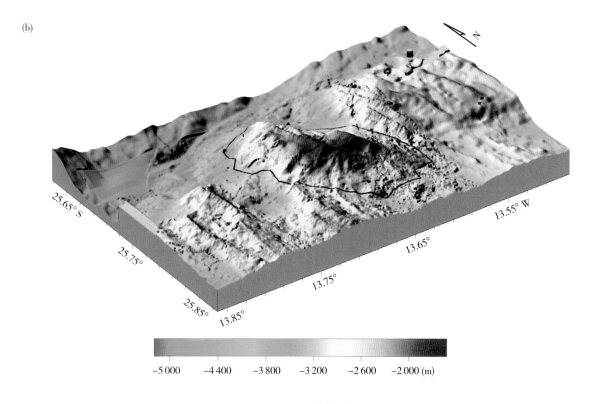

图 3-17　方舟海山

(a) 地形图（等深线间隔 200 m）；(b) 三维图

Fig.3-17　Fangzhou Seamount

(a) Bathymetric map (the contour interval is 200 m); (b) 3-D topographic map

3.2.16　洵美海丘
Xunmei Hill

中文名称 Chinese Name	洵美海丘 Xunmei Haiqiu	英文名称 English Name	Xunmei Hill	所在大洋 Ocean or Sea	南大西洋 South Atlantic Ocean
发现情况 Discovery Facts	此海丘于 2011 年由中国科考船"大洋一号"在执行 DY125-22 航次调查时发现。 This hill was discovered by the Chinese R/V *Dayang Yihao* during the DY125-22 cruise in 2011.				
命名历史 Name History	在我国大洋航次调查报告和 GEBCO 地名辞典中未命名。 This hill has not been named in Chinese cruise reports or GEBCO gazetteer.				
特征点坐标 Coordinates	26°01.20'S，13°51.64'W			长 (km)×宽 (km) Length (km)× Width (km)	10×9
最大水深（m） Max Depth（m）	2 800	最小水深（m） Min Depth（m）	2 494	高差（m） Total Relief（m）	306
地形特征 Feature Description	洵美海丘发育在洋中脊内部裂谷扩张中心处，属于该一级脊段的岩浆活动中心，为内部裂谷隆起高地。海丘形态近似圆形，长宽分别为 10 km 和 9 km，高差 306 m（图 3–18）。 Xunmei Hill is located in the expanding center of the inner rift in the mid-ocean ridge. It belongs to the magma activity center of this ridge segment and is a highland of the inner rift. This hill has a nearly round shape. The length is 10 km and the width is 9 km, which makes the total relief being 306 m (Fig.3–18).				
命名释义 Reason for Choice of Name	"洵美"出自《诗经·国风·静女》"自牧归荑，洵美且异。匪女之为美，美人之贻"。意为姑娘从郊野采来茅草芽送我作为信物，真是美好新异。并不是茅草芽有多美，而是因为美人所赠。刻画了情人互赠信物，情意缠绵的情景。洵美，美好之意。 "Xunmei" comes from a poem named *Jingnü* in *Shijing · Guofeng. Shijing* is a collection of ancient Chinese poems from 11th century B.C. to 6th century B.C. "The maiden in the fields brings me exotic grass. I love what nature yields when it comes from the lass." The man in the poem describes his rendezvous with his sweet heart. The maiden gives him a gift as a token of her tender love. "Xunmei" means truly elegant.				

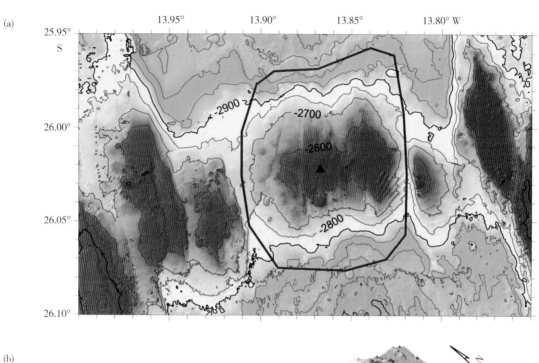

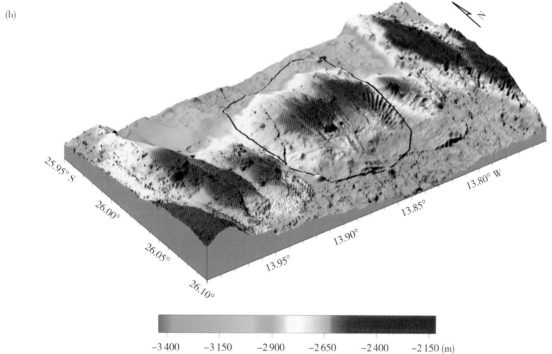

图 3-18　洵美海丘

(a) 地形图（等深线间隔 100 m）；(b) 三维图

Fig.3-18　Xunmei Hill

(a) Bathymetric map (the contour interval is 100 m); (b) 3-D topographic map

3.2.17 淘美 1 号热液区
Xunmei-1 Hydrothermal Field

中文名称 Chinese Name	淘美 1 号热液区 Xunmei-1 Reyequ	英文名称 English Name	Xunmei-1 Hydrothermal Field	所在大洋 Ocean or Sea	南大西洋 South Atlantic Ocean
发现情况 Discovery Facts	此热液区于 2011 年由中国科考船"大洋一号"在执行 DY125-22 航次调查时发现，通过海底观察及取样证实。 该热液区位于南大西洋洋中脊中央裂谷中的洼地，围岩为玄武岩，水深约 2 565 m。 This hydrothermal field was discovered by the Chinese R/V *Dayang Yihao* during the DY125-22 cruise in 2011. It was confirmed by undersea observation and samples. The Xunmei-1 Hydrothermal Field is located in the depression of the central rift valley in the South Atlantic Ridge with the water depth of about 2 565 m. Its surrounding rocks are basalt.				
命名历史 Name History	我国大洋 22 航次现场调查报告中曾命名 26°S 热液区。 It was once named "26°S Hydrothermal Field" in the Chinese 22 cruise reports.				
特征点坐标 Coordinates	26°01.27'S，13°51.13'W			水深 (m) Depth (m)	2 565
命名释义 Reason for Choice of Name	采用热液区附近的"淘美海丘"的名字命名，命名为"淘美 1 号热液区"（图 3–19）。 It was named Xunmei-1 Hydrothermal Field after the nearby Xunmei Hill（Fig.3–19）.				

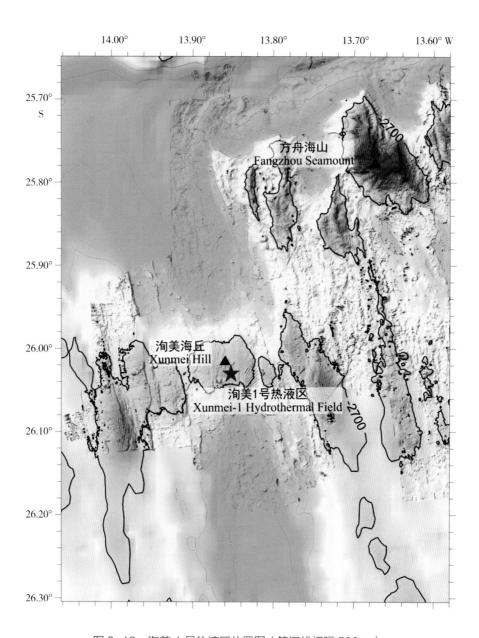

图 3-19　洵美 1 号热液区位置图（等深线间隔 500 m）

Fig.3-19　Bathymetric map of the Xunmei-1 Hydrothermal Field (The contour interval is 500 m)

3.2.18 文王海脊
Wenwang Ridge

中文名称 Chinese Name	文王海脊 Wenwang Haiji		英文名称 English Name	Wenwang Ridge	所在大洋 Ocean or Sea	南大西洋 South Atlantic Ocean
发现情况 Discovery Facts	此海脊于 2017 年由中国科考船"向阳红 01"在执行测量任务时调查发现。 This ridge was discovered by the Chinese R/V *Xiangyanghong* 01 during the survey in 2017.					
命名历史 Name History	由我国命名为文王海脊,于 2018 年提交 SCUFN 审议通过。 This feature was named Wenwang by China and the name was approved by SCUFN in 2018.					
特征点坐标 Coordinates	22°44.1′S,13°11.0′W				长 (km)×宽 (km) Length (km)× Width (km)	63×21
最大水深(m) Max Depth(m)	4 730	最小水深(m) Min Depth(m)		1 970	高差(m) Total Relief(m)	2 760
地形特征 Feature Description	文王海脊位于凯风海山北侧,两者由一条转换断层分割,发育在裂谷西侧离轴区域。 Wenwang Ridge is located in the north of Kaifeng Seamount. The two seamounts are cut by a transform fault. It develops at off-axis area west of the valley.					
命名释义 Reason for Choice of Name	文王姬昌,周朝的创建者,中国历史上著名的贤君,在天文、航行等方面亦有杰出贡献(图 3–20)。 Jichang, called Wenwang, was the founder of Zhou Dynasty. He was a famous sage in China history, who also made outstanding contributions in Astronomy and Navigation (Fig.3–20).					

(a)

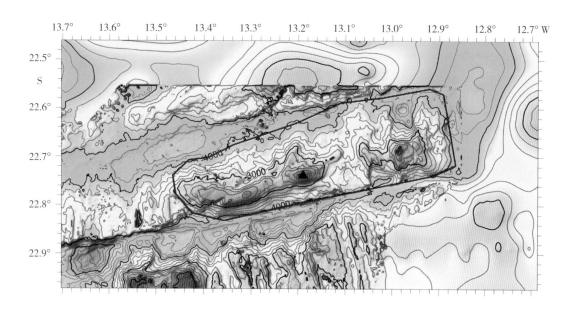

(b)

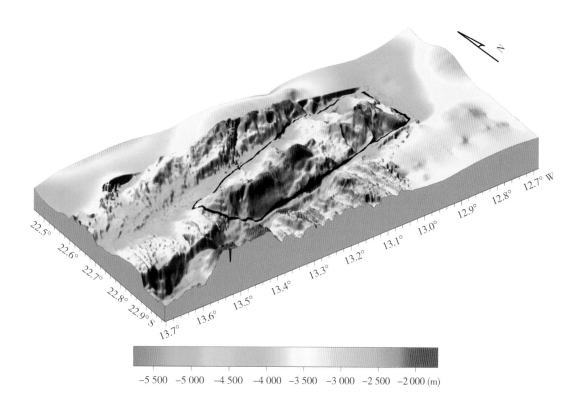

图 3-20　文王海脊

(a) 地形图（等深线间隔 200 m）；(b) 三维图

Fig.3-20　Wenwang Seamount

(a) Bathymetric map (the contour interval is 200 m); (b) 3-D topographic map

第4章
印度洋地理实体

Chapter 4
Undersea Features in the
Indian Ocean

第1节 命名概况

印度洋是地球上第三大洋，位于亚洲、南极洲、大洋洲和非洲之间，东南部和西南部分别与太平洋和大西洋相接，南部与南大洋相连。印度洋海底地貌同样分为大陆架、大陆坡、大陆隆、大洋中脊和大洋盆地5大类基本地貌单元，其中"入"字形展布的洋中脊是海底最显著的地貌单元。

印度洋大陆架的平均宽度比大西洋狭窄，大陆架水深一般不超过200 m。大陆坡地形较太平洋和大西洋平缓，规模也较小，宽度一般只有20～50 km。大陆边缘地貌突出的特点是大陆隆或海台较多而且分布广。另外，海沟和水下冲积锥在印度洋海底地貌也较为突出，其中在孟加拉湾有一巨大的恒河水下冲积锥，其面积达200万平方千米，在此水下冲积锥上部直到河口，还形成宽达13～36 km的水下溺谷，溺谷在水下冲积锥上再分散成树状扇形谷，延伸上千千米。海沟主要发育在安达曼群岛－努沙登加拉群岛等系列岛链之南，这一岛弧－海沟系具有平行的双重岛弧，两列岛弧之间为水深3 000～4 000 m的海沟，外侧为爪哇海沟。爪哇海沟宽50～100 km，长4 500 km，海沟最深点7 450 m。

印度洋"入"字形的大洋中脊系统由4条洋中脊组成，自北向南分别为卡尔斯伯格脊、中印度洋脊、西南印度洋脊以及东南印度洋脊。卡尔斯伯格脊是印度洋中脊系统的西北分支，走向北西—南东向，西起欧文断裂带，东至66°E附近与中印度洋脊相接，长约1 500 km，宽约800 km，水深1 800～3 600 m，平均相对高差2 100 m。中印度洋脊是印度洋中脊的北部分支，大致从阿姆斯特丹岛向北延伸，平均宽740～930 km左右，一般高出两侧海盆1 300～2 500 m，个别脊峰露出海面而形成岛屿，例如罗德里格斯岛、阿姆斯特丹岛和圣保罗岛等。西南印度洋脊在罗德里格斯岛附近与中印度洋脊相接，向西南延伸，在15°W附近与大西洋中脊相连，相对高差1 500～3 000 m。东南印度洋脊是印度洋中脊的东南分支，相对高差3 000 m左右，中脊区宽约500 km。东南印度洋海岭发育的转换断层数量较少，中脊外侧两翼地形以阶梯形状缓慢下降向深海盆地过渡。

印度洋中脊将印度洋分成为东、西、南三大海域。东部海域被近南北向延伸的90°E海岭分隔为中印度洋海盆和西澳大利亚海盆。中印度洋海盆南北纵贯，北部为恒河水下冲积锥所掩覆的锡兰深海平原，几乎全部为印度河水下冲积锥所填积。西澳大利亚海盆北部直接与爪哇海沟相接，其间分布一系列东北—西南向的海丘，东南部被一些海岭、海丘和海台分割，海底地貌错综复杂。印度洋西部海域由于马达加斯加岛的中间突起，把大洋海盆分成索马里海盆和马达加斯加海盆。在马达加斯加岛与马斯克林海台之间，有许多火山形成的海山和在火山基座上发育成的珊瑚岛，这些海山和珊瑚岛之间，发育深海平原和断层性的海沟。印度洋南部海域被凯尔盖朗海底高原分隔为3个海盆，即为克罗泽海盆、大西洋－印度洋海盆以及澳大利亚－南极海盆（图4-1）。

Section 1　Overview of the Naming

The Indian Ocean is the world's third largest ocean. It is located between Asia, Antarctica, Oceania and Africa. It connects the Pacific Ocean and Atlantic Ocean in the southeast and southwest respectively, and the Southern Ocean in the south. The seabed geomorphology of the Indian Ocean can also be divided into 5 categories of fundamental geomorphic units, which are continental shelf, continental slope, continental rise, mid-ocean ridge and ocean basin. Among them, the λ-shaped distributed mid-ocean ridges are the most significant undersea features.

The average width of the Indian Ocean Continental Shelf is narrower than the Atlantic's, with the water depth of less than 200 m generally. The topography of continental slope is smoother than that of the Pacific Ocean and the Atlantic Ocean with smaller scale, and the width usually being only 20–50 km. The prominent features of marginal continent geomorphology are a lot of widely distributed continental rises or plateaus. In addition, trenches and underwater alluvial cones are also prominent among the Indian Ocean seabed geomorphology. Among them, there is a huge Ganges underwater alluvial cone in the Bay of Bengal with an area of 2 million square kilometers. The upper part of this underwater alluvial cone reaches as far as the estuary and also forms the underwater liman with width of 13–36 km. The liman then disperses into a tree-like fan valley on the underwater alluvial cone, extending thousands of kilometers. Trenches mainly develop in the south of Andaman Islands-Nusa Tenggara Islands and other island chains. This island arc-trench system has a parallel double arc, between which there is a trench with the water depth of 3 000–4 000 m. Java Trench is on the outside. The Java Trench has the width of 50–100 km, length of 4 500 km with the deepest point being 7 450 m.

The λ-shaped Indian Ridge system consists of 4 ridges, which are Carlsberg Ridge, Central Indian Ridge, Southwest Indian Ridge, and Southeast Indian Ridge. The Carlsberg Ridge is the northwestern branch of the Indian Ridge system, running NW to SE, from Owen Fracture Zone in the west to nearly 66° E in the east. The length is about 1 500 km and the width is about 800 km. The water depth is 1 800–3 600 m, The average relative relief is about 2 100 m. The Central Indian Ridge is the northern branch of Indian Ridge system, generally extending from Amsterdam Island to the north direction, with the average width being around 740–930 km, generally 1 300–2 500 m higher than the sea basins on the both sides. A few ridge summits reach out of the sea surface, forming islands, such as Rodrigues Island, Amsterdam Island and St. Paul Island. The Southwest Indian Ridge connects the Central Indian Ridge near Rodrigues Island, running to SW, and connecting with the Mid-Atlantic Ridge near 15° W, relative total relief being 1 500–3 000 m. The Southeast Indian Ridge is the southeastern branch of Indian Ridge system, relative total relief being around 3 000 m, width of the ridge area being about 500 km. There are rare transform faults on the Southeast Indian Ridge and the ridge is in transition to deep-sea basin on the both side in stair-step shaped.

The Indian Ridge divides the Indian Ocean into the eastern, western and southern three large part of the ocean. The eastern sea is split into the Central Indian Basin and the West Australia Basin by the roughly N to S running 90° E ridge. The Central Indian Basin runs across from north to south. The northern part is the Ceylon abyssal plain covered by the Ganges underwater alluvial cone, almost all filled by the Indus River underwater alluvial cone. The West Australia Basin connects the Java Trench directly in the north with many NE to SW running hills distributing in it. The southern part's seabed geomorphology is really complicated because it was cut by some ridges, hills and plateaus. The western sea area of Indian Ocean divides the ocean basin into Somali Basin and Madagascar Basin because of the central protuberance of Madagascar Island. Between Madagascar Island and Mascarene Plateau, there are many volcanic seamounts and coral islands developed on the bases of volcanoes. There develops abyssal plains and fault trenches between these seamounts and coral islands. The southern sea area of Indian Ocean is split into 3 sea basins by undersea Kerguelen Plateau, which are Crozet Basin, Atlantic-Indian Basin and Australian-Antarctic Basin (Fig. 4–1).

图 4-1　印度洋海底地理实体命名区域示意图

Fig.4-1　The Indian Ocean undersea features naming regions

第 2 节　西北印度洋洋中脊地理实体和热液区

西北印度洋洋中脊包括卡尔斯伯格脊和希巴洋脊。其中，卡尔斯伯格脊位于印度洋洋中脊西北段，是印度洋洋中脊体系最年轻的一支，近 NW—SE 向走向，西起欧文断裂带，与红海 – 亚丁湾相连，东至 66° E 附近，与中印度洋脊相接；长约 1500 km，宽约 800 km，水深约 1800 ～ 4500 m，是非洲板块与印度 – 澳大利亚板块之间的离散型板块边界，分隔阿拉伯海盆和东索马里海盆。

卡尔斯伯格脊全扩张速率为 24.6 ～ 34.4 mm/a（Dyment, 1998），属于慢速扩张洋脊，扩张速率表现为西北段较东南段慢；脊轴地形较为崎岖，发育较宽的深大裂谷（宽 13 ～ 26 km，深 1.5 ～ 3.0 km）。卡尔斯伯格脊发育 5 条大型断裂带（李四光断裂带、竺可桢断裂带、徐霞客断裂带、宝船断裂带及郦道元断裂带），将洋中脊分为 5 个一级洋脊段，并以众多非转换不连续带划分出 15 个二级洋脊段。洋中脊北段岩浆活动弱，发育大规模拆离断层，并形成多处大洋核杂岩，并出露地幔类岩石（Han et al., 2012）。该洋中脊发育始于 58 Ma 左右，是非洲板块与印度板块差异运动、留尼旺热点活动及中印度洋脊延伸作用的共同结果，并随板块运动至现今位置。

在西北印度洋区域，共命名地理实体 16 个，热液区 3 个。其中海山 6 个，包括玉磬海山、排箫海山、庸鼓海山、烈祖海山、温恭海山和客怡海山；海丘 2 个，大禧海丘和龙灯海丘；断裂带 4 个，包括李四光断裂带、竺可桢断裂带、徐霞客断裂带和郦道元断裂带；海脊 3 个，包括万舞海脊、卧蚕海脊和玄鸟海脊；海山群 1 个，天龙海山群；热液区 3 个，包括卧蚕 1 号热液区、卧蚕 2 号热液区和天休 1 号热液区（图 4-2）。

Section 2 Undersea Features and Hydrothermal Fields in the Mid-ocean Ridges in the Northwest Indian Ocean

The mid-ocean ridges in Northwest Indian Ocean include the Carlsberg Ridge and the Sheba Ridge. The Carlsberg Ridge is located in the northwestern part of the Indian Ridge and is the youngest one in the Indian Ridge system. It runs nearly NW to SE and starts from Owen Fracture Zone in the west, connecting the Red Sea and Gulf of Aden, to near 66° E, connecting the Central Indian Ridge in the east. It has the length of about 1 500 km, width of about 800 km, and water depth of about 1 800– 4 500 m. It is the discrete plate boundary between the Australian Plate and the African Plate, separating Arabian Basin and East Somalia Basin.

The full spreading rate of the Carlsberg Ridge is 24.6–34.4 mm/a, belonging to the slow-spreading ridges. The northwestern part spreads more slowly than southwestern part; the topography of ridge axis is rugged, with wide deep gaps (width of 13–26 km, depth of 1.5–3.0 km). The Carlsberg Ridge develops 5 large fracture zones (Lisiguang Fracture Zone, Chukochen Fracture Zone, Xuxiake Fracture Zone, Baochuan Fracture Zone and Lidaoyuan Fracture Zone), dividing the ridge into 5 first order segments, 14 secondary order segments according to a number of non-transform discontinuities. The northern part of the ridge has weak magma activity, developing large-scale detachment faults and forming several oceanic core complexes and exposing mantle rocks (Han et al., 2012). This ridge development began around 58 Ma and it is the result of difference movements of the African Plate and the Indian Plate, the Reunion hotspot activities and the Central Indian Ridge expansion, and reaches the present position with the moving of the plates.

In total, 16 undersea features and 3 hydrothermal fields have been named by China in the region of the Northwest Indian Ocean. Among them, there are 6 seamounts, including Yuqing Seamount, Paixiao Seamount, Yonggu Seamount, Liezu Seamount, Wengong Seamount and Keyi Seamount; 2 hill, including Daxi Hill and Longdeng Hill; 4 fracture zones, including Lisiguang Fracture Zone, Chukochen Fracture Zone, Xuxiake Fracture Zone and Lidaoyuan Fracture Zone; 3 ridges, including Wocan Ridge, Xuanniao Ridge and Wanwu ridge; 1 seamounts, which is Tianlong Seamounts; 3 hydrothermal fields, including Wocan-1 Hydrothermal Field, Wocan-2 Hydrothermal Field and Tianxiu-1 Hydrothermal Field (Fig. 4–2).

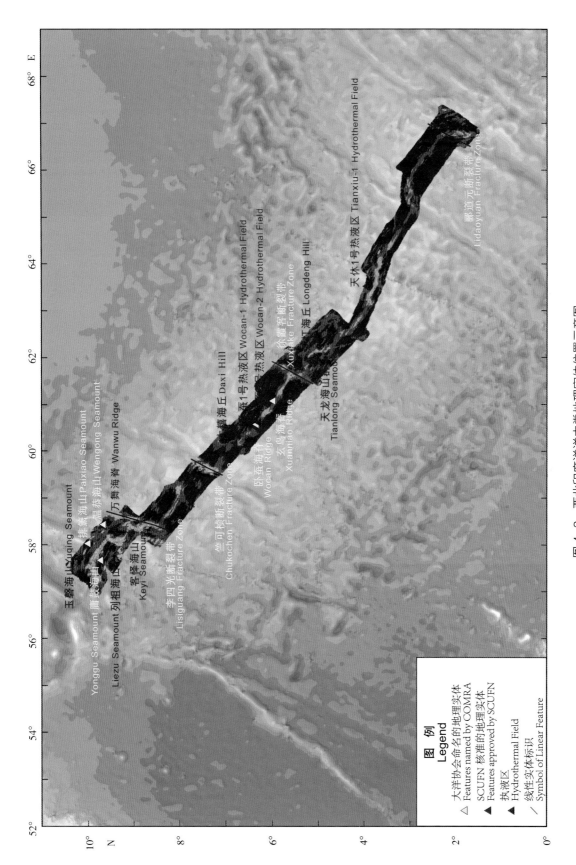

图 4-2　西北印度洋洋中脊地理实体位置示意图

Fig.4-2　Locations of the undersea features in the Northwest Indian Ridge

4.2.1　玉磬海山
Yuqing Seamount

中文名称 Chinese Name	玉磬海山 Yuqing Haishan	英文名称 English Name	Yuqing Seamount	所在大洋 Ocean or Sea	西北印度洋 Northwest Indian Ocean
发现情况 Discovery Facts	此海山于 2012 年 5 月在由中国科考船"李四光号"在执行 DY125-24 航次时调查发现。 This seamount was discovered by the Chinese R/V *Lisiguanghao* during the DY125-24 cruise in May, 2012.				
命名历史 Name History	由我国命名为玉磬海山，于 2016 年提交 SCUFN 审议通过。 Yuqing Seamount was named by China and approved by SCUFN in 2016.				
特征点坐标 Coordinates	10°11.55′N，57°42.40′E			长 (km) × 宽 (km) Length (km) × Width (km)	22 × 10
最大水深（m） Max Depth（m）	3 200	最小水深（m） Min Depth（m）	1 100	高差（m） Total Relief（m）	2 100
地形特征 Feature Description	该海山属断块山，呈长条状，长宽分别为 22 km 和 10 km。海山顶部水深 1 100 m，山麓水深 3 200 m，最大高差 2 100 m（图 4–3）。 Yuqing Seamount is a kind of block mountain which has an elongated shape. The length is 22 km and the width is 10 km. The top depth of the seamount is 1 100 m and the piedmont depth is 3 200 m, which makes the total relief being 2 100 m (Fig.4–3).				
命名释义 Reason for Choice of Name	"玉磬"出自《诗经·商颂·那》"既和且平，依我磬声"。《那》是《商颂》的第一篇，是殷商后代祭祀先祖的颂歌，主要表现的是祭祀祖先时的音乐舞蹈活动，以乐舞的盛大来表示对先祖的尊崇，以此求取祖先之神的庇护佑助。文中涉及了四种乐器，鞉鼓、管、磬、镛。其中磬为中国古代玉石制作的乐器。 "Yuqing" comes from a poem named *Nuo* in *Shijing · Shangsong*. *Shijing* is a collection of ancient Chinese poems from 11th century B.C. to 6th century B.C. "The music sounds harmonious and fine to the rhythm of the sounding chime." *Nuo* is the first poem in *Shangsong* and it is a poem by the descendants of Shang to pay tribute to their ancestors. The music consists of four kinds of instruments that is Gu (drums), Guan (flute), Qing (an ancient instrument of China made of jade) and Yong (a large bell).				

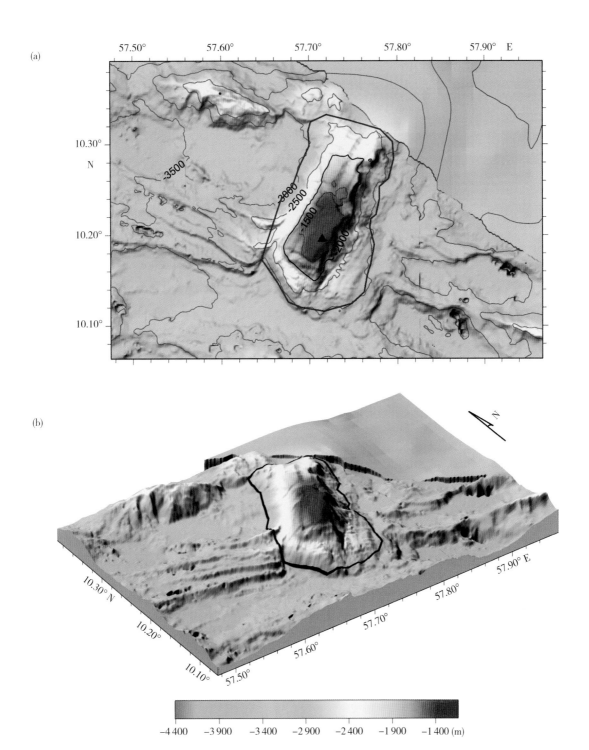

图 4-3　玉磬海山

(a) 地形图（等深线间隔 500 m）；(b) 三维图

Fig.4-3　Yuqing Seamount

(a) Bathymetric map (the contour interval is 500 m); (b) 3-D topographic map

4.2.2 排箫海山
Paixiao Seamount

中文名称 Chinese Name	排箫海山 Paixiao Haishan	英文名称 English Name	Paixiao Seamount	所在大洋 Ocean or Sea	西北印度洋 Northwest Indian Ocean
发现情况 Discovery Facts	此海山于 2012 年 5 月由中国科考船"李四光号"在执行 DY125-24 航次时调查发现。 This seamount was discovered by the Chinese R/V *Lisiguanghao* during the DY125-24 cruise in May, 2012.				
命名历史 Name History	在我国大洋航次调查报告和 GEBCO 地名辞典中未命名。 This seamount has not been named in Chinese cruise reports or GEBCO gazetteer.				
特征点坐标 Coordinates	10°00.38′N，58°00.20′E		长(km)×宽(km) Length (km)× Width (km)		21×18
最大水深（m） Max Depth（m）	3 400	最小水深（m） Min Depth（m）	1 650	高差（m） Total Relief（m）	1 750
地形特征 Feature Description	该海山属断块山，呈龟背状，长宽分别为 21 km 和 18 km。海山顶部水深 1 650 m，山麓水深 3 400 m，最大高差 1 750 m（图 4-4）。 Paixiao Seamount is a kind of block mountain which has a shape like the back of a turtle. The length is 21 km and the width is 18 km. The top depth of the seamount is 1 650 m and the piedmont depth is 3 400 m, which makes the total relief being 1 750 m (Fig.4-4).				
命名释义 Reason for Choice of Name	"排箫"出自《诗经·商颂·那》中"鞉鼓渊渊，嘒嘒管声"。《那》是《商颂》的第一篇，是殷商后代祭祀先祖的颂歌，主要表现的是祭祀祖先时的音乐舞蹈活动，以乐舞的盛大来表示对先祖的尊崇，以此求取祖先之神的庇护佑助。文中涉及了四种乐器，鞉鼓、管、磬、镛。此句意思是拨浪鼓儿响咚咚，箫管声声多清亮。管为箫管，是中国古代一种乐器。根据海山的形状，取名为"排箫"。 "Paixiao" comes from a poem named *Nuo* in *Shijing · Shangsong*. *Shijing* is a collection of ancient Chinese poems from 11th century B.C. to 6th century B.C. "The drums and tambourines resound with shrill of the flute." *Nuo* is the first poem in *Shangsong* and it is a poem by the descendants of Shang to pay tribute to their ancestors. The music consists of four kind of instruments that is Gu (drums), Guan (flute), Qing (an ancient instrument of China made of jade) and Yong (A large bell). "Paixiao" are a group of flutes. The seamount is named "Paixiao" since it look like a group of flutes.				

(a)

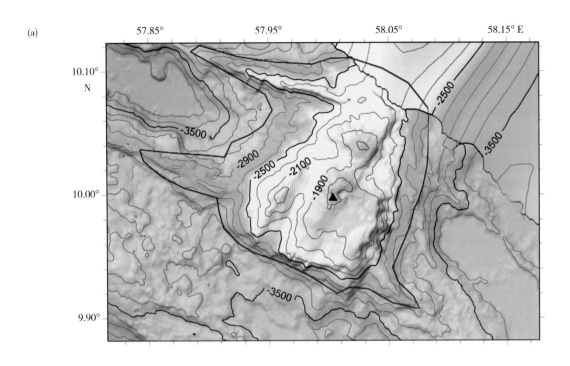

(b)

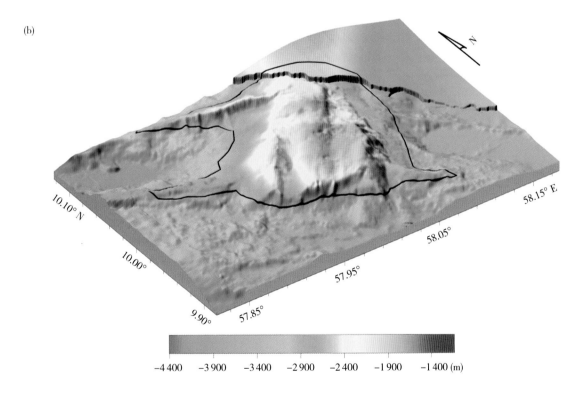

图 4-4　排箫海山

(a) 地形图 （等深线间隔 200 m）；(b) 三维图

Fig.4-4　Paixiao Seamount

(a) Bathymetric map (the contour interval is 200 m); (b) 3-D topographic map

4.2.3 庸鼓海山
Yonggu Seamount

中文名称 Chinese Name	庸鼓海山 Yonggu Haishan	英文名称 English Name	Yonggu Seamount	所在大洋 Ocean or Sea	西北印度洋 Northwest Indian Ocean
发现情况 Discovery Facts	此海山于 2012 年 5 月由中国科考船"李四光号"在执行 DY125-24 航次时调查发现。 This seamount was discovered by the Chinese R/V *Lisiguanghao* during the DY125-24 cruise in May, 2012.				
命名历史 Name History	在我国大洋航次调查报告和 GEBCO 地名辞典中未命名。 This seamount has not been named in Chinese cruise reports or GEBCO gazetteer.				
特征点坐标 Coordinates	9°43.52′N，57°38.46′E			长(km) × 宽(km) Length (km) × Width (km)	34 × 28
最大水深（m） Max Depth（m）	3 100	最小水深（m） Min Depth（m）	1 400	高差（m） Total Relief（m）	1 700
地形特征 Feature Description	该海山包括 2 座山峰，长宽分别为 34 km 和 28 km。海山顶部水深 1 400 m，山麓水深 3 100 m，最大高差 1 700 m（图 4–5）。 Yonggu Seamount contains two summits. The length is 34 km and the width is 28 km. The top depth of the seamount is 1 400 m and the piedmont depth is 3 100 m, which makes the total relief is 1 700 m (Fig.4–5).				
命名释义 Reason for Choice of Name	"庸鼓"出自《诗经·商颂·那》"庸鼓有斁，万舞有奕"。《那》是《商颂》的第一篇，是殷商后代祭祀先祖的颂歌，主要表现的是祭祀祖先时的音乐舞蹈活动，以乐舞的盛大来表示对先祖的尊崇，以此求取祖先之神的庇护佑助。文中涉及了四种乐器，鞉鼓、管、磬、镛。庸指钟，意为敲钟击鼓，场面宏大。 "Yonggu" comes from a poem named *Nuo* in *Shijing·Shangsong*. *Shijing* is a collection of ancient Chinese poems from 11th century B.C. to 6th century B.C. "The sounds of bells and drums rise high; The grand performance is on display." *Nuo* is the first poem in *Shangsong* and it is a poem by the descendants of Shang to pay tribute to their ancestors. The music consists of four kind of instruments that is Gu (drums), Guan (flute), Qing (an ancient instrument of China made of jade) and Yong (A large bell). "Yonggu" means bells and drums.				

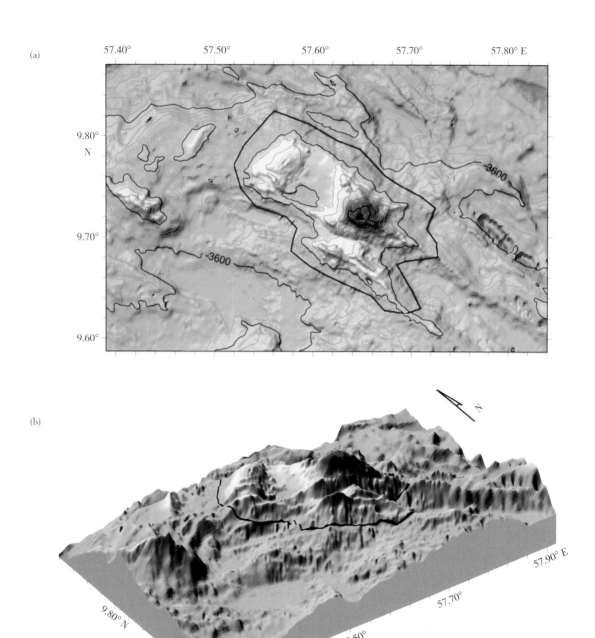

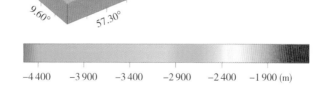

图 4-5　庸鼓海山

(a) 地形图（等深线间隔 200 m）；(b) 三维图

Fig.4-5　Yonggu Seamount

(a) Bathymetric map (the contour interval is 200 m); (b) 3-D topographic map

4.2.4 烈祖海山
Liezu Seamount

中文名称 Chinese Name	烈祖海山 Liezu Haishan	英文名称 English Name	Liezu Seamount	所在大洋 Ocean or Sea	西北印度洋 Northwest Indian Ocean
发现情况 Discovery Facts	此海山于 2012 年 5 月由中国科考船"李四光号"在执行 DY125-24 航次时调查发现。 This seamount was discovered by the Chinese R/V *Lisiguanghao* during the DY125-24 cruise in May, 2012.				
命名历史 Name History	由我国命名为烈祖海山，于 2016 年提交 SCUFN 审议通过。 Liezu Seamount was named by China and approved by SCUFN in 2016.				
特征点坐标 Coordinates	9°29.97′N，57°35.79′E			长(km)×宽(km) Length (km)× Width (km)	21 × 12
最大水深（m） Max Depth（m）	2 900	最小水深（m） Min Depth（m）	1 180	高差（m） Total Relief（m）	1 720
地形特征 Feature Description	该海山属断块海山，呈长条状，其东侧为断崖，长宽分别为 21 km 和 12 km。海山顶部水深 1 180 m，山麓水深 2 900 m，最大高差 1 720 m（图 4–6）。 Liezu Seamount is a kind of block mountain which has an elongated shape. The eastern side is the cliff. The length is 21 km and the width is 12 km. The top depth of the seamount is 1 180 m and the piedmont depth is 2 900 m, which makes the total relief being 1 720 m (Fig.4–6).				
命名释义 Reason for Choice of Name	"烈祖"出自《诗经·商颂·那》"奏鼓简简，衎我烈祖"。《那》是《商颂》的第一篇，主要表现的是祭祀祖先时的音乐舞蹈活动，以乐舞的盛大来表示对先祖的尊崇，以此求取祖先之神的庇护佑助。烈祖指中国古代的功烈之祖。该句意为鼓儿敲起咚咚响，使先祖心情欢畅。 "Liezu" comes from a poem named *Nuo* in *Shijing · Shangsong*. *Shijing* is a collection of ancient Chinese poems from 11th century B.C. to 6th century B.C. "The drums resound so eloquent; To please our former Kings and queens." *Nuo* is the first poem in *Shangsong* and it is a poem by the descendants of Shang to pay tribute to their ancestors. "Liezu" means the meritorious ancestors of China.				

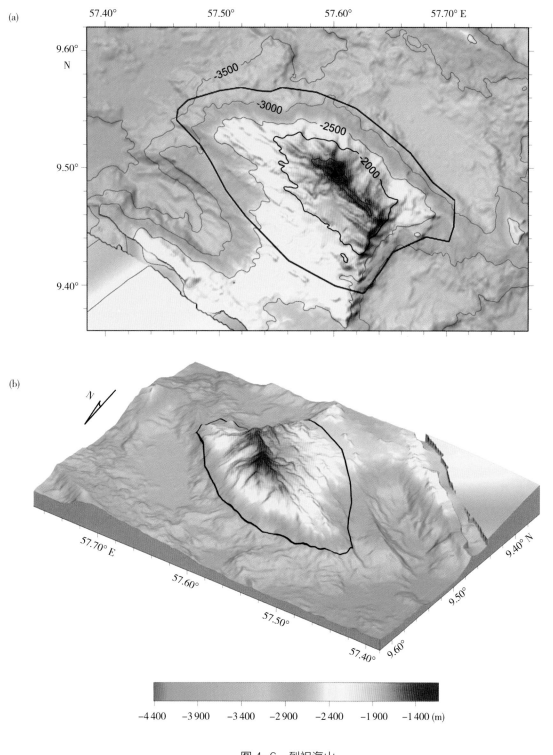

图 4-6 烈祖海山

(a) 地形图（等深线间隔 500 m）；(b) 三维图

Fig.4-6 Liezu Seamount

(a) Bathymetric map (the contour interval is 500 m); (b) 3-D topographic map

4.2.5 温恭海山
Wengong Seamount

中文名称 Chinese Name	温恭海山 Wengong Haishan	英文名称 English Name	Wengong Seamount	所在大洋 Ocean or Sea	西北印度洋 Northwest Indian Ocean
发现情况 Discovery Facts	此海山于 2012 年 5 月由中国测量船"李四光号"在执行 DY125-24 航次时调查发现。 This seamount was discovered by the Chinese R/V *Lisiguanghao* during the DY125-24 cruise in May, 2012.				
命名历史 Name History	在我国大洋航次调查报告和 GEBCO 地名辞典中未命名。 This seamount has not been named in Chinese cruise reports or GEBCO gazetteer.				
特征点坐标 Coordinates	9°39.30′N，58°24.12′E			长 (km) × 宽 (km) Length (km) × Width (km)	28 × 17
最大水深（m） Max Depth（m）	3 400	最小水深（m） Min Depth（m）	2 080	高差（m） Total Relief（m）	1 320
地形特征 Feature Description	该海山呈不规则状，地形相对平缓，由 2 座山峰组成，其长宽分别为 28 km 和 17 km。海山顶部水深 2 080 m，山麓水深 3 400 m，最大高差 1 320 m（图 4–7）。 Wengong Seamount has an irregular shape with relatively smooth topography. It consists of two summits. The length of Wengong Seamount is 28 km and the width is 17 km. The top depth is 2 080 m and the piedmont depth is 3 400 m, which makes the total relief being 1 320 m(Fig.4–7).				
命名释义 Reason for Choice of Name	"温恭"出自《诗经·商颂·那》中的诗句"温恭朝夕，执事有恪"。《那》是《商颂》的第一篇，是殷商后代祭祀先祖的颂歌，主要表现的是祭祀祖先时的音乐舞蹈活动，以乐舞的盛大来表示对先祖的尊崇，以此求取祖先之神的庇护佑助。该句意为温和恭敬、小心谨慎做事情。 "Wengong" comes from a poem named *Nuo* in *Shijing · Shangsong*. *Shijing* is a collection of ancient Chinese poems from 11th century B.C. to 6th century B.C. "Mild and pious day and night, they held the service time and again." *Nuo* is the first poem in *Shangsong* and it is a poem by the descendants of Shang to pay tribute to their ancestors. "Wengong" means being respectful and modest.				

(a)

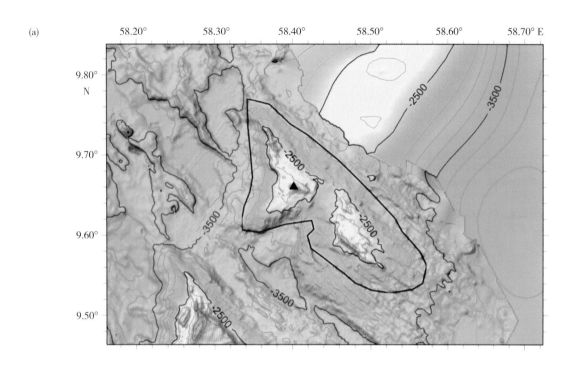

(b)

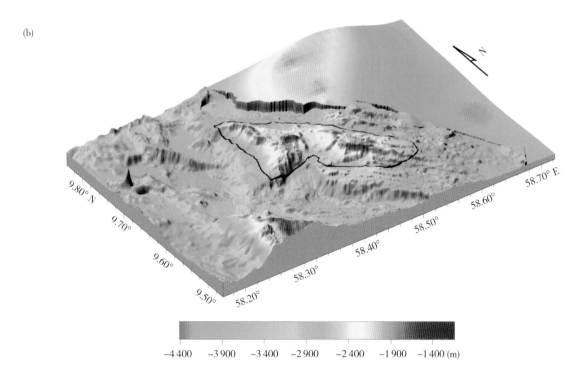

−4 400　−3 900　−3 400　−2 900　−2 400　−1 900　−1 400 (m)

图 4-7　温恭海山

(a) 地形图（等深线间隔 200 m）；(b) 三维图

Fig.4-7　Wengong Seamount

(a) Bathymetric map (the contour interval is 200 m); (b) 3-D topographic map

4.2.6　万舞海山
Wanwu Seamount

中文名称 Chinese Name	万舞海山 Wanwu Haishan		英文名称 English Name	Wanwu Seamount	所在大洋 Ocean or Sea	西北印度洋 Northwest Indian Ocean
发现情况 Discovery Facts	此海山于 2012 年 5 月由中国测量船"李四光号"在执行 DY125-24 航次时调查发现。 This Seamount was discovered by the Chinese R/V *Lisiguanghao* during the DY125-24 cruise in May, 2012.					
命名历史 Name History	由我国命名为万舞海山，于 2016 年提交 SCUFN 审议通过。 Wanwu Seamount was named by China and approved by SCUFN in 2016.					
特征点坐标 Coordinates	9°27.18′N，58°19.44′E				长(km)×宽(km) Length (km) × Width (km)	22 × 13
最大水深（m） Max Depth（m）	3 500	最小水深（m） Min Depth（m）	1 300	高差（m） Total Relief（m）	2 200	
地形特征 Feature Description	海山形态狭长，走向北西，长宽分别为 22 km 和 13 km。海山顶部水深 1 300 m，山麓水深 3 500 m，最大高差 2 200 m（图 4-8）。 This seamount has an elongated shape and runs NW to SE. The length of Wanwu Seamount is 22 km and the width is 13 km. The top depth is 1 300 m and the piedmont depth is 3 500 m, which makes the total relief being 2 200 m (Fig.4-8).					
命名释义 Reason for Choice of Name	"万舞"出自《诗经·商颂·那》"庸鼓有斁，万舞有奕"，意为众人跳舞。《那》是《商颂》的第一篇，是殷商后代祭祀先祖的颂歌，主要表现的是祭祀祖先时的音乐舞蹈活动，以乐舞的盛大来表示对先祖的尊崇，以此求取祖先之神的庇护佑助。该句意为钟鼓齐鸣，场面盛大。 "Wanwu" comes from a poem named *Nuo* in *Shijing · Shangsong*. *Shijing* is a collection of ancient Chinese poems from 11th century B.C. to 6th century B.C. "The sounds of bells and drums rise high; The grand performance is on display." *Nuo* is the first poem in *Shangsong* and it is a poem by the descendants of Shang to pay tribute to their ancestors. "Wanwu" means grand performance with bells and drums rising together.					

(a)

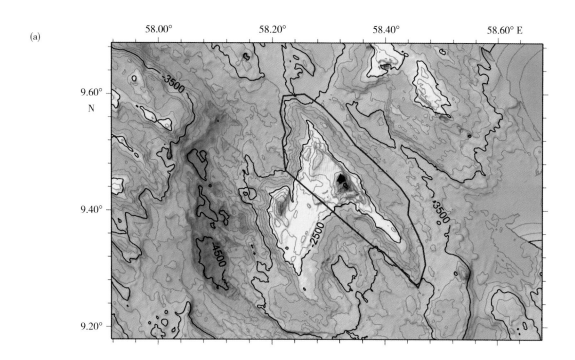

(b)

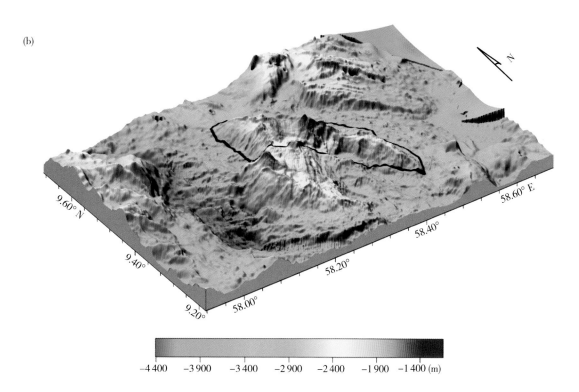

图 4-8　万舞海山

(a) 地形图（等深线间隔 200 m）；(b) 三维图

Fig.4-8　Wanwu Seamount

(a) Bathymetric map (the contour interval is 200 m); (b) 3-D topographic map

4.2.7　客怿海山
Keyi Seamount

中文名称 Chinese Name	客怿海山 Keyi Haishan	英文名称 English Name	Keyi Seamount	所在大洋 Ocean or Sea	西北印度洋 Northwest Indian Ocean
发现情况 Discovery Facts	此海山于 2012 年 5 月由中国测量船"李四光号"在执行 DY125-24 航次时调查发现。 This seamount was discovered by the Chinese R/V *Lisiguanghao* during the DY125-24 cruise in May, 2012.				
命名历史 Name History	由我国命名为客怿海山,于 2013 年提交 SCUFN 审议通过。 Keyi Seamount was named by China and approved by SCUFN in 2013.				
特征点坐标 Coordinates	9°03.70′N,58°13.10′E			长(km)×宽(km) Length (km) × Width (km)	24×19
最大水深(m) Max Depth(m)	3 800	最小水深(m) Min Depth(m)	1 300	高差(m) Total Relief(m)	2 500
地形特征 Feature Description	该海山属断块山,形状不规则,长宽分别为 24 km 和 19 km。海山顶部水深 1 300 m,山麓水深 3 800 m,最大高差 2 500 m(图 4–9)。 Keyi Seamount is a kind of block mountain which has an irregular shape. The length is 24 km and the width is 19 km. The top depth is 1 300 m and the piedmont depth is 3 800 m, which makes the total relief being 2 500 m (Fig.4–9).				
命名释义 Reason for Choice of Name	"客怿"出自《诗经·商颂·那》中的诗句"我有嘉客,亦不夷怿"。《那》是《商颂》的第一篇,是殷商后代祭祀先祖的颂歌,主要表现的是祭祀先祖时的音乐舞蹈活动,以乐舞的盛大来表示对先祖的尊崇,以此求取祖先之神的庇护佑助。意为有宾客到来,大家都很快乐,反映了中华民族好客的优良传统。 "Keyi" comes from a poem named *Nuo* in *Shijing · Shangsong*. *Shijing* is a collection of ancient Chinese poems from 11th century B.C. to 6th century B.C. "Here my worthy guests stand by, all of them happy and gay." *Nuo* is the first poem in *Shangsong* and it is a poem by the descendants of Shang to pay tribute to their ancestors. "Keyi" means that everyone is very happy when the guests come. This reflects hospitable tradition of Chinese.				

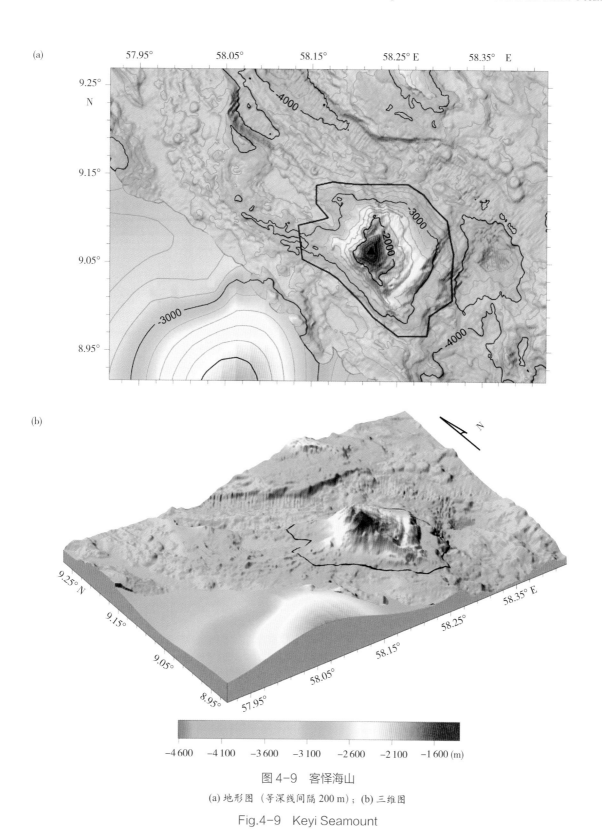

图 4-9 客怿海山

(a) 地形图（等深线间隔 200 m）；(b) 三维图

Fig.4-9 Keyi Seamount

(a) Bathymetric map (the contour interval is 200 m); (b) 3-D topographic map

4.2.8 李四光断裂带
Lisiguang Fracture Zone

中文名称 Chinese Name	李四光断裂带 Lisiguang Duanliedai	英文名称 English Name	Lisiguang Fracture Zone	所在大洋 Ocean or Sea	西北印度洋 Northwest Indian Ocean
发现情况 Discovery Facts	李四光断裂带于2012年5月由中国测量船"李四光号"在执行DY125-24航次时调查发现。 This fracture zone was discovered by the Chinese R/V *Lisiguanghao* during the DY125-24 cruise in May, 2012.				
命名历史 Name History	在我国大洋航次调查报告和GEBCO地名辞典中未命名。 This fracture zone has not been named in Chinese cruise reports or GEBCO gazetteer.				
特征点坐标 Coordinates	8°38.69′N，58°25.79′E			长(km) × 宽(km) Length (km) × Width (km)	170 × 20
最大水深（m） Max Depth（m）	4 400	最小水深（m） Min Depth（m）	3 300	高差（m） Total Relief（m）	1 100
地形特征 Feature Description	李四光断裂带属转换断层，呈带状展布，与卡尔斯伯格脊走向垂直，近SW—NE向，长宽分别为170 km和20 km。该断裂带最浅处水深3 300 m，断裂带最深处水深4 400 m，最大高差1 100 m（图4–10）。 Lisiguang Fracture Zone is a kind of transform fault, running like a band nearly NE to SW and perpendicular to the strike of Carlsberg Ridge. The length is 170 km and the width is 20 km. The minimum depth is 3 300 m and the maximum depth is 4 400 m, which makes the total relief being 1100 m (Fig.4–10).				
命名释义 Reason for Choice of Name	李四光（公元1889—1971年）系中国著名地质学家，地质力学理论的创始人，中国现代地球科学和地质工作的主要领导人和奠基人之一，为中国石油工业的发展作出了重要贡献，2009年当选为100位新中国成立以来感动中国人物之一。该地理实体命名为"李四光断裂带"，以纪念李四光先生。 Li Siguang (A.D. 1889–1971), a famous Chinese geologist, was the builder of geomechanics theory. He made outstanding contributions to changing the situation of "oil deficiency" in the country, enabling the large-scale development of oil fields to raise the country to the ranks of the world's major oil producers. This feature zone is named "Lisiguang Fracture Zone", in memory of Li Siguang.				

(a)

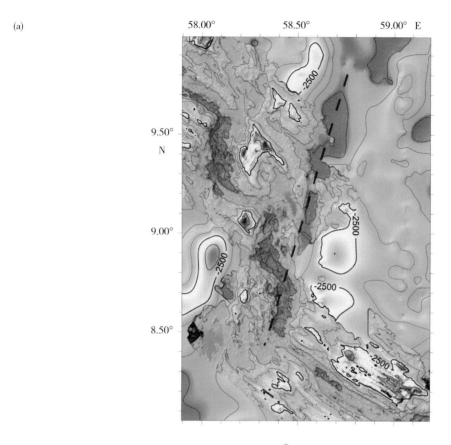

(b)

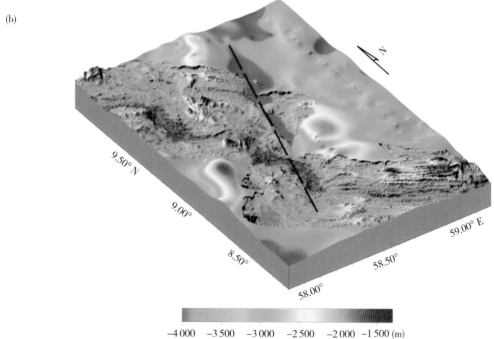

图 4-10　李四光断裂带

(a) 地形图（等深线间隔 500 m）；(b) 三维图

Fig.4-10　Lisiguang Fracture Zone

(a) Bathymetric map (the contour interval is 500 m); (b) 3-D topographic map

4.2.9 竺可桢断裂带
Chukochen Fracture Zone

中文名称 Chinese Name	竺可桢断裂带 Zhukezhen Duanliedai	英文名称 English Name	Chukochen Fracture Zone	所在大洋 Ocean or Sea	西北印度洋 Northwest Indian Ocean
发现情况 Discovery Facts					
命名历史 Name History					
特征点坐标 Coordinates	7°23.50′N，59°37.89′E			长(km)× 宽(km) Length (km) × Width (km)	70 × 25
最大水深（m） Max Depth（m）	4 700	最小水深（m） Min Depth（m）	2 500	高差（m） Total Relief（m）	2 200
地形特征 Feature Description					
命名释义 Reason for Choice of Name					

发现情况 / Discovery Facts: 竺可桢断裂带于2012年5月由中国科考船"李四光号"在执行DY125-24航次时调查发现。
This fracture zone was discovered by the Chinese R/V *Lisiguanghao* during the DY125-24 cruise in May 2012.

命名历史 / Name History: 在我国大洋航次调查报告和GEBCO地名辞典中未命名。
This fracture zone has not been named in Chinese cruise reports or GEBCO gazetteer.

地形特征 / Feature Description: 竺可桢断裂带属转换断层，呈带状，与卡尔斯伯格脊走向垂直，近SW—NE向，长宽分别为70 km和25 km。断裂带最浅处水深2 500 m，断裂带最深处水深4 700 m，最大高差2 200 m（图4–11）。
Chukochen Fracture Zone is a kind of transform fault, running like a band nearly NE to SW and perpendicular to the strike of Carlsberg Ridge. The length is 70 km and the width is 25 km. The minimum depth is 2 500 m and the maximum depth is 4 700 m, which makes the total relief being 2 200 m (Fig.4–11).

命名释义 / Reason for Choice of Name: 竺可桢（公元1890—1974年）是当代著名地理学家、气象学家和教育家，中国近代地理学和气象学的奠基者，对中国气候的形成、特点、区划及变迁等和对地理学以及自然科学史都有深刻的研究。该地理实体命名为"竺可桢断裂带"，以纪念竺可桢先生的重大贡献。
Chu Kochen (A.D.1890–1974) is a famous Chinese geographer, meteorologist and educationalist, one of the founder of geography and meteorology of China. He has investigated the formation, characteristics, division and changes of Chinese climate as well as the geography and history of natural science. The feature zone is named after Chu Kochen, in memory of his great contributions.

(a)

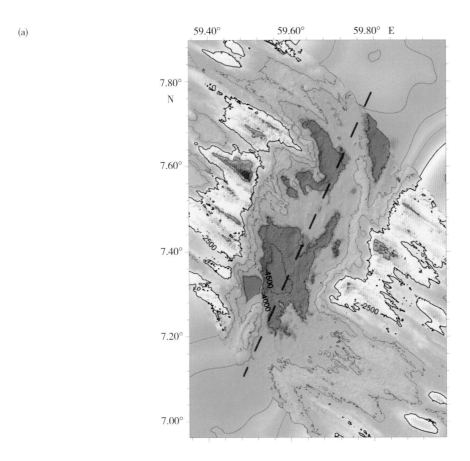

(b)

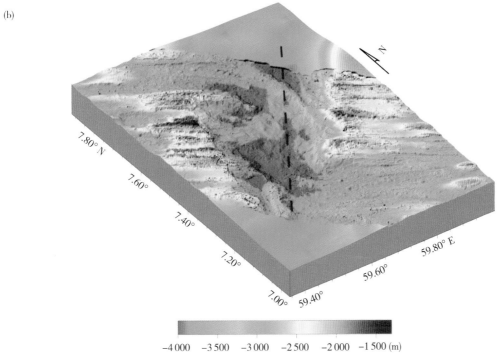

图 4-11　竺可桢断裂带

(a) 地形图（等深线间隔 500 m）；(b) 三维图

Fig.4-11　Chukochen Fracture Zone

(a) Bathymetric map (the contour interval is 500 m); (b) 3-D topographic map

4.2.10 卧蚕海脊
Wocan Ridge

中文名称 Chinese Name	卧蚕海脊 Wocan Haiji		英文名称 English Name	Wocan Ridge	所在大洋 Ocean or Sea	西北印度洋 Northwest Indian Ocean
发现情况 Discovery Facts	此海脊于 2012 年 5 月由中国科考船中国测量船"李四光号"在执行 DY125-24 航次时调查发现。 This ridge was discovered by the Chinese R/V *Lisiguanghao* during the DY125-24 cruise in May, 2012.					
命名历史 Name History	在我国大洋航次调查报告和 GEBCO 地名辞典中未命名。 This ridge has not been named in Chinese cruise reports or GEBCO gazetteer.					
特征点坐标 Coordinates	6°21.17′N，60°32.46′E				长(km)×宽(km) Length (km)× Width (km)	34×6
最大水深（m） Max Depth（m）	3 500	最小水深（m） Min Depth（m）	2 850	高差（m） Total Relief（m）		650
地形特征 Feature Description	该海脊位于卡尔斯伯格脊中央裂谷，属新火山脊，呈线状，与卡尔斯伯格脊走向一致为 SE—NW 向，长 34 km，宽 6 km。该海脊顶部水深 2 850 m，断裂带底部水深 3 500 m，最大高差 650 m（图 4–12）。 Wocan Ridge is located in the central rift valley of Carlsberg Ridge and is a kind of new volcanic ridge. It extends like a line and is parallel with the strike of Carlsberg Ridge, which runs NW to SE. The length is 34 km and the width is 6 km. The top depth is 2 850 m and the bottom depth of the fracture zone is 3 500 m, which makes the total relief being 650 m (Fig.4–12).					
命名释义 Reason for Choice of Name	蚕在人类经济生活及文化历史上有重要地位，约在 4 000 多年前中国已开始养蚕和利用蚕丝。卧蚕海脊俯视平面形态似卧蚕，故以此命名。 "Wocan" means silkworm in Chinese. Silkworm has an important position in human economy and civilization. Chinese began to rear silkworm and produce silk for more than 4 000 years ago. This ridge is named "Wocan" since it has an overlook plane shape like a silkworm.					

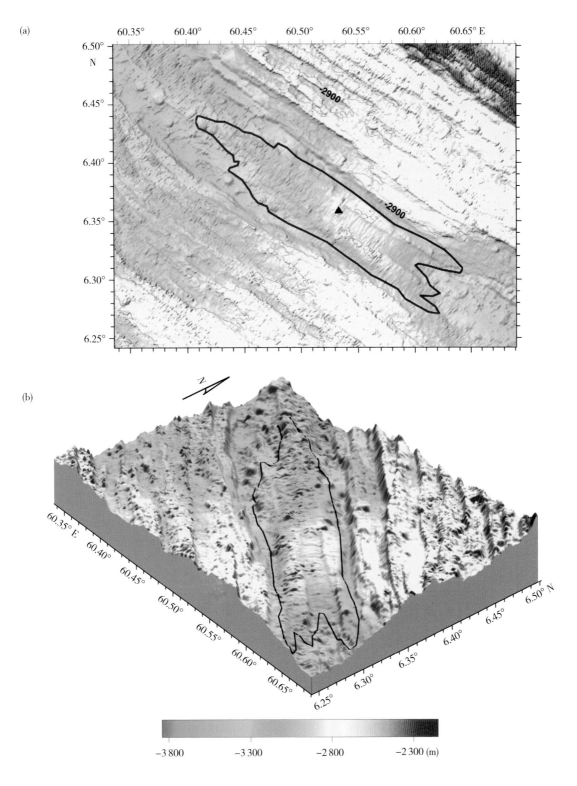

图 4-12　卧蚕海脊

(a) 地形图（等深线间隔 200 m）；(b) 三维图

Fig.4-12　Wocan Ridge

(a) Bathymetric map (the contour interval is 200 m); (b) 3-D topographic map

4.2.11　卧蚕 1 号热液区、卧蚕 2 号热液区
Wocan-1 Hydrothermal Field and Wocan-2 Hydrothermal Field

中文名称 Chinese Name	卧蚕 1 号热液区 卧蚕 2 号热液区 Wocan-1 Reyequ Wocan-2 Reyequ	英文名称 English Name	Wocan-1 Hydrothermal Field Wocan-2 Hydrothermal Field	所在大洋 Ocean or Sea	西北印度洋 Northwest Indian Ocean
发现情况 Discovery Facts	此热液区于 2013 年 5 月由中国科考船"竺可桢号"在执行 DY125-28 航次调查时，通过海底观察及水体热液异常探测证实。 　　该热液区位于洋中脊中央裂谷新火山脊（卧蚕海脊）之上，由东西两个热液区组成，西侧为卧蚕 1 号热液区，水深 2 950 m；东侧为卧蚕 2 号热液区，水深 3 110 m；两者相距约 1.7 km。 　　The hydrothermal fields were discovered by the Chinese R/V *Chukochenhao* during the DY125-28 cruise in May, 2013. They were confirmed by undersea observation and hydrothermal anomaly detection. 　　The hydrothermal fields are located in the new volcanic ridge (Wocan Ridge) of the central rift valley of the mid-ocean ridge. They consist of the eastern and western hydrothermal field. The western one is Wocan-1 Hydrothermal Field with the water depth of 2 950 m. The eastern one is Wocan-2 Hydrothermal Field with the water depth of 3 110 m. The distance between them is about 1.7 km.				
命名历史 Name History	原我国大洋调查报告中用名"CR1-2 热液区"。 It was named "CR1-2 Hydrothermal Field" in the Chinese cruise reports.				
特征点坐标 Coordinates	6°21.92′N，60°31.57′E（卧蚕 1 号 / Wocan-1） 6°22.73′N，60°30.38′E（卧蚕 2 号 / Wocan-2）		水深 (m) Depth (m)		2 950 ~ 3 110
命名释义 Reason for Choice of Name	以此热液区所在的地理实体卧蚕海脊的名字命名，为方便使用，命名为卧蚕 1 号热液区和卧蚕 2 号热液区（图 4–13）。 　　The hydrothermal fields were named after the undersea feature Wocan Ridge right there. For convenience, they were named Wocan-1 Hydrothermal Field and Wocan-2 Hydrothermal Field (Fig.4–13).				

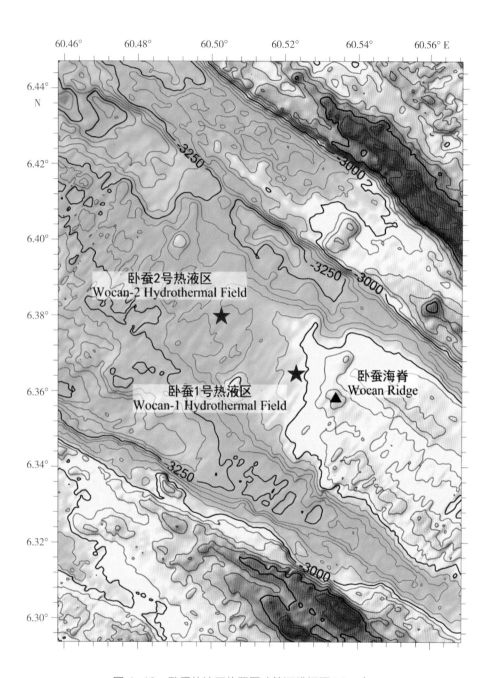

图 4-13　卧蚕热液区位置图（等深线间隔 50 m）

Fig.4-13　Bathymetric map of the Wocan Hydrothermal Field (The contour interval is 50 m)

4.2.12 玄鸟海脊
Xuanniao Ridge

中文名称 Chinese Name	玄鸟海脊 Xuanniao Haiji	英文名称 English Name	Xuanniao Ridge	所在大洋 Ocean or Sea	西北印度洋 Northwest Indian Ocean
发现情况 Discovery Facts	此海脊于 2012 年 5 月由中国测量船"李四光号"在执行 DY125-24 航次时调查发现。 This ridge was discovered by the Chinese R/V *Lisiguanghao* during DY125-24 cruise in May, 2012.				
命名历史 Name History	在我国大洋航次调查报告和 GEBCO 地名辞典中未命名。 This ridge has not been named in Chinese cruise reports or GEBCO gazetteer.				
特征点坐标 Coordinates	5°56.68′N，61°02.26′E			长(km)×宽(km) Length (km)× Width (km)	34×7
最大水深（m） Max Depth（m）	3 700	最小水深（m） Min Depth（m）	3 200	高差（m） Total Relief（m）	500
地形特征 Feature Description	该海脊位于卡尔斯伯格脊中央裂谷，属新火山脊，呈线状，与卡尔斯伯格脊走向一致为 SE—NW 向，长 34 km，宽 7 km。海脊顶部水深 3 200 m，断裂带底部水深 3 700 m，最大高差 500 m（图 4–14）。 Xuanniao Ridge is located in the central rift valley of Carlsberg Ridge and is a kind of new volcanic ridge. It extends like a line and is parallel with the strike of Carlsberg Ridge, which runs NW to SE. The length is 34 km and the width is 7 km. The top depth of the ridge is 3 200 m and the bottom depth is 3 700 m, which makes the total relief being 500 m (Fig.4–14).				
命名释义 Reason for Choice of Name	"玄鸟"出自《诗经·商颂·玄鸟》"天命玄鸟，降而生商，宅殷土芒芒"。此诗是祭祀殷高宗武丁的颂歌，中国古人称燕子为玄鸟，传说有娀氏之女简狄吞燕卵而怀孕生契，契建商。该句意为上天命令燕子降，来到人间生育商王，居住在广袤的殷地。 "Xuanniao" comes from a poem named *Xuanniao* in *Shijing · Shangsong*. *Shijing* is a collection of ancient Chinese poems from 11th century B.C. to 6th century B.C. "From eggs of God-sent swallow Qi sprang, fore-father of the House of Shang; On this land of Yin inhabited his gang." "Xuanniao" means swallows in ancient China. The legend said that Yousong's daughter Jiandi gave birth to Qi, who established Shang Dynasty, after she ate an egg of swallow.				

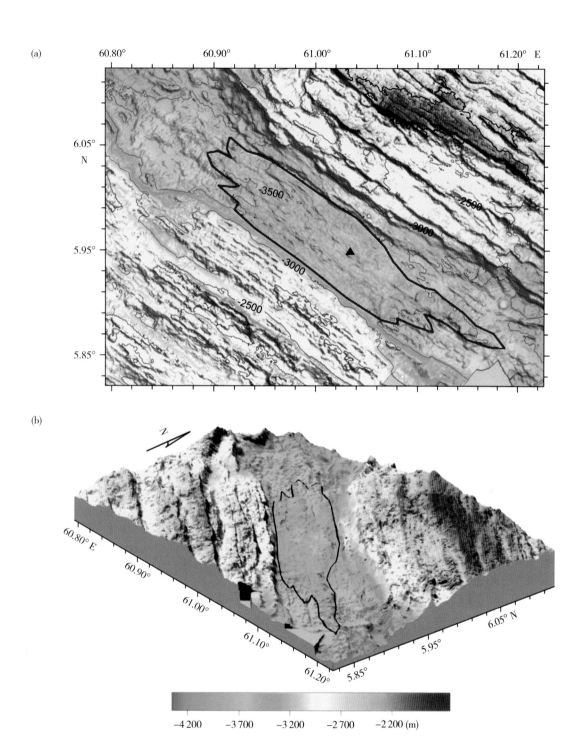

图 4-14　玄鸟海脊

(a) 地形图（等深线间隔 500 m）；(b) 三维图

Fig.4-14　Xuanniao Ridge

(a) Bathymetric map (the contour interval is 500 m); (b) 3-D topographic map

4.2.13 徐霞客断裂带
Xuxiake Fracture Zone

中文名称 Chinese Name	徐霞客断裂带 Xuxiake Duanliedai	英文名称 English Name	Xuxiake Fracture Zone	所在大洋 Ocean or Sea	西北印度洋 Northwest Indian Ocean
发现情况 Discovery Facts	徐霞客断裂带于2012年5月由中国测量船"李四光号"在执行DY125-24航次时调查发现。 This fracture zone was discovered by the Chinese R/V *Lisiguanghao* during the DY125-24 cruise in May, 2012.				
命名历史 Name History	在我国大洋航次调查报告和GEBCO地名辞典中未命名。 This fracture zone has not been named in Chinese cruise reports or GEBCO gazetteer.				
特征点坐标 Coordinates	5°35.62′N，61°39.11′E			长(km)×宽(km) Length (km)×Width (km)	100×17
最大水深（m） Max Depth（m）	4 200	最小水深（m） Min Depth（m）	2 600	高差（m） Total Relief（m）	1 600
地形特征 Feature Description	徐霞客断裂带属转换断层，呈带状，与卡尔斯伯格脊走向垂直，近SW—NE向，长宽分别为100 km和17 km。断裂带最浅处水深2 600 m，断裂带最深处水深4 200 m，最大高差1 600 m（图4–15）。 Xuxiake Fracture Zone is a kind of transform fault, running like a band nearly NE to SW and perpendicular to the strike of Carlsberg Ridge. The length is 100 km and the width is 17 km. The minimum depth of the fracture zone is 2 600 m and the maximum depth is 4 200 m, which makes the total relief being 1 600 m (Fig.4–15).				
命名释义 Reason for Choice of Name	徐霞客（公元1587—1641年），名弘祖，字振之，号霞客，是我国明代著名的地理学家、旅行家。他先后游历了相当于今江苏、安徽等十六省，足迹遍及大半个中国，著有《徐霞客游记》。该断裂带以徐霞客的名字命名，以纪念徐霞客在地理学方面的伟大成就。 Xu Xiake (A.D. 1587–1641) was a Chinese travel writer and geographer of the Ming Dynasty, known best for his famous geographical treatise, and noted for his bravery and humility. He traveled throughout more than 16 provinces in China, documenting his travels extensively. The records of his travels were compiled posthumously in his famous travel books. The fracture zone is named after Xu Xiake, in memory of his significant achievements in geography.				

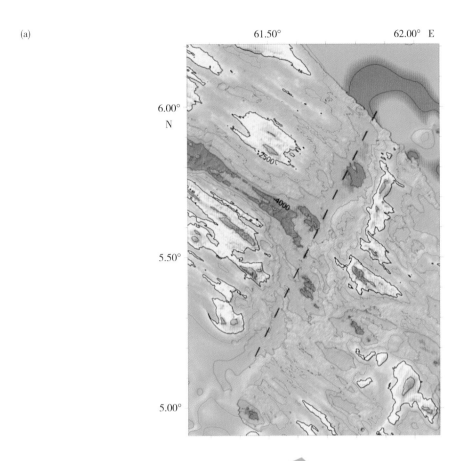

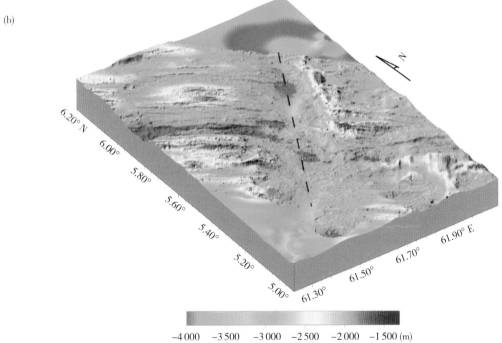

图 4-15　徐霞客断裂带

(a) 地形图（等深线间隔 500 m）；(b) 三维图

Fig.4-15　Xuxiake Fracture Zone

(a) Bathymetric map (the contour interval is 500 m); (b) 3-D topographic map

4.2.14 天休 1 号热液区
Tianxiu-1 Hydrothermal Field

中文名称 Chinese Name	天休 1 号热液区 Tianxiu-1 Reyequ	英文名称 English Name	Tianxiu-1 Hydrothermal Field	所在大洋 Ocean or Sea	西北印度洋 Northwest Indian Ocean
发现情况 Discovery Facts	此热液区于 2012 年 5 月由中国科考船"大洋一号"在执行 DY125-26 航次调查时，通过海底观察及水体热液异常探测证实。该热液区位于洋中脊中央裂谷南侧壁底之上，水深约 3 400 m。 This hydrothermal Field was discovered by the Chinese R/V *Dayang Yihao* during the DY125-26 cruise in May, 2012. It was confirmed by undersea observation and hydrothermal anomaly detection. This hydrothermal field is located in the bottom of the southern wall of the central rift valley of the mid-ocean ridge, with the water depth of about 3 400 m.				
命名历史 Name History	原我国大洋调查报告中曾用名为 3°42′N 热液区。 It was named "3°42′N Hydrothermal Field" in the former Chinese cruise reports.				
特征点坐标 Coordinates	3°41.76′N，63°50.36′E			水深 (m) Depth(m)	3 400
命名释义 Reason for Choice of Name	"天休"出自《诗经·商颂·长发》，"受小球大球，为下国缀旒，何天之休。"意为接受小法和大法，表率诸侯做榜样，承蒙老天赐福祥。"天休"即上天赐福。科考队员将该热液区的发现比喻犹如天降福祥，故以此命名（图 4-16）。 "Tianxiu" comes from a poem named *Changfa* in *Shijing · Shangsong*. *Shijing* is a collection of ancient Chinese poems from 11th century B.C. to 6th century B.C. "He followed statutes great and small and became a model for all with heaven's blessing on his side". "Tianxiu" means gods' blessings. This hydrothermal field is named "Tianxiu-1 Hydrothermal Field" since the scientist regard the discovery of this hydrothermal field just like blessings from gods (Fig.4–16).				

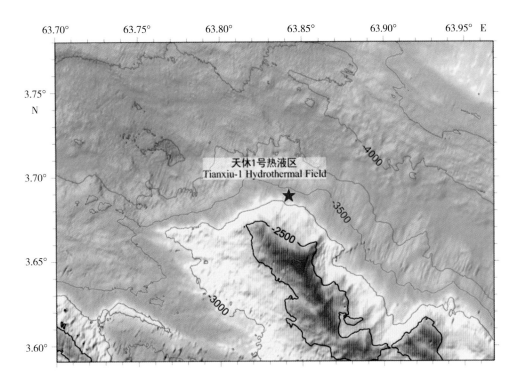

图 4-16　天休 1 号热液区位置图（等深线间隔 500 m）

Fig.4-16　Bathymetric map of the Tianxiu-1 Hydrothermal Field

(The contour interval is 500 m)

4.2.15 郦道元断裂带
Lidaoyuan Fracture Zone

中文名称 Chinese Name	郦道元断裂带 Lidaoyuan Duanliedai	英文名称 English Name	Lidaoyuan Fracture Zone	所在大洋 Ocean or Sea	西北印度洋 Northwest Indian Ocean
发现情况 Discovery Facts	郦道元断裂带于2012年5月由中国测量船"李四光号"在执行DY125-24航次时调查发现。 This fracture zone was discovered by the Chinese R/V *Lisiguanghao* during the DY125-24 cruise in May, 2012.				
命名历史 Name History	在我国大洋航次调查报告和GEBCO地名辞典中未命名。 This fracture zone has not been named in Chinese cruise reports or GEBCO gazetteer.				
特征点坐标 Coordinates	2°17.09′N，66°47.41′E		长(km)×宽(km) Length (km)× Width (km)		150×20
最大水深（m） Max Depth（m）	4 000	最小水深（m） Min Depth（m）	2 000	高差（m） Total Relief（m）	2 000
地形特征 Feature Description	郦道元断裂带属转换断层，呈带状，与卡尔斯伯格脊走向垂直，近 SW—NE 向，长宽分别为150 km 和20 km。断裂带最浅处水深2 000 m，断裂带最深处水深4 000 m，最大高差2 000 m（图4–17）。 Lidaoyuan Fracture Zone is a kind of transform fault, running like a band nearly NE to SW and perpendicular to the strike of Carlsberg Ridge. The length is 150 km and the width is 20 km. The minimum depth is 2 000 m and the maximum depth is 4 000 m, which makes the total relief being 2 000 m (Fig.4–17).				
命名释义 Reason for Choice of Name	郦道元（约公元470—527年），字善长，北朝北魏地理学家，游历秦岭、淮河以北和长城以南广大地区，考察河道沟渠，收集有关的风土民情、历史故事和神话传说，撰有《水经注》四十卷。该断裂带以郦道元的名字命名，以纪念郦道元在地理学方面的伟大成就。 Li Daoyuan (about A.D. 470–527) was a famous geographer of Beiwei Dynasty. He visited the North of Qinling Mountains, Huaihe River and the South of the Great Wall, and collected the culture, history and myths. He is known as the author of a famous book on geography. The fracture zone is named after Li Daoyuan, in memory of his significant achievements in geography.				

(a)

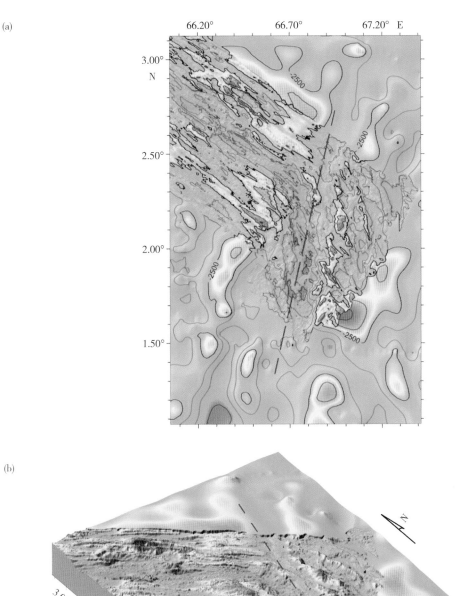

(b)

−4 000　−3 500　−3 000　−2 500　−2 000　−1 500 (m)

图 4-17　郦道元断裂带

(a) 地形图（等深线间隔 500 m）；(b) 三维图

Fig.4-17　Lidaoyuan Fracture Zone

(a) Bathymetric map (the contour interval is 500 m); (b) 3-D topographic map

4.2.16　大糦海丘
Daxi Hill

中文名称 Chinese Name	大糦海丘 Daxi Haiqiu	英文名称 English Name	Daxi Hill	所在大洋 Ocean or Sea	西北印度洋 Northwest Indian Ocean
发现情况 Discovery Facts	此海丘于 2012 年 5 月由中国测量船"李四光号"在执行测量任务时调查发现。 This hill was discovered by the Chinese R/V *Lisiguanghao* during the survey in May, 2012.				
命名历史 Name History	由我国命名为大糦海丘，于 2017 年提交 SCUFN 审议通过。 This feature was named Daxi Hill by China and the name was approved by SCUFN in 2017.				
特征点坐标 Coordinates	06°48.1'N, 60°10.5'E			长(km) × 宽(km) Length (km) × Width (km)	10 × 5
最大水深（m） Max Depth（m）	4 000	最小水深（m） Min Depth（m）	3 300	高差（m） Total Relief（m）	700
地形特征 Feature Description	大糦海丘位于西北印度洋，客怿海山东南方向 330 km，该海丘属新火山脊，呈长条状（图 4–18）。 Daxi Hill is located in the Northwest Indian Ocean and at 330 km southeast to Keyi Seamount. This ridge belongs to new volcanic ridge and has an elongated shape (Fig.4–18).				
命名释义 Reason for Choice of Name	"大糦"出自《诗经·商颂·玄鸟》中的诗句"龙旂十乘，大糦是承"，意指非常丰富的食物。这是因为大糦热液区热液活动较为频繁，为盲虾、螃蟹等生物提供了充足的营养，好似龙宫盛馔。 "Daxi" comes from a poem named *Xuanniao* in *Shijing · Shangsong. Shijing* is a collection of ancient Chinese poems from 11th century B.C. to 6th century B.C. "Ten carts with Dragon Flags are full of food." "Daxi" means abundant food. Because the activity of Daxi Hydrothermal Field is more frequent, that provides adequate nutrition to the shrimp, crab and other creatures around, like a feast. Daxi Hill was named after Daxi Hydrothermal Field right here.				

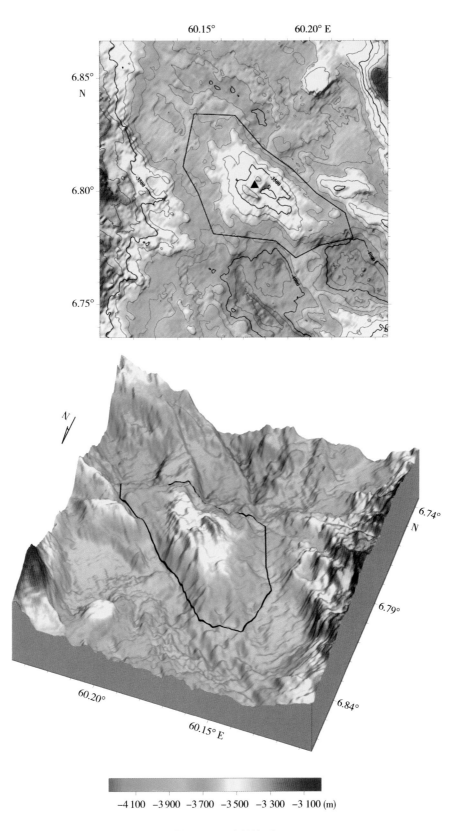

图 4-18　大糦海丘

(a) 地形图（等深线间隔 100 m）；(b) 三维图

Fig.4-18　Daxi Hill

(a) Bathymetric map (the contour interval is 100 m); (b) 3-D topographic map

4.2.17　天龙海山群
Tianlong Seamounts

中文名称 Chinese Name	天龙海山群 Tianlong Haishanqun	英文名称 English Name	Tianlong Seamounts	所在大洋 Ocean or Sea	西北印度洋 Northwest Indian Ocean
发现情况 Discovery Facts	此海山群于 2012 年由中国科考船"海洋 18"船在执行测量任务时调查发现。 This seamounts was discovered by the Chinese R/V *Haiyang* 18 during the survey in 2012.				
命名历史 Name History	由我国命名为天龙海山群，于 2018 年提交 SCUFN 审议通过。 This feature was named Tianlong Seamounts by China and the name was approved by SCUFN in 2018.				
特征点坐标 Coordinates	4°59.7′N，62°02.5′E 5°04.4′N，61°59.0′E 5°04.8′N，61°49.0′E 5°10.0′N，61°53.5′E			长(km) × 宽(km) Length (km) × Width (km)	53 × 28
最大水深（m） Max Depth（m）	4 377	最小水深（m） Min Depth（m）	1 748	高差（m） Total Relief（m）	2 629
地形特征 Feature Description	该海山群属于断块山，由洋底核杂岩（Oceanic Core Complex）组成，状如卧龙（图 4–19）。 Tianlong Seamounts is a blockgebirge, which is composed of Oceanic Core Complex. It seems like a crouched dragon (Fig.4–19).				
命名释义 Reason for Choice of Name	此海山群形似一条龙，故名。 The seamounts is named Tianlong because it looks like a dragon.				

(a)

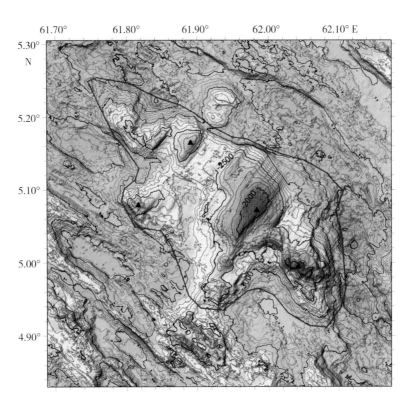

(b)

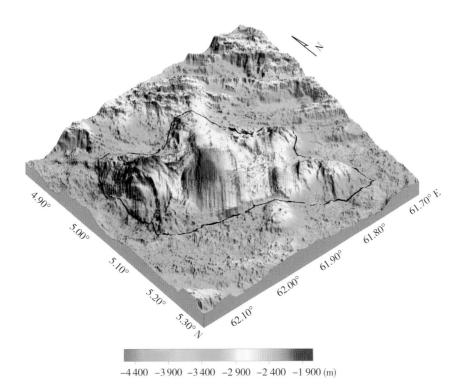

−4 400　−3 900　−3 400　−2 900　−2 400　−1 900 (m)

图 4-19　天龙海山群

(a) 地形图（等深线间隔 100 m）；(b) 三维图

Fig.4-19　Tianlong Seamounts

(a) Bathymetric map (the contour interval is 100 m); (b) 3-D topographic map

4.2.18　龙灯海丘
Longdeng Hill

中文名称 Chinese Name	龙灯海丘 Longdeng Haiqiu	英文名称 English Name	Longdeng Hill	所在大洋 Ocean or Sea	西北印度洋 Northwest Indian Ocean
发现情况 Discovery Facts	此海丘于 2012 年由中国科考船"海洋 18"船在执行测量任务时调查发现。 This Hill was discovered by the Chinese R/V *Haiyang* 18 during the survey in 2012.				
命名历史 Name History	由我国命名为龙灯海丘，于 2018 年提交 SCUFN 审议通过。 This feature was named Longdeng Hill by China and the name was approved by SCUFN in 2018.				
特征点坐标 Coordinates	5°13.8'N，61°55.6'E			长 (km)×宽 (km) Length (km)×Width (km)	10×8.7
最大水深（m） Max Depth（m）	3 451	最小水深（m） Min Depth（m）	2 608	高差（m） Total Relief（m）	843
地形特征 Feature Description	该海丘为火山成因，呈圆锥状（图 4–20）。 Longdeng Hill is volcanogenic and cone-shaped (Fig.4–20).				
命名释义 Reason for Choice of Name	邻近天龙海山群，形似为天龙指路的明灯，故名。 This hill is close to Tianlong Seamounts and looks like a guiding light for Tianlong, so it is named Longdeng, which means light for the Tianlong.				

(a)

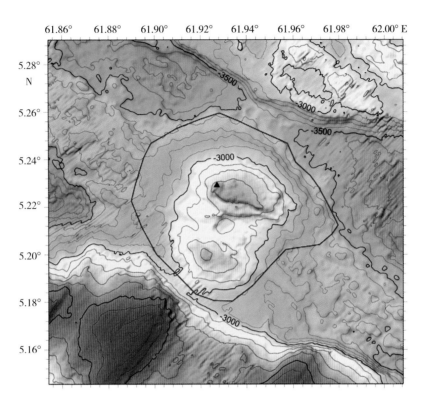

(b)

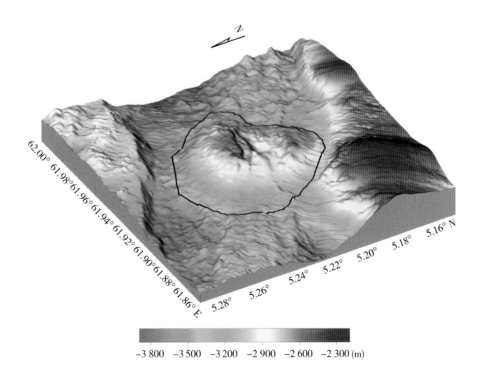

图 4-20　龙灯海丘

(a) 地形图（等深线间隔 100 m）；(b) 三维图

Fig.4-20　Longdeng Hill

(a) Bathymetric map (the contour interval is 100 m); (b) 3-D topographic map

第 3 节　西南印度洋洋中脊地理实体和热液区

西南印度洋中脊（SWIR）全长约 8 000 km，西起布维三联点（Bouvet Triple Junction，BTJ，55°S，00°40′W），东至罗德里格斯三联点（Rodrigues Triple Junction，RTJ，25°30′S，70°E），构成了非洲板块、索马里板块和南极洲板块之间主要的板块边界。西南印度洋中脊作为世界上超慢速扩张洋中脊，整条洋中脊的扩张速率基本一致，接近 13 ～ 16 mm/a（Mendel et al., 1997; Georgen et al., 2001），西段的扩张速度略快于东段（Chu and Gordon, 1999）。沿西南印度洋中脊走向，清楚可见结构上明显的变化，根据其几何形态和扩张历史，可将 SWIR 划分为七段（Georgen et al., 2001）。我国在西南印度洋开展的调查工作主要集中在 49°—70°E 之间，洋中脊走向北东。海底地形形态总体表现为与走向平行的隆洼相间，指示了与洋脊走向垂直的周期扩张运动，洋脊被数条近 N—S 的转换断层或构造错断切割。扩张裂谷多呈中心对称分布，裂谷宽几到几十千米，最宽达 30 千米，裂谷深几百米到上千米，中央裂谷内地形复杂，发育谷中谷，谷底地形不平坦，发育小海山或海山脊，而在裂谷壁上多发育有台地或丘状突起，部分地段裂谷被断裂所切割，发生扭转。

在西南印度洋洋中脊区域，共命名地理实体 9 个，热液区 14 处。其中海山 6 个，包括乔岳海山、天作海山、骏惠海山、天成海山、汪大渊海山、封演海山；海脊 2 个，包括崇牙海脊和思文海脊；海丘 1 个，断桥海丘。区域内 14 处热液区分别是龙旂 1 号热液区、龙旂 2 号热液区、断桥 1 号热液区、玉皇 1 号热液区、长白 1 号热液区、天作 1 号热液区、天成 1 号热液区、苏堤 1 号热液区、白堤 1 号热液区、东龙井 1 号热液区、西龙井 1 号热液区、栈桥 1 号热液区、骏惠 1 号热液区和百神 1 号热液区（图 4-21）。

Section 3 Undersea Features and Hydrothermal Fields in the Southwest Indian Ridge

The Southwest Indian Ridge (SWIR) has a total length of about 8 000 km, starting from the Bouvet Triple Junction (Bouvet Triple Junction, BTJ, 55° S, 00° 40′ W) in the west, to the Rodrigues Triple Junction (Rodrigues Triple Junction, RTJ , 25° 30′ S, 70° E), constituting the boundary between the African Plate and the Antarctic Plate. As one of the world's slowest spreading ridges, the spreading rate of the whole Southwest Indian Ridge is basically consistent, nearly 13–16 mm/a (Mendel et al., 1997; Georgen et al., 2001), with the western part expanding faster than the eastern part (Chu and Gordon, 1999). Obvious changes can be clearly seen in structure along the strike of the Southwest Indian Ridge. SWIR can be divided into 7 segments according to its geometry and spreading history (Georgen et al., 2001). Chinese investigations carried out in the Southwest Indian Ocean mainly focus between 49°–70° E, with the ridge running NE to SW. The overall seabed topography and morphology is featured by rises and depressions parallel with the strike, which indicates the periodic spreading motion perpendicular to the ridge strike. The ridge is cut by several nearly N to S running transform faults or tectonic dislocation. The expansion rift mostly has symmetric distribution, with the width of several to dozens of kilometers, its maximum width being 30 km, and the depth of a few hundreds to thousands of meters. Inside the axial rift, topography is complicated and it develops valleys inside the valleys. The topography is not flat at the bottom, which develops small seamounts or ridges, while developing platforms or dome-shaped protuberance on the rift wall. Some sections of the rift valley are cut by the fractures and thus twisted.

In total, 9 undersea features and 14 hydrothermal fields have been named by China in the region of the Southwest Indian Ridge. Among them, there are 6 seamounts, including Qiaoyue Seamount, Tianzuo Seamount, Junhui Seamount, Tiancheng Seamount; Wangdayuan Seamount and Fengyan Seamount; 2 ridges, including Chongya Ridge and Siwen Ridge; 1 hill, Duanqiao Hill. The 14 hydrothermal fields in this area are respectively Longqi-1 Hydrothermal Field, Longqi-2 Hydrothermal Field, Duanqiao-1 Hydrothermal Field, Yuhuang-1 Hydrothermal Field, Changbai-1 Hydrothermal Field, Tianzuo-1 Hydrothermal Field, Tiancheng-1 Hydrothermal Field, Sudi-1 Hydrothermal Field, Baidi-1 Hydrothermal Field, Donglongjing-1 Hydrothermal Field, Xilongjing-1 Hydrothermal Field, Zhanqiao-1 Hydrothermal Field, Junhui-1 Hydrothermal Field and Baishen-1 Hydrothermal Field (Fig. 4-21).

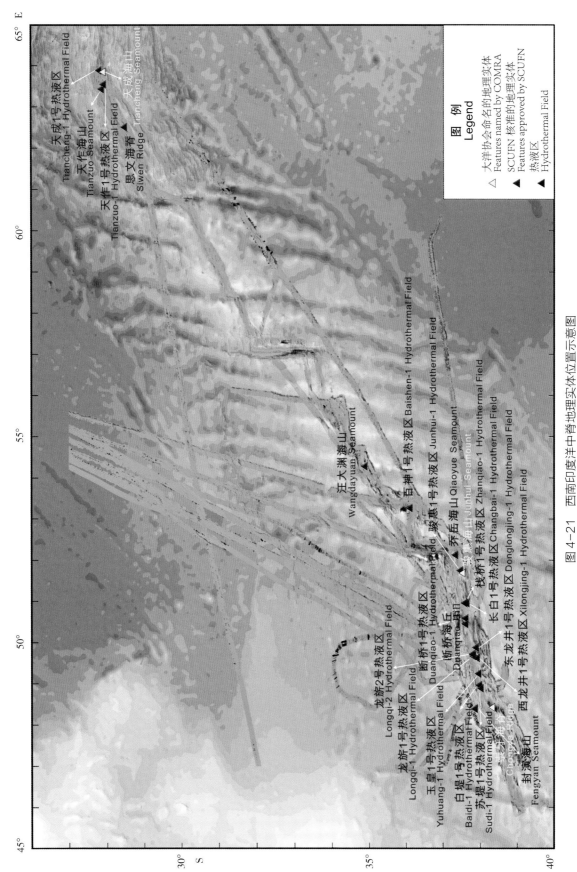

图 4-21　西南印度洋中脊地理实体位置示意图

Fig.4-21　Locations of the undersea features in the Southwest Indian Ridge

4.3.1 崇牙海脊
Chongya Ridge

中文名称 Chinese Name	崇牙海脊 Chongya Haiji		英文名称 English Name	Chongya Ridge	所在大洋 Ocean or Sea	西南印度洋 Southwest Indian Ocean
发现情况 Discovery Facts	此海脊于 2011 年由中国科考船 "大洋一号" 在执行 DY125-22 航次时调查发现。 This ridge was discovered by the Chinese R/V *Dayang Yihao* during the DY125-22 cruise in 2011.					
命名历史 Name History	在我国大洋航次调查报告和 GEBCO 地名辞典中未命名。 This ridge has not been named in Chinese cruise reports or GEBCO gazetteer.					
特征点坐标 Coordinates	38°29.12′S，48°23.99′E				长 (km) × 宽 (km) Length (km) × Width (km)	50 × 15
最大水深（m） Max Depth（m）	2 500	最小水深（m） Min Depth（m）	900		高差（m） Total Relief（m）	1 600
地形特征 Feature Description	该海脊位于西南印度洋洋中脊中央裂谷，呈长条状，沿轴近 SW—NE 走向展布，长轴长约 50 km。海山顶部水深 900 m，山麓水深 2 500 m，高差 1 600 m（图 4–22）。 Chongya Ridge is located in the central rift valley of Southwest Indian Ridge. It has an elongated shape and runs NE to SW along the axis. The length of major axis is 50 km. The top depth of this ridge is 900 m and the piedmont depth is 2 500 m, which makes the total relief being 1 600 m (Fig.4–22).					
命名释义 Reason for Choice of Name	"崇牙" 出自《诗经·周颂·有瞽》中的诗句 "设业设虡，崇牙树羽。" 这首诗描述了在周王朝宗庙的大庭上，排列着很多乐器，当乐器齐备和乐师到齐的时候就开始合奏，声音洪亮和谐，先祖神灵前来欣赏，参祭的贵客聆听了全部演奏。崇牙指用来悬挂乐器的木齿。 "Chongya" comes from a poem named *Yougu* in *Shijing · Zhousong*. *Shijing* is a collection of ancient Chinese poems from 11th century B.C. to 6th century B.C. "The music-stands have been arrayed, with hooks and plumes displayed." The poem is a song by the king in Zhou Dynasty when music is played to pay tribute to the ancestors. When the musicians and the instruments are ready, the music will start. "Chongya" means prominent wood for hanging musical instruments.					

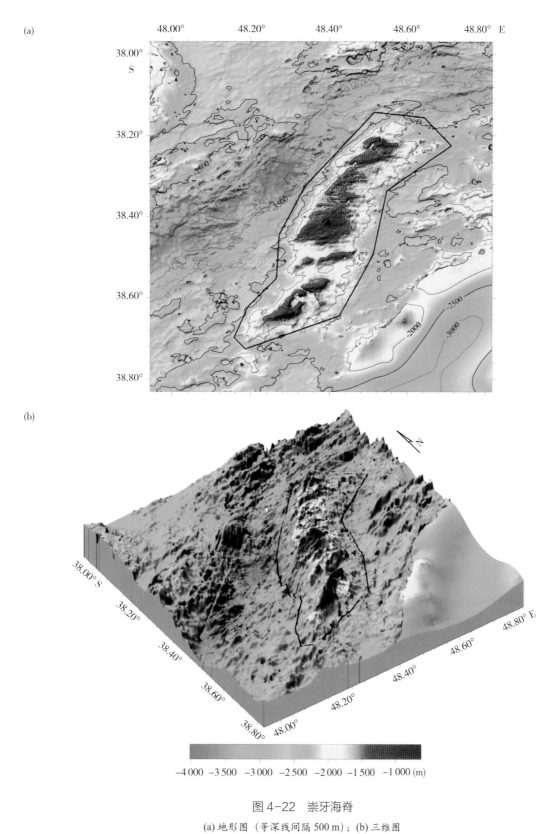

图 4-22　崇牙海脊

(a) 地形图（等深线间隔 500 m）；(b) 三维图

Fig.4-22　Chongya Ridge

(a) Bathymetric map (the contour interval is 500 m); (b) 3-D topographic map

4.3.2 苏堤 1 号热液区
Sudi-1 Hydrothermal Field

中文名称 Chinese Name	苏堤 1 号热液区 Sudi-1 Reyequ	英文名称 English Name	Sudi-1 Hydrothermal Field	所在大洋 Ocean or Sea	西南印度洋 Southwest Indian Ocean
发现情况 Discovery Facts	此热液区于 2014 年由中国科考船"大洋一号"在执行 DY125-30 航次调查时，通过海底观察和水体热液异常探测发现。 该热液区位于西南印度洋洋中脊中央裂谷南侧斜坡之上，围岩为玄武岩，水深约2 100 m。 This hydrothermal field was discovered by the Chinese R/V *Dayang Yihao* during the DY125-30 cruise in 2014. Sudi-1 Hydrothermal Field is located in the southern slope of the central rift valley of the Southwest Indian Ridge. The surrounding rocks are basalt and the water depth is about 2 100 m.				
命名历史 Name History	在 DY125-30 航次现场报告中曾用名为"苏堤热液区"，英文曾用名为"Su Causeway Hydrothermal Field"。 It was once named "Sudi Reyequ" in Chinese and "Su Causeway Hydrothermal Field" in English in the Chinese DY125-30 ocean cruise field report.				
特征点坐标 Coordinates	38°09.72′S，48°35.72′E			水深 (m) Depth (m)	2 100
命名释义 Reason for Choice of Name	苏堤和白堤为杭州西湖上的两个著名的长堤，苏堤是杭州西湖十景之首"苏堤春晓"的所在地。是诗人苏轼（苏东坡）任杭州知州时，疏浚西湖，利用浚挖的淤泥构筑并历经后世演变而形成的。本热液区与本区东北侧发现的白堤 1 号热液区遥相呼应故命名为苏堤 1 号热液区（图 4–23）。 Sudi and Baidi are two famous causeways in the West Lake in Hangzhou, China. Sudi is the place where the Su Causeway, the No.1 of the Ten Sceneries of the West Lake, locates. It echoes to the Baidi-1 Hydrothermal Field at the northeastern side to this area and thus was named Sudi-1 Hydrothermal Field (Fig.4–23).				

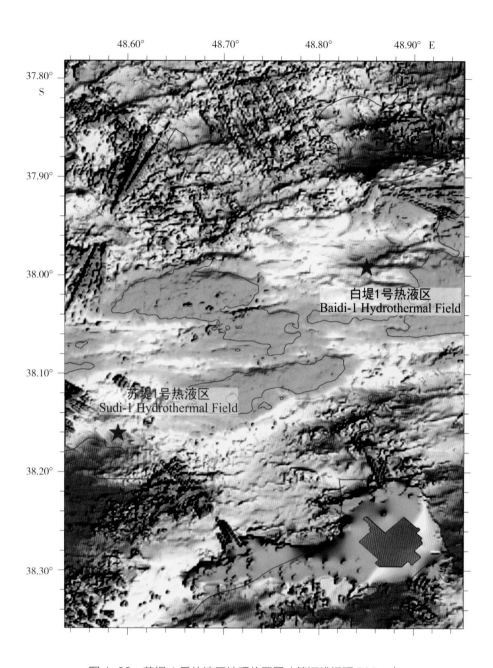

图 4-23　苏堤 1 号热液区地理位置图（等深线间隔 200 m）

Fig.4-23　Bathymetric map of the Sudi-1 Hydrothermal Field

(The contour interval is 200 m)

4.3.3 白堤1号热液区
Baidi-1 Hydrothermal Field

中文名称 Chinese Name	白堤1号热液区 Baidi-1 Reyequ	英文名称 English Name	Baidi-1 Hydrothermal Field	所在大洋 Ocean or Sea	西南印度洋 Southwest Indian Ocean
发现情况 Discovery Facts	此热液区于2014年由中国科考船"大洋一号"在执行DY125-30航次调查时，通过海底观察和水体热液异常探测发现。 该热液区位于西南印度洋洋中脊中央裂谷新生洋脊之上，围岩为玄武岩，水深约2 300 m。 This hydrothermal field was discovered by the Chinese R/V *Dayang Yihao* during the DY125-30 cruise in 2014. Baidi-1 Hydrothermal Field is located in the newly born ridge of the central rift valley of the Southwest Indian Ridge. The surrounding rocks are basalt and the water depth is about 2 300 m.				
命名历史 Name History	在DY125-30航次现场报告中曾用名为"白堤热液区"，英文曾名为"Bai Causeway Hydrothermal Field"。 It was once named "Baidi Reyequ" in Chinese and "Bai Causeway Hydrothermal Field" in English in the Chinese DY125-30 ocean cruise field report.				
特征点坐标 Coordinates	37°59.51′S，48°51.20′E			水深 (m) Depth (m)	2 300
命名释义 Reason for Choice of Name	苏堤和白堤为杭州西湖上的两个著名的长堤，白堤是连接杭州市区和西湖风景区的纽带，东起"断桥残雪"，经锦带桥向西，止于"平湖秋月"。此热液区与本区西南侧发现的苏堤1号热液区遥相呼应，故命名为白堤1号热液区（图4-24）。 Sudi and Baidi are two famous causeways in the West Lake in Hangzhou, China. The Baidi Causeway, on the northern shore of the West Lake, links the downtown of Hangzhou and the West Lake. It echos to the Sudi-1 Hydrothermal Field at the southwestern side of this area and thus was named Baidi-1 Hydrothermal Field (Fig.4-24).				

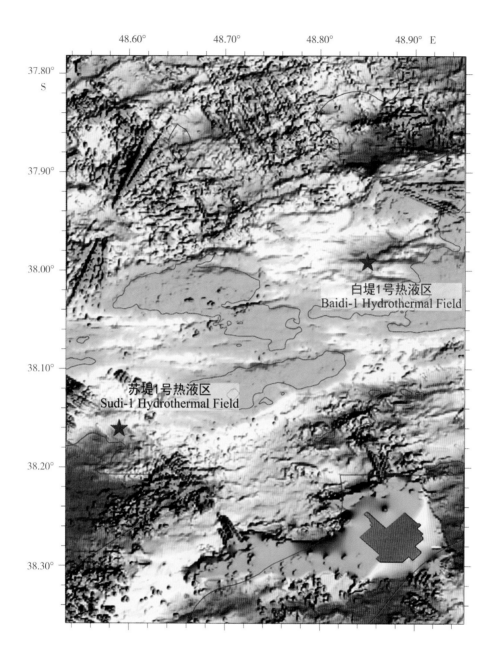

图 4-24　白堤 1 号热液区地理位置图（等深线间隔 200 m）

Fig.4-24　Bathymetric map of the Baidi-1 Hydrothermal Field

(The contour interval is 200 m)

4.3.4 玉皇 1 号热液区
Yuhuang-1 Hydrothermal Field

中文名称 Chinese Name	玉皇 1 号热液区 Yuhuang-1 Reyequ	英文名称 English Name	Yuhuang-1 Hydrothermal Field	所在大洋 Ocean or Sea	西南印度洋 Southwest Indian Ocean
发现情况 Discovery Facts	— 此热液区于 2009 年 3—4 月由中国科考船"大洋一号"在执行 DY115-21 航次调查时，通过海底观察及取样验证。 该热液区位于西南印度洋洋中脊中央裂谷南侧离轴火山的平缓坡地上，围岩为玄武岩，水深约 1 600 m。 This hydrothermal field was discovered by the Chinese R/V *Dayang Yihao* during the DY115-21 cruise from March to April, 2009. This hydrothermal field is located in the smooth slope of the off-axis volcano, west to the central rift valley of the Southwest Indian Ridge. The surrounding rocks are basalt. The water depth is about 1 600 m.				
命名历史 Name History	在航次现场调查报告中曾用名为"玉皇山热液区"，英文曾用名为"Jade Emperor Mountain"。 It was once named "Yuhuangshan Hydrothermal Field" in Chinese and "Jade Emperor Mountain Hydrothermal Field" in English in the Chinese cruise reports.				
特征点坐标 Coordinates	37°56.31′S，49°15.88′E			水深 (m) Depth (m)	1 600
命名释义 Reason for Choice of Name	该热液区采用中国杭州名山玉皇山命名，该山位于杭州西湖与钱塘江之间，原名"龙山"，远望如巨龙横卧，雄姿俊法，风起云涌时，但见湖山空阔，江天浩瀚，境界壮伟高远，史称"万山之祖"。与凤凰山首尾相连，有"龙飞凤舞"的美称。是我国道教全真派圣地，也是新西湖十景之一"玉皇飞云"所在地（图 4–25）。 This hydrothermal field was named after Yuhuang Mountain, a famous mountain in Hangzhou, China. This mountain is located between the West Lake and Qiantang River. It was named "Longshan" before, and it is Chinese Taoism Quanzhen shrine and also the location of "Yu Huang Fei Yun", one of the new Ten Sceneries of the West Lake (Fig.4–25).				

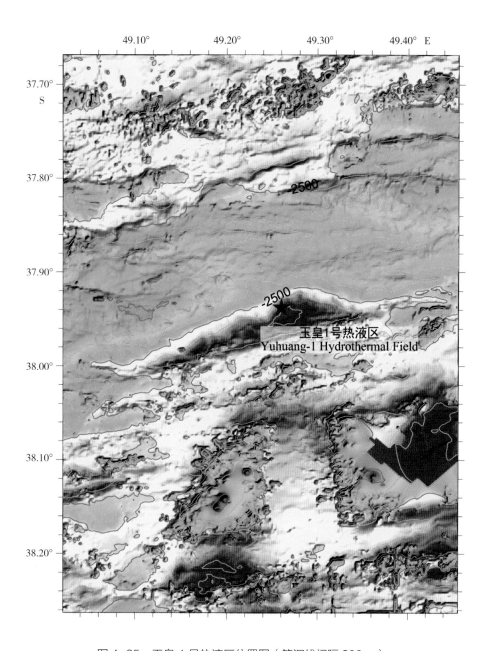

图 4-25　玉皇 1 号热液区位置图（等深线间隔 200 m）

Fig.4-25　Bathymetric map of the Yuhuang-1 Hydrothermal Field

(The contour interval is 200 m)

4.3.5　龙旂 1 号热液区、龙旂 2 号热液区
Longqi-1 Hydrothermal Field, Longqi-2 Hydrothermal Field

中文名称 Chinese Name	龙旂 1 号热液区 龙旂 2 号热液区 Longqi-1 Reyequ Longqi-2 Reyequ	英文名称 English Name	Longqi-1 Hydrothermal Field Longqi-2 Hydrothermal Field	所在大洋 Ocean or Sea	西南印度洋 Southwest Indian Ocean
发现情况 Discovery Facts	龙旂 1 号热液区于 2005 年执行中国大洋 DY115-17 航次调查时，发现水体热液异常，2007 年 DY115-19 航次调查时，通过海底观察和取样证实，DY115-20、DY125-30 航次开展了进一步调查。 龙旂 2 号热液区于 2014 年由中国科考船"大洋一号"DY125-30 航次在该区域调查发现。 龙旂 1 号热液区位于洋中脊小型非转换断层错断与中脊裂谷正断裂交汇点，中轴裂谷东南斜坡的丘状突起正地形上，围岩为玄武岩，水深约为 2 700 m；龙旂 2 号热液区位于龙旂 1 号热液区的南侧，围岩为玄武岩，水深约为 2 600 m。 Longqi-1 Hydrothermal Field was detected hydrothermal anomaly by the Chinese R/V *Dayang Yihao* during the DY115-17 cruise in 2005. It was confirmed by undersea observation and sampling during the DY115-19 cruise in 2007. Further investigation were carried out during the DY115-20 and DY125-30 cruises. Longqi-2 Hydrothermal Field was discovered by the Chinese R/V *Dayang Yihao* during the DY125-30 cruise in 2014. Longqi-1 Hydrothermal Field is located in the intersection point of the small non-transform fault dislocation and the normal fault of the central rift valley of the mid-ocean ridge, the dome shaped normal topography of southeastern slope of the central rift valley. The surrounding rocks are basalt and the water depth is about 2 700 m. Longqi-2 Hydrothermal Field is located in the south to Longqi-1. The surrounding rocks are basalt and the water depth is about 2 600 m.				
命名历史 Name History	在 DY115-17 航次至 DY115-20 航次期间曾用名"A 区"或"49.6°E 热液区"，英国科学家在 2012 年曾用名为"Dragon Area"。 It was named "A section" or "49.6°E Hydrothermal Field" during the Chinese DY115-17 cruise to DY115-20 cruise. It was once named Dragon Area by the scientists of the UK in 2012.				
特征点坐标 Coordinates	37°46.98′S，49°39.00′E（龙旂 1 号 / Longqi-1） 37°48.20′S，49°39.72′E（龙旂 2 号 / Longqi-2）		水深 (m) Depth (m)		2 600 ~ 2 700
命名释义 Reason for Choice of Name	这是中国科考队首次发现热液区，也是西南印度洋脊上发现的首个热液区。以中华民族的图腾"龙"进行命名。后续航次在该区域附近发现的热液区，以 1 号、2 号顺序命名（图 4-26）。 This is the first time that Chinese expedition team discovered a hydrothermal field, and it is the first hydrothermal field that discovered on the Southwest Indian Ridge. It was named after the Chinese totem "Dragon". The hydrothermal fields discovered by the following cruises in this area were named in the order 1 and 2 (Fig.4-26).				

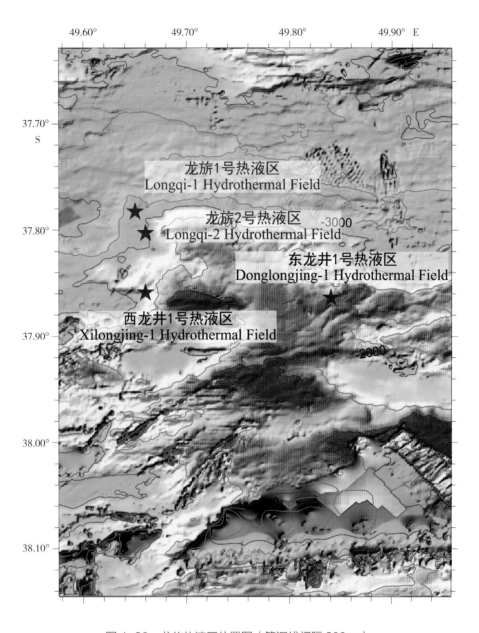

图 4-26　龙旂热液区位置图（等深线间隔 200 m）

Fig.4-26　Bathymetric map of the Longqi Hydrothermal Field

(The contour interval is 200 m)

4.3.6 西龙井 1 号热液区
Xilongjing-1 Hydrothermal Field

中文名称 Chinese Name	西龙井 1 号热液区 Xilongjing-1 Reyequ	英文名称 English Name	Xilongjing-1 Hydrothermal Field	所在大洋 Ocean or Sea	西南印度洋 Southwest Indian Ocean
发现情况 Discovery Facts	此热液区于 2014 年由中国科考船"大洋一号"在执行 DY125-30 航次调查时，通过海底观察、水体热液异常探测及取样证实。 此热液区位于洋中脊中央裂谷南侧一火山口的底部斜坡，围岩为玄武岩，水深约 2 200 m。 This hydrothermal field was discovered by the Chinese R/V *Dayang Yihao* during the DY125-30 cruise in 2014. It was confirmed by undersea observation, hydrothermal anomaly detection and samples. Xilongjing-1 Hydrothermal Field is located in the slope of the bottom of a crater in the central rift valley of the Southwest Indian Ridge. The surrounding rocks are basalt and the water depth is about 2 200 m.				
命名历史 Name History	在 DY125-30 航次现场报告中曾用名为"西龙井热液区"，英文曾用名为"Dragon Well West"。 It was once named as "Xilongjing Reyequ" in Chinese and "Dragon Well West" in English in the Chinese DY125-30 ocean cruise field report.				
特征点坐标 Coordinates	37°51.54′S，49°39.79′E			水深 (m) Depth (m)	2 200
命名释义 Reason for Choice of Name	龙井是一种绿茶，位列中国十大名茶之首，原产于杭州西湖一带。本区发现东西两个热液区，故分别命名为东龙井 1 号热液区和西龙井 1 号热液区（图 4-27）。 Longjing is kind of green tea and it ranks first in Chinese top ten tea, which is grown on the hills around the West Lake in Hangzhou. Two hydrothermal fields were discovered in this area and are named Donglongjing-1 Hydrothermal Field and Xilongjing-1 Hydrothermal Field respectively (Fig.4-27).				

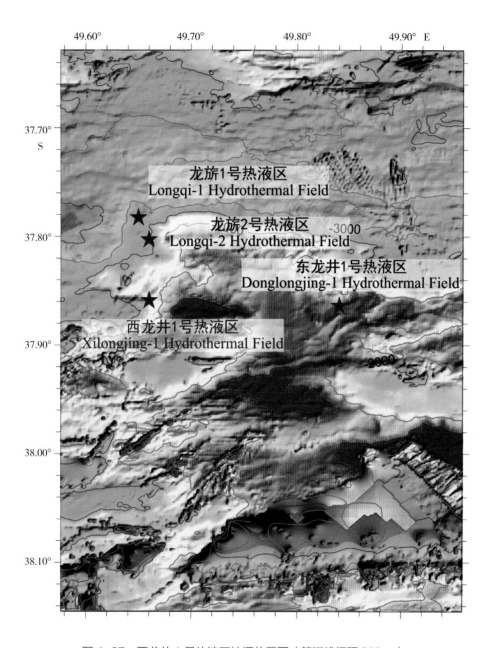

图 4-27　西龙井 1 号热液区地理位置图（等深线间隔 200 m）

Fig.4-27　Bathymetric map of the Xilongjing-1 Hydrothermal Field

(The contour interval is 200 m)

4.3.7 东龙井 1 号热液区
Donglongjing-1 Hydrothermal Field

中文名称 Chinese Name	东龙井 1 号热液区 Donglongjing-1 Reyequ	英文名称 English Name	Donglongjing-1 Hydrothermal Field	所在大洋 Ocean or Sea	西南印度洋 Southwest Indian Ocean
发现情况 Discovery Facts	此热液区于 2014 年由中国科考船"大洋一号"在执行 DY125-30 航次调查时，通过海底观察及取样证实为碳酸盐热液区。 该热液区位于西南印度洋洋中脊中央裂谷南侧，围岩为玄武岩，水深约 1 700 m。 This hydrothermal field was discovered by the Chinese R/V *Dayang Yihao* during the DY125-30 cruise in 2014. It was confirmed to be carbonate hydrothermal field by undersea observation and samples. Donglongjing-1 Hydrothermal Field is located in the south to the central rift valley of the Southwest Indian Ridge. The surrounding rocks are basalt and the water depth is 1 700 m.				
命名历史 Name History	在 DY125-30 航次现场报告中曾用名为"东龙井热液区"，英文曾用名为"Dragon Well East"。 It was once named as "Donglongjing Reyequ" in Chinese and "Dragon Well East" in English in the Chinese DY125-30 ocean cruise field report.				
特征点坐标 Coordinates	37°51.89′S，49°50.57′E			水深 (m) Depth (m)	1 700
命名释义 Reason for Choice of Name	龙井是一种绿茶，位列中国十大名茶之首，原产于杭州西湖一带。本区发现东西两个热液区，故分别命名为东龙井 1 号热液区和西龙井 1 号热液区（图 4-28）。 Longjing is kind of green tea and it ranks first in Chinese top ten tea, which is grown on the hills around the West Lake in Hangzhou. Two hydrothermal fields were discovered in this area and are named Donglongjing-1 Hydrothermal Field and Xilongjing-1 Hydrothermal Field respectively (Fig.4-28).				

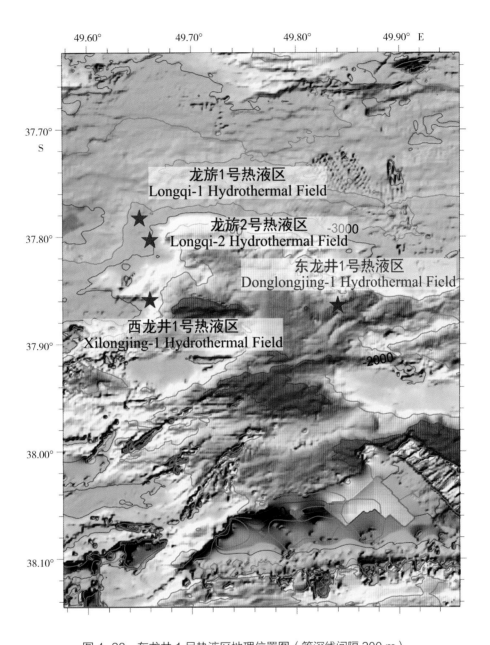

图 4-28　东龙井 1 号热液区地理位置图（等深线间隔 200 m）

Fig.4-28　Bathymetric map of the Donglongjing-1 Hydrothermal Field

(The contour interval is 200 m)

4.3.8 断桥1号热液区
Duanqiao-1 Hydrothermal Field

中文名称 Chinese Name	断桥1号热液区 Duanqiao-1 Reyequ	英文名称 English Name	Duanqiao-1 Hydrothermal Field	所在大洋 Ocean or Sea	西南印度洋 Southwest Indian Ocean
发现情况 Discovery Facts	此热液区于 2008 年 5 月由中国科考船"大洋一号"在执行 DY115-20 航次调查时,通过海底观察及取样证实。 该热液区位于西南印度洋洋中脊轴部高地上,围岩为玄武岩,总体水深约 1 700 m。 This hydrothermal field was discovered by the Chinese R/V *Dayang Yihao* during the DY115-20 cruise in May, 2008. It was confirmed by undersea observation and samples. This hydrothermal field is located in the axial elevation of the Southwest Indian Ridge. The surrounding rocks are basalt and the water depth is about 1 700 m.				
命名历史 Name History	在 DY115-20 航次及其他相关文献曾用名为"50.5°E 热液区"。DY125-30 航次现场报告中曾用名"断桥热液区"。 It was once named "50.5°E Hydrothermal Field" in the Chinese DY115-20 cruise report and other related documents. It was once named "Duanqiao Hydrothermal Field" in the Chinese DY125-30 ocean cruise field report.				
特征点坐标 Coordinates	37°39.45′S,50°28.02′E			水深 (m) Depth (m)	1 700
命名释义 Reason for Choice of Name	断桥是杭州西湖十景"断桥残雪"的所在地。汉族民间爱情传说《白蛇传》的故事即发生于此。传说白娘子与许仙断桥相会,为断桥景物增添了浪漫色彩。该热液区发育于两侧相对低洼而中部相对隆起的地貌部位,形似沟通南北两侧的桥梁,故以此命名(图 4–29)。 Duanqiao is the location of "Duan Qiao Can Xue" among the Ten Sceneries of the West Lake in Hangzhou, China. This hydrothermal field developed from the geomorphic part with two sides being low and central section being rised, which looked like bridges connecting south and north. That is why it got its name (Fig.4–29).				

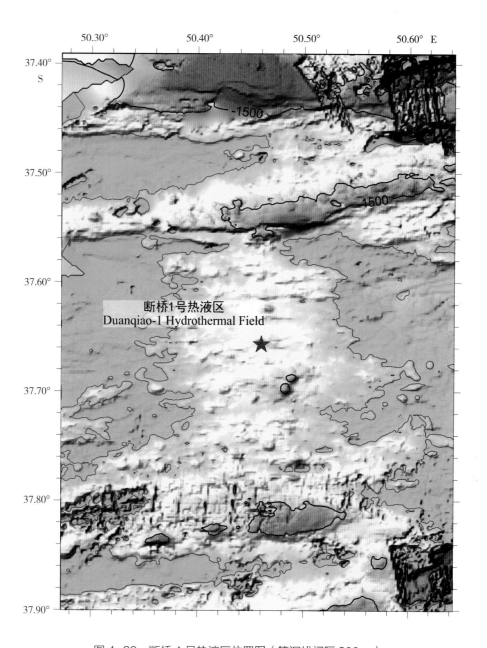

图 4-29　断桥 1 号热液区位置图（等深线间隔 200 m）

Fig.4-29　Bathymetric map of the Duanqiao-1 Hydrothermal Field

(The contour interval is 200 m)

4.3.9 栈桥 1 号热液区
Zhanqiao-1 Hydrothermal Field

中文名称 Chinese Name	栈桥 1 号热液区 Zhanqiao-1 Reyequ	英文名称 English Name	Zhanqiao-1 Hydrothermal Field	所在大洋 Ocean or Sea	西南印度洋 Southwest Indian Ocean
发现情况 Discovery Facts	此热液区于 2014 年由中国科考船"大洋一号"在执行 DY125-30 航次调查时,通过海底观察、水体热液异常探测和取样证实。 　　该热液区位于西南印度洋洋中脊中央裂谷南侧斜坡之上,围岩为玄武岩,水深约 2 500 m。 This hydrothermal field was discovered by the Chinese R/V *Dayang Yihao* during the DY125-30 cruise in 2014. It was confirmed by undersea observation, hydrothermal anomaly detection and samples. Zhanqiao-1 Hydrothermal Field is located in the slope south to the central rift valley of the Southwest Indian Ridge. The surrounding rocks are basalt and the water depth is 2 500 m.				
命名历史 Name History	在 DY125-30 航次现场报告中曾用名为"栈桥热液区",英文曾用名为"Landing Stage Hydrothermal Field"。 It was once named as "Zhanqiao Reyequ" in Chinese and "Landing Stage Hydrothermal Field" in English in the Chinese DY125-30 ocean cruise field report.				
特征点坐标 Coordinates	37°33.27′S,50°59.81′E			水深 (m) Depth(m)	2 500
命名释义 Reason for Choice of Name	栈桥是青岛海滨著名景点之一,全长 440 米,南端筑有回澜阁,是观海看景的绝佳之处之一(图 4–30)。 Zhanqiao is a famous site in Qingdao, China, which extends to the Qingdao Bay as long as 440 m. It is a perfect site for sea view in Qingdao (Fig.4–30).				

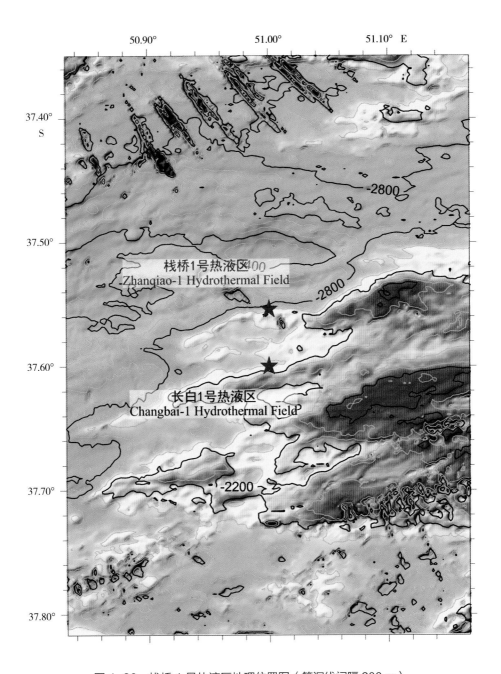

图 4-30　栈桥 1 号热液区地理位置图（等深线间隔 200 m）

Fig.4-30　Bathymetric map of the Zhanqiao-1 Hydrothermal Field

(The contour interval is 200 m)

4.3.10 长白1号热液区
Changbai-1 Hydrothermal Field

中文名称 Chinese Name	长白1号热液区 Changbai-1 Reyequ	英文名称 English Name	Changbai-1 Hydrothermal Field	所在大洋 Ocean or Sea	西南印度洋 Southwest Indian Ocean
发现情况 Discovery Facts	此热液区于 2008 年由中国科考船"大洋一号"在执行 DY115-20 航次调查时，通过海底观察及取样证实该区域为碳酸盐区。 该热液区位于西南印度洋洋中脊中央裂谷南侧的斜坡，围岩为玄武岩，水深 2 200 m 至 2 600 m。 This hydrothermal field was discovered by the Chinese R/V *Dayang Yihao* during the DY115-20 cruise in 2008. It was confirmed to be carbonate area by undersea observation and samples. This hydrothermal field is located in the slope south to the central rift valley of the Southwest Indian Ridge. The surrounding rocks are basalt. The water depth is 2 200 m to 2 600 m.				
命名历史 Name History	DY115-20 航次曾使用"51°E 热液区"名称。DY125-30 航次现场报告中曾使用"长白山热液区"名称（陶春辉等，2014）。 It was once named "51°E Hydrothermal Field" in the Chinese DY115-20 ocean cruise. It was once named "Changbaishan Hydrothermal Field" in the Chinese DY125-30 ocean cruise field report (Tao et al., 2014).				
特征点坐标 Coordinates	37°36.00′S，51°00.00′E			水深 (m) Depth (m)	2 200 ~ 2 600
命名释义 Reason for Choice of Name	通过视像系统发现该区域海底覆盖了大范围的白色碳酸盐沉积，故根据其颜色命名为"长白1号热液区"（图 4–31）。 This area is covered by wide range of carbonate deposits with white color, from the visible video system. Bai means white color in Chinese. So it was named "Changbai-1 Hydrothermal Field" according to the color (Fig.4–31).				

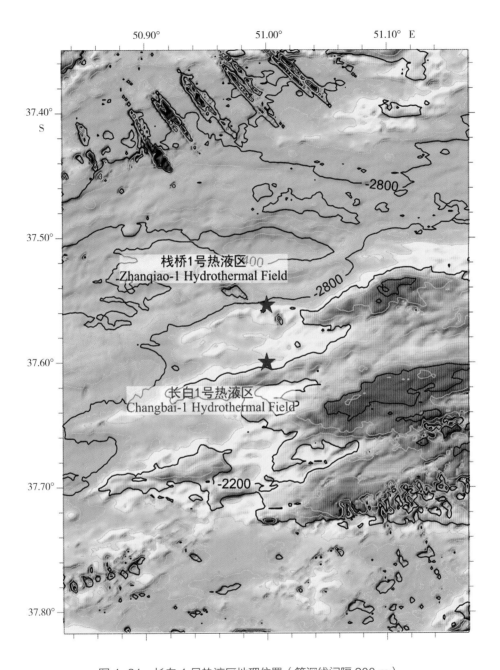

图 4-31　长白 1 号热液区地理位置（等深线间隔 200 m）

Fig.4-31　Bathymetric map of the Changbai-1 Hydrothermal Field

(The contour interval is 200 m)

4.3.11 骏惠海山
Junhui Seamount

中文名称 Chinese Name	骏惠海山 Junhui Haishan	英文名称 English Name	Junhui Seamount	所在大洋 Ocean or Sea	西南印度洋 Southwest Indian Ocean
发现情况 Discovery Facts	此海山于 2009 年由中国科考船"大洋一号"在执行 DY115-21 航次时调查发现。 This seamount was discovered by the Chinese R/V *Dayang Yihao* during the DY115-21 cruise in 2009.				
命名历史 Name History	在我国大洋航次调查报告和 GEBCO 地名辞典中未命名。 This seamount has not been named in Chinese cruise reports or GEBCO gazetteer.				
特征点坐标 Coordinates	37°30.03′S，51°42.00′E		长(km)×宽(km) Length (km) × Width (km)		23 × 14
最大水深（m） Max Depth（m）	3 200	最小水深（m） Min Depth（m）	1 300	高差（m） Total Relief（m）	1 900
地形特征 Feature Description	该海山位于西南印度洋洋中脊中央裂谷，东西宽约 23 km，南北 14 km，北侧紧邻近东西向海沟，高差约 1 900 m（图 4-32）。 Junhui Seamount is located in the central rift valley of the Southwest Indian Ridge. The width is 23 km from east to west and 14 km from north to south. It is adjacent to a nearly E to W running trench on the north. The total relief is 1 900 m (Fig.4-32).				
命名释义 Reason for Choice of Name	"骏惠"出自《诗经·周颂·维天之命》中的诗句，"骏惠我文王，曾孙笃之。""骏惠"指极力遵从。此诗旨在赞美周文王上应天命，品德纯美；文王德业泽被后代，后代当遵其遗教，发扬光大。 "Junhui" comes from a poem named *Weitianzhiming* in *Shijing · Zhousong*. *Shijing* is a collection of ancient Chinese poems from 11th century B.C. to 6th century B.C. "Endowed with favours plus Lord Wen's fame, his followers will remain the same." "Junhui" means sparing no effort to obey. This poem means that we should follow the noble character of Lord Wen.				

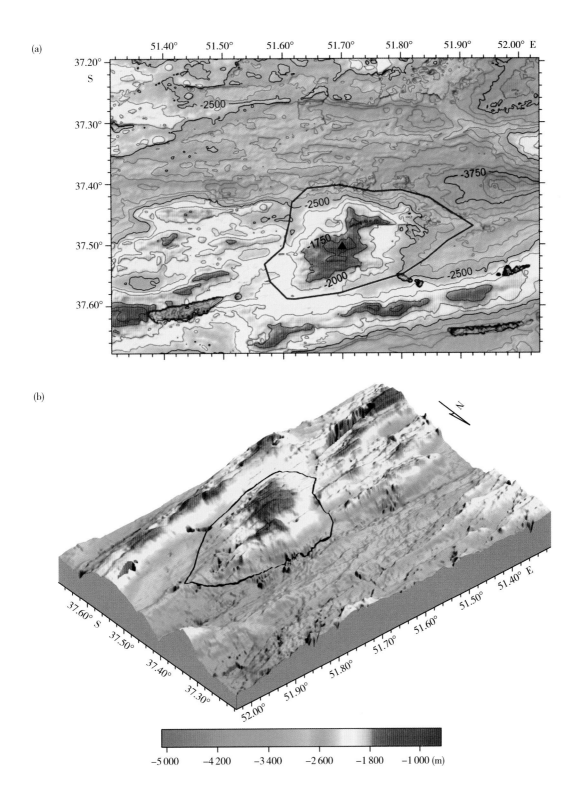

图 4-32　骏惠海山

(a) 地形图（等深线间隔 250 m）；(b) 三维图

Fig.4-32　Junhui Seamount

(a) Bathymetric map (the contour interval is 250 m); (b) 3-D topographic map

4.3.12 骏惠 1 号热液区
Junhui-1 Hydrothermal Field

中文名称 Chinese Name	骏惠 1 号热液区 Junhui-1 Reyequ	英文名称 English Name	Junhui-1 Hydrothermal Field	所在大洋 Ocean or Sea	西南印度洋 Southwest Indian Ocean
发现情况 Discovery Facts	此热液区于 2009 年由中国科考船"大洋一号"在执行 DY115-20 航次调查时，通过海底观察及水体热液异常探测发现。 该热液区位于西南印度洋洋中脊中央裂谷南侧，围岩为玄武岩，水深约 1 600 m。 This hydrothermal field was discovered by the Chinese R/V *Dayang Yihao* during the DY115-20 cruise in 2009. It was confirmed by undersea observation and hydrothermal anomaly detection. Junhui-1 Hydrothermal Field is located in south to the central rift valley of the Southwest Indian Ridge. The surrounding rocks are basalt and the water depth is about 1 600 m.				
命名历史 Name History	在我国大洋航次调查报告中未命名。 This hydrothermal field has not been named in the Chinese cruise reports.				
特征点坐标 Coordinates	37°27.94′S，51°43.91′E			水深 (m) Depth (m)	1 600
命名释义 Reason for Choice of Name	采用热液区附近的"骏惠海山"的名字命名，命名为"骏惠 1 号热液区"（图 4-33）。 This hydrothermal field is named Junhui-1 Hydrothermal Field after the nearby Junhui Seamount (Fig.4-33).				

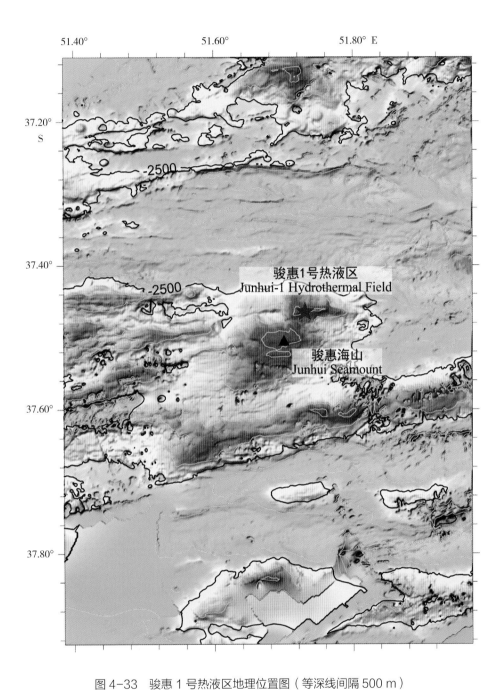

图 4-33　骏惠 1 号热液区地理位置图（等深线间隔 500 m）

Fig.4-33　Bathymetric map of the Junhui-1 Hydrothermal Field

(The contour interval is 500 m)

4.3.13 乔岳海山
Qiaoyue Seamount

中文名称 Chinese Name	乔岳海山 Qiaoyue Haishan	英文名称 English Name	Qiaoyue Seamount	所在大洋 Ocean or Sea	西南印度洋 Southwest Indian Ocean
发现情况 Discovery Facts	乔岳海山于 2009 年 2 月由中国科考船"大洋一号"在执行 DY115-21 航次时调查发现。 This seamount was discovered by the Chinese R/V *Dayang Yihao* during the DY115-21 cruise in February, 2009.				
命名历史 Name History	由我国命名为"乔岳海山",于 2012 年提交 SCUFN 审议通过。 Qiaoyue Seamount was named by China and approved by SCUFN in 2012.				
特征点坐标 Coordinates	37°20.00′S,52°07.00′E			长(km)×宽(km) Length (km) × Width (km)	43 × 24
最大水深（m） Max Depth（m）	4 200	最小水深（m） Min Depth（m）	1 400	高差（m） Total Relief（m）	2 800
地形特征 Feature Description	乔岳海山位于西南印度洋洋中脊中央裂谷,东西宽约 43 km,南北 24 km,海山顶部水深最浅 1 400 m,山麓水深 4 200 m,高差约 2 800 m。山体陡峭（图 4–34）。 Qiaoyue Seamount is located in the central rift valley of the Southwest Indian Ridge. The width is 43 km from east to west and 24 km from north to south. The top depth of this seamount is 1 400 m and the piedmont depth is 4 200 m, which makes the total relief being 2 800 m. It has steep slopes (Fig.4–34).				
命名释义 Reason for Choice of Name	"乔岳"取自《诗经·周颂·般》中的诗句"陟其高山,嶞山乔岳,允犹翕河"。此篇旨在赞美周朝的美丽河山,乔岳表示高大的山峰,形容海山形态高大。 "Qiaoyue" comes from a poem named *Ban* in *Shijing · Zhousong. Shijing* is a collection of ancient Chinese poems from 11th century B.C. to 6th century B.C. "When he looks at what lies down below, he sees the towering ridges long and narrow." *Ban* is a poem about Lord Wu who made offerings to the mountains and rivers when he was on a tour around the country. Qiaoyue means tall mountains. It was used to describe the seamount here.				

(a)

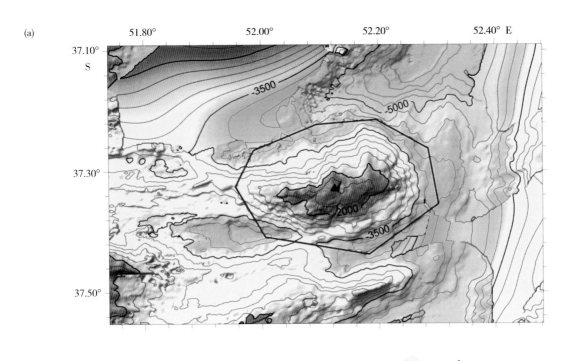

(b)

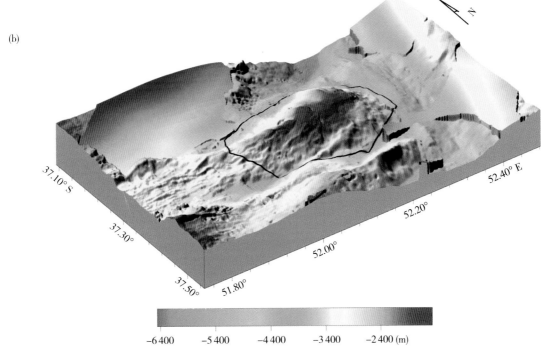

图 4-34　乔岳海山

(a) 地形图（等深线间隔 300 m）；(b) 三维图

Fig.4-34　Qiaoyue Seamount

(a) Bathymetric map (the contour interval is 300 m); (b) 3-D topographic map

4.3.14 百神 1 号热液区
Baishen-1 Hydrothermal Field

中文名称 Chinese Name	百神 1 号热液区 Baishen-1 Reyequ	英文名称 English Name	Baishen-1 Hydrothermal Field	所在大洋 Ocean or Sea	西南印度洋 Southwest Indian Ocean
发现情况 Discovery Facts	此热液区于 2007 年由中国科考船"大洋一号"在执行 DY115-20 航次调查时，通过海底观察和水体热液异常探测发现。 该热液区位于洋中脊中央裂谷北侧新生洋脊斜坡之上，水深约 2 400 m。 This hydrothermal field was discovered by the Chinese R/V *Dayang Yihao* during the DY115-20 cruise in 2007. It was confirmed by undersea observation and hydrothermal anomaly detection. Baishen-1 Hydrothermal Field is located in the slope of the new volcanic ridge north to the central rift valley of the Southwest Indian Ridge. The water depth is 2 400 m.				
命名历史 Name History	在我国大洋航次调查报告中未命名。 This hydrothermal field has not been named in the Chinese cruise reports.				
特征点坐标 Coordinates	36°06.07′S，53°18.00′E		水深 (m) Depth (m)		2 400
命名释义 Reason for Choice of Name	"百神"出自《诗经·周颂·时迈》"怀柔百神，及河乔岳，允王维后"。此篇为歌颂武王列封诸侯，威震四方，安抚百神，偃武修文，从而发扬光大祖先功业。百神指各方神灵（图 4–35）。 "Baishen" comes from a poem named *Shimai* in *Shijing · Zhousong*. *Shijing* is a collection of ancient Chinese poems from 11th century B.C. to 6th century B.C. "The host of gods he will subdue, the rills and hills subdued, too; Great indeed is our Lord Wu." This is a poem by Lord Wu to pay tribute to the various deities. Baishen means various deities (Fig.4–35).				

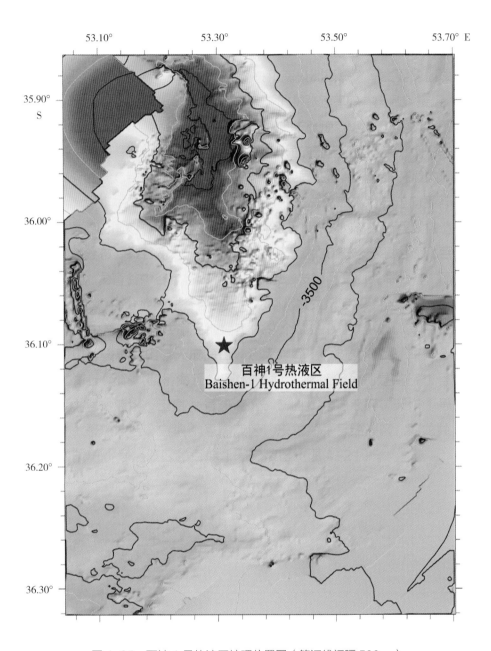

图 4-35　百神 1 号热液区地理位置图（等深线间隔 500 m）

Fig.4-35　Bathymetric map of the Baishen-1 Hydrothermal Field

(The contour interval is 500 m)

4.3.15 思文海脊
Siwen Ridge

中文名称 Chinese Name	思文海脊 Siwen Haiji		英文名称 English Name	Siwen Ridge	所在大洋 Ocean or Sea	西南印度洋 Southwest Indian Ocean
发现情况 Discovery Facts	此海脊于 2012 年由中国科考船"大洋一号"在执行 DY125-26 航次时调查发现。 Siwen Ridge was discovered by the Chinese R/V *Dayang Yihao* during the DY125-26 cruise in 2012.					
命名历史 Name History	该海脊由我国命名为思文海脊,于 2014 年提交 SCUFN 审议通过。 Siwen Ridge was named by China and approved by SCUFN in 2014.					
特征点坐标 Coordinates	28°30.50′S,62°29.70′E			长(km)× 宽(km) Length (km)× Width (km)		13 × 49
最大水深(m) Max Depth(m)	4 750	最小水深(m) Min Depth(m)	2 700	高差(m) Total Relief(m)		2 050
地形特征 Feature Description	该海脊位于洋中脊中央裂谷,山体呈长条状,高差约 2 050 m(图 4–36)。 Siwen Ridge is located in the central rift valley of the Southwest Indian Ridge and has an elongated shape. The total relief is about 2 050 m (Fig.4–36).					
命名释义 Reason for Choice of Name	"思文"出自《诗经·周颂·思文》"思文后稷,克配彼天"。"思文"指纪念、回忆,"后稷"指周人的始祖,此诗指后人想起后稷当初治理国家的功德,可与上天相比。以"思文"命名此海脊,表示后人不忘祖先的伟大功绩。 "Siwen" comes from a poem named *Siwen* in *Shijing · Zhousong. Shijing* is a collection of ancient Chinese poems from 11th century B.C. to 6th century B.C. "Oh, virtuous Houji, you're great as Heaven." "Siwen" means commemorating and remembering. It means that the descendants of the Zhou regarded the achievements and merits of Houji, who was the leader of their ancestors, as comparable to deity. The ridge is named Siwen in order to showcase that Chinese will not forget the great achievements of their ancestors.					

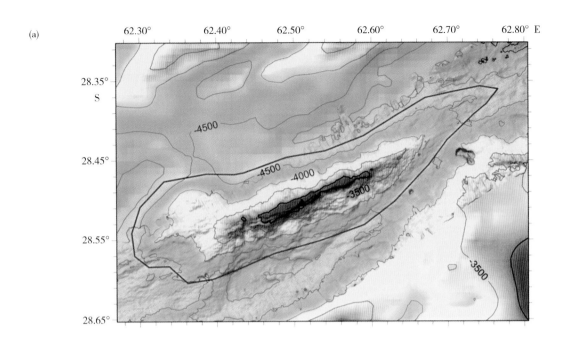

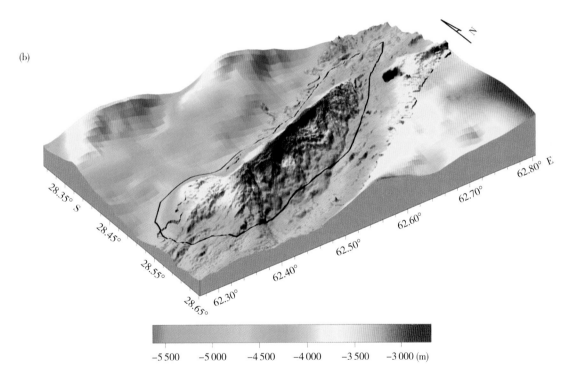

图 4-36 思文海脊

(a) 地形图 （等深线间隔 500 m）；(b) 三维图

Fig.4-36 Siwen Ridge

(a) Bathymetric map (the contour interval is 500 m); (b) 3-D topographic map

4.3.16 天作海山
Tianzuo Seamount

中文名称 Chinese Name	天作海山 Tianzuo Haishan	英文名称 English Name	Tianzuo Seamount	所在大洋 Ocean or Sea	西南印度洋 Southwest Indian Ocean
发现情况 Discovery Facts					
命名历史 Name History					
特征点坐标 Coordinates	27°53.20′S，63°27.80′E			长(km)×宽(km) Length (km)× Width (km)	48×28
最大水深（m） Max Depth（m）	5 200	最小水深（m） Min Depth（m）	2 550	高差（m） Total Relief（m）	2 650
地形特征 Feature Description					
命名释义 Reason for Choice of Name					

发现情况 / Discovery Facts:
此海山于 2009 年 2 月由中国科考船 "大洋一号" 在执行 DY115-21 航次时调查发现。
This seamount was discovered by the Chinese R/V *Dayang Yihao* during the DY115-21 cruise in February, 2009.

命名历史 / Name History:
由我国命名为天作海山，于 2013 年提交 SCUFN 审议通过。
Tianzuo Seamount was named by China and approved by SCUFN in 2013.

地形特征 / Feature Description:
该海山位于洋中脊中央裂谷，高差约 2 650 m（图 4–37）。
Tianzuo Seamount is located in the central rift valley of the Southwest Indian Ridge. The total relief is about 2 650 m (Fig.4–37).

命名释义 / Reason for Choice of Name:
"天作" 出自《诗经·周颂·天作》"天作高山，大王荒之。" 意思是高耸的岐山自然形成，大王带领国民勇于开拓。2009 年中国科学家在此海山首次发现了与超基性岩相关的多金属硫化物矿点，是中国大洋科考的重大突破，此海山命名为天作海山，以纪念此发现。

"Tianzuo" comes from a poem named *Tianzuo* in *Shijing · Zhousong*. *Shijing* is a collection of ancient Chinese poems from 11th century B.C. to 6th century B.C. "The heaven made the mountain grand; Lord Tai reclaimed the barren land". Chinese scientists first found polymetallic sulfide mines related to ultrabasic rocks on this seamount in 2009. It is a major breakthrough of Chinese ocean scientific survey. This seamount is named Tianzuo seamount in the memory of this discovery.

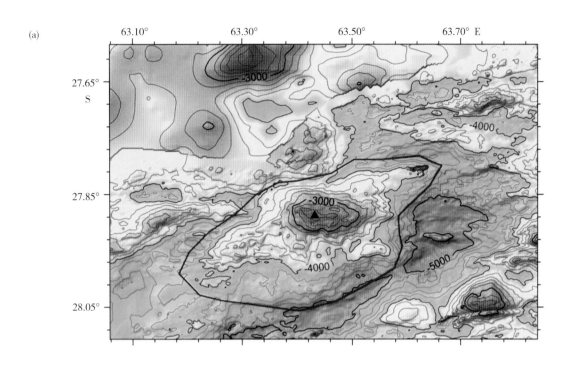

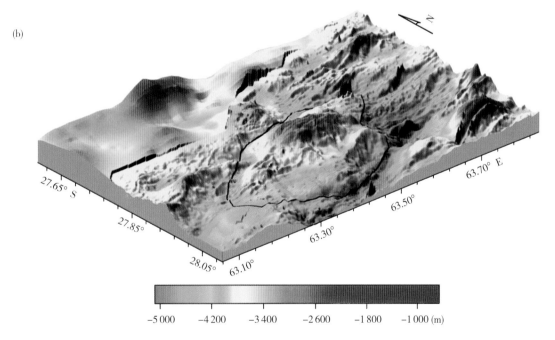

图 4-37　天作海山

(a) 地形图（等深线间隔 200 m）；(b) 三维图

Fig.4-37　Tianzuo Seamount

(a) Bathymetric map (the contour interval is 200 m); (b) 3-D topographic map

4.3.17 天成海山
Tiancheng Seamount

中文名称 Chinese Name	天成海山 Tiancheng Haishan	英文名称 English Name	Tiancheng Seamount	所在大洋 Ocean or Sea	西南印度洋 Southwest Indian Ocean
发现情况 Discovery Facts	此海山于 2009 年由中国科考船"大洋一号"在执行 DY115-21 航次时调查发现。 This seamount was discovered by the Chinese R/V *Dayang Yihao* during the DY115-21 cruise in 2009.				
命名历史 Name History	在我国大洋航次调查报告和 GEBCO 地名辞典中未命名。 This seamount has not been named in Chinese cruise reports or GEBCO gazetteer.				
特征点坐标 Coordinates	27°48.90′S，63°53.70′E			长 (km)×宽 (km) Length (km)× Width (km)	21×12
最大水深（m） Max Depth（m）	4 000	最小水深（m） Min Depth（m）	2 000	高差（m） Total Relief（m）	2 000
地形特征 Feature Description	该海山位于西南印度洋洋中脊中央裂谷，南北宽约 21 km，东西 12 km，高差约 2 000 m。边坡陡峭，南侧形成断崖（图 4–38）。 Tiancheng Seamount is located in the central rift valley of the Southwest Indian Ridge. The width is 21 km from north to south and 12 km from east to west. The total relief is 2 000 m. The slopes are steep, and the southern side of this seamount forms a cliff (Fig.4–38).				
命名释义 Reason for Choice of Name	"天成"出自《诗经·周颂·昊天有成命》中的诗句"昊天有成命，二后受之"。此诗为祭祀周王的作品，诗句意指文王武王接受天命，尽心治理国家，天下安定。 "Tiancheng" comes from a poem named *Haotianyouchengming* in *Shijing · Zhousong*. *Shijing* is a collection of ancient Chinese poems from 11th century B.C. to 6th century B.C. "The heaven issued lofty orders; Lord Wen and Wu received the orders." The poem means that Lord Wen and Wu accepted God's command, dedicated to govern the country, in order to keep the world peaceful and stable.				

(a)

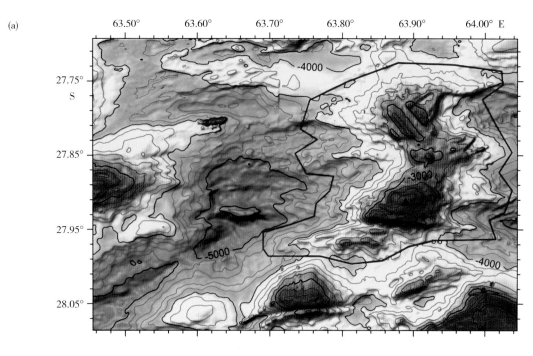

(b)

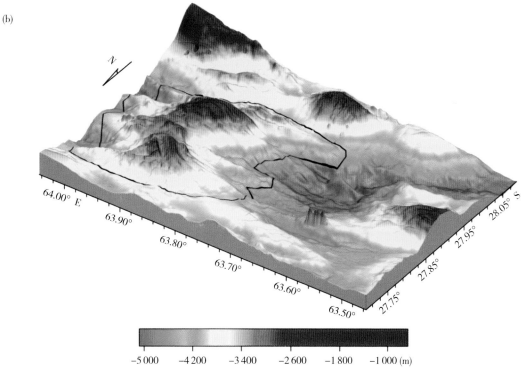

图 4-38　天成海山

(a) 地形图（等深线间隔 200 m）；(b) 三维图

Fig.4-38　Tiancheng Seamount

(a) Bathymetric map (the contour interval is 200 m); (b) 3-D topographic map

4.3.18　天作 1 号热液区和天成 1 号热液区
Tianzuo-1 Hydrothermal Field and Tiancheng-1 Hydrothermal Field

中文名称 Chinese Name	天作 1 号热液区 天成 1 号热液区 Tianzuo-1 Reyequ Tiancheng-1 Reyequ	英文名称 English Name	Tianzuo-1 Hydrothermal Field Tiancheng-1 Hydrothermal Field	所在大洋 Ocean or Sea	西南印度洋 Southwest Indian Ocean
发现情况 Discovery Facts	两个热液区于 2009 年由中国科考船 "大洋一号" 在执行 DY115-20 航次调查时，通过海底观察及取样证实。 两个热液区位于西南印度洋洋中脊中央裂谷新生洋脊之上，水深 2 900 ～ 3 800 m， These two hydrothermal fields were discovered by the Chinese R/V *Dayang Yihao* during the DY115-20 cruise in 2009. They were confirmed by undersea observation and samples. The two hydrothermal fields are located in the new volcanic ridge of the central rift valley of the Southwest Indian Ridge. The water depth is 2 900 m to 3 800 m.				
命名历史 Name History	在我国大洋航次调查报告中未命名。 It has not been named in the Chinese cruise reports.				
特征点坐标 Coordinates	27°57.04′S，63°32.48′E（天作 1 号 / Tianzuo-1） 27°51.10′S，63°55.40′E（天成 1 号 / Tiancheng-1）			水深 (m) Depth (m)	2 900 ～ 3 800
命名释义 Reason for Choice of Name	根据两个热液区的地理位置，采用相邻的海底地理实体天作海山、天成海山的专名分别对两个热液区命名（图 4-39）。 According to their locations, these two hydrothermal fields were named after the adjacent Tianzuo Seamount and Tiancheng Seamount (Fig.4-39).				

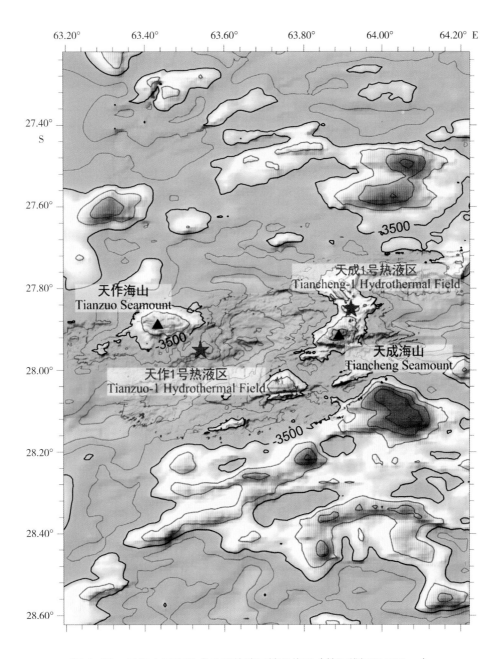

图 4-39　天作 1 号和天成 2 号热液区地理位置（等深线间隔 500 m）

Fig.4-39　Bathymetric map of the Tianzuo-1/Tiancheng-1 Hydrothermal Field

(The contour interval is 500 m)

4.3.19 汪大渊海山
Wangdayuan Seamount

中文名称 Chinese Name	汪大渊海山 Wangdayuan Haishan	英文名称 English Name	Wangdayuan Seamount	所在大洋 Ocean or Sea	西南印度洋 Southwest Indian Ocean
发现情况 Discovery Facts					

发现情况 Discovery Facts	此海山于 2010 年和 2016 年两次由中国科考船"大洋一号"在执行测量任务时调查发现。 This seamount was discovered by the Chinese R/V *Dayang Yihao* during the survey in 2010 and 2016.
命名历史 Name History	由我国命名为汪大渊海山，于 2017 年提交 SCUFN 审议通过。 This feature was named Wangdayuan Seamount by China and the name was approved by SCUFN in 2017.

特征点坐标 Coordinates	34°53.60′S, 54°16.77′E		长(km)×宽(km) Length (km)× Width (km)	21×14
最大水深（m） Max Depth（m）	3 500	最小水深（m） Min Depth（m）	850	高差（m） Total Relief（m）	2 650

地形特征 Feature Description	此海山位于西南印度洋脊（图 4–40）。 This seamount is located in the Southwest Indian Ridge (Fig.4–40).
命名释义 Reason for Choice of Name	汪大渊（1311 年～？），字焕章，江西南昌人，元代杰出的民间航海旅行家，曾两次进行远洋考察，远至东南亚、南亚、西亚、印度洋、地中海等广大海域。他根据亲身经历著有《岛夷志略》，涉及亚、非、欧三大洲 220 多个国家与地区，是考察元代远洋活动的重要文献。以汪大渊命名此海山，以此纪念其对远洋考察的贡献。 Wang Dayuan (A.D. 1311~?), whose another name taken at the age of twenty was Huanzhang, lived in Nanchang, Jiangxi Province, China. He was an outstanding folk navigator and traveler who had carried out two far-sea explorations, as far as Southeast Asia, South Asia, West Asia, Indian Ocean, Mediterranean Sea and so on. According to personal experience, he wrote a famous book, which involved more than 220 nations and regions of Asia, Africa and Europe. It is an important literature to study far-sea activities of Yuan Dynasty. The seamount is named after "Wangdayuan", to commemorate his contributions to far-sea explorations.

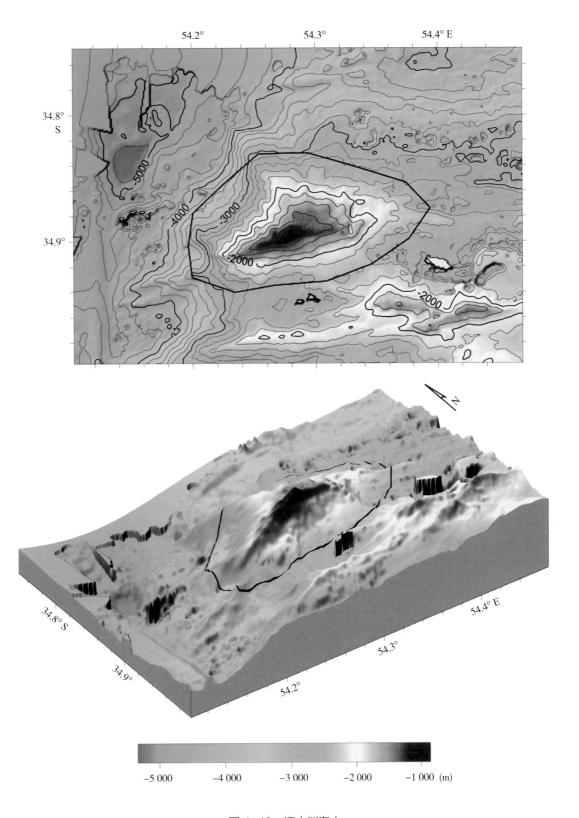

图 4-40　汪大渊海山

(a) 地形图（等深线间隔 200 m）；(b) 三维图

Fig.4-40　Wangdayuan Seamount

(a) Bathymetric map (the contour interval is 200 m); (b) 3-D topographic map

4.3.20 封演海山
Fengyan Seamount

中文名称 Chinese Name	封演海山 Fengyan Haishan	英文名称 English Name	Fengyan Seamount	所在大洋 Ocean or Sea	西南印度洋 Southwest Indian Ocean
发现情况 Discovery Facts	此海山于 2010 年由中国科考船"大洋一号"在执行测量任务时调查发现。 This seamount was discovered by the Chinese R/V *Dayang Yihao* during the survey in 2010.				
命名历史 Name History	由我国命名为封演海山，于 2017 年提交 SCUFN 审议通过。 This feature was named Fengyan Seamount by China and the name was approved by SCUFN in 2017.				
特征点坐标 Coordinates	37°52.94′S, 48°24.32′E			长（km）× 宽（km） Length (km) × Width (km)	27 × 16
最大水深（m） Max Depth（m）	2 500	最小水深（m） Min Depth（m）	1 100	高差（m） Total Relief（m）	1 400
地形特征 Feature Description	此海山位于西南印度洋脊，呈长条形，近东西走向（图 4–41）。 This seamount is located in the Southwest Indian Ridge. It has an elongated shape and extends towards nearly E-W (Fig.4–41).				
命名释义 Reason for Choice of Name	封演（生卒年不详），中唐时人，撰《说潮》一文，首次提出了月亮和海水"潜相感"（相互作用）而引发潮汐的观点，清晰地论述一月之中海潮变化的规律。以封演命名此海山，以此纪念其对潮汐研究的贡献。 Feng Yan (age unknown) lived in Middle Tang Dynasty. He wrote a famous book which first proposed the viewpoint that the moon's interaction with sea water each other causes tide, which clearly discussed the tide variation law during a month. The seamount is named after "Fengyan" to memorize his contributions to study on tide.				

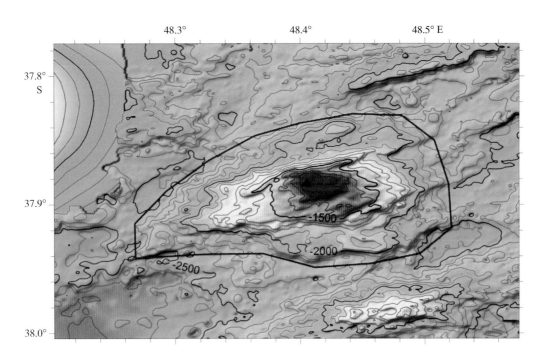

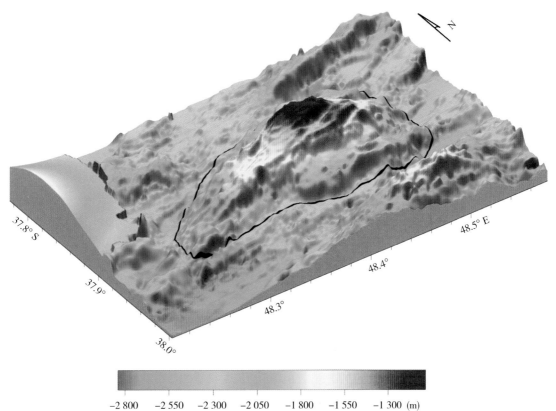

图 4-41　封演海山

(a) 地形图（等深线间隔 100 m）；(b) 三维图

Fig.4-41　Fengyan Seamount

(a) Bathymetric map (the contour interval is 100 m); (b) 3-D topographic map

4.3.21 断桥海丘
Duanqiao Hill

中文名称 Chinese Name	断桥海丘 Duanqiao Haiqiu	英文名称 English Name	Duanqiao Hill	所在大洋 Ocean or Sea	西南印度洋 Southwest Indian Ocean
发现情况 Discovery Facts	此海丘于 2016 年由中国科考船"向阳红"10 船在执行测量任务时调查发现。 This hill was discovered by the Chinese R/V *Xiangyanghong* 10 during the survey in 2016.				
命名历史 Name History	由我国命名为断桥海丘，于 2017 年提交 SCUFN 审议通过。 This feature was named Duanqiao Hill by China and the name was approved by SCUFN in 2017.				
特征点坐标 Coordinates	37°37.42′S, 50°27.79′E			长(km)× 宽(km) Length (km) × Width (km)	2 × 1.5
最大水深（m） Max Depth（m）	1 800	最小水深（m） Min Depth（m）	1 500	高差（m） Total Relief（m）	300
地形特征 Feature Description	此海丘位于西南印度洋脊，近似圆形（图 4–42）。 This hill is located at Southwest Indian Ridge and has a near circular shape (Fig.4–42).				
命名释义 Reason for Choice of Name	2008 年，在该区域发现一处海底热液区，命名为断桥热液区，该热液区发育于两侧相对低洼而中部相对隆起的地貌部位，形似沟通南北两侧的桥梁，2016 年，基于 AUV 地形探测，在热液区附近发现该海丘，命名为断桥海丘。发现此热液区时，调查船的首席科学家的家乡在中国杭州，于是以其家乡著名景点名称"断桥"命名。 A undersea hydrothermal field was found in this region in 2008 and was named "Duanqiao Hydrothermal Field". It develops in the geomorphic unit which is low-lying in both sides and ridgy in the middle relatively, just like a bridge connecting north and south. This hill was found near the hydrothermal field based on AUV topography exploration in 2016 and named "Duanqiao Hill". When it was found, hometown of the chief scientist of the R/V is in Hangzhou ,China. So it was named after a famous tour site called "Duanqiao" in his hometown.				

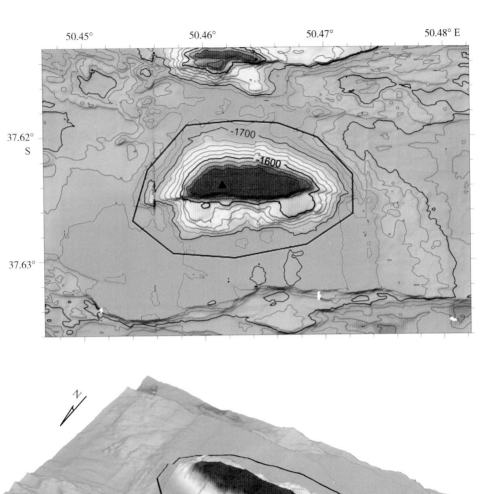

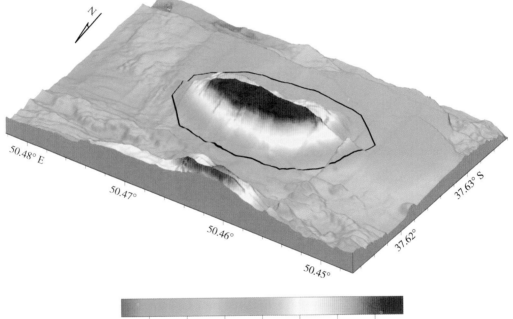

−1 850　−1 800　−1 750　−1 700　−1 650　−1 600　−1 550　(m)

图 4-42　断桥海丘

(a) 地形图（等深线间隔 20 m）；(b) 三维图

Fig.4-42　Duanqiao Hill

(a) Bathymetric map (the contour interval is 20 m); (b) 3-D topographic map

第 4 节　东南印度洋地理实体

东南印度洋中脊始于罗德里格斯三联点，向东南方向延伸，止于西南太平洋麦夸里（Macquarie）三联点，走向北西西—南东东，长约 8 200 km，宽 400 ～ 1 600 km，最深处约 4 700 m，分隔澳大利亚板块和南极洲板块（Lelieved et al., 2001）。洋中脊向东与东太平洋海隆相接，沟通印度洋和太平洋洋中脊。

东南印度洋中脊属中速扩张洋中脊（Hellebrand et al., 2002），现今洋中脊全扩张速率从靠近罗德里格斯三联点的 57.5 mm/a，增加到阿姆斯特丹和圣保罗岛处的 68 mm/a，然后逐渐增加到最大的 76.5 mm/a（50°S，114°E 附近）。随后扩张速率缓慢下降，在乔治亚断裂附近下降到 72 mm/a。东南印度洋中脊的扩张速率是非对称的（Sempéré and Cochran, 1997）。

澳大利亚南极不协调区（Australia Antarctic Discordance, AAD）位于东南印度洋中脊的 120°—128°E 段，其主要地形特征表现为深大的裂谷、起伏不平的地形和空间上靠近断裂带。

东南印度洋区域，共命名地理实体 1 个，即阳阿海脊（图 4-43）。

Section 4　Undersea Features in the Southeast Indian Ocean

The Southeast Indian Ridge starts from the Rodrigues Triple Junction, extending towards the southeast and ends at the Macquarie Triple Junction in Southwest Pacific Ocean. The ridge runs NWW to SEE with the length of 8 200 km and width of 400–1 600 km. The maximum depth of the ridge is about 4 700 m and it separates the Australian Plate and the Antarctic Plate (Lelieved et al., 2001). The mid-ocean ridge connects the East Pacific Rise in the east and links up the Indian Ocean and the Pacific Ridge.

The Southeast Indian Ridge is a kind of medium-spreading ridges (Hellebrand et al., 2002). The current full spreading rate of the mid-ocean ridge increases from 57.5 mm/a near the Rodrigues Triple Junction to 68 mm/a at Amsterdam Island and St. Paul Island, then increases to the maximum 76.5 mm/a (near 50°S, 114°E). It decreases slowly to 72 mm/a near Georgia Fracture. The spreading rate of the Southeast Indian Ridge is asymmetric (Sempéré and Cochran, 1997).

The Australia Antarctic Discordance (AAD) is located in the 120°–128°E segment of the Southeast Indian Ridge, whose major topography is featured deep and large rifts, fluctuant terrain and closed to fracture zone spatially.

In total, one undersea feature has been named by China in the region of the Southeast Indian Ocean, Yang'e Ridge (Fig. 4–43).

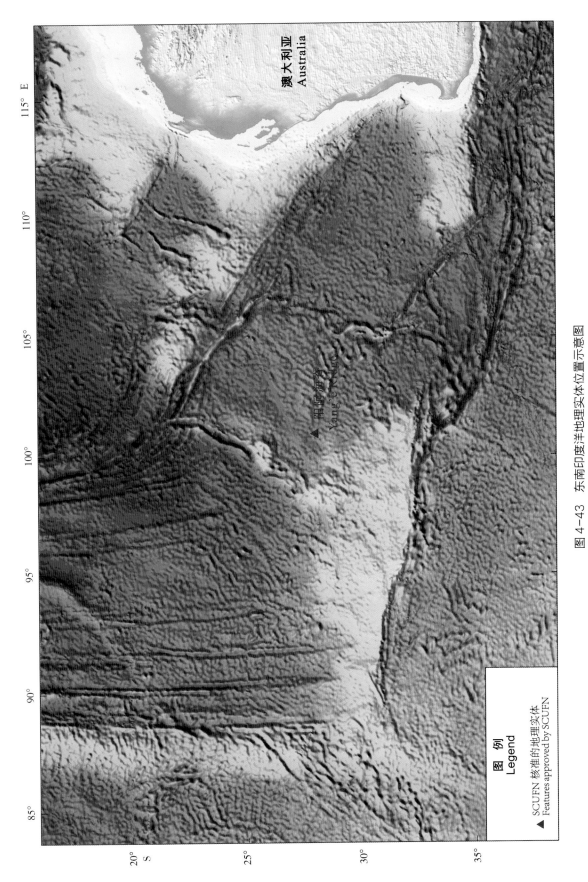

图 4-43　东南印度洋地理实体位置示意图

Fig.4-43　Location of the undersea feature in the Southeast Indian Ocean

4.4.1 阳阿海脊
Yang'e Ridge

中文名称 Chinese Name	阳阿海脊 Yang'e Haiji		英文名称 English Name	Yang'e Ridge	所在大洋 Ocean or Sea	东南印度洋 Southeast Indian Ocean
发现情况 Discovery Facts	此海山于 2014 年 7 月由中国测量船 "竺可桢号" 在执行测量任务时调查发现。 This seamount was discovered by the Chinese R/V *Chukochenhao* in July, 2014.					
命名历史 Name History	由我国命名为 "阳阿海脊"，于 2015 年提交 SCUFN 审议通过。 Yang'e Ridge was named by China and approved by SCUFN in 2015.					
特征点坐标 Coordinates	28°00.0′S，100°54.3′E			长(km)×宽(km) Length (km) × Width (km)		32×9
最大水深（m） Max Depth（m）	4 789	最小水深（m） Min Depth（m）	3 432	高差（m） Total Relief（m）		1 357
地形特征 Feature Description	此海脊位于沃顿海盆南侧，呈条带状近南北向延伸，顶部水深 3 432 m（图 4-44）。 This ridge is located in the south to Wharton Basin with a stripped shape, running nearly N to S. The top water depth is 3 432 m (Fig.4-44).					
命名释义 Reason for Choice of Name	阳阿：古代神话传说中的山名，朝阳初升时所经之处。出自屈原的《楚辞·九歌》"与女沐兮咸池，晞女髮兮阳之阿"，指对美丽女子的景仰之情。以此命名海脊，表现古人丰富的情感生活。 Yang'e, namely the place where the rising sun passes at dawn, is a legendary hill in an ancient Chinese fairy tale. The title is taken from *Chu Ci* (The Poetry of Chu State), The poem expresses the admiration for fair women. The reason why Yange'e is quoted to designate this ridge is to demonstrate the delicate inner life of ancient Chinese people.					

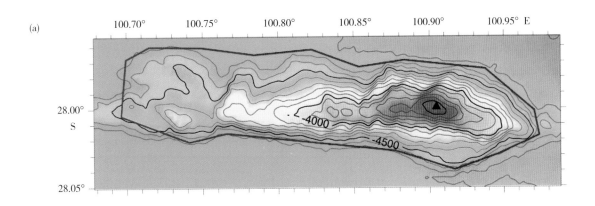

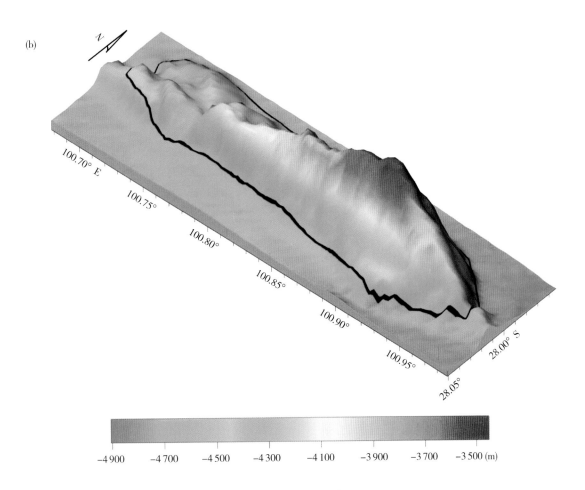

图 4-44　阳阿海脊

(a) 地形图（等深线间隔 100 m）；(b) 三维图

Fig.4-44　Yang'e Ridge

(a) Bathymetric map (the contour interval is 100 m); (b) 3-D topographic map

参考文献
References

陈永顺 . 2003 . 海底扩张和大洋中脊动力学问题概述 // 张有学主编 . 地球的结构、演化和动力学 . 北京：高等
　　教育出版社 , 283–317.

李兵 . 2014. 南大西洋中脊 14°S 内角热液区成矿作用研究 . 中国科学院大学博士学位论文 , 1–122.

陶春辉 , 李怀明 , 金肖兵 , 等 . 2014. 西南印度洋脊的海底热液活动和硫化物勘探 . 科学通报 , (19): 1812–1822.

汪榕培 , 任秀桦 . 1995. 诗经 (中英文版). 沈阳 : 辽宁教育出版社 .

Chu D, Gordon R G. 1999. Evidence for motion between Nubia and Somalia along the Southwest Indian ridge. Nature,
　　398(6722): 64–67.

Devey C W, German C, Haase K,et al.2010.The relationships between volcanism, tectonism, and hydrothermal activity on
　　the southern equatorial Mid–Atlantic Ridge. Geophysical Monograph Series, 188:133–152.

Dick H J B, Lin J, Schouten H. 2003. An ultraslow–spreading class of ocean ridge. Nature, 426(6965):405–412.

Dyment J. 1998. Evolution of the Carlsberg Ridge between 60 and 45 Ma: Ridge propagation, spreading asymmetry, and
　　the Deccan–Reunion hotspot. Journal of Geophysical Research: Solid Earth, 103: 24067–24084.

Flood P G. 1999. Development of northwest Pacific guyots: general results from Ocean Drilling Program legs 143 and 144.
　　The Island Arc, 8: 92–98.

Georgen J E, Lin J, and Dick H J B. 2001. Evidence from gravity anomalies for interactions of the Marion and Bouvet
　　hotspots with the Southwest Indian Ridge: Effects of transform offsets, Earth Planet. Sci. Lett., 187, 283–300.

Georgen J E, Kurz M D, Dick H J B, et al. 2003. Low 3He/4He ratios in basalt glasses from the western Southwest Indian
　　Ridge (10–24°E). Earth and Planetary Science Letters, 206(3–4): 509–528.

Han X, Wu Z, Qiu B. 2012. Morphotectonic characteristics of the northern part of the Carlsberg Ridge near the Owen
　　Fracture Zone and the occurrence of oceanic core complex formation. AGU Fall Meeting, San Francisco, abstract
　　OS13B–1722.

Hannington M, Ronde C. and Petersen S. 2005. Seafloor tectonics and submarine hydrothermal systems. Economic
　　Geology (Economic Geology 100th Anniversary Volume), 111–141.

Haymon R M, Fornari D J, Edwards M H, et al. 1991. Hydrothermal vent distribution along the East Pacific Rise
　　crest (9°09′–9°54′ N) and its relationship to magmatic and tectonic processes on fast–spreading mid–ocean ridges[J].
　　Earth and Planetary Science Letters, 104(2): 513–534.

Hellebrand E, Snow J E, Hoppe P, et al. 2002. Garnet–field melting and late–stage refertilization in 'Residual' abyssal
　　peridotites from the Central Indian Ridge. Journal of Petrology, 43(12): 2305–2338.

Kaneda K，Kodaira S，Nishizawa A，et al. 2010. Structural evolution of preexisting oceanic crust through intraplate
　　igneous activities in the Marcus–Wake seamount chain. Geochemistry Geophysics Geosystems, 11(10): 69–71.

Lelieveld J, Crutzen P J, Ramanathan V, et al. 2001. The Indian Ocean experiment: widespread air pollution from South and Southeast Asia. Science, 291(5506): 1031–1036.

Menard H W. 1966. Fracture zones and offsets of the East Pacific Rise. Journal of Geophysical Research, 71(2): 682–685.

Mendel V, Sauter D, Parson L, et al. 1997. Segmentation and morphotectonic variations along a super slow–spreading center: The Southwest Indian Ridge (57°–70°E). Marine Geophysical Researches, 19: 505–533.

Nemoto K, Kroenke L W. 1985. Sio Guyot: a complex volcanic edifice in the western Mid–Pacific Mountains[J]. Geomarine letters, 5(2): 83–89.

Sager W W, Duncan R A, Handschumacher D W. 1993. Paleomagnetism of the Japanese and Marcus–Wake seamounts, Western Pacific Ocean// Pringle M S, Sager W W, Sliter W V. et al. The Mesozoic Pacific: Geology, Tectonics, and Volcanism. American Geophysical Union, Washington, D. C: 401–435.

Sempéré J, Cochran J R. 1997. The Southeast Indian Ridge between 88°E and 118°E: Variations in crustal accretion at constant spreading rate. Journal of Geophysical Research: Solid Earth, 102(B7): 15489–15505.

Winterer E L and Metzler C V. 1984. Origin and subsidence of guyots in Mid–Pacific Mountains. Journal of Geophysical Research, 89(B12): 9969–9979.

附录：名录简表

Appendix: Simplification of Undersea Feature Names

序 号 S/N	名 称 Name	所在大洋 Ocean or Sea	特征点坐标 Coordinates	页 码 Page
1	日昇平顶海山 Risheng Guyot	西太平洋 West Pacific Ocean	20°42.60′N，127°44.10′E	18
2	日潭海丘 Ritan Hill	西太平洋 West Pacific Ocean	21°09.40′N，127°45.20′E	20
3	月潭海脊 Yuetan Ridge	西太平洋 West Pacific Ocean	20°54.16′N，128°06.01′E	22
4	海东青海山 Haidongqing Seamount	西太平洋 West Pacific Ocean	18°41.70′N，125°34.40′E	24
5	天保海山 Tianbao Seamount	西太平洋 West Pacific Ocean	17°54.10′N，123°17.20′E	26
6	静好海山 Jinghao Seamount	西太平洋 West Pacific Ocean	18°18.50′N，123°16.60′E	28
7	翠翘海丘 Cuiqiao Hill	西太平洋 Western Pacific Ocean	16°45.5′N，124°34.3′E	30
8	睢鸠海山群 Jujiu Seamount	西太平洋 Western Pacific Ocean	17°05.5′N，124°59.2′E	32
9	槐序海山 Huaixu Seamount	西太平洋 Western Pacific Ocean	16°59.9′N，129°29.3′E	34
10	首阳海山 Shouyang Seamount	西太平洋 Western Pacific Ocean	17°31.6′N，128°45.5′E	36
11	立春海山 Lichun Seamount	西太平洋 Western Pacific Ocean	14°08.0′N，132°45.2′E	38
12	惊蛰海山 Jingzhe Seamount	西太平洋 Western Pacific Ocean	14°41.7′N，134°34.1′E	40
13	春分海丘 Chunfen Hill	西太平洋 Western Pacific Ocean	14°42.1′N，134°31.0′E	42
14	清明海山 Qingming Seamount	西太平洋 Western Pacific Ocean	15°10.1′N，133°55.2′E	44
15	谷雨海山 Guyu Seamount	西太平洋 Western Pacific Ocean	15°43.6′N，131°40.8′E	46
16	小满海丘 Xiaoman Hill	西太平洋 Western Pacific Ocean	15°32.5′N，134°32.9′E	48

序 号 S/N	名 称 Name	所在大洋 Ocean or Sea	特征点坐标 Coordinates	页 码 Page
17	芒种海盆 Mangzhong Basin	西太平洋 Western Pacific Ocean	15°21.0′N，134°12.2′E	50
18	飒沓海山 Sata Seamount	西太平洋 Western Pacific Ocean	18°14.8′N，126°17.5′E	52
19	骐骆平顶海山群 Qiluo Guyots	西太平洋 West Pacific Ocean	20°15.50′N，149°39.20′E （维骐平顶海山 / Weiqi Guyot） 20°00.80′N，150°09.20′E （维骆平顶海山 / Weiluo Guyot） 19°40.15′N，150°46.65′E （劳里平顶海山 / Lowrie Guyot）	56
20	维骐平顶海山 Weiqi Guyot	西太平洋 West Pacific Ocean	20°15.50′N，149°39.20′E	58
21	维骆平顶海山 Weiluo Guyot	西太平洋 West Pacific Ocean	20°00.80′N，150°09.20′E	60
22	劳里平顶海山 Lowrie Guyot	西太平洋 West Pacific Ocean	19°40.15′N，150°46.65′E	62
23	鹿鸣平顶海山 Luming Guyot	西太平洋 West Pacific Ocean	20°02.20′N，151°38.64′E	64
24	赫姆勒平顶海山 Hemler Guyot	西太平洋 West Pacific Ocean	19°45.77′N，151°45.00′E	66
25	戈沃罗夫平顶海山 Govorov Guyot	西太平洋 West Pacific Ocean	17°58.85′N，151°05.88′E	68
26	斯科尔尼亚科瓦平顶海山 Skornyakova Guyot	西太平洋 West Pacific Ocean	16°51.92′N，149°53.80′E	70
27	戈尔金平顶海山 Gordin Guyot	西太平洋 West Pacific Ocean	16°56.78′N，150°47.82′E	72
28	伊利切夫平顶海山 Il'ichev Guyot	西太平洋 West Pacific Ocean	16°55.00′N，152°04.98′E	74
29	佩加斯平顶海山 Pegas Guyot	西太平洋 West Pacific Ocean	15°36.56′N，152°05.28′E	76
30	采薇海山群 Caiwei Seamounts	西太平洋 West Pacific Ocean	15°40.00′N，155°10.00′E （采薇平顶海山 / Caiwei Guyot） 15°18.00′N，155°00.00′E （采杞平顶海山 / Caiqi Guyot） 15°21.07′N，154°36.05′E （采菽海山 / Caishu Seamount）	78
31	采薇平顶海山 Caiwei Guyot (Pallada Guyot)	西太平洋 West Pacific Ocean	15°40.00′N，155°10.00′E	80

序　号 S/N	名　称 Name	所在大洋 Ocean or Sea	特征点坐标 Coordinates	页　码 Page
32	采菽海山 Caishu Seamount	西太平洋 West Pacific Ocean	15°21.07′N，154°36.05′E	82
33	采杞平顶海山 Caiqi Guyot	西太平洋 West Pacific Ocean	15°18.00′N，155°00.00′E	84
34	嘉偕平顶海山群 Jiaxie Guyots	西太平洋 West Pacific Ocean	12°52.20′N，156°48.60′E （维嘉平顶海山 / Weijia Guyot） 112°13.20′N，156°22.80′E （维偕平顶海山 / Weixie Guyot） 12°31.80′N，156°20.63′E （维祯平顶海山 / Weizhen Guyot）	86
35	维嘉平顶海山 Weijia Guyot (Ita Mai Tai Guyot)	西太平洋 West Pacific Ocean	12°52.20′N，156°48.60′E	88
36	维祯平顶海山 Weizhen Guyot	西太平洋 West Pacific Ocean	12°31.80′N，156°20.63′E	90
37	维偕平顶海山 Weixie Guyot (Gelendzhik Guyot)	西太平洋 West Pacific Ocean	12°13.20′N，156°22.80′E	92
38	布塔科夫平顶海山 Butakov Guyot	西太平洋 West Pacific Ocean	11°26.99′N，156°35.94′E	94
39	南台平顶海山群 Nantai Guyots	西太平洋 West Pacific Ocean	20°00.79′N，156°34.80′E （巴蒂泽平顶海山 / Badiza Guyot） 20°23.32′N，155°58.62′E （德茂平顶海山 / Demao Guyot） 20°53.83′N，156°14.34′E （邦基平顶海山 / Bangji Guyot） 20°55.82′N，157°09.54′E （南台平顶海山 / Nantai Guyot） 21°05.60′N，158°28.92′E （阿诺德平顶海山 / Arnold Guyot）	98
40	邦基平顶海山 Bangji Guyot	西太平洋 West Pacific Ocean	20°53.83′N，156°14.34′E （最高峰 / peak） 20°38.71′N，156°08.64′E	100
41	德茂平顶海山 Demao Guyot	西太平洋 West Pacific Ocean	20°23.32′N，155°58.62′E	102
42	巴蒂泽平顶海山 Batiza Guyot	西太平洋 West Pacific Ocean	20°00.79′N，156°34.80′E	104
43	南台平顶海山 Nantai Guyot	西太平洋 West Pacific Ocean	20°55.82′N，157°09.54′E	106

序　号 S/N	名　称 Name	所在大洋 Ocean or Sea	特征点坐标 Coordinates	页　码 Page
44	阿诺德平顶海山 Arnold Guyot	西太平洋 West Pacific Ocean	21°10.30′N，157°38.94′E 21°05.60′N，158°28.92′E （最高峰 / peak）	108
45	徐福海山群 Xufu Seamounts	西太平洋 West Pacific Ocean	19°57.80′N，157°27.30′E （瀛洲海山 / Yingzhou Seamount） 19°46.30′N，157°22.80′E （方丈平顶海山 / Fangzhang Guyot） 19°32.30′N，157°56.00′E （徐福平顶海山 / Xufu Guyot） 19°12.30′N，158°14.00′E （蓬莱海山 / Penglai Seamount）	110
46	瀛洲海山 Yingzhou Seamount	西太平洋 West Pacific Ocean	19°57.80′N，157°27.30′E	112
47	方丈平顶海山 Fangzhang Guyot	西太平洋 West Pacific Ocean	19°46.30′N，157°22.80′E	114
48	徐福平顶海山 Xufu Guyot	西太平洋 West Pacific Ocean	19°32.30′N，157°56.00′E	116
49	蓬莱海山 Penglai Seamount	西太平洋 West Pacific Ocean	19°12.30′N，158°14.00′E	118
50	大成海山群 Dacheng Seamounts	西太平洋 West Pacific Ocean	21°32.00′N，159°32.00′E （拉蒙特平顶海山 / Lamont Guyot） 21°41.80′N，160°40.30′E （大成平顶海山 / Dacheng Guyot） 21°10.94′N，160°29.37′E （蝴蝶海山 / Hudie Seamount） 21°12.15′N，160°43.41′E （百合平顶海山 / Baihe Guyot）	120
51	拉蒙特平顶海山 Lamont Guyot	西太平洋 West Pacific Ocean	21°32.00′N，159°32.00′E	122
52	大成平顶海山 Dacheng Guyot	西太平洋 West Pacific Ocean	21°41.80′N，160°40.30′E	124
53	蝴蝶海山 Hudie Seamount	西太平洋 West Pacific Ocean	21°10.94′N，160°29.37′E	126
54	百合平顶海山 Baihe Guyot	西太平洋 West Pacific Ocean	21°12.15′N，160°43.41′E	128
55	牛郎平顶海山 Niulang Guyot	西太平洋 West Pacific Ocean	20°22.80′N，160°45.40′E	130
56	织女平顶海山 Zhinyu Guyot	西太平洋 West Pacific Ocean	19°39.20′N，160°09.40′E	132

序 号 S/N	名 称 Name	所在大洋 Ocean or Sea	特征点坐标 Coordinates	页 码 Page
57	湛露海山群 Zhanlu Seamounts	西太平洋 West Pacific Ocean	20°01.92′N，161°41.99′E （显允海山 / Xianyun Seamount） 19°51.23′N，161°56.03′E （令德海山 / Lingde Seamount） 19°53.01′N，162°45.97′E （湛露平顶海山 / Zhanlu Guyot） 19°32.52′N，162°34.80′E （令仪海山 / Lingyi Seamount） 19°08.25′N，162°07.58′E （恺悌平顶海山 / Kaiti Guyot） 19°37.09′N，162°13.74′E （麦克唐奈平顶海山 / McDonnell Guyot）	134
58	显允海山 Xianyun Seamount	西太平洋 West Pacific Ocean	20°01.92′N，161°41.99′E	136
59	令德海山 Lingde Seamount	西太平洋 West Pacific Ocean	19°51.23′N，161°56.03′E	138
60	麦克唐奈平顶海山 McDonnell Guyot	西太平洋 West Pacific Ocean	19°37.09′N，162°13.74′E	140
61	湛露平顶海山 Zhanlu Guyot	西太平洋 West Pacific Ocean	19°53.01′N，162°45.97′E	142
62	令仪海山 Lingyi Seamount	西太平洋 West Pacific Ocean	19°32.52′N，162°34.80′E	144
63	恺悌平顶海山 Kaiti Guyot	西太平洋 West Pacific Ocean	19°08.25′N，162°07.58′E	146
64	凤鸣海山 Fengming Seamount	西太平洋 West Pacific Ocean	22°05.5′N，160°15.6′E	148
65	纳济莫夫平顶海山群 Nazimov Guyots	西太平洋 West Pacific Ocean	15°09.98′N，162°52.61′E （南 / South） 16°06.4′N，162°59.88′E （北 / North）	152
66	潜鱼平顶海山 Qianyu Guyot	中太平洋 Central Pacific Ocean	22°58.40′N，175°38.50′E	158
67	斯干平顶海山群 Sigan Guyots	中太平洋 Central Pacific Ocean	21°44.72′N，175°00.85′E （如竹平顶海山 / Ruzhu Guyot） 21°57.03′N，175°32.57′E （如松平顶海山 / Rusong Guyot） 21°41.40′N，176°23.31′E （如翼平顶海山 / Ruyi Guyot） 21°10.56′N，176°27.42′E （如翚平顶海山 / Ruhui Guyot）	160

序 号 S/N	名 称 Name	所在大洋 Ocean or Sea	特征点坐标 Coordinates	页 码 Page
68	如竹平顶海山 Ruzhu Guyot	中太平洋 Central Pacific Ocean	21°43.90′N，175°03.59′E	162
69	如松平顶海山 Rusong Guyot	中太平洋 Central Pacific Ocean	21°57.03′N，175°32.57′E	164
70	如翼平顶海山 Ruyi Guyot	中太平洋 Central Pacific Ocean	21°41.40′N，176°23.31′E	166
71	如翚平顶海山 Ruhui Guyot	中太平洋 Central Pacific Ocean	21°10.56′N，176°27.42′E	168
72	鹤鸣平顶海山群 Heming Guyots	中太平洋 Central Pacific Ocean	19°44.88′N，171°54.88′E （紫檀平顶海山 / Zitan Guyot） 20°09.14′N，172°03.04′E （九皋平顶海山 / Jiugao Guyot） 20°16.27′N，172°08.64′E （乐园平顶海山 / Leyuan Guyot）	170
73	乐园平顶海山 Leyuan Guyot	中太平洋 Central Pacific Ocean	20°16.27′N，172°08.64′E	172
74	九皋平顶海山 Jiugao Guyot	中太平洋 Central Pacific Ocean	20°09.14′N，172°03.04′E	174
75	紫檀平顶海山 Zitan Guyot	中太平洋 Central Pacific Ocean	19°44.88′N，171°54.88′E	176
76	鸿雁平顶海山群 Hongyan Guyots	中太平洋 Central Pacific Ocean	19°54.41′N，172°55.21′E （刘徽平顶海山 / Liuhui Guyot） 19°51.13′N，173°10.58′E （法显平顶海山 / Faxian Guyot） 20°01.98′N，173°33.86′E （贾耽平顶海山 / Jiadan Guyot） 19°40.52′N，173°35.40′E （义净平顶海山 / Yijing Guyot） 19°34.21′N，173°57.53′E （窦叔蒙平顶海山 / Doushumeng Guyot） 20°20.00′N，174°10.00′E （尤因海山 / Ewing seamount）	178
77	刘徽平顶海山 Liuhui Guyot	中太平洋 Central Pacific Ocean	19°54.41′N，172°55.21′E	180
78	法显平顶海山 Faxian Guyot	中太平洋 Central Pacific Ocean	19°51.13′N，173°10.58′E	182
79	贾耽平顶海山 Jiadan Guyot	中太平洋 Central Pacific Ocean	20°1.98′N，173°33.86′E	184

序 号 S/N	名 称 Name	所在大洋 Ocean or Sea	特征点坐标 Coordinates	页 码 Page
80	义净平顶海山 Yijing Guyot	中太平洋 Central Pacific Ocean	19°40.52′N，173°35.40′E	186
81	尤因海山 Ewing Seamount	中太平洋 Central Pacific Ocean	20°19.80′N，174°13.10′E	188
82	窦叔蒙平顶海山 Doushumeng Guyot	中太平洋 Central Pacific Ocean	19°34.21′N，173°57.53′E	190
83	希格平顶海山 HIG Guyot	中太平洋 Central Pacific Ocean	19°10.00′N，173°15.00′E	192
84	白驹平顶海山群 Baiju Guyots	中太平洋 Central Pacific Ocean	17°52.00′N，178°12.00′E （路易斯·阿加西斯平顶海山 / Louis Agassiz Guyot） 17°54.00′N，178°33.00′E （亚历山大·阿加西斯平顶海山 / Alexander Agassiz Guyot） 17°53.90′N，178°58.70′E （白驹平顶海山 / Baiju Guyot）	194
85	路易斯·阿加西斯平顶海山 Louis Agassiz Guyot	中太平洋 Central Pacific Ocean	17°52.00′N，178°12.00′E	196
86	亚历山大·阿加西斯平顶海山 Alexander Agassiz Guyot	中太平洋 Central Pacific Ocean	17°54.00′N，178°33.00′E	198
87	白驹平顶海山 Baiju Guyot	中太平洋 Central Pacific Ocean	17°53.90′N，178°58.70′E	200
88	无羊海山群 Wuyang Seamounts	中太平洋 Central Pacific Ocean	18°31.00′N，179°36.00′W （艾利森平顶海山 / Allison Guyot） 18°03.52′N，179°22.64′W （麾肱海山 / Huigong Seamount） 18°49.80′N，179°16.80′W （牧来平顶海山 / Mulai Guyot） 18°40.28′N，178°58.00′W （阿池海脊 / Echi Ridge） 18°06.50′N，178°42.50′W （维鱼平顶海山 / Weiyu Guyot） 18°42.60′N，178°42.50′W （簑笠平顶海山 / Suoli Guyot） 18°19.70′N，178°15.90′W （犉羊海山 / Chunyang Seamount） 18°01.40′N，178°24.00′W （年丰平顶海山 / Nianfeng Guyot）	202

序 号 S/N	名 称 Name	所在大洋 Ocean or Sea	特征点坐标 Coordinates	页 码 Page
89	艾利森平顶海山 Allison Guyot	中太平洋 Central Pacific Ocean	18°31.00′N，179°36.00′W	204
90	牧来平顶海山 Mulai Guyot	中太平洋 Central Pacific Ocean	18°49.80′N，179°16.80′W	206
91	阿池海脊 Echi Ridge	中太平洋 Central Pacific Ocean	18°40.28′N，178°58.00′W	208
92	蓑笠平顶海山 Suoli Guyot	中太平洋 Central Pacific Ocean	18°42.60′N，178°42.50′W	210
93	犉羊海山 Chunyang Seamount	中太平洋 Central Pacific Ocean	18°19.70′N，178°15.90′W	212
94	年丰平顶海山 Nianfeng Guyot	中太平洋 Central Pacific Ocean	18°01.40′N，178°24.00′W	214
95	维鱼平顶海山 Weiyu Guyot	中太平洋 Central Pacific Ocean	18°06.50′N，178°42.50′W	216
96	麾肱海山 Huigong Seamount	中太平洋 Central Pacific Ocean	18°03.52′N，179°22.64′W	218
97	彤弓海山群 Tonggong Seamounts	中太平洋 Central Pacific Ocean	13°46.60′N，165°40.02′W （彤弓海山 / Tonggong Seamount） 14°13.80′N，165°51.60′W （北彤弓海山 / Beitonggong Seamount） 13°42.60′N，165°27.00′W （东彤弓海山 / Dongtonggong Seamount） 13°16.20′N，165°28.02′W （南彤弓海山 / Nantonggong Seamount）	222
98	北彤弓海山 Beitonggong Seamount	中太平洋 Central Pacific Ocean	14°13.80′N，165°51.60′W	224
99	彤弓海山 Tonggong Seamount	中太平洋 Central Pacific Ocean	13°46.60′N，165°40.02′W	226
100	东彤弓海山 Dongtonggong Seamount	中太平洋 Central Pacific Ocean	13°42.60′N，165°27.00′W	228
101	南彤弓海山 Nantonggong Seamount	中太平洋 Central Pacific Ocean	13°16.20′N，165°28.02′W	230

序　号 S/N	名　称 Name	所在大洋 Ocean or Sea	特征点坐标 Coordinates	页　码 Page
102	谷陵海山群 Guling Seamounts	中太平洋 Central Pacific Ocean	10°57.10′N，170°22.20′W （西谷海山 / Xigu Seamount） 10°47.90′N，170°07.00′W （西陵海山 / Xiling Seamount） 10°44.80′N，169°36.50′W （东陵海山 / Dongling Seamount） 10°44.60′N，169°12.70′W （东谷海山 / Donggu Seamount）	232
103	西谷海山 Xigu Seamount	中太平洋 Central Pacific Ocean	10°57.10′N，170°22.20′W	234
104	西陵海山 Xiling Seamount	中太平洋 Central Pacific Ocean	10°47.90′N，170°07.00′W	236
105	东陵海山 Dongling Seamount	中太平洋 Central Pacific Ocean	10°44.80′N，169°36.50′W	238
106	东谷海山 Donggu Seamount	中太平洋 Central Pacific Ocean	10°44.60′N，169°12.70′W	240
107	柔木海山群 Roumu Seamounts	中太平洋 Central Pacific Ocean	10°17.60′N，167°59.40′W （南水杉海山 / Nanshuishan Seamount） 10°25.90′N，167°42.50′W （西柔木海山 / Xiroumu Seamount） 10°36.30′N，167°27.90′W （北柔木海山 / Beiroumu Seamount） 10°33.80′N，168°18.80′W （西水杉海山 / Xishuishan Seamount） 10°46.50′N，167°29.60′W （银杉海山 / Yinshan Seamount） 10°33.70′N，168°00.20′W （水杉海山 / Shuishan Seamount） 10°22.80′N，167°25.90′W （柔木海山 / Roumu Seamount）	242
108	西水杉海山 Xishuishan Seamount	中太平洋 Central Pacific Ocean	10°33.80′N，168°18.80′W	244
109	水杉海山 Shuishan Seamount	中太平洋 Central Pacific Ocean	10°33.70′N，168°00.20′W	246
110	南水杉海山 Nanshuishan Seamount	中太平洋 Central Pacific Ocean	10°17.60′N，167°59.40′W	248
111	西柔木海山 Xiroumu Seamount	中太平洋 Central Pacific Ocean	10°25.90′N，167°42.50′W	250

序　号 S/N	名　称 Name	所在大洋 Ocean or Sea	特征点坐标 Coordinates	页　码 Page
112	柔木海山 Roumu Seamount	中太平洋 Central Pacific Ocean	10°28.40′N，167°20.30′W 10°21.30′N，167°01.60′W 10°23.20′N，167°10.10′W 10°22.80′N，167°25.90′W （峰顶 / top）	252
113	北柔木海山 Beiroumu Seamount	中太平洋 Central Pacific Ocean	10°36.30′N，167°27.90′W	254
114	银杉海山 Yinshan Seamount	中太平洋 Central Pacific Ocean	10°46.50′N，167°29.60′W	256
115	鉴真海岭 Jianzhen Ridge	东太平洋 East Pacific Ocean	10°43.67′N，152°05.64′W 10°44.41′N，151°45.00′W 10°45.16′N，151°06.18′W	262
116	甘雨海山 Ganyu Seamount	东太平洋 East Pacific Ocean	10°30.20′N，152°24.20′W	264
117	茑萝海丘群 Niaoluo Hills	东太平洋 East Pacific Ocean	10°19.03′N，152°01.08′W	266
118	郑和海岭 Zhenghe Ridge	东太平洋 East Pacific Ocean	10°13.00′N，153°11.00′W 10°14.38′N，152°52.32′W 10°06.93′N，152°12.12′W 10°06.07′N，151°26.88′W 10°07.00′N，151°12.00′W	268
119	巩珍圆海丘群 Gongzhen Knolls	东太平洋 East Pacific Ocean	10°33.07′N，154°27.60′W 10°31.12′N，154°32.88′W 10°29.66′N，154°28.20′W 10°29.34′N，154°32.88′W 10°28.91′N，154°25.38′W 10°27.61′N，154°30.84′W 10°25.34′N，154°25.26′W （巩珍圆海丘 / Gongzhen Knoll）	270
120	巩珍圆海丘 Gongzhen Knoll	东太平洋 East Pacific Ocean	10°25.34′N，154°25.26′W	272
121	维熊圆海丘 Weixiong Knoll	东太平洋 East Pacific Ocean	10°16.65′N，154°31.20′W	274
122	维罴圆海丘 Weipi Knoll	东太平洋 East Pacific Ocean	10°18.92′N，153°51.72′W	276
123	维貙海丘 Weihui Hill	东太平洋 East Pacific Ocean	9°58.62′N，154°00.00′W	278
124	维蛇海丘 Weishe Hill	东太平洋 East Pacific Ocean	9°56.89′N，154°21.54′W	280

序 号 S/N	名 称 Name	所在大洋 Ocean or Sea	特征点坐标 Coordinates	页 码 Page
125	达奚通圆海丘 Daxitong Knoll	东太平洋 East Pacific Ocean	10°04.45′N，154°42.78′W	282
126	魏源海山 Weiyuan Seamount	东太平洋 East Pacific Ocean	9°48.40′N，154°31.80′W	284
127	张炳熹海岭 Zhangbingxi Ridge	东太平洋 East Pacific Ocean	9°46.35′N，154°17.64′W 9°43.01′N，153°49.68′W 9°39.93′N，152°38.22′W 9°36.96′N，151°53.88′W 9°42.26′N，151°06.60′W	286
128	清高海山 Qinggao Seamount	东太平洋 East Pacific Ocean	9°43.00′N，153°49.70′W	288
129	芳伯海山 Fangbo Seamount	东太平洋 East Pacific Ocean	9°40.10′N，152°38.50′W	290
130	斯翼海丘 Siyi Hill	东太平洋 East Pacific Ocean	9°28.43′N，154°03.78′W	292
131	斯寝海山 Siqin Seamount	东太平洋 East Pacific Ocean	9°27.46′N，153°34.44′W	294
132	郑庭芳海山群 Zhengtingfang Seamounts	东太平洋 East Pacific Ocean	9°06.46′N，154°13.68′W 9°00.52′N，154°04.14′W （峰顶 / top）	296
133	陈伦炯海山群 Chenlunjiong Seamounts	东太平洋 East Pacific Ocean	9°08.50′N，153°41.60′W （长庚海山 / Changgeng Seamount） 9°05.59′N，153°24.06′W （峰顶 /top） 9°03.86′N，153°03.90′W 8°56.10′N，153°47.30′W （景福海丘 / Jingfu Hill） 8°53.93′N，153°40.62′W	298
134	长庚海山 Changgeng Seamount	东太平洋 East Pacific Ocean	9°08.50′N，153°41.60′W	300
135	景福海丘 Jingfu Hill	东太平洋 East Pacific Ocean	8°56.10′N，153°47.30′W	302
136	如霆海山 Ruting Seamount	东太平洋 East Pacific Ocean	8°31.03′N，148°21.18′W	304
137	苏洵圆海丘 Suxun Knoll	东太平洋 East Pacific Ocean	8°10.35′N，146°43.20′W	306
138	苏轼海丘 Sushi Hill	东太平洋 East Pacific Ocean	8°12.62′N，146°38.70′W	308

序 号 S/N	名 称 Name	所在大洋 Ocean or Sea	特征点坐标 Coordinates	页 码 Page
139	苏辙圆海丘 Suzhe Knoll	东太平洋 East Pacific Ocean	8°12.18′N，146°30.54′W	310
140	王勃圆海丘 Wangbo Knoll	东太平洋 East Pacific Ocean	7°49.67′N，146°06.42′W	312
141	杨炯海丘 Yangjiong Hill	东太平洋 East Pacific Ocean	7°49.72′N，146°01.45′W	314
142	卢照邻圆海丘 Luzhaolin Knoll	东太平洋 East Pacific Ocean	7°50.53′N，145°55.32′W	316
143	骆宾王圆海丘 Luobinwang Knoll	东太平洋 East Pacific Ocean	7°47.99′N，145°52.56′W	318
144	天祐圆海丘 Tianhu Knoll	东太平洋 East Pacific Ocean	8°28.30′N，145°44.70′W	320
145	天祐南海底洼地 Tianhunan Depression	东太平洋 East Pacific Ocean	8°24.66′N，145°45.42′W 8°18.60′N，145°43.92′W 8°09.09′N，145°42.60′W 8°00.06′N，145°41.82′W （最深点 / deepest point） 7°52.85′N，145°39.96′W	322
146	朱应海山 Zhuying Seamount	东太平洋 East Pacific Ocean	8°41.00′N，144°12.60′W	324
147	朱应西海底洼地 Zhuyingxi Depression	东太平洋 East Pacific Ocean	8°58.16′N，144°19.26′W 8°53.72′N，144°22.20′W 8°48.87′N，144°23.16′W （最深点 / deepest point）	326
148	朱应北海底洼地 Zhuyingbei Depression	东太平洋 East Pacific Ocean	8°47.09′N，144°11.46″W （最深点 / deepest point） 8°55.97′N，144°14.82′W 9°02.27′N，144°13.62′W	328
149	嘉卉圆海丘 Jiahui Knoll	东太平洋 East Pacific Ocean	8°29.80′N，144°24.30′W	330
150	楚茨海山 Chuci Seamount	东太平洋 East Pacific Ocean	7°49.20′N，144°31.70′W	332

序 号 S/N	名 称 Name	所在大洋 Ocean or Sea	特征点坐标 Coordinates	页 码 Page
151	楚茨海山链 Chuci Seamount Chain	东太平洋 East Pacific Ocean	7°47.51′N，144°41.16′W 7°49.20′N，144°31.70′W （楚茨海山 / Chuci Seamount） 7°50.86′N，144°20.16′W 7°57.12′N，144°16.20′W 7°58.52′N，144°07.74′W 7°59.22′N，144°00.66′W 7°57.33′N，143°52.20′W	334
152	亿庾海山 Yiyu Seamount	东太平洋 East Pacific Ocean	8°25.19′N，143°08.70′W	336
153	怀允海山 Huaiyun Seamount	东太平洋 East Pacific Ocean	9°22.32′N，142°26.76′W	338
154	启明海山 Qiming Seamount	东太平洋 East Pacific Ocean	8°20.30′N，142°16.50′W	340
155	启明南海底洼地 Qimingnan Depression	东太平洋 East Pacific Ocean	7°51.88′N，142°12.18′W 8°04.37′N，142°16.26′W （最深点 / deepest point） 8°19.73′N，142°18.30′W	342
156	圆鼓圆海丘 Yuangu Knoll	东太平洋 East Pacific Ocean	10°26.7′N，154°18.0′W	344
157	刺螺平顶海山群 Ciluo Guyots	东太平洋 East Pacific Ocean	9°17.0′N，158°31.9′W 9°12.4′N，157°59.2′W	346
158	颂首平顶海山 Fenshou Guyot	东太平洋 East Pacific Ocean	3°01.28′N，101°55.21′W	352
159	西鸳鸯海丘 Xiyuanyang Hill	东太平洋 East Pacific Ocean	3°10.23′N，103°02.18′W	354
160	东鸳鸯海丘 Dongyuanyang Hill	东太平洋 East Pacific Ocean	3°09.76′N，103°00.65′W	356
161	白茅海脊 Baimao Ridge	东太平洋 East Pacific Ocean	3°21.56′N，100°04.92′W 3°21.71′N，100°00.98′W 3°21.33′N，99°56.84′W	358
162	白云海丘 Baiyun Hill	东太平洋 East Pacific Ocean	2°59.21′N，100°37.59′W	360
163	硕人海丘 Shuoren Hill	东太平洋 East Pacific Ocean	2°33.50′N，102°59.54′W	362
164	啸歌海丘 Xiaoge Hill	东太平洋 East Pacific Ocean	1°55.27′N，103°08.95′W	364

序 号 S/N	名 称 Name	所在大洋 Ocean or Sea	特征点坐标 Coordinates	页 码 Page
165	太白海脊 Taibai Ridge	东太平洋 East Pacific Ocean	1°12.69′N，102°07.42′W 1°19.07′N，101°52.46′W 1°24.45′N，101°40.41′W	366
166 ~ 172	太白热液区 Taibai Hydrothermal Fields	东太平洋 East Pacific Ocean	1°36.90′N，102°15.60′W （太白北 1 号 / Taibaibei-1） 1°39.60′N，102°16.38′W （太白北 2 号 / Taibaibei-2） 1°40.80′N，102°16.20′W （太白北 3 号 / Taibaibei-3） 1°44.40′N，102°17.40′W （太白北 4 号 / Taibaibei-4） 1°03.60′N，102°10.20′W （太白南 1 号 / Taibainan-1） 1°02.40′N，102°10.80′W （太白南 2 号 / Taibainan-2） 1°00.00′N，102°11.40′W （太白南 3 号 / Taibainan-3）	368
173	太白海渊 Taibai Deep	东太平洋 East Pacific Ocean	1°33.60′N，101°13.20′W	370
174	宝石 1 号热液区 Baoshi-1 Hydrothermal Field	东太平洋 East Pacific Ocean	1°13.20′S，101°29.40′W	372
175	那居圆海丘 Nuoju Knoll	东太平洋 East Pacific Ocean	0°38.36′N，101°25.43′W	374
176	依蒲海丘 Yipu Hill	东太平洋 East Pacific Ocean	0°30.62′N，101°02.97′W	376
177	白华海丘 Baihua Hill	东太平洋 East Pacific Ocean	0°29.63′N，102°56.10′W	378
178	维翰海山 Weihan Seamount	东太平洋 East Pacific Ocean	0°05.20′S，101°24.20′W	380
179	鸟巢海丘 Niaochao Hill	东太平洋 East Pacific Ocean	1°22.00′S，102°27.50′W	382
180 ~ 185	鸟巢热液区 Niaochao Hydrothermal Field	东太平洋 East Pacific Ocean	0°50.22′S，102°27.00′W （鸟巢北 1 号 / Niaochaobei-1） 1°22.08′S，102°27.00′W （鸟巢 /Niaochao） 2°01.20′S，102°37.20′W （鸟巢南 1 号 / Niaochaonan-1） 2°04.62′S，102°36.60′W （鸟巢南 2 号 / Niaochaonan-2） 2°09.06′S，102°38.40′W （鸟巢南 3 号 / Niaochaonan-3） 2°13.14′S，102°39.00′W （鸟巢南 4 号 / Niaochaonan-4）	384

序 号 S/N	名 称 Name	所在大洋 Ocean or Sea	特征点坐标 Coordinates	页 码 Page
186	徽猷 1 号热液区 Huiyou-1 Hydrothermal Field	东太平洋 East Pacific Ocean	3°06.26′S，102°33.18′W	386
187	角弓 1 号热液区 Jiaogong-1 Hydrothermal Field	东太平洋 East Pacific Ocean	5°18.00′S，106°28.92′W	388
188 ~ 189	鱼藻 1 号热液区 鱼藻 2 号热液区 Yuzao-1 Hydrothermal Field Yuzao-2 Hydrothermal Field	东太平洋 East Pacific Ocean	6°03.60′S，106°50.76′W （鱼藻 1 号 / Yuzao-1） 6°04.80′S，106°50.94′W （鱼藻 2 号 / Yuzao-2）	390
190	驺虞海脊 Zouyu Ridge	南大西洋 South Atlantic Ocean	13°10.02′S，14°25.80′W 13°19.80′S，14°24.00′W 13°32.04′S，14°23.40′W	400
191 ~ 192	驺虞 1 号热液区 驺虞 2 号热液区 Zouyu-1 Hydrothermal Field Zouyu-2 Hydrothermal Field	南大西洋 South Atlantic Ocean	13°17.15′S，14°24.79′W （驺虞 1 号 / Zouyu-1） 13°15.26′S，14°24.77′W （驺虞 2 号 / Zouyu-2）	402
193	太极 1 号热液区 Taiji-1 Hydrothermal Field	南大西洋 South Atlantic Ocean	13°35.42′S，14°31.01′W	404
194	如玉海脊 Ruyu Ridge	南大西洋 South Atlantic Ocean	13°56.13′S，14°19.40′W	406
195	采蘩海山 Caifan Seamount	南大西洋 South Atlantic Ocean	14°03.10′S，14°21.10′W	408
196 ~ 197	采蘩 1 号热液区 采蘩 2 号热液区 Caifan-1 Hydrothermal Field Caifan-2 Hydrothermal Field	南大西洋 South Atlantic Ocean	14°02.87′S，14°22.79′W （采蘩 1 号热液区 / Caifan-1 Hydrothermal Field） 14°02.18′S，14°21.83′W （采蘩 2 号热液区 / Caifan-2 Hydrothermal Field）	410
198	唐棣海脊 Tangdi Ridge	南大西洋 South Atlantic Ocean	14°28.09′S，13°40.64′W	412
199	安吉海丘 Anji Hill	南大西洋 South Atlantic Ocean	14°52.20′S，13°22.50′W	414
200	德音裂谷 Deyin Gap	南大西洋 South Atlantic Ocean	15°15.76′S，13°21.86′W	416
201	德音 1 号热液区 Deyin-1 Hydrothermal Field	南大西洋 South Atlantic Ocean	15°09.94′S，13°21.38′W	418
202	宵征海山 Xiaozheng Seamount	南大西洋 South Atlantic Ocean	16°12.9′S，13°06.5′W	420

序　号 S/N	名　称 Name	所在大洋 Ocean or Sea	特征点坐标 Coordinates	页　码 Page
203	凯风海山 Kaifeng Seamount	南大西洋 South Atlantic Ocean	22°56.6′S，13°25.9′W	422
204	赤狐海丘 Chihu Hill	南大西洋 South Atlantic Ocean	23°44.65′S，13°21.94′W	424
205	黑乌海丘 Heiwu Hill	南大西洋 South Atlantic Ocean	23°49.26′S，13°15.78′W	426
206	方舟海山 Fangzhou Seamount	南大西洋 South Atlantic Ocean	25°46.8′S，13°38.8′W	428
207	洵美海丘 Xunmei Hill	南大西洋 South Atlantic Ocean	26°01.20′S，13°51.64′W	430
208	洵美1号热液区 Xunmei-1 Hydrothermal Field	南大西洋 South Atlantic Ocean	26°01.27′S，13°51.13′W	432
209	文王海脊 Wenwang Ridge	南大西洋 South Atlantic Ocean	22°44.1′S，13°11.0′W	434
210	玉磬海山 Yuqing Seamount	西北印度洋 Northwest Indian Ocean	10°11.55′N，57°42.40′E	446
211	排箫海山 Paixiao Seamount	西北印度洋 Northwest Indian Ocean	10°00.38′N，58°00.20′E	448
212	庸鼓海山 Yonggu Seamount	西北印度洋 Northwest Indian Ocean	9°43.52′N，57°38.46′E	450
213	烈祖海山 Liezu Seamount	西北印度洋 Northwest Indian Ocean	9°29.97′N，57°35.79′E	452
214	温恭海山 Wengong Seamount	西北印度洋 Northwest Indian Ocean	9°39.30′N，58°24.12′E	454
215	万舞海山 Wanwu Seamount	西北印度洋 Northwest Indian Ocean	9°27.18′N，58°19.44′E	456
216	客怿海山 Keyi Seamount	西北印度洋 Northwest Indian Ocean	9°03.70′N，58°13.10′E	458
217	李四光断裂带 Lisiguang Fracture Zone	西北印度洋 Northwest Indian Ocean	8°38.69′N，58°25.79′E	460
218	竺可桢断裂带 Chukochen Fracture Zone	西北印度洋 Northwest Indian Ocean	7°23.50′N，59°37.89′E	422
219	卧蚕海脊 Wocan Ridge	西北印度洋 Northwest Indian Ocean	6°21.17′N，60°32.46′E	464

序 号 S/N	名 称 Name	所在大洋 Ocean or Sea	特征点坐标 Coordinates	页 码 Page
220 ~ 221	卧蚕 1 号热液区 卧蚕 2 号热液区 Wocan-1 Hydrothermal Field Wocan-2 Hydrothermal Field	西北印度洋 Northwest Indian Ocean	6°21.92′N，60°31.57′E （卧蚕 1 号 / Wocan-1） 6°22.73′N，60°30.38′E （卧蚕 2 号 / Wocan-2）	466
222	玄鸟海脊 Xuanniao Ridge	西北印度洋 Northwest Indian Ocean	5°56.68′N，61°02.26′E	468
223	徐霞客断裂带 Xuxiake Fracture Zone	西北印度洋 Northwest Indian Ocean	5°35.62′N，61°39.11′E	470
224	天休 1 号热液区 Tianxiu-1 Hydrothermal Field	西北印度洋 Northwest Indian Ocean	3°41.76′N，63°50.36′E	472
225	郦道元断裂带 Lidaoyuan Fracture Zone	西北印度洋 Northwest Indian Ocean	2°17.09′N，66°47.41′E	474
226	大糦海丘 Daxi Hill	西北印度洋 Northwest Indian Ocean	06°48.1′N，60°10.5′E	476
227	天龙海山群 Tianlong Seamounts	西北印度洋 Northwest Indian Ocean	4°59.7′N，62°02.5′E 5°04.4′N，61°59.0′E 5°04.8′N，61°49.0′E 5°10.0′N，61°53.5′E	478
228	龙灯海丘 Longdeng Hill	西北印度洋 Northwest Indian Ocean	5°13.8′N，61°55.6′E	480
229	崇牙海脊 Chongya Ridge	西南印度洋 Southwest Indian Ocean	38°29.12′S，48°23.99′E	486
230	苏堤 1 号热液区 Sudi-1 Hydrothermal Field	西南印度洋 Southwest Indian Ocean	38°09.72′S，48°35.72′E	488
231	白堤 1 号热液区 Baidi-1 Hydrothermal Field	西南印度洋 Southwest Indian Ocean	37°59.51′S，48°51.20′E	490
232	玉皇 1 号热液区 Yuhuang-1 Hydrothermal Field	西南印度洋 Southwest Indian Ocean	37°56.31′S，49°15.88′E	492
233 ~ 234	龙旂 1 号热液区 龙旂 2 号热液区 Longqi-1 Hydrothermal Field Longqi-2 Hydrothermal Field	西南印度洋 Southwest Indian Ocean	37°46.98′S，49°39.00′E （龙旂 1 号 / Longqi-1） 37°48.20′S，49°39.72′E （龙旂 2 号 / Longqi-2）	494
235	西龙井 1 号热液区 Xilongjing-1 Hydrothermal Field	西南印度洋 Southwest Indian Ocean	37°51.54′S，49°39.79′E	496
236	东龙井 1 号热液区 Donglongjing-1 Hydrothermal Field	西南印度洋 Southwest Indian Ocean	37°51.89′S，49°50.57′E	498

序　号 S/N	名　称 Name	所在大洋 Ocean or Sea	特征点坐标 Coordinates	页　码 Page
237	断桥 1 号热液区 Duanqiao-1 Hydrothermal Field	西南印度洋 Southwest Indian Ocean	37°39.45′S，50°28.02′E	500
238	栈桥 1 号热液区 Zhanqiao-1 Hydrothermal Field	西南印度洋 Southwest Indian Ocean	37°33.27′S，50°59.81′E	502
239	长白 1 号热液区 Changbai-1 Hydrothermal Field	西南印度洋 Southwest Indian Ocean	37°36.00′S，51°00.00′E	504
240	骏惠海山 Junhui Seamount	西南印度洋 Southwest Indian Ocean	37°30.03′S，51°42.00′E	506
241	骏惠 1 号热液区 Junhui-1 Hydrothermal Field	西南印度洋 Southwest Indian Ocean	37°27.94′S，51°43.91′E	508
242	乔岳海山 Qiaoyue Seamount	西南印度洋 Southwest Indian Ocean	37°20.00′S，52°07.00′E	510
243	百神 1 号热液区 Baishen-1 Hydrothermal Field	西南印度洋 Southwest Indian Ocean	36°06.07′S，53°18.00′E	512
244	思文海脊 Siwen Ridge	西南印度洋 Southwest Indian Ocean	28°30.50′S，62°29.70′E	514
245	天作海山 Tianzuo Seamount	西南印度洋 Southwest Indian Ocean	27°53.20′S，63°27.80′E	516
246	天成海山 Tiancheng Seamount	西南印度洋 Southwest Indian Ocean	27°48.90′S，63°53.70′E	518
247 ~ 248	天作 1 号热液区 天成 1 号热液区 Tianzuo-1 Hydrothermal Field Tiancheng-1 Hydrothermal Field	西南印度洋 Southwest Indian Ocean	27°57.04′S，63°32.48′E （天作 1 号 / Tianzuo-1） 27°51.10′S，63°55.40′E （天成 1 号 / Tiancheng-1）	520
249	汪大渊海山 Wangdayuan Seamount	西南印度洋 Southwest Indian Ocean	34°53.60′S，54°16.77′E	522
250	封演海山 Fengyan Seamount	西南印度洋 Southwest Indian Ocean	37°52.94′S，48°24.32′E	524
251	断桥海丘 Duanqiao Hill	西南印度洋 Southwest Indian Ocean	37°37.42′S，50°27.79′E	526
252	阳阿海脊 Yang'e Ridge	东南印度洋 Southeast Indian Ocean	28°00.0′S，100°54.3′E	530

索 引
Index

名　称 Name	页　码 Page	名　称 Name	页　码 Page
黑乌海丘 Heiwu Hill	426	骏惠海山 Junhui Seamount	506
鸿雁平顶海山群 Hongyan Guyots	178	凯风海山 Kaifeng Seamount	422
蝴蝶海山 Hudie Seamount	126	恺悌平顶海山 Kaiti Guyot	146
怀允海山 Huaiyun Seamount	338	客怿海山 Keyi Seamount	458
槐序海山 Huaixu Seamount	34	拉蒙特平顶海山 Lamont Guyot	122
麾肱海山 Huigong Seamount	218	劳里平顶海山 Lowrie Guyot	62
徽猷 1 号热液区 Huiyou-1 Hydrothermal Field	386	乐园平顶海山 Leyuan Guyot	172
嘉卉圆海丘 Jiahui Knoll	330	李四光断裂带 Lisiguang Fracture Zone	460
嘉偕平顶海山群 Jiaxie Guyots	86	立春海山 Lichun Seamount	38
贾耽平顶海山 Jiadan Guyot	184	郦道元断裂带 Lidaoyuan Fracture Zone	474
鉴真海岭 Jianzhen Ridge	262	烈祖海山 Liezu Seamount	452
角弓 1 号热液区 Jiaogong-1 Hydrothermal Field	388	令德海山 Lingde Seamount	138
惊蛰海山 Jingzhe Seamount	40	令仪海山 Lingyi Seamount	144
景福海丘 Jingfu Hill	302	刘徽平顶海山 Liuhui Guyot	180
静好海山 Jinghao Seamount	28	龙灯海丘 Longdeng Hill	480
九皋平顶海山 Jiugao Guyot	174	龙旂 1 号热液区 龙旂 2 号热液区 Longqi-1 Hydrothermal Field Longqi-2 Hydrothermal Field	494
雎鸠海山群 Jujiu Seamount	32		
骏惠 1 号热液区 Junhui-1 Hydrothermal Field	508	卢照邻圆海丘 Luzhaolin Knoll	316

名　称 Name	页　码 Page
伊利切夫平顶海山 Il'ichev Guyot	74
依蒲海丘 Yipu Hill	376
义净平顶海山 Yijing Guyot	186
亿庾海山 Yiyu Seamount	336
银杉海山 Yinshan Seamount	256
瀛洲海山 Yingzhou Seamount	112
庸鼓海山 Yonggu Seamount	450
尤因海山 Ewing Seamount	188
鱼藻 1 号热液区 鱼藻 2 号热液区 Yuzao-1 Hydrothermal Field Yuzao-2 Hydrothermal Field	390
玉皇 1 号热液区 Yuhuang-1 Hydrothermal Field	492
玉磬海山 Yuqing Seamount	446
圆鼓圆海丘 Yuangu Knoll	344
月潭海脊 Yuetan Ridge	22
栈桥 1 号热液区 Zhanqiao-1 Hydrothermal Field	502

名　称 Name	页　码 Page
湛露海山群 Zhanlu Seamounts	134
湛露平顶海山 Zhanlu Guyot	142
张炳熹海岭 Zhangbingxi Ridge	286
郑和海岭 Zhenghe Ridge	268
郑庭芳海山群 Zhengtingfang Seamounts	296
织女平顶海山 Zhinyu Guyot	132
朱应北海底洼地 Zhuyingbei Depression	328
朱应海山 Zhuying Seamount	324
朱应西海底洼地 Zhuyingxi Depression	326
竺可桢断裂带 Chukochen Fracture Zone	422
紫檀平顶海山 Zitan Guyot	176
驺虞 1 号热液区 驺虞 2 号热液区 Zouyu-1 Hydrothermal Field Zouyu-2 Hydrothermal Field	402
驺虞海脊 Zouyu Ridge	400